Erosion and Growth of Solids Stimulated by Atom and Ion Beams

NATO ASI Series

Advanced Science Institutes Series

A Series presenting the results of activities sponsored by the NATO Science Committee, which aims at the dissemination of advanced scientific and technological knowledge, with a view to strengthening links between scientific communities.

The Series is published by an international board of publishers in conjunction with the NATO Scientific Affairs Division

A	Life Sciences	Plenum Publishing Corporation
B	Physics	London and New York
C	Mathematical and Physical Sciences	D. Reidel Publishing Company Dordrecht and Boston
D	Behavioural and Social Sciences	Martinus Nijhoff Publishers Dordrecht/Boston/Lancaster
E	Applied Sciences	
F	Computer and Systems Sciences	Springer-Verlag Berlin/Heidelberg/New York
G	Ecological Sciences	

Series E: Applied Sciences – No. 112

Erosion and Growth of Solids Stimulated by Atom and Ion Beams

edited by

G. Kiriakidis

Physics Department
University of Crete
Heraklion, Crete
Greece

G. Carter

Department of Electronic
and Electrical Engineering
University of Salford
Salford, U.K.

J.L. Whitton

Physics Department
Queen's University
Kingston
Canada

1986 **Martinus Nijhoff Publishers**
Dordrecht / Boston / Lancaster
Published in cooperation with NATO Scientific Affairs Division

Proceedings of the NATO Advanced Study Institute on Erosion and Growth of Solids Stimulated by Atom and Ion Beams, Heraklion, Crete, Greece, September 16-27, 1985

Library of Congress Cataloging in Publication Data

ISBN-13: 978-94-010-8468-0   e-ISBN-13: 978-94-009-4422-0
DOI:10.1007/978-94-009-4422-0

Distributors for the United States and Canada: Kluwer Academic Publishers, 190 Old Derby Street, Hingham, MA 02043, USA

Distributors for the UK and Ireland: Kluwer Academic Publishers, MTP Press Ltd, Falcon House, Queen Square, Lancaster LA1 1RN, UK

Distributors for all other countries: Kluwer Academic Publishers Group, Distribution Center, P.O. Box 322, 3300 AH Dordrecht, The Netherlands

PREFACE

The members of the organising Committee and their
colleagues have, for many years been investigating the evolution of the fascinating surface features which develop
during sputtering erosion of solids. Such experimental,
theoretical and computational studies have also been carried
out in many international laboratories and, as well as much
commonality and agreement, substantial disagreements were
unresolved. In view of the increasing importance of such
processes in technological applications such as microlithographic etching for the patterning of solid state devices and
in fusion technology it was felt opportune to hold a meeting
in this area. Furthermore the use of energetic atomic and ion
fluxes is also becoming of increasing importance in assisting
or modifying the growth of thin films in a number of important
industrial processes and it was therefore rational to combine
the study of both erosional and growth processes in a single
meeting.

These proceedings include 16 invited review and 15 oral or
poster presented contributions to the NATO Advanced Study
Institute on the "Erosion and Growth of Solids Stimulated by
Atom and Ion Beams". The review contributions span the range
from the fundamental concepts of ballistic sputtering, and how
this influences surface morphology evolution, through processes
involving entrapment of incident species to mechanisms involved
in the use of chemically reactive ion species. Further reviews
outline the influence of energetic irradiation upon surface
growth by atomic deposition whilst others discuss technological
applications of both areas of growth and erosion. These general,
state of the art discussions, are supported by detailed
accounts of relevant studies in the contributed papers.

The proceedings should be of particular interest to those
involved in the morphological, composition and structural
modifications of solid surfaces in such diverse application
areas as microelectronics, coating technology and fusion
reactor development.

We would like to express our particular appreciation to the
lecturers and all the participants for their continuous
friendly and effective cooperation throughout the meeting. Also,
our secretary Ms Hara Parasyri who carried much of the
organizational burden with great efficiency. Last but not least
we gratefully acknowledge financial support by the Scientific
Affairs Division of NATO and by the following sponsors:

VIII

University of Crete, Greece
Greek Ministry of Research,
Vacuum Science Workshop (VSW), U.K.,
National Science Foundation, U.S.A.,
IBM, U.S.A.,
Bell-Northern Research Ltd, Canada,
Vacuum Generators (VG), U.K.

G. Kiriakidis
G. Carter
J. Whitton

Heraklion, December 1985

TABLE OF CONTENTS

PHYSICAL SPUTTERING OF ELEMENTAL METALS AND SEMICONDUCTORS

Wolfgang O. HOFER
Kernforschungsanlage Jülich GmbH[*], IGV, D-5170 Jülich, F.R. Germany

1. INTRODUCTORY AND HISTORICAL REMARKS

This overview is not intended as yet another contribution to existing reviews on the physics of sputtering and its history. In this respect, the interested reader is referred to the excellent recent monographs and conference proceedings given in Ref.1 to 5. The intention of this article is to give an introduction to physical sputtering, guided by the interplay between collisional and "thermal" sputtering which has obscured the understanding of the phenomenon since its discovery. It will become apparent that angular distributions of sputtered particles have always played a decisive role in identifying the relevant mechanisms, energy distributions have allowed quantifying them, and mass distributions now urge further efforts towards understanding collective processes.

Sputtering of solids is the emission of surface atoms upon impact of energetic radiation. If we restrict ourselves to metals and semiconductors, "radiation" means particle radiation, first and foremost ion radiation. The phenomenon was discovered in studies of electrical gas discharges about 130 years ago. It took half a century before the nature of the particles involved - ions and electrons in the plasma, to name the modern terms - became clear /6,7/, but from this time on the understanding of sputtering (of cathodes) developed much faster. During the first decade of this century STARK suggested two seemingly opposing mechanisms of surface-atom ejection; it has required intensive research up until recently before a clear answer in favour of one or the other could be given. STARK perceived cathodic erosion by ion impact as either an evaporation/sublimation process from a hot micro-spot /6/, or as a result of a succession of binary collisions during which part of the initial (inwardly-directed) momentum is directed back towards surface atoms /8/.

A thermal mechanism for sputtering was first proposed by HITTORF /9/ who understood cathodic erosion as plain sublimation from cathodes heated in the gas discharge - thus overriding the statement of his former teacher PLÜCKER that particle emission from cathodes is not coupled to the discharge power /10/. GRANQUIST /11/ eventually disproved sputtering as a bulk sublimation process by showing that the yield from externally heated cathodes is largely independent of temperature. - KOHLSCHÜTTER /12/ reviewed the early development up to 1912 and concluded by renouncing his own model in favour of STARK's.

The concept of a hot spot developing around the projectile's point of impact was pursued in various areas of radiation/condensed-matter interaction: by DESSAUER in radiation biology ("Punktwärmetheorie", /13/), by KAPITZA in ion-induced electron emission /14/ and by v. HIPPEL in sputtering /15/.

2. INFORMATION FROM ANGULAR DISTRIBUTIONS

In 1935 SEELIGER and SOMMERMEIER /16/ performed the first angular distribution studies on polycrystalline and liquid metals. Their results resembled KNUDSEN's cosine-distribution from evaporating liquids so closely that this finding was considered as corroboration of the evaporation-from-a-hot-spot model /17-19/.

It was 20 years before WEHNER's discovery of anisotropic emission from single crystals changed the situation again. In 1955 WEHNER had found that enhanced emission occurs along close-packed lattice directions such as <110> in fcc and <111> in bcc lattices /20/. During the following 5 years this effect was confirmed by several researchers over a broad range of impact and target parameters /21-25/. Preferred ejection along close-packed lattice rows is established now over five orders of magnitude of projectile energy, for metals as well as semiconductors and for heavy-ion as well as electron bombardment; it is thus a general irradiation effect in crystalline solids and, since sublimation from single crystals was found to be of a structureless cosine-shape /26,27/, it precludes thermal sputtering's being of a dominant nature.

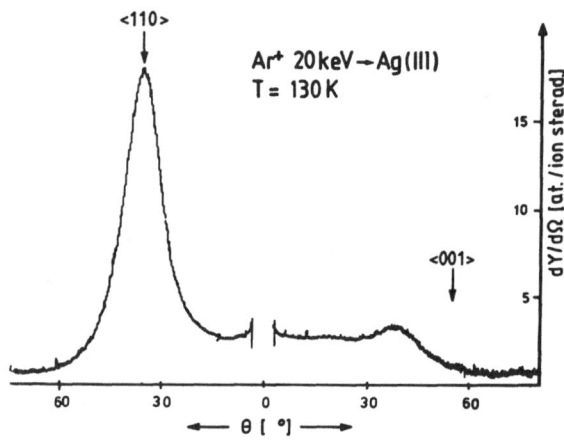

Fig. 1: Anisotropic emission from single crystals.
Emission distribution in the (1$\bar{1}$0) azimuth from (111)Ag under irradiation with 20 keV Ar$^+$ ions. The main ejection coincides with the <110> lattice direction while the weaker maximum is shifted inwards from the <100> direction, probably due to unbalanced surface collision /29/. From LINDERS et al. /28/.

The discovery of anisotropic emission from single crystals introduced another sputtering concept, that of emission by focusing collision sequences. The controversies over this mechanism have led to the situation that for 20 years single crystal investigations concentrated on particular ejection kinematics (direct or assisted focusing, surface ejection, primary knock-on etc.) /20-38/. Information on the broad areas of physical sputtering, which are: light-ion sputtering, linear cascade sputtering and sputtering from collision spikes, was primarily deduced from polycrystalline targets.

2.1 Emission from Single Crystals

In 1957 SILSBEE /39/ pointed out the possibility of a lattice influence on the energy dissipation of the recoils. In particular, he demonstrated by applying the hard sphere approximation how momentum focusing along <110> in fcc lattices could be accomplished, provided the interaction potential is sufficiently strong. WEHNER's findings of 1955/56 were not considered as evidence enough for the existence of focusing collision sequences, mainly because of the low (10^2 eV) projectile energies in these experiments. With the confirmation of preferential ejection at higher (10^4 eV) energies /21-25/, however, this restraint no longer applied and focusons became the general interpretation for ejection along close-packed directions.
In the hard core approximation it is easy to derive an upper energy limit for collimated recoil momentum transport. This energy is called the focusing energy and is given by

$$E_f = 2A \exp\left(-\frac{D}{2a}\right) \tag{2.1}$$

where A ($A \sim Z^n$, n > 0) and a are the parameters of the Born-Mayer potential, and D is the equilibrium interatomic spacing in the row, see Fig.2. Focusing energy values calculated from Eq.2.1 lie about a factor 2 too high compared with more accurate calculations /40,41/, but E_f's deduced from energy distribution measurements are even larger than the hard-core-approximation values. Hence we consider the following focusing energies for the <110> direction as acceptable compromises Al:5eV, Cu:25eV, Au:50eV.

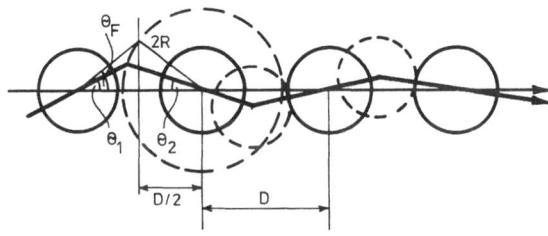

Fig. 2: Focusing collision sequence along a close-packed lattice direction. In the hard core approximation momentum alignment is possible only when the collision radius (2R) is larger than the half equilibrium distance, i.e. the collision occurs before the saddle-point of the potential of the surrounding lattice.

Rather than arguing about more accurate E_f values, in discussions on the role of focusing collision sequences, it is more constructive to consider also the focuson energy loss, ΔE, to the neighbouring lattice. Both E_f and ΔE increase with the interaction potential (e.g. A in Eq.2.1) and both are, of course, fully subject to its large uncertainties in the 1 to 100 eV energy regime. Since, however, E_f (together with the surface binding energy E_b) determines the fraction of the recoil spectrum contributing to focusing collision sequences, and ΔE determines their length, E_f and ΔE affect the contribution of focusons in opposite ways. The far-reaching potential of Au-atoms, for instance, results in a larger E_f than for Al, but this causes a high ΔE for Au as well. $E_f/\Delta E$, the decisive quantity as regards the con-

4

tribution of focusing collision sequences to sputtering, is therefore not too different for the aforementioned metals and in no way precludes focuson contributions in Al /42,30/.

For some time, focusing collision sequences were not only regarded as the origin of anisotropic emission from single crystals but also as the main ejection mechanism in general. This transport of energy along correlated collision sequences conflicted with the notion of random collision cascades as being the prime transport mechanism. Thus, for the decision on the starting basis for theoretical treatments, two events in the mid sixties were important:
- SMITH and co-workers /31,32/ stated that even for a pronounced "spot pattern" (this is the deposit distribution anisotropic emission generates on collector plates in the vicinity of the sputtered target) the majority of ejected material is contained in the random, cosine-like background;
- LEHMANN & SIGMUND /43/, and HARRISON et al. /44/ pointed out that anisotropic emission need not necessarily be a focuson effect. The former authors suggested that it might be a consequence of a selective influence of the surface binding energy on the low-energy part of the recoil spectrum, Fig.3. HARRISON et al. found in computer simulations anisotropic emission in the absence of the intersection of the surface by focusing collision sequences.

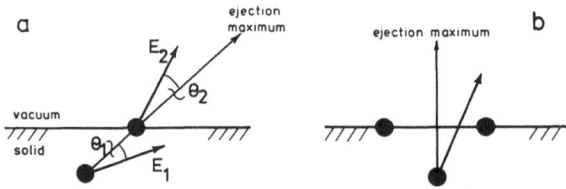

Fig. 3: Preferred ejection in low-index lattice directions according to LEHMANN & SIGMUND /43/. In the left-hand case surface binding imposes a selection criterion on the impact angles, on the right-hand potential minima between symmetric surface atom arrangements constitute "windows" for low-energy particles. Compared to these selection mechanisms, it is of secondary importance in this model whether or not the collision with the surface atom is focusing ($\theta_2 < \theta_1$). Both mechanisms work only for that part of the recoil energy spectrum which lies near the surface binding energy.

This so-called surface model for anisotropic emission is consistent with the conception of sputtering as being determined by kinetic energy imparted to surface atoms by collision cascades /45,46/. While most of the subsequent angular distribution investigations on single crystals were devoted to determining the respective contributions of the mechanisms causing anisotropic emission as well as their fraction in total emission, sputtering theory based on collision cascades meanwhile found its main support in its good agreement with experimental data of the total yield dependence on the energy, mass and incident angle of the projectile. In SIGMUND's theory of the sputtering of amorphous matter, energy dissipation within the cascade volume is treated by linear Boltzmann equations. Such a linearized transport-theoretical description requires dilute cascades, i.e. cascades where energy can be distributed to an ever increasing number of atoms which were at rest before the collision. Any such dilute or "linear" cascade will eventually transform into a state in which practically all atoms in the slowing-down volume are in motion. Such an ensemble of energetic atoms behaves rather like a high-pressure gas. It is often referred to as a (collision or "thermal") spike.

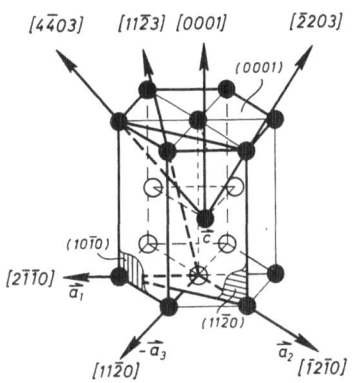

Fig. 4: Hcp lattices are ideally suited for checking the LEHMANN-SIGMUND ejection mechanism; they show both close-packed lattice rows, <1120>, and pair arrangements without a third partner below the surface, <2023>. Furthermore, the <0001> lattice direction is a typical case for assisted focusing collision sequences, where an atom is steered by a ring of three neighbouring atoms onto the next atom in the line.

For an investigation of the LEHMANN-SIGMUND model for anisotropic emission, hexagonal close-packed (hcp) monocrystals are the ideal targets: they have both lattice directions allowing momentum focusing, and pairs of atoms where only the surface ejection (LS) model would give an anisotropic component, Fig.4. The existence of preferred ejection from the Zn (0001) plane has been known since the early stages of spot pattern generation /25/; it was in fact mentioned by LEHMANN and SIGMUND as an argument against focusons. But a clear assignment of the respective contributions and degrees of anisotropy required a direct comparison between the <1120> and the <2023> emission distribution. This was carried out by HOFER /30/. Figure 5, which is taken from this work, clearly shows the capacity of the surface ejection model. Since, furthermore, the Mg <1120> emission is identical with the Al <110> emission it can be concluded that for low-Z elements
- the prime anisotropy-generating mechanism along close-packed lattice directions is the L-S mechanism;
- the contribution of focusing collision sequences to total emission is negligible.

The reason why these investigations were carried out on low-Z crystals is also noteworthy: all medium-to-high-Z hcp metals showed strong ejection along <0001> /29,30/. This preferred emission, which is of the kind shown in Fig.3b, is highly disturbing for <2023> investigations. For this reason Mg was chosen, where <0001> emission is absent; the reason for this in turn is the openness of the lattice - or the weak interaction potential of its atoms; the "surface filter" is too weak. Hence low E_f, low ΔE, and the absence of <0001> emission from Mg all have the same reason: "the small core size compared to the cell size" /39/ for low-Z crystals. Again we caution, however, not to discard lightly focusing collisions in these lattices! $E_f > E_b$ must hold here, too, otherwise none of the models would explain the rather sharp spot width; and, as mentioned above, $E_f/\Delta E$ is as large in Al as in Au or Cu.

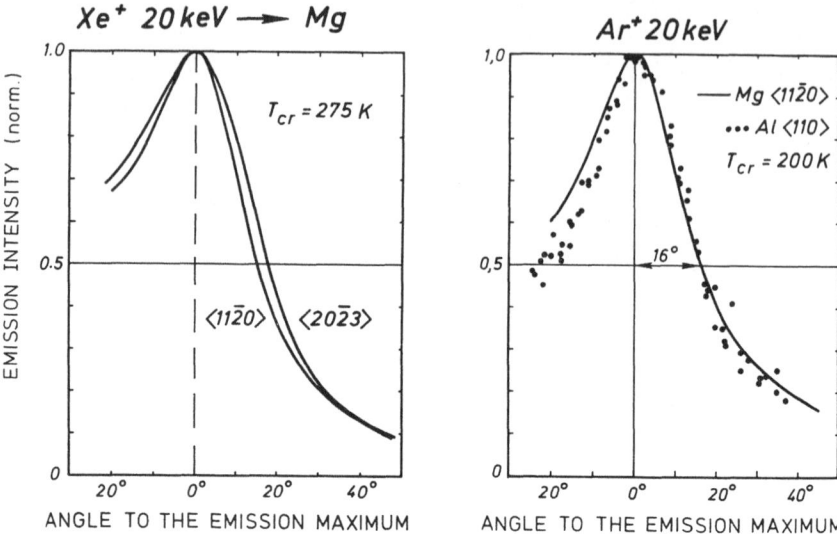

Fig. 5: *Comparison of preferred emission from low-Z single crystals.*
a) *<1120> and <2023> emission distributions from hcp Mg.*
   For <2023> only the LS-mechanism, Fig. 3, can cause preferred emission while for <1120> also focusons can contribute. Their contribution to anisotropicity is obviously small, but not negligible.

b) *<1120>emission from hcp Mg and <110> emission from fcc Al. Both lattice rows are closest-packed and allow focusing collision sequences. From HOFER /30/.*

The existence of <0001> ejection for medium and high-Z hcp crystals together with the total absence of preferred <111> emission from fcc crystals allows another important conclusion. Since in the two-layer model (LS mechanism) of Fig. 3 these two directions are identical and the difference comes only with the third layer (layer sequence in hcp: ABAB ......, in fcc: ABCABC ...), the mere existence of potential minima in the top surface layer is not sufficient to explain why there is no preferred <111> emission; rather the atom to be ejected has to receive its momentum in a directed way such as by virtue of the focusing capacity of the surrounding lattice. Such an assisted focusing emission was first described by NELSON and THOMPSON /23/. It is particularly pronounced for <100> in fcc and <0001> in hcp lattices.

At the high-Z end of the stable nuclides, Au has been investigated in great detail /23,33-38/. While the early work was naturally interpreted in terms of the focuson model, SCHULZ and SIZMANN /35,36/ found no feature in extensive studies of the temperature influence on <110> ejection which was at variance with the surface ejection model. SZYMCZAK /37,38/ recently investigated emission from the (111) surface by varying the projectile energy

over a broad regime (0.1 to 270 keV). Similarly to Ref.32, he tried to un-
ravel the anisotropic component from the random part by fitting analytic
functions to the measured distribution. Such a procedure is not unproblem-
atical because a reliable random-fit requires knowing the emission distribution over the full upper half-space /32,28/, otherwise the fit has to be
made at an angular region which is most strongly influenced by surface
structures; furthermore, a true distinction between random and surface-
deflected focusons requires energy-dispersive measurements, see Chap.3. But
it is certainly acceptable to deduce from Fig. 6 that even for high-Z elements the majority of particles is ejected randomly. The anisotropic component, however, is no longer negligible. Since the authors conclude that
the <110> ejection is mostly due to focusons, the latter conclusion holds
also for these.

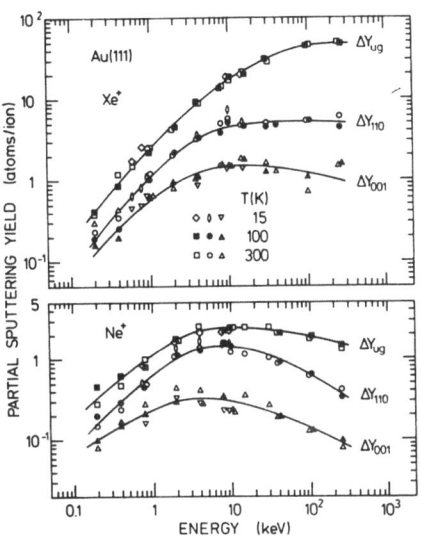

Fig. 6: Partial sputtering yields from an Au(111) surface bombarded with
Ne⁺ and Xe⁺ ions. ΔY: random emission, ΔY₁₁₀: preferred <110> emissions,
ΔY₁₀₀: preferred <100> emissions. There is hardly any influence of tem-
perature on partial yields between 15 and 300 K. From SZYMCZAK /38/.

The temperature of the target has very little influence on the sputtering of metals. This is true for total and partial yields, as well as for
the angular distribution of the random part of the emission, see
ROOSENDAAL's and HOFER's reviews in Ref.3, and Ref.30-38,47,48. The width
of preferred ejection in closest packed directions increases of course with
temperature (<110> in fcc: /33, 35-38/, <11$\bar{2}$0> in hcp: /30/) due to
enhanced lattice vibration, but this has little influence on the yield.

There was confusion for some time about whether or not sputtering near
the melting point would show an exponential yield increase and whether this

8

phenomenon could be attributed to thermal spikes /49,50/. Emission from surface areas which are intersected by such high density collision cascades was conceived as a sublimation/evaporation process which should therefore depend exponentially on temperature. This is, however, not the case. Even in situations where we can safely assume spike effects to determine emission, the angular distributions and their integrals remain only weakly dependent on bulk temperature /47/, Fig.7. Such a high-density cascade is met in heavy molecule ion bombardment: the molecule disintegrates at the first nuclear collision in the solid, the fragments deposit their energy in essentially the same cascade volume and one has energy densities where the majority of atoms is indeed in motion ($\equiv$ spike) /51,52/. Non-linearity between the numbers of impinging and ejected atoms is then observed.

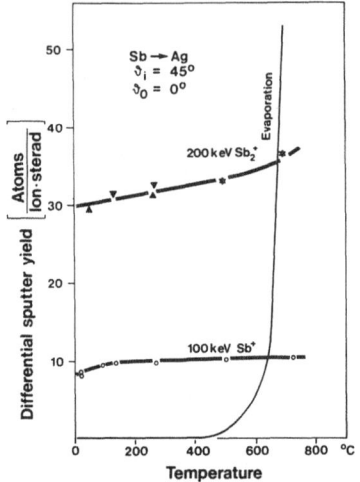

Fig. 7: Temperature dependence of the differential yield of polycrystalline silver for 100 keV per atom Sb$^+$ bombardment. From HOFER et al. /47/.

Semiconductors behave differently from metals as regards radiation damage and sputtering. Their covalent binding prevents annealing of radiation damage below about 300°C. At this amorphous/crystalline transition the sputtering yield may vary by more than a factor of 2 /53/ and the angular distribution changes from a structureless form of Eq.2.2 to an anisotropic emission distribution /54/. Semiconductors therefore lend themselves particularly well to investigations of random matter and have been chosen in this sense for angular emission distribution studies /55-57/.

Anisotropic emission from monocrystals can be observed down to energies near the ejection threshold /20,58-60/, Fig.8. In this regime, none of the aforementioned mechanisms applies since the penetration depth of the projectiles is only 1 to 2 monolayers. Ejection is then accomplished only by direct projectile/target atom collisions as the higher generations of recoils have too little energy. In this respect, low-energy sputtering is similar to light-ion sputtering.

Light-ion sputtering, i.e. sputtering by hydrogen or helium isotopes, is mostly due to primary recoils. Owing to the small energy transfer by these projectiles, energy is deposited rather locally /61-63/. This results in a predominating influence of the projectile/target atom collision on the ejection process. Specific ejection kinematics, Fig.9, are to be considered in interpretations of total (Y) and differential (dY/dΩ, dY/dE) sputtering yields. For this reason the dependencies of these quantities on the projectile's mass, energy and incident angle are much stronger than in the cascade sputtering regime, Fig.10-13. Experimental as well as computer investigations have shown that it is the direct ejection of surface atoms by (back)scattered projectiles that is the preponderant ejection mechanism of light-ion sputtering.

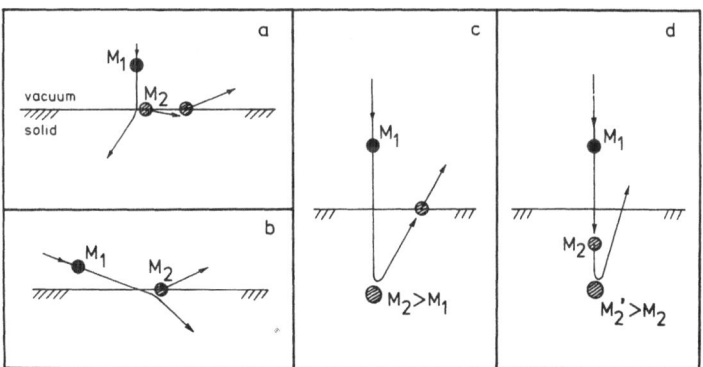

Fig. 8: <110> emission distributions from an Au(100) surface bombarded by 50 eV inert gas ions. Ejection occurs due to backscattered projectiles. From VAN VEEN and FLUIT /60/.

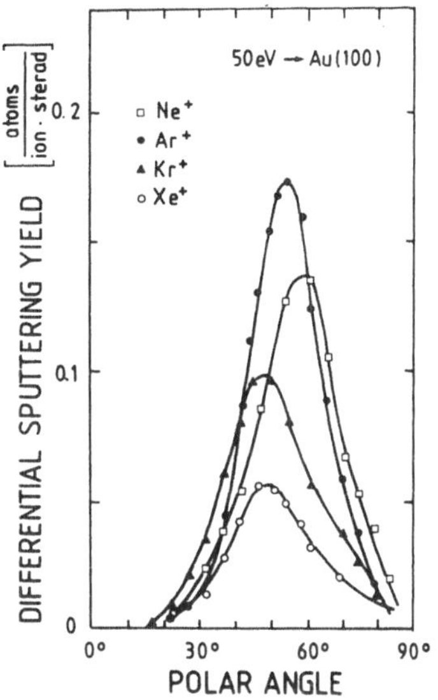

Fig. 9: Collission kinematics of low-energy projectile impacts leading to ejection of surface atoms. a + b: Ejection of primary (b) and higher-order (a: secondary) recoils, c: Ejection by backscattered projectiles, d: Ejection of atoms recoiling at heavier constituents in the target.

2.2 Emission from Polycrystals

Despite the great interest single crystals had attracted since WEHNER's discovery in 1955, ejection distributions from polycrystalline solids continued to be the subject of experimental and theoretical investigations. This is mostly because the theories for light-ion, linear cascade and spike sputtering are all for random media. But there is also great interest in angular distributions in applied fields such as plasma-surface interaction and thin film production /55,65-68/.

In the case of sputtering by linear collision cascades and elastic collision spikes the emission distribution is largely independent of mass, energy and incident angle of the projectiles, see e.g. /55,56,67/ and HOFER's review in Ref.3, vol. III. They are all rotationally symmetric with respect to the surface normal and conform to

$$\frac{dY}{d\Omega}(\theta_0) \sim \cos^{\nu_0}\theta_0 \qquad 1 \le \nu_0 < 2 \quad , \tag{2.2}$$

where θ_0 is the polar angle of the out-going particle. Such behaviour is expected from the mechanism of (linear and non-linear) cascades and is reproduced by computer simulations /63,63a/.

In the low-energy regime of medium to heavy ion sputtering i.e. $m_1 > 4$, $E_1 < 1$ keV, the specific collisions shown in Fig.9 determine emission. At perpendicular projectile incidence, emission therefore takes place primarily at large polar angles (so-called "under-cosine" distributions) while at oblique incidence emission is mainly in the opposite, "specular", direction - similar to, but by far not as pronounced as for light-ion sputtering, Fig.11. As mentioned above, light-ion sputtering is primarily an effect of projectile/ target atom collision. Hence at perpendicular incidence ejection is due to backscattered projectiles. For this reason, the angular distributions are of the over-cosine type /66/, exactly opposite to angular distributions for heavier ions /54-58/. At oblique incidence the projectiles can eject surface atoms already on their way into the solid, which causes the dramatic increase of both the total, Fig.10b, and especially the differential yield in the specular direction, Fig.11. The direct ejection of surface atoms is also apparent in the energy distribution, where they cause a shift to higher energies and a hump at the high-energy tail of the spectrum, see Chap.3.

The contribution of all the ejection mechanisms based on a special constellation of the collision partners, namely
- ejection by primary recoils (Fig. 9a),
- ejection of primary recoils (Fig. 9b,c), and
- ejection by backscattered projectiles (Fig. 9c)
becomes more and more a second-order effect as the collision cascade develops. This is primarily a matter of high enough projectile energy or, more properly, of high enough energy transfer to primary recoils.

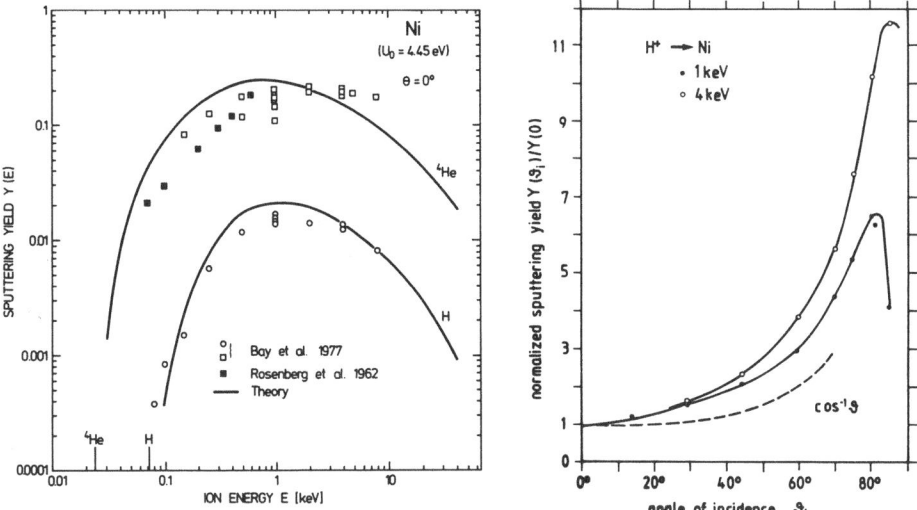

Fig. 10: *Dependence of the total yield on energy (a) and angle of in-cidence (b) of light-ion projectiles. The theoretical curves in (a) were calculated by LITTMARK and FEDDER /62/, assuming ejection to be entirely due to backscattered H⁺-ions. Fig. 10b is from BAY and BOHDANSKY /64/ and shows much stronger yield enhancement than is known in the linear cascade sputtering regime.*

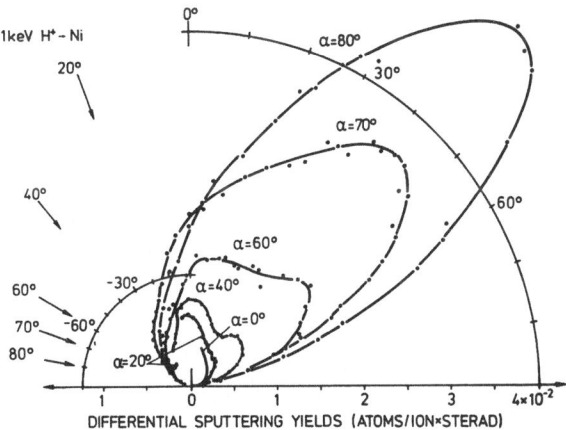

Fig. 11: *Angular distribution of Ni sputtered by hydrogen ions of various incident angles. Preferential forward emission at grazing angle is due to primary recoil ejection (cf. Fig. 9b, 10b). From BAY et al. /65/.*

2.3 Influence of Surface Topography

The development of surface structures on crystalline solids is a major problem in all sputtering investigations where fluences in excess of 10^{18} ions/cm^2 are applied. With polycrystalline targets the situation is worse because the stochastic orientation of the grains causes discontinuities in erosion speed at the grain boundaries. These discontinuities are likely to induce surface structures such as ridges and pyramids - in addition to the characteristic structures developing on the individual single crystalline grains. A polycrystalline surface subjected to intense particle irradiation will therefore be covered with arrays of differently oriented structures, mostly of the facet type, the orientation of the structures within the array (grain surface) often evincing astonishing regularity /69-71/.

Any deviation from a flat target surface must give rise to changes in yield and angular distribution, irrespective of the type of orientation of the structure. Because of the increase in yield caused by oblique incidence (cf. Fig. 10a)

$$Y(\theta_i) = Y(0) \cdot \cos^{-\nu_i}(\theta_i) \quad \text{for } 0 \lesssim \theta_i \lesssim 70^0 \quad \frac{1}{2} \lesssim \nu_i \lesssim 3 \quad (2.3)$$

structured surfaces lead to enhanced emission; but recapture of ejected particles at these ridges, facets or cones may compensate this effect. LITTMARK and HOFER /72/ considered this problem theoretically in the cascade-sputtering regime and found the total yield in all cases of Eq.2.3 to be larger than for flat surfaces. Only in the case of steep, needle-like cone structures (incidence angle $\theta_i > 70°$) will the recapture of released particles prevail and the yield therefore be smaller /73/.

Another noteworthy result was found for the angular distribution from polycrystals which develop facets. The emission distribution from such surfaces was found to be barely distinguishable from a cosine function, although the genuine distribution from a flat surface may be anything but of cosine shape. Surface structures appear to have a randomizing effect on the emission distribution. The genuine emission distribution may thus not be so well in keeping with isotropic theory as would appear from cosine-distributions obtained from structured surfaces. Careful electron microscopic control is thus a necessary accompaniment to angular distribution measurements on crystalline solids in order to guarantee that the observed distribution is not feigned by surface topography. If information on collision processes is the aim of the investigation, it is in any case preferable to conduct such investigations on amorphous solids such as semiconductors, which become amorphous when the irradiation is performed at moderate temperatures (< 300°C) /55-57/.

3. INFORMATION FROM ENERGY DISTRIBUTIONS

The energy distribution of sputtered particles affords vital information on the mechanism of sputtering and the collision processes on which it is based. Thermal spike components, the theoretically predicted E^{-2} high-energy tail, focusing energies, and the refractive action of the surface barrier have been identified by dY/dE investigations. Owing to their experimental complexity, information on energy distributions came rather late in the history of sputtering, and the number of dY/dE measurements is by far smaller than for dY/dΩ, /74-92/ and Ref.3 Vol. III.

3.1 Sputtered Ions

Measurements of the energy distribution of sputtered ions are easier to perform than the analysis of neutrals, because problems associated with post-ionization for particle detection do not arise. Such ion spectra are of limited value in elucidating sputtering mechanisms, however, because particles undergo charge-changing processes during their passage through the surface. These processes are dwell-time, thus velocity dependent and therefore strongly influence the low-energy part of the spectrum. This effect may be described phenomenologically by

$$N^+(E)dE \sim N(E) \cdot P^+(E) \; dE \qquad (3.1)$$

where $N(E)$ is the energy spectrum of sputtered neutrals, $N^+(E)$ that of sputtered ions, and $P^+(E)$ the probability that an ionized particle leaves the surface in this charge state /93-95/. Since $dP^+/dE > 0$, the measured ion distribution $N^+(E)$ differs from the energy distribution of neutrals in that the maximum is shifted to higher energies and the high energy tail is more pronounced. As will be discussed later, the same effect occurs with electronically excited (metastable) neutrals. Moreover, since ions constitute only a very small fraction ($< 10^{-3}$) of the flux of particles sputtered from clean metal surfaces, they may not be representative of the majority flux. Information on the main ejection mechanisms must, therefore, be deduced from spectra of neutral particles. There are, however, several cases where ion spectra are the only source of information, the most prominent examples being the energy distribution of sputtered clusters /92/ and that of the different components in alloys. Both fields are a domain of Secondary Ion Mass Spectrometry (SIMS).

3.2. Sputtered Neutrals

Energy analysis of sputtered neutral particles can most suitably be performed by
- time-of-flight (TOF) methods, and
- Doppler-shift methods using tuneable lasers.

The first, and also the most extensive, investigations of sputtered neutrals used the TOF technique. This is particularly true of the pioneering work of THOMPSON and co-workers. A high-speed spinning rotor was used for time-resolved registration of particles sputtered by a chopped ion beam /78/. The particle density condensed at the rotor rim reflects the velocity distribution of the sputtered flux; it was determined by autoradiography, but other trace analysis techniques can be used as well. Fig.12 is taken from this work.

The energy distribution of particles conforms to the relation (often referred to as the Thompson-formula)

$$N(E) \sim \frac{E}{(E + E_b)^3} \qquad (3.2)$$

The theoretical derivation of this relation is based on two assumptions:
- linear collision cascades; for such cascade recoils the energy distribution falls off with E^{-2}, see the overviews of ROBINSON and SIGMUND in Ref.3, Vol. I. This internal recoil spectrum is transformed into Eq.3.2 by virtue of a
- planar surface barrier ($\sim E_b \cos^{-2}\theta$)); this work-function type binding force leaves the momentum component parallel to the surface unaffected, while the normal component is reduced by an amount equivalent to the binding energy /78/.

The maximum of Eq.3.2 lies at half the sublimation energy $(\hat{E} = E_b/2)$, a result which has been verified so often in the linear cascade regime that it became customary to characterize energy spectra by the purely thermodynamical quantity E_b. On the other hand, the confirmation of Eq.3.2 is an argument against the rotationally symmetric barrier used in the LS-model for anisotropic ejection, Chap.2.1. It is conceivable that these models can be reconciled.

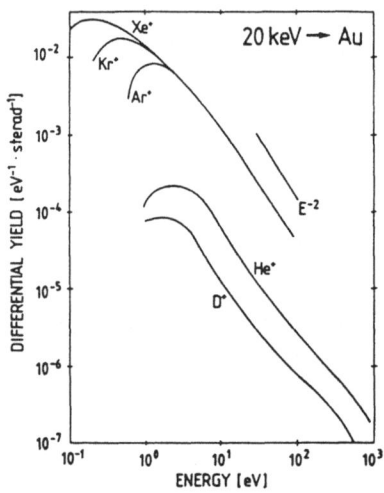

Fig. 12: Energy distribution of sputtered Au atoms from a polycrystalline target. The mass of the projectiles was varied over such a large range that light-ion sputtering, linear cascade sputtering, and sputtering by collision spikes were covered. $\theta_i = 45°$, $\theta_o = 75°$. From AHMAD et al. /81/.

In reference to Fig.12, linear cascades are represented there by Ar^+ bombardment. There are deviations, however, for both substantially lighter (D^+) and heavier (Xe^+) projectiles. For light-ion sputtering neither the position of the maximum nor the high-energy tail conforms to predictions from linear cascade theory. This is explained by light-ion sputtering's being due to scattered or backscattered projectiles, see Chap.2. For this reason, the energy spectra also depend on the projectile's energy and incident angle /88,91/, Fig.13a,b; precisely this is not the case in the linear cascade regime /79-81/.

For heavy-ion bombardment, on the other hand, the high energy part of the spectrum in Fig.12 falls off with E^{-2} as is characteristic of linear cascades, but the low-energy portion of the spectrum is more pronounced. This is ascribed to collision spikes. Emission of atoms from spikes is conceived as a kind of evaporation from surface areas with temperatures of $3-10 \times 10^3$ K. In extreme cases temperatures up to 2×10^4 K have been reported /96/. There is still some controversy as to the magnitude of the partial spike yield deduced from unfolded energy spectra, as well as its dependence on target temperature. Neither the high spike yields nor the exceedingly large spike temperatures reported in /96/ have ever been reproduced. Contrary to the total yield, however, there seems to be some influence of the target temperature on the partial spike yield in energy spectra /79/.

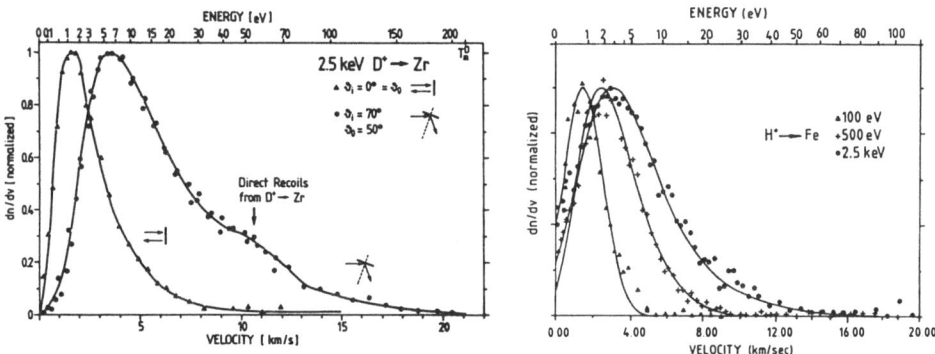

Fig. 13: Velocity distribution of sputtered ground state atoms from poly-crystalline Fe and Zr targets sputtered by light ions. The distributions are shifted towards higher energies in

a) *with increasing projectile energy owing to higher (primary) recoil energies. Average velocities are at 100 eV: 1.8 km/s, 500 eV: 4 km/s, 2500 eV: 6.2 km/s. From BAY and SCHWEER /91/.*

b) *with increasing incident angle owing to direct ejection of surface atoms. Average velocities: 6.5 and 11.5 km/s. From BAY and BERRES /88/.*

The results shown in Figs.13,14 were obtained by the Doppler-shift method, see also BETZ et al., this Study Institute. It is based on the shift of the excitation frequency by the velocity (Doppler effect) and is carried out by measuring the fluorescence yield as a function of the detuning of a (CW or pulsed) laser. The advantage of this method is that it can distinguish between atoms of different excitation, charge, and binding states. It thus allows discrimination of sputtered clusters (agglomerates of ejected atoms) which may falsify the low energy part of the spectrum, and it opens up a hitherto inaccessible field of spectroscopy, namely that of metastable neutrals, Fig.14.

16

3.3 Excitation and Charge State

During their egress from the surface the particles experience electronic transitions. The probability of excitation/ionization falls sharply with increasing excitation energy, whence ionization is a rare event ($< 10^{-3}$) compared to ejection as an excited neutral. Excited neutral particles either de-excite spontaneously, in which case they can be studied by spectroscopy of the light emitted immediately in front of the target surface, or they remain in metastable states. These excited metastable neutrals have recently attracted great interest because laser induced fluorescence has just come into the position of being able to determine their population and energy distribution. It was only in this decade that it was realized that excited neutrals may amount to up to 50 % of the sputtered neutral flux from clean metal surfaces /86/. The actual fraction depends on the projectile energy, and even more so on reactive gas coverage /84-90/. With oxygen exposure, for instance, the energy spectra change drastically: the ground state neutrals are reduced in intensity (a reduction factor of 50 was found for 15 keV $Ar^+ \rightarrow Cr$, and of 120 for 1 keV $Ar^+ \rightarrow Ti$) and show a deficiency in the low-energy part of the spectrum; the metastables show a shift of the spectra towards higher energies which increases with the excitation level, Fig. 14. In good agreement with observations on secondary ions, de-excitation or neutralization (see Chap.3.1) increases with the dwell time of exiting particles in the surface region. The survival probability P(E) of an atom in a particular excitation state therefore falls with the velocity $v_\perp$ normal to the surface, a dependence of the form

$$P(E) \sim \exp\left(-\frac{cons.}{v_\perp}\right) \tag{3.3}$$

often being in at least qualitative agreement with observation. P(E) can be generalized to include ionized states in Eq.3.1 as well.

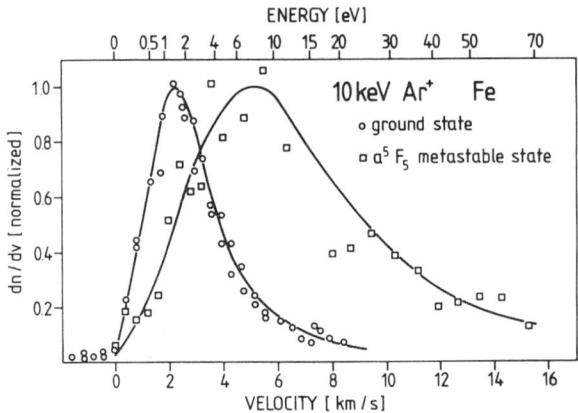

Fig. 14: Velocity distribution of sputtered ground state and metastable excited state (ΔE = 0.6 eV) neutrals from polycrystalline iron. The average velocity of the metastables is 19.6 km/s, a factor of 2 larger than the ground state neutrals. While the distribution of ground state atoms agrees with Eq. 3.2, there is no way to fit such a function with a thermodynamically reasonable binding energy. From SCHWEER and BAY /87/.

4. COLLECTIVE EMISSION

4.1 Cluster Emission

When around 1960 the first mass resolved flux measurements of positive and neutral particles were carried out, it became evident that a large fraction of particles is ejected as agglomerates of target atoms, clusters as they are called /98-101/. For many elements the intensity of cluster ions is so large that it can be utilized for production of composite ion beams which cannot otherwise be generated, e.g. Al_{17}^+, V_{14}^+, Cu_{11}^+, U_5^+, $(V_3 Nb_4)^+$ /102,103/. Owing to this large intensity, clusters can also be used for chemical analysis of composite materials since their stoichiometry correlates with that of the target (SIMS or SNMS[*] analysis) /104-107/.

One should not under-estimate the fraction of bound species in the flux of sputtered particles. For secondary ions the number of atoms contained in agglomerates can easily amount up to 40 % of the charged flux. In the case shown in Fig. 15, for instance, summation over only the first three clusters (n = 2,3,4) yields about as many atoms ejected in clusters as in the atomic state. The key question, however, is whether information obtained with charged clusters can also be transferred to neutral clusters. Neutral diatomic fractions of 20 to 40 % can be deduced from published work on Cu and Ag /100, 105, 107/, and recent measurements of GNASER & HOFER suggest similarly high fractions for Ge. But it appears doubtful whether this result can be generalized /108/.

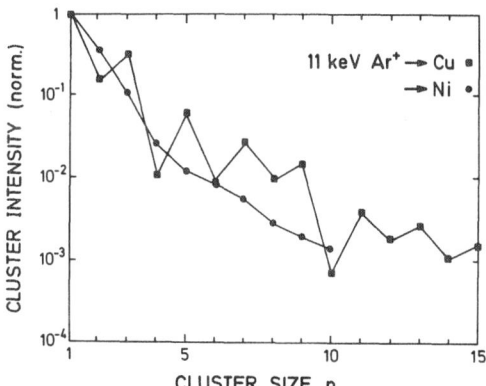

Fig. 15: Intensity ratio of clusters sputtered from polycrystalline copper and nickel. The appearance of oscillations in the abundance ratio for clusters with change in parity of the binding electrons (here Cu_n^+) reflects the importance of stability on cluster ejection. It should be noted that the number of atoms ejected in charged clusters is comparable in magnitude to the number of ejected atomic ions. From RODRIGUEZ-MURCIA and BESKE /115/.

[*] Secondary Ion Mass Spectrometry
Sputtered Neutrals Mass Spectrometry

Cluster emission is one of the least understood fields in ion-solid interaction. Even the basic ejection mechanism is not established on a generally accepted basis. Two models exist which take quite opposite views: collective ejection, and statistical agglomeration of independently emitted atoms in the vacuum. Collective emission of surface atoms which constitute the cluster after the ejection event was the general consensus up to about 1975, when some measurements (on diatomic clusters), statistical calculations, and computer simulations suggested that clusters may also form by association above the surface /105,106,109-111/.

There is presently no decisive information obtainable from theory. Owing to the small energy of clusters /92/, the evolution of collision cascades has to be traced down to energies well below 10 eV. This is a regime where attractive forces also have to be included and the binary collision approximation is too crude. Plural particle interaction, however, has as yet been treated only by computer simulation. These calculations suffer from severe statistical problems due to the appreciable computation time required per emission event. Moreover, cluster stability after emission is an intricate issue /112-117/. Electronic, vibrational and rotational excitation may prevent the formation of a bound state. The importance of cluster stability is reflected impressively in the oscillating Cu_n^+ abundance distribution, Fig.15, which was first reported in sputtering by BLAISE & SLODZIAN (Cr_n^+, Cu_n^+, /113/) and HORTIG & MÜLLER (Ag_n^-, /102/). The effect was known at that time from spark source mass spectrometry of group IV elements, see /111/ and references therein; also the explanation in terms of the parity of molecular orbitals is the same /112, 114/.

Owing to these complications in quantitative descriptions of cluster models one can do no more than rule out certain processes by general physical arguments. There is first the vanishingly small probability that clusters larger than dimers (n > 2) form by statistical association after the ejection event. Note that Fig.15 shows ejection probabilities of 10^{-2} up to n = 9. In a directed flux of atoms this can be accomplished only in correlated emission. This reduces the controversy to n = 2 clusters. For these, ejection from lattice sites further apart than next nearest neighbours can be ruled out because otherwise binding of the two atoms cannot be accomplished. Binding requires the relative kinetic energy to be smaller than the potential well. This rules out association of atoms which have not been nearest or next nearest neighbours in the solid. Similar to this argument which restricts the allowed area from where sputtered atoms may stem if they are to form a cluster, there is also a restriction which narrows down the perpendicular dimension: the last chance for the two atoms to dissipate excess energy is in the interaction region with the surface; this is again of the order of 5 Å. As these necessary conditions restrict the allowed volume to about 5x5x5 Å it is very suggestive indeed to perceive the emission event as a collective one - even more so as there are also restrictions in time /92/: in order that particles can interact above the surface, the maximum time interval between two momentum transfer processes must not be larger than some 10^{-14} s. This is a very small fraction of the lifetime of a collision cascade.

To summarize, cluster emission thus appears to be an event at the very-low-energy stage of the cascade, where non-binary interactions favour simultaneous ($\Delta t < 10^{-13}$ s) knock-on processes near and in the surface, and where soft collisions upon egress from the surface reduce the relative kinetic energy of the particles to be bound in the cluster or molecule. These cluster components need not necessarily stem from nearest neighbour sites in the lattice, although this becomes a more and more stringent condition with increasing cluster size. In all probability, nonlinear effects in high-density cascades near the surface will promote cluster emission.

4.2 Surface Craters and Sub-Cascades

An extreme case of non-linearity and collective emission has been reported by MERKLE & JÄGER /118/. In electron-microscopy investigations of defect structures created by heavy-ion bombarded Au MERKLE observed - in addition to collapsed depleted zones in the bulk - surface craters of about 50 Å in diameter. About 10^3 atoms have to be removed to form such a crater. The number of these craters increased sharply with energy and was particularly pronounced for molecular ion (Bi$_2^+$) bombardment Fig.16. At an energy of 100 keV the material removed just from these craters- there are presumably many more of sub-microscopic size - had reached the yield value calculated from linear cascades. This is the same energy range where BAY et al. /119/ have measured enhanced Au-selfsputtering yields which they have attributed to collision spikes. MERKLE follows this explanation but points out that it is not so much the average (linear) cascade size which determines high-energy (> 100 keV) sputtering, but the individual sub-cascade as long as it is generated only near enough to the surface. These sub-cascades are the zones which constitute the spike, and their distribution in space must be known in order to allow a more accurate calculation of sputtering yields. This problem is coupled to the often addressed fluctuations in energy deposition in individual cascades; for further details in connection with cluster emission see also /120/.

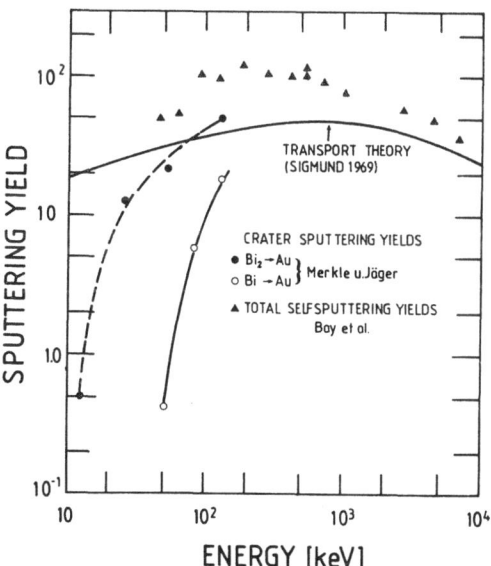

Fig. 16: Energy dependence of the total (Δ) and partial (o) sputtering yield of polycrystalline gold due to Au$^+$, and Bi$^+$, Bi$_2^+$ bombardment, respectively. From MERKLE & JÄGER /118/.

4.3 Microparticle and "Chunk" Emission

Microparticle emission had already been observed in the first erosion investigations of cathodes in glow discharges. PLÜCKER /10/ reported luminous objects leaving the cathode without any influence of electric or magnetic fields on their trajectory. HITTORF /9/ gave an explanation in terms of cone formation: after prolonged bombardment of Al-cathodes he observed needle or hair-like structures which, owing to their reduced thermal contact to the bulk of the cathode, start melting at their tips and thereby emit macroscopic particles. Such an explanation can also be found in recent work on exfoliation and blistering, see SCHERZER's review in Ref.3, Vol.II.

Emission of μm-sized particles containing up to 10^{10} atoms was recently also found in neutron sputtering, where it caused the sputtering yields to exceed the physically reasonable values ($\sim 10^{-5}$ atoms/neutron) by orders of magnitude /121/. The concern which this high erosion caused in the thermonuclear fusion community triggered off intensive research in several laboratories. The final consensus was that chunk emission in neutron sputtering is due to loosely bound particles on "technical surfaces" /122/.

5. SUMMARY

Sputtering of elemental metals and semiconductors is well understood in the regimes of specific knock-on ejection, i.e. light-ion as well as near-threshold sputtering, and linear collision cascades. Theory is often in remarkably good quantitative agreement with experiment, in spite of large uncertainties in the interaction potential at those energies which dominate in the emission spectrum.

The lack of knowledge of interatomic forces, in both the repulsive and the attractive regions, is the main obstacle to a better understanding of the emission and stability of clusters. Their contribution to total ejection is not negligible.

Understanding of sputtering by collision spikes also suffers from uncertainties in the low-energy interaction region, but the main problem to solve in this regime is the energy transport across the spike/solid interface as this determines the evolution of the spike in space and time as well as energy confinement in the cascade volume. Sputtering by spikes is also experimentally not satisfactorily established; it is still not clear to what extent the low-energy part of energy spectra is influenced by non-linear effects, the lattice temperature, and clusters.

Acknowledgements

I should like to thank M.M. Ferguson for her continual help during the preparation of the manuscript and H.H. Andersen, R. Kelly and J.L. Whitton for valuable comments.

* EURATOM Association

REFERENCES

1. Behrisch R.: Festkörperzerstäubung durch Ionenbeschuß, in Ergebn. Exakt Naturwiss. 35 (1964) 295
2. Kaminsky M.: Atomic and Ionic Impact Phenomena on Metal Surfaces, Springer-Verlag Berlin, Heidelberg, New York (1965)
3. Behrisch R. (ed.): Sputtering by Particle Bombardment
   Vol. I: Physical Sputtering of Single-Element Solids (1981)
   Vol. II: Sputtering of Alloys and Compounds, Electron and Neutron Sputtering, Surface Topography (1983)
   Vol. III: in preparation (2001)
4. Behrisch R., W. Heiland, W. Poschenrieder, P. Staib, H. Verbeek (Eds.): Ion Surface Interaction, Sputtering and Related Phenomena, Gordon & Breach, London, New York Paris (1973)
   Also published as volume 18/19 in Radiation Effects
5. Varga P., G. Betz, F.P. Viehböck (Eds.): Symposium on Sputtering, Perchtoldsdorf/Vienna (1980)
   Published by IAP, Technische Universität Wien, Austria
6. Stark J.: Die Elektrizität in Gasen, Barth, Leipzig (1902)
   "In erster Linie haben wir die Zerstäubung als Verdampfung zu betrachten und auf die hohe Temperatur an der Oberfläche der Kathode zurückzuführen. Es können die Metallmoleküle da, wo positive Ionen auftreffen, infolge von deren riesiger kinetischer Energie für kurze Zeit unmittelbar nach dem Auftreffen eine sehr hohe Temperatur von den positiven Ionen übernehmen, infolge davon in den Gasraum hineinverdampfen ..."

   "We must regard sputtering primarily as evaporation caused by the high temperature at the surface of the cathode. Where the positive ions strike, the metal molecules can acquire from them, as a result of their huge kinetic energy, a very high temperature which persists for a short time directly after the impact; the metal particles consequently evaporate into the vacuum ... " p. 434
7. Thomson J.J., G.P. Thomson: Conduction of Electricity through Gases, Cambridge Univ. Pr., Vol. I (1928), Vol. II (1933)
8. Stark J.: Z. Elektrochem. 14 (1908) 752, and 15 (1909) 509
   "Ein Atomstrahl (Kanalstrahl) besitzt eine große kinetische Energie; beim Auftreffen auf ein Atom an der Oberfläche eines festen oder flüssigen Körpers kann er an dieses einen Teil seiner kinetischen Energie durch Stoß abgeben, mag er hierbei reflektiert werden oder nicht. Indem das getroffene Atom eine Geschwindigkeitskomponente in der Richtung des auftreffenden Atomstrahls von diesem übernimmt, wird es in gewissem Sinne selbst zu einem geladenen oder ungeladenen Strahle und kann, indem es auf seinem Wege auf ein anderes Atom trifft, an diesem eine Reflexion erfahren: und erhält es hierbei eine Geschwindigkeitskomponente senkrecht zur Oberfläche nach außen, so kann es aus dieser heraus in den Gasraum übertreten ..."

   "An atomic ray (canal ray) possesses a large kinetic energy; during impact with an atom on the surface of a solid or fluid body, it can give up a part of its kinetic energy to the atom in the collision, whether or not thereby undergoing reflection. The atom struck acquires a component of velocity in the direction of the incident atomic beam, it will itself in a sense become a charged or uncharged ray and can experience reflection by striking another atom in its path; should it thus acquire a velocity component outwards normal to the surface, it can exit from the surface into the vacuum ..." p. 754

22

9. Hittorf W.: Ann. Physik u. Chemie (Lpzg.) 21 (1884) 90
10. Plücker J.: ibid. 104 (1858) 113
11. Granquist G.: Öfvers. Svenska Vet. Akad. Förh. 55 (1898) 709
12. Kohlschütter V.: Die Zerstäubung durch Kanalstrahlen, Jahrb. d. Radio-
    aktivität u. Elektronik 9 (1912) 355
13. Dessauer Fr.: Z. Physik 23 (1923) 38
14. Kapitza P.: Phil Mag. 45 (1923) 989 see, however, also A. Becker: Ann.
    Physik 75 (1924) 217 and Ref. 18, where a more critical view of thermal
    electron emission is taken.
15. Hippel A. v.: Ann. Physik 81 (1926) 1043
16. Seeliger R., K. Sommermeier: Z. Physik 93 (1935) 692
17. Sommermeyer K.: Ann. Physik 25 (1936) 481
18. Morgulis N.D.: Z. Eksper. Teoret. Fiz. 9 (1939) 1484 (in Russian)
19. Townes C.H.: Phys. Rev. 65 (1944) 319
20. Wehner G.K.: J. Appl. Phys. 26 (1955) 1056, and
    Phys. Rev. 102 (1956) 690
21. Yurasova V.E.: Sov. Phys.-Techn.Phys. 3 (1958) 1806
    Yurasova V.E., N. Pleshivtsev, I. Orfanov: Sov. Phys.-JETP 37 (1960) 689
    Yurasova V.E., I.G. Sirotenko: ibid. 14 968 (1962)
22. Thompson M.W.: Phil. Mag. 4 (1959) 139
23. Nelson R.S., M.W. Thompson: Proc. Roy. Soc. A 259 (1961) 458
24. Molchanov V.A., V.G. Tel'kovskii, V.M. Chickerov:
    Sov. Phys.-Doklady 6 (1961) 486
25. Perovic B.: Bull. Inst. Nucl. Sci. "Boris Kidrich" 11 (1961) 37
26. Cooper C.B., J. Comas: J. Appl. Phys. 36 (1965) 2891
27. Chen G.P., Besocke K., S. Berger, W.O. Hofer: to be published
28. Linders J., H. Niedrig, M. Sternberg: Nucl. Instr. Meth.
    Phys. Res. B2 (1984) 649
29. Robinson M.T., A.L. Southern: J. Appl. Phys. 38 (1967) 2969
    Robinson M.T.: ibid. 40 (1969) 4982
    Robinson M.T., A.L. Southern: ibid. 39 (1968) 3463
30. Hofer W.O.: Thesis Universität München 1972, and Ref. 4 p. 7
31. Olson N.Th., H.P. Smith: AIAA-Journal 4 (1966) 916
    Olson N.Th., H.P. Smith: Phys. Rev. 157 (1967) 241
32. Musket R.G., H.P. Smith: J. Appl. Phys. 39 (1968) 3579
    Higgins T.B., N.Th. Olson, H.P. Smith: J. Appl. Phys. 39 (1968) 4849
33. Nelson R.S., M.W. Thompson, H. Montgomery: Phil. Mag. 7 (1962) 1385
34. Chapman G.E., J.C. Kelly: Austral. J. Phys. 20 (1967) 283
35. Schulz F., R. Sizmann: Proc. 8th Int. Conf. Ion. Phen. Gases,Vienna,
    1967, p. 35
36. Schulz F.: Thesis Techn. Universität München 1967
37. Szymczak W., K. Wittmaack: Nucl. Instr. Meth. 194 (1982) 561
38. Szymczak W.: Thesis, Universität München 1985
39. Silsbee R.H.: J. Appl. Phys. 28 (1957) 1246
40. Lehmann Chr., G. Leibfried: Z. Physik 162 (1971) 203
41. Robinson M.T.: Ref. 2 Vol. I p. 73-218
42. Nelson R.S., R. v. Jan: Phil Mag. 17 (1968) 1017
43. Lehmann Chr., P. Sigmund: phys. stat. sol. 16 (1966) 507
44. Harrison D.E., J.P. Johnson, N.S. Levy: Appl. Phys. Lett. 8 (1966) 33,
    and J. Appl. Phys. 39 (1968) 3742
45. Brandt W., R. Laubert: Nucl. Instr. Meth. 47 (1967) 201
46. Sigmund P.: Phys. Rev. 184 (1969) 383
47. Besocke K., S. Berger, W.O. Hofer, U. Littmark: Radiat. Eff. 66 (1982) 35
    Hofer W.O., K. Besocke, B. Stritzker: Appl. Phys. A 30 (1983) 83
48. Sigmund P., M. Szymonski: Appl. Phys. A 33 (1984) 141

49. Nelson R.S.: Phil. Mag. 11 (1965) 291
50. Kelly R.: Radiat. Eff. 32 (1977) 91
51. Andersen H.H., H.L. Bay: Radiat. Eff. 19 (1973) 139
52. Sigmund P.: Appl. Phys. Lett. 25 (1974) 169
53. Sommerfeldt H., E.S. Mashkova, V.A. Molchanov: Radiat. Eff. 9 (1971) 267,
    and Phys. Lett. 38 A (1972) 237
    Holmén G.: Radiat. Eff. 24 (1975) 7
54. Anderson G.S., G.K. Wehner: J. Appl. Phys. 34 (1963) 3492, and
    Surface Sci. 2 (1964) 367
    Anderson G.S.: J. Appl. Phys. 37 (1966) 3455
55. Tsuge H., S. Esho: J. Appl. Phys. 52 (1981) 4391
56. Okutani T., M. Shikata, S. Ishimura, R. Shimizu: J. Appl. Phys. 51
    (1980) 2884
57. Andersen H.H., B. Stenum, T. Sørensen, H.J. Whitlow: Nucl. Instr. Meth.
    Phys. Res. B 6 (1985) 459
58. Koedam M., Thesis Reijksuniversiteit Utrecht 1961
59. Weijsenfeld C.H.: Thesis Reijksuniversiteit Utrecht 1966
60. Veen van A., J.M. Fluit: Nucl. Instr. Meth. 170 (1980) 341, and
    Thesis Reijksuniversiteit Utrecht 1979
61. Littmark, U., G. Maderlechner: SPIG 1976 Dubrovnik Jugoslavia, p. 136
62. Fedder S., U. Littmark: Nucl. Instr. Meth. 194 (1982) 607
63. Biersack J.P., W. Eckstein: Appl. Phys. A 34 (1984) 73
63a.Hautala M., H.J. Whitlow: Nucl. Instr. Meth. Phys. Res. B6 (1985) 466
64. Bay H.L., J. Bohdansky: Appl. Phys. 19 (1979) 421
65. Bay H.L., J. Bohdansky, W.O. Hofer, J. Roth: Appl. Phys. 21 (1980) 327
66. Hofer W.O., H.L. Bay, P.J. Martin: J. Nucl. Instr. 76/77 (1978) 156
67. Rödelsperger K., A. Scharmann: Z. Physik B 28 (1977) 37, and
    Nucl. Instr. Meth. 132 (1976) 355
68. Motohiro T., Y. Taga, K. Nakajima: Surf. Sci. 118 (1982) 66
69. Hermanne, N. Ref. 4 p. 161
70. Alexander V., H.J. Lippold, H. Niedrig: Radiat. Eff. 56 (1981) 241, and
    Ref. 5, p. 622
71. Carter G., B. Navinsek, J.L. Whitton: Ref. 3 Vol. II p. 231
72. Littmark U., W.O. Hofer: J. Mater. Sci. 13 (1978) 2577
73. Panitz J.K.G., D.J. Sharp: J. Vac. Sci. Technol. 17 (1980) 282
74. Thompson M.W., R.S. Nelson: Phil. Mag. 7 (1962) 2015
75. Thompson M.W.:Phys. Lett. 6 (1963) 24
76. Stuart R.V., G.K. Wehner: J. Appl. Phys. 35 (1964) 1819, and
    40 (1969) 803
77. Hulpke E., Ch. Schlier: Z. Physik 207 (1967) 294
78. Thompson M.W., B.W. Farmery, P.A. Newson: Phil. Mag 18 (1968) 361
    Thompson M.W.: ibid. 18 (1968) 377
    Farmery B.W., M.W. Thompson: ibid. 18 (1968) 415
79. Chapman G.E., B.W. Farmery, M.W. Thompson, I.H. Wilson: Radiat. Eff. 13
    (1972) 121
80. Reid I.H., M.W. Thompson, B.W. Farmery: Phil. Mag. A 42 (1980) 151
81. Ahmad S., B.W. Farmery, M.W. Thompson: Phil. Mag. A 44 (1981) 1387
82. Ahmad S., M.W. Thompson: Phil. Mag. A 50 (1984) 299
83. Husinsky W., G. Betz, I. Girgis: J. Vac. Sci. Technol. A 2 (1984) 698
84. Husinsky W., G. Betz, I. Girgis, F. Viehböck, H.L. Bay: J. Nucl. Mater.
    128 & 129 (1984) 577
85. Bay H.L., W. Berres, E. Hintz: Nucl. Instr. Meth. 194 (1982) 555
86. Schweer B., H.L. Bay: in Proc. IV Int. Conf. on Solid Surfaces and
    IIIrd Eur. Conf. on Surf. Sci., ed. by D.A. Degras ans M. Costa
    (Société Francaise du Vide, Paris 1980) p. 1349

24

87. Schweer B., H.L. Bay: Appl. Phys. A 29 (1982) 53
    Dullni E.: Appl. Phys. A38 (1985) 131
88. Bay H.L., W. Berres: Nucl. Instr. Meth. Phys. Res. B 2 (1984) 606
89. Yu M.L., D. Grischkowsky, A.C. Blatant: Phys. Rev. Lett. 48 (1982) 427
90. Wright R.B., M.J. Pellin, D.M. Gruen: Surf. Sci. 110 (1981) 151
    Young C.E., W.F. Calaway, M.J. Pellin, D.M. Gruen: J. Vac. Sci. Techn.
    A2 (1984) 693
91. Bay H.L., B. Schweer: in Sympos. on Surf. Sci. Obertraun, Austria 1985,
    ed. by G. Betz et al. TU Wien p. 147
92. Staudenmaier G.: Radiat. Eff. 13 (1972) 87
93. Wittmaack K. in Inelastic Ion-Surface Collis., eds.: N.H. Tolk et al.,
    Academic Press 1977 p. 153; review paper.
94. Veksler V.I.: Sov. Phys. JETP 11 (1960) 235
95. Vasile M.J.: Phys. Rev. B 29 (1984) 3785
96. Szymonski M., A.E. de Vries: Phys. Lett. 63A (1977) 359, and
    J. Phys. D, 11 (1978) 751
97. de Vries A.E.: Ref. 5 p. 256
98. Honig R.E.: J. Appl. Phys. 29 (1958) 549
99. Krohn, V.E.: J. Appl. Phys. 33 (1962) 3523
100. Woodyard J.R., C.B. Cooper: J. Appl. Phys. 35 (1964) 1107
     Woodyard J.R.: 15th Ann. Conf. Mass Spec. Appl. Top., Denver (1967)
     p.254
101. Hofer W.O.: Nucl. Instr. Meth. 170 (1980) 275, review paper
102. Hortig G., M. Müller: Z. Physik 221 (1969) 119
103. Thum F., W.O. Hofer: Surf. Sci. 90 (1979) 331
104. Schou J., W.O. Hofer: Appl. Surf. Sci. 10 (1982) 383
105. Oechsner H., W. Gerhard: Surf. Sci. 44 (1974) 480
106. Gerhard W.: Z. Physik B 22 (1975) 41
107. Oechsner H., E. Stumpe: Appl. Phys. 14 (1977) 43
108. Gnaser H., W.O. Hofer: to be published
109. Können G.P., A. Tip, A.E. de Vries: Radiat. Eff. 21 (1974) 269,
     and 26 (1975) 23
110. Bitenskii I.S., E.S. Parilis: Sov. Phys. Tech. Phys. 23 (1978) 1104
111. Harrison D.E., C.B. Delaplain: J. Appl. Phys. 47 (1976) 2252
112. Dörnenburg E., H. Hintenberger: Z. Naturforschg. 14a (1959) 765,
     and 16a (1961) 532
113. Blaise G., G. Slodzian: C.R. Acad. Sc. Paris, Ser. B 266 (1968) 1525
114. Joyes P., J. Phys. Chem. Solids 32 (1971) 1269
115. Rodriguez-Murcia H., H.E. Beske: Adv. in Mass Spectrom. 7 (1978) 598,
     and Report Jül-1292 (1976)
116. Dzhemaliev N.Kh., R.I. Kurbanov: Izv. Akad. Nauk SSSR, Ser. Fiz. 43
     (1979) 606
117. Snowdon K.: Nucl. Instr. Meth. B   in press
118. Merkle K.L., W. Jäger: Phil. Mag. A 44 (1981) 741
     Merkle K.L.: 35th Ann. Proc. Electron Microscopy Soc. Amer., Boston,
     Mass. (1977), ed.: G.W. Bailey, p. 36
119. Bay H.L., H.H. Andersen, W.O. Hofer, O. Nielsen: Nucl. Instr. Meth.
     133 (1976) 301
120. Sigmund P.: Ref. 93 p. 121
121. Kaminsky M., J. Peavey, S. Das: Phys. Rev. Lett. 32 (1974) 599,
     and J. Nucl. Mater. 53 (1974) 162
122. Behrisch R.: Nucl. Instr. Meth. 132 (1976) 293, and Ref. 3, Vol. II p.
     179, review articles

ION IMPLANTATION MECHANISMS AND RELATED COMPUTATIONAL ISSUES

Omer F. Goktepe
White Oak Laboratory, Naval Surface Weapons Center
Silver Spring, Maryland 20903-5000 USA

INTRODUCTION

Ion implantation modifies the electric and atomic structures of solids and it potentially has various important scientific and technological applications. However, for a successful application of ion implantation it is necessary to understand well the physical and chemical mechanisms involved. At present, the low-energy mechanisms of the atomic collisions phenomena in solids is perhaps the least understood domain of ion implantation. This is partly due to the complexity of the mechanisms involved at low energies in which the energetic atoms and their cascade atoms end the "slowing-down" process and begin the "thermalization" process where the various chemical forces dominate the purely collisional physical forces in a much larger time frame. Also, at present, the analytically available solution for the transport of charged particles does not adequately represent the low-energy region.

The slowing down of fast particles is commonly studied by using the Boltzmann transport equation in its familiar "forward" form. The counterpart of this situation is solving the transport equation in the "adjoint" or "backward" form. Such an equation was first obtained by Lindhard(1) from the probabilistic considerations and later Sigmund typified, solved in the asymptotic region and applied it extensively to sputtering(2) and other atomic collision effects(3,4). The solution for the adjoint form of the transport equation differs from the one for the forward form in that the energy variable is reversed, i.e., it is the initial rather than the final energies which are the independent variables. Therefore, the asymptotic solution which is valid in the region far from the high-energy source region in the forward form corresponds to the asymptotic solution which is valid in the region far from the low-energy "response" region in the adjoint form.

After the description of ion implantation phenomena, the little known methods of solution, namely the forward and adjoint modes of the transport equation and their utilizations are emphasized. Due to difficulties in obtaining the exact solutions, their asymptotic solutions' validity is discussed. It is noted that the asymptotic solution of the adjoint equation which is presently available, and valid for the region far from the low-energy region cannot adequately incorporate the low-energy effects such as defect or displacement mechanisms.

In section (A) the general mechanisms of ion implantation such as the slowing-down process, sputtering, atomic mixing and the little understood defect and displacement mechanisms are discussed. In section (B) a general discussion of the forward-adjoint formulation as applied to the Boltzmann transport equation is presented. In section (C) the fundamentals of the Monte Carlo simulation method are introduced as a scientific tool. In the final section (D), an example of a multiple-energy ion implantation simulation result which reveals some non-trivial fluence-dependent features is shown.

A) DESCRIPTION OF ION IMPLANTATION PHENOMENA

As a beam of energetic ions penetrates a solid, the kinetic energy is dissipated by two processes, namely the electronic interactions and the nuclear collisions. These two processes may be taken as independent of each other to a good approximation. In the electronic interactions the particle loses energy in a nearly continuous process, where atomic excitation, ionization and charge exchange dominate. In the nuclear stopping process, except at very low energies, the interaction between the beam atom and the target atoms is of an isolated or binary collision type and results in a scattering, in which the incident beam atom suffers a loss of energy and a change in the direction of its motion. Consequently the trajectory of the beam atom inside the target is mainly determined by a sequence of nuclear collisions with target atoms. Finally, the beam atom is either completely stopped in the target or emerges from the surface with some residual energy.

The beam ions that are stopped inside the solid may either be permanently trapped or may migrate to the surface by way of diffusion and segregation. A simple model of implantation based on only collisional mechanisms excluding diffusion and segregation has been adapted from ref (5) and it is shown with the solid lines in Fig. 1 at different fluences. In this model it is further assumed that (i) there is no range-shortening due to implanted ions in the target, (ii) total sputtering yield of the target is constant and no preferential sputtering takes place, and (iii) there is no atomic or recoil mixing effect. Therefore, a Gaussian range distribution for the implants and a constant surface recession velocity due to sputter erosion as a function of fluence are assumed.

Starting from a low fluence level as shown in Fig. 1a the trapped beam atoms continuously build up a concentration profile in the target. After the receding surface crosses the portions of implant profile (1b), there is a competition between implant collection and implant removal. Since the bulk of the solid sample offers a large sink for beam atoms and since under the constant ion bombardment these atoms may be subsequently released by erosion of the solid matrix, the concentration of implanted beam atoms will build up until equilibrium is reached (1c).

One of the complications of the ion implantation mechanism is the cascade effect due to the generation of recoil atoms. In general, each incident beam atom produces direct and indirect displacement of atoms of the solid from their lattice sites. The energetic target atom displaced by the projectile, i.e., primary knock-on atom, may then be treated as a secondary projectile, and it too, in the course of its trajectory, may have sufficiently energetic interactions with other target atoms to produce further displacements and a second generation of knock-on atoms.

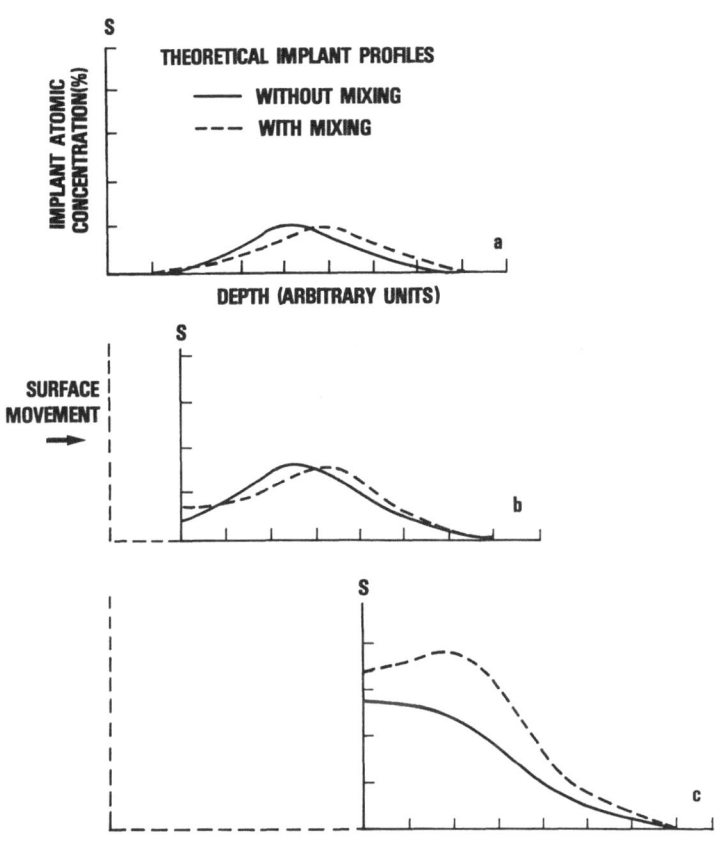

Figure 1. Steps to reach the steady-state implant depth profile due to
the model of ref. (5) (solid line) and with the contribution of atomic
mixing calculated in ref. (6) (dashed line), (a) implant concentration
builds up and the target surface moves inward, (b) continuous implant
collection and removal proceeds, (c) the implant saturation profile is
reached.

Therefore, the initial energy of the primary knock-on atom struck by the
incident beam atom is dissipated through a multiple displacement sequence
known as an atomic displacement cascade. Since the energetic target atoms
may also displace the trapped beam atoms from their positions, the cascade
contains both scattering and recoiling target atoms of original and
implanted kind. The production of the atomic displacement cascade process
continues until all incident beam and target atoms are brought to rest
through a series of nuclear and electronic stopping events with other
atoms of the solid or until they escape the target surface by overcoming
the surface binding energy. Theoretical calculations (6) of the atomic

mixing effect under simplifying assumptions, i.e., matrix atoms are not displaced from their lattice sites and the masses of the implant and matrix atoms do not differ substantially so that the flux of implantation species is determined in a single-element like medium, are shown by the dotted lines in Fig. 1. Similar mixing results were also obtained in ref. (7) generally showing that the mixing broadens the high fluence profiles and increases the concentration of implant on the surface as well as in the bulk. However, at high influences the build up of the surface concentration proceeds much more slowly with mixing than without mixing (8).

Other atomic displacement effects which complicate our understanding of ion implantation are the preferential sputtering and the range shortening effects. In compounds or alloys, one would expect, qualitatively, that the kinetic energy of a bombarding ion will not be shared by the different atomic species according to the composition only, but will also depend on other factors such as the interaction potential, the atomic mass ratios, etc. Non-stoichiometric effects have been observed in sputtering (9,10,11). The range shortening effect is due to a stopping power increase from building up the implant collection. In the physical system this implant collection will produce stresses followed by relaxation or expansion of the solid. This effect can be accounted for analytically by relaxing the solid to a nominal atomic density; one such procedure is given in ref. (7).

Besides the atomic displacement effects, the energy dissipation by the incident ion may result in electronic excitation effects, such as photon, electron, etc., emission. The way this electronic energy loss is expended in the medium determines the so-called "damage energy" available for producing defects by atomic displacement. It is suggested (12) that in defect production calculations the electronic energy loss be considered only down to the energies below which a moving atom cannot displace the target atom from its lattice site. Better understanding of the defect (i.e., vacancies, dangling bonds, voids, density fluctuations, etc.) production mechanisms is needed because many properties of amorphous as well as crystalline solids are affected by defects and can be defect controlled (13).

The slowing-down of ions (or atoms) and target atoms in solids is properly the domain of atomic collision physics. The collision cascades end in an aggregate of point defects in a solid. Following the slowing down of cascades, the processes of solid state physics are dominant (radiation enhanced diffusion, radiation induced segregation, nucleation, precipitation, void formation, etc.) The time scale here can be much larger than the 10^{-13} - 10^{-11} seconds for the slowing-down of cascades. Very little is known (14,15,16) about the transition between the slowing-down of cascade atoms and the final state of a configuration of point defects, i.e., the calming-down of a highly excited, disordered lattice to thermal equilibrium.

B) ANALYTICAL METHODS

The linear transport theory has been applied to atomic collision cascades in an elemental target to calculate the sputtering yield (2) and a binary target to investigate the role of the mass difference in cascade development (17). In the following, some of the assumptions used to simplify and/or linearize the transport equation will be stated, followed by a discussion of the methods of solution in the operator form of the linear transport equation.

Major complications in the analytical formulation of atomic collision cascades are: (a) the complexity of the atomic cross sections which are highly anisotropic and vary with the interaction potential used, (b) the need to consider the fluxes of each recoiling target species along with that of the projectile, (c) the need for a better displacement model for target atoms, (d) finite geometry effects, and (e) uncertainties in electronic losses.

Common assumptions made in the transport calculations are: (a) binary atom collisions only are considered, (b) all atomic collisions obey the laws of classical mechanics, (c) excitation of electrons or ionization of atoms enters only as a source of energy loss, but does not influence the collision dynamics, (d) an infinite medium is assumed, (e) one of two colliding atoms is initially at rest, and (f) three mutually connected assumptions in regard to the spatial structure of the target are applied, i.e., homogeneity, isotropy and randomness. Then, the linear transport equation for the slowing down of charged particles i in a medium composed of an arbitrary number of different scattering species j is written as (18,19)

$$L_1 \, \phi_i(\underline{p}) + I_{ij}(\underline{p}) = S_i(\underline{p}) \tag{1}$$

where $\underline{p}$ is the generic phase-space coordinate including space, energy and direction, ϕ_i is the particle (beam ion or recoil atom) flux distribution, L_1 is an operator which represents losses due to "drift" out of phase space, elastic collisions, and slowing down by electronic interactions, I_{ij} is the collision integral term and contains gains due to the atoms scattered in from higher energies, Fig. 2a, and the recoils scattered by moving atoms, Fig. 2b, and S_i is the external source term.

In Equation 1, the term I_{ij} is the coupling term for the fluxes of the beam atom and/or recoil species. In a single particle flux, an energetic atom of species i with kinetic energy E "scattering" into the range (E, E + dE) when in collision with a target atom of species j is considered, as shown in Fig. 2a. However, the "recoiling" lattice atoms which receive sufficient energy can also contribute to the process which generates a flux of cascade atoms. Such an energetic atom of species j, energy E_j' which gives rise to a recoiling i-species in the range (E, E + dE) is shown in Fig. 2b. It is noted that the cascade process has a multiplication effect which takes place practically from the incident beam energies down to the displacement threshold energies which can be curtailed only by the binding energy losses during the "recoiling" process of lattice atoms. This cascade behavior alters the normal collision term in Eq. (1) in a non-trivial way and requires special consideration (20).

It is assumed that, in principle, the set of linearly coupled equations in Eq. 1 can be transformed into a more manageable set of separate equations for each specie by the appropriate operations of calculus. Then, the equations for the cascade species can be represented

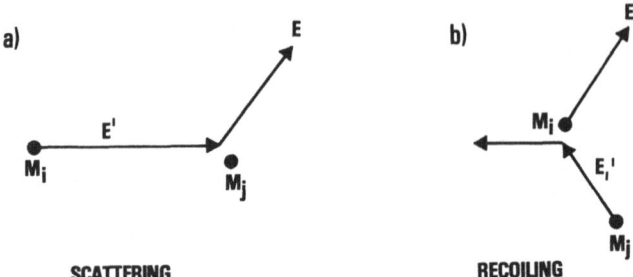

Figure 2. a) Elastic scattering of an energetic atom of species i with kinetic energy E' into (E, E + dE) when in collision with a target atom of species j, and b) recoiling mechanism of a target atom of species i, energy, E, resulting from the elastic collision with an energetic atom of species j and energy E_1'.

$$L\phi(\underline{p}) = S(\underline{p}) \tag{2}$$

in the operator form where L is a linear Boltzmann operator including integral or differential operators, or both, and the source term, S, now also including the recoiling target atoms. Equation (2) is usually defined in a certain domain

$$\underline{r} \in \mathcal{D} \quad \text{(finite space)}$$

$$E_o < E < E_1$$

$$t_i < t < t_f$$

where E_o is the lower limit of energy, E_1 is the upper limit of energy, t_i is the initial time, and t_f is the final time. The solution of Eq. (2) is subject to boundary conditions at the external boundaries of $\mathcal{D}$, and initial conditions at $t = t_i$.

The adjoint Boltzmann equation is defined as

$$L^* \phi^*(\underline{p}) = S^*(\underline{p}) \tag{3}$$

where ϕ^* is the adjoint function, S^* is the adjoint source, and L^* is the adjoint Boltzmann operator defined by

$$\langle \phi^*, L\phi \rangle = \langle \phi, L^* \phi^* \rangle \tag{4}$$

where the brackets mean the scalar product. The solutions of adjoint equations are subject to the boundary condition at the external surface of the space and final conditions at $t = t_f$.

The energy variable in the equation used by Sigmund and the one used in a typical neutron transport equation are reversed. In other words, they are semi-adjoint equations (this may cause problems when the cross sections become spatial and time dependent[21]). Sigmund[2] called his equation "forward", but it will be called "adjoint" here.[1]

The definition of the adjoint source S^* is rather arbitrary, and of course, the physical meaning of ϕ^* changes depending on the definition of S^*. This point will be clear when the physical quantities which are the effects of interest to us are defined.

The basic result of a solution to a problem is a flux-integrated quantity or more simply a "response" which can be defined as the functional

$$R(\phi) \equiv \int_{\underline{p}} \Sigma_R (\underline{p}) \phi (\underline{p}) d\underline{p} \equiv \langle \Sigma_R, \phi \rangle \tag{5}$$

Here, $\phi(\underline{p})$ is the flux solution of the Boltzmann transport equation, $\Sigma_R(\underline{p})$ is the response function which relates the flux to the integrated response being studied and $d\underline{p} = dr\ dEd\Omega$. The response of interest may be sputtering, entrapment, dose, scalar fluence, energy deposition, ionization, reaction rate, etc., which can be an end mechanism of an atom or an effect induced by the atom.

One approach to calculate the "response" is to solve Eq. 2 and to introduce the solution into Eq. (5). In the forward formulation as in Eq. 2, the computation of ω at the domain of the response function must proceed step-by-step starting at the source condition (in energy, position, direction, time) as in the neutron slowing down calculations, and then this solution must be introduced into Eq. (5).

However, an alternative approach to calculate the "response" is to solve the adjoint equation (3) with $S^* = \Sigma_R(\underline{p})$ and then calculate

$$R(\phi^*) = \langle S, \phi^* \rangle \equiv \int_{\underline{p}} S(\underline{p})\phi^*(\underline{p})d\underline{p} \tag{6}$$

---

[1]Technically, the definition of "forward" and "adjoint" for a specific application is rather arbitrary. In Sigmund's equation the initial energy is the variable and the final energy is a parameter, which is reversed in the neutron slowing down case. Two ways of determining the right terminology are: 1) it is a "forward" equation if it looks "forward" from t to t + Δt; 2) if the initial variable is fixed, it is "forward"; if the final variable is fixed, it is "adjoint".

The equivalence of Eqs. (5) and (6) has been shown analytically in several references, i.e., references (22) and (23), and therefore will not be shown here. However, this can be seen from Eq. (4) by direct substitution of the source and response functions, or by Green's function considerations.[2]

At this point, some further comments about the forward and adjoint solutions are appropriate. The quantity of interest in the adjoint solution, Eq. (6), is more sensitive to different "forward" source parameters as is Sigmund's solution where the initial parameters are treated as variables. On the other hand, in the forward solution, Eq. (5), it is more sensitive to response function parameters. It is just those effects of low energy mechanisms, i.e., domain of response function, which are presently not well understood.

If it were possible to obtain exact solutions for the forward or adjoint equation, the equivalence of Eqs. (5) and (6) would hold, and, of course, we would not worry about the sensitivity of solutions to the forward or adjoint formulation. In this case, the solution of forward flux in the domain of response function, and the solution of adjoint flux in the domain of source function would be exact. Then the integrations over the source or response functions would give the effect of interest ("response"). However, this is not the case and, in general, the exact solutions are difficult to obtain analytically. In some simplifying cases, such as the neutron slowing down problem where a detector response is determined by a very high absorption rate at the low energy region and the energy distribution of the source neutrons effectively fall to zero at some low energy so that the asymptotic solution (far from the source energies) in low energies can be found, the calculation of response can be done with reasonably good accuracy.

---

[2]In Green's function formulation the identity represented by Eqs. (5) and (6) can be written as

$$R(\phi) = \int d\underline{p}' \, \Sigma_R(\underline{p}') \int d\underline{p}'' \, G(\underline{p}' \to \underline{p}'') \, S(\underline{p}'')$$

$$= \int d\underline{p} \, S(\underline{p}) \int d\underline{p}'' \, G(\underline{p} \to \underline{p}'') \Sigma_R(\underline{p}'')$$

$$= R(\phi^*)$$

where $G(\underline{p}' \to \underline{p}'')$ is the flux at $\underline{p}'$ due to the delta function particle source at $\underline{p}''$ and $G(\underline{p} \to \underline{p}'')$ is the adjoint flux which represents the importance of the point impulse source at $\underline{p}$ in increasing the response at $\underline{p}''$.

However, as pointed out earlier, in the case of flux of recoil atoms in the cascade there are source contributions by the recoiling target atoms (Fig. 2b) practically at all energies from the displacement threshold up to the incident beam energy which has been well documented by computer simulation calculations (24). The asymptotic solution for the adjoint flux obtained in reference (2) for E >> U (where U is of the order of a few eV and E is greater than 100-200 eV) may not adequately represent the collisional mechanisms in the energy range below 100-200 eV's. Using "bombarding ions" or "primary recoils" (25) as a first approximation for the source of "cascade recoils", because of the arguments given above, will not be adequate.

The insensitivity of the adjoint asymptotic solution to the low-energy collision can be improved by using the forward equation (19, 26) or by using numerical methods (27). The importance of the low-energy mechanisms cannot be overemphasized because both the computer simulations (28-30) and a deterministic solution (31) of the transport equations show that the results are very much dependent on the low-energy parameters, i.e., displacement threshold energies used for the solution of equations. The adjoint flux evaluated "exactly" would have the advantage of incorporating the low-energy displacement mechanisms. However, the asymptotic solution cannot represent any displacement model because of the energy region it is defined in, e.g., far from the energy domain of the low-energy displacement effects.

C) SIMULATION METHOD

Computer simulation is a very useful technique to explain various physical phenomena of atomic collisions in solids which defy precise analytical treatment. Simulation is a sampling experiment over time whenever the "model" contains at least one stochastic variable; the word "model" emphasizes that we are not studying the object of interest but only a representation of it (32). There are "deterministic" simulations which do not use random numbers; however, atomic collision applications require a technique of sampling stochastic variables using random numbers which is referred to as "Monte Carlo" simulation. In the next section an example will illustrate the direct simulation of particle history which is called "analog Monte Carlo", although the "adjustable" parameters used commonly in analog simulations, i.e., reducing the mean free path, could bias the results. Some of the assumptions of the analytical calculations are also used in the Monte Carlo method, but to a lesser degree. The "analog Monte Carlo" method is also referred to as the "binary Monte Carlo" method because of the binary atomic collision approximation. In another Monte Carlo method which is referred to as the "molecular dynamics" method (33,34), the multiple collisions of atoms under their mutual forces are allowed. By this method more realistic solutions at low energies can be obtained; but, it suffers from the limitations on the incident energies and number of histories.

The Monte Carlo method has a number of distinct advantages over present analytical formulations which have been discussed in the previous section. Namely, having the possibility of simulating in both forward (as done so far) and adjoint formulations, the Monte Carlo method allows more rigorous treatment of elastic scattering, explicit consideration of surfaces and interfaces, and easy determination of energy and angular distributions (35). However, since simulation involves experimentation over time, the technique can be utilized best for the study of the dynamic behavior of atomic collisions. For this reason the Monte Carlo simulation method is the most promising one to study the dynamic changes in ion implantation (36-38), atomic mixing (24,39,40), preferential sputtering (30,41-43) and radiation damage mechanisms.

When the simulation model contains stochastic variables, even a simple model produces results which usually cannot be expressed analytically; this holds even more for the study of "transient" behavior than "steady-state" behavior. If we are only interested in the result after a long time period, limit theorems may yield "steady-state" solutions which may be expressed analytically, provided the variables do not interact in a complicated way. It should be mentioned that analytical expressions can be incorporated into the simulation model, if one wishes, usually as a variance reduction and cost cutting measure.

The simulation models are based on our thorough knowledge of the behavior of the system elements. An incomplete understanding of the physical situation affects analytical formulations as well. However, the tendency to simplify analytical models to obtain a general solution is customary and not as severe a limitation in the analytical methods as it is in the simulation method. It is because in simulation, the results are very much input parameter and model dependent in lieu of analytically expressible general solutions. But, if the inputs and the models used in the analytical calculations and Monte Carlo simulations are exactly the same, the expected (or average) answer of the Monte Carlo calculation of the particle transport satisfies the integral transport equation, and if the number of simulated particles is sufficiently large, then the estimated answer approaches the exact solution of the integral equation (44).

Finally, a comment is given concerning why Monte Carlo is regarded as "different". There are several reasons for this. First, it is a stochastic method. This is different from the conventional deterministic solutions to which people are more accustomed. Second and even more important, one can interpret the use of Monte Carlo in two different ways. It may be considered as a stochastical method to solve deterministic problems, which means in this case, the integral transport equation. But one may also look at it as a stochastical method to simulate stochastical processes, without requiring analytical expressions to process. Mainly for this last reason, the Monte Carlo method will remain as a very important solution technique.

D) EXAMPLE OF MULTIPLE-ENERGY ION IMPLANTATION

As stated earlier, the analytical equations for particle transport are complicated by being integro-differential equations which can be solved analytically only in some very special cases. Also, in spite of the recent development of large computers, the numerical solutions such as the discrete ordinates to solve directly the full Boltzmann equation are presently limited to two-dimensional and comparatively simple problems. However, the Monte Carlo method used in this section does not have the limitations of the analytical and numerical methods. More details of the results shown here can be found in reference (36) and the EVOLVE code which has been used to obtain them is described in detail in reference (45).

Here, the results will be shown of pure copper target implantations and post-implantation of a previously implanted copper target. The effects of the sputtering rates and dynamic target changes for different irradiation energies and fluences on the implant concentration profiles are shown. Statistical error associated with each data point is difficult to estimate because of continuous change in the target composition in dynamic simulation. However, the standard deviation is estimated to be of the order of 6-10 percent up to the medium penetration depths, increasing in deeper regions because of the difficulty in penetration and consequent reduction in the sample size.

Figure 3 shows the 10 keV and 20 keV gold implantation profiles in an initially pure copper target at three different fluence levels. Also, the partial sputtering yields of copper and gold as a function of the number of layers sputtered are shown in the figure. The fluence levels corresponding to the implant profiles shown are marked by arrows. In the conversion of these simulations, 1000 histories corresponded to the fluence of 5.5×10^{15} Au^+ ions cm^{-2}.

The numerals inside the concentration profiles indicate the total number of the implanted gold atoms, showing that the 10 keV gold bombardment of copper yields a higher number of implants than the 20 keV bombardment in a shallower depth. The shifts of the mean ranges of the implanted gold atoms towards the surface as a result of the movements of the surface with increased influences are seen in 10 keV but not in 20 keV bombardment. It was found that less sputtering was required to reach the steady state when 10 keV gold bombardment was used compared to the 20 keV one. Sputtering of an amount of target material more than the thickness of the altered layer was required in both cases to reach the steady state. It required a sputtering of the 138$\overset{0}{A}$ depth of target (46 layers) in 10 keV gold bombardment and a slightly higher depth (about 60 layers) in the 20 keV gold bombardment to reach the steady state of gold implant concentration. The preferential sputtering factor for copper and gold is nearly unity in keV bombardment and the maximum concentration of the gold implant on the surface at steady-state is inversely proportional to the total sputtering yield ($Y \approx 8.5$ for 10 keV and $Y \approx 12$ for 20 keV) as predicted in reference (7).

In Fig. 3 the static TRIM (46) results have also been shown to compare with the steady-state results of EVOLVE. In the static TRIM (which was the version available), only the slowing down of the incident ions in a pure target is taken into account, the reflection of the incident ions, the atomic cascades in the target or the sputtering processes are not. In comparing the steady-state EVOLVE results with the static TRIM results, an inward broadening of the back of the gold profile in EVOLVE is observed in both 10 and 20 keV beam energies. This effect can be attributed to the recoil mixing mechanism. The EVOLVE profiles are broader in front also, as expected from the surface movement which brings the portion of the implant profile to the surface, as shown in Fig. 1. Broadening of the implant profile in front and higher accumulation of implants in the bulk are expected from the atomic mixing effects as shown

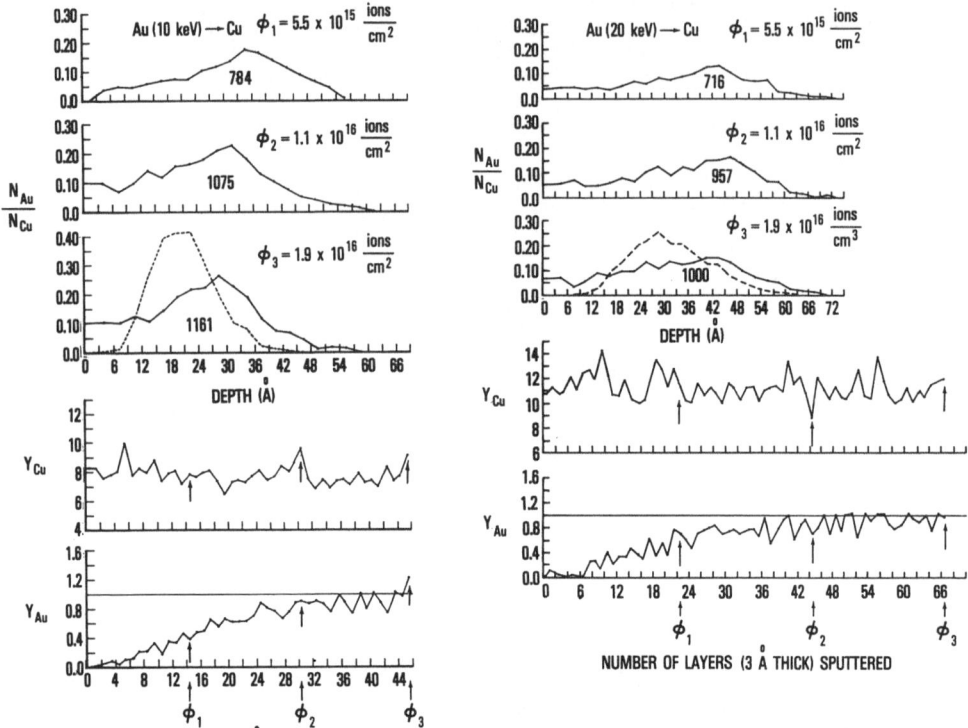

Figure 3. 10 keV and 20 keV gold implantations into copper. Gold concentration profiles at three different fluence levels and the partial sputtering yields of copper and gold as the number of layers sputtered are shown. The numericals inside the concentration profiles indicate the total number of the implanted gold atoms. The dotted line profile is the static TRIM results.

in Fig. 1. However, EVOLVE results also show that the high sputtering rate suppresses the recoil mixing effects in front of the profile. These results are, in general, in good agreement with observations of the experimental profiles obtained in 40 keV germanium bombardment of a silicon target in reference (6).

Ion implantations at selected energies and fluences are commonly believed to yield concentration distributions which can be approximated by the sum of the individual distributions. However, the computer simulations which take into account the dynamic changes in the target show different results. As seen in Fig. 4 the presence of the previously implanted ions influence the distribution of those that follow.

Figure 4 shows the results of the 20 keV post-implantation of gold atoms on the previously implanted (at 10 keV) gold profile and the results of the 10 keV post-implantation of gold atoms on the previously implanted (at 20 keV) gold profile. The results show marked differences. In the case of the 20 keV post-implantation we observe a rapid loss of the previously implanted gold atoms and then a slow build up of them, but not reaching the previous 10 keV concentration levels. In the case of the 10 keV post-implantation there is a rapid build-up of gold atoms while the maximum range of the implant profile is reduced by sputtering, therefore increasing the surface and bulk concentration of gold atoms. After the partial sputtering yield of gold becomes larger than unity as a result of the sputtering of the layers with a high concentration of gold, the total

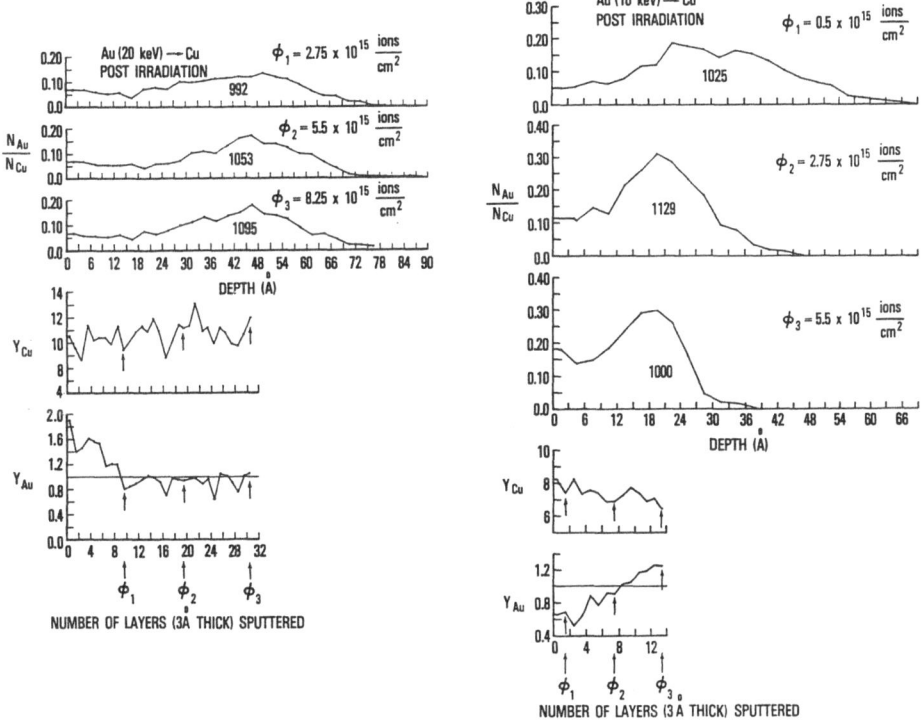

Figure 4. 20 keV and 10 keV gold post-implantation of the previously implanted copper samples shown in Fig. 3.

number of implanted gold atoms begins to decrease. A very large number of layers has to be sputtered before steady state gold implant concentration is obtained.

CONCLUSION

Emphasis has been given to the analytical approaches and the computational methods to describe the ion implantation mechanisms. Due to the fact that more and more experimental and computational data reveal the importance of the low-energy effects, the limitations of the presently available solution to describe such collisional effects have been discussed.

Unlike the analytical methods, the solutions of the transport equation for the charged particles are available by the Monte Carlo method in the forward form. They are exact solutions; therefore, they do not sacrifice the low-energy effects mechanisms as does the adjoint asymptotic solution which is analytically available. Monte Carlo solutions are important in obtaining exact solutions which are not available otherwise. Also, they are useful to compare the exact with the available analytical solution(s). Another advantage of the Monte Carlo solution is that it does not require an analytical expression with which to work. All one needs is the interaction cross-sections to describe the physical phenomena; the formulation of the problem is done by the logical decisions and the conditions imposed during the random walk of the charged particle. The particle can travel from the "source" to the "response" domain (forward method) or from the "response" to the "source" domain (adjoint or backward method) and the solution obtained is the evaluation of the "integral" form of Boltzmann's transport equation rather than the commonly known "integro-differential" form of it. Finally, an example has been presented of Monte Carlo solution for multiple-energy ion implantation which is not easily obtained analytically.

Acknowledgement. This work was supported by the Independent Research Program of the White Oak Laboratory, U. S. Department of the Navy. It is my pleasure to acknowledge many stimulating discussions on this topic with Drs. D. G. Simons and D. J. Land.

REFERENCES:

1. J. Lindhard, V. Nielsen, M. Scharff and P. V. Thomsen, K. Danske Vidensk. Selsk; Mat.-Fys. Meddr. 33 (1968) 1.
2. P. Sigmund, Phys. Rev. 184 (1969) 383; 187 (1969) 768.
3. P. Sigmund, M. T. Matthies and D. L. Philips, Rad. Effects 11 (1971) 39.
4. P. Sigmund, Rev. Roum. Phys. 17 (1972) 823, 969, 1079.
5. G. Carter, J. N. Baruah and W. A. Grant, Rad. Effects, 16 (1972) 107.
6. A. Gras-Marti, J. J. Jimenez-Rodriguez, J. Peon-Fernandez, M. Rodriguez-Vidal, N. P. Tognetti, G. Carter, M. J. Nobes and D. G. Armour, Vacuum, 32 (1982) 433.
7. U. Littmark and W. O. Hofer, Nucl, Instr. and Meth. 170 (1980) 177.

8. K. Wittmaack, Nucl. Instr. and Meth. B7/8 (1985) 779.
9. H. H. Andersen in: Ion Implantation and Beam Processing, eds., J. S. Williams and J. M. Poate (Academic Press Australia) Chapter 6 (1984) 127.
10. H. H. Andersen in: SPIG 1980, ed. M. Matic, Boris Kidric Inst. Nucl. Sci. (Beograd, 1980) 421.
11. Z. L. Liau and J. W. Mayer in: Treatise on Material Science and Technology, ed. J. K. Hirvonen, (Academic Press, Inc.) Vol. 18 (1980) 17.
12. M. T. Robinson and O. S. Oen, Jour. of Nucl. Mat. 110 (1982) 147.
13. S. R. Elliot, Physics of Amorphous Materials (Longman Inc., New York) 1984.
14. W. L. Johnson. Y. T. Cheng, M. Van Rossum and M-A. Nicolet, Nucl. Instr. and Meth. B7/8 (1985) 657.
15. H. Weidersich, Nucl. Instr. and Meth. B7/8 (1985) 1.
16. R. Kelly, Rad. Effects, 80 (1984) 273.
17. N. Andersen and P. Sigmund, Kgl. Danske Vid. Selsk, Mat. Fys. Medd, 39 (1974) No. 3.
18. G. Bell and S. Glasstone, (1970) Nuclear Reactor Theory, Van Nostrand, New York.
19. M. M. R. Williams, Progress in Nucl. Ener., 3 (1979) 1.
20. M. M. R. Williams, J. Phys. A; Math. Gen. 9 (1976) 771.
21. M. M. R. Williams, Annals of Nucl. Ener., 5 (1978) 149.
22. J. Spanier and E. M. Gelbard, Monte Carlo Principles and Neutron Transport Problems, (Addison-Wesley Publ. Co.) 1969.
23. H. Kahn, Applications of Monte Carlo, Rand Corporation (1954).
24. M. L. Roush, F. Davarya, O. F. Goktepe and T. D. Andreadis, Nucl. Inst. and Meth. 209/210 (1983) 67.
25. P. Sigmund in: SIMS IV, eds., A. Benninghoven, J. Okano, R. Shimizu, and H. W. Werner (Springer-Verlag Berlin) (1984) 2.
26. M. Urbassek, Nucl. Instr. and Meth. B4 (1984) 356.
27. T. J. Hoffmann, H. L. Dodds, Jr., M. T. Robinson and D. K. Holmes, Nucl. Sci. and Engin. 68 (1978) 204.
28. O. Goktepe, T. D. Andreadis, M. Rosen, G. P. Mueller and M. L. Roush, Nucl. Instr. and Meth. (in press).
29. M. Rosen, G. P. Mueller, M. L. Roush, T. D. Andreadis and O. F. Goktepe, Nucl. Instr. and Meth. (in press).
30. O. F. Goktepe and M. L. Roush, Nucl. Instr. and Meth. B7/8 (1985) 803.
31. G. P. Mueller, Nucl. Instr. and Meth. 170 (1980) 389.
32. J. P. C. Kleijnen, Statistical Techniques in Simulation, (Marcel Dekker, Inc. New York, 1974).
33. D. Harrison, Rad. Effects 70 (1983) 1.
34. J. R. Beeler, Computer Experiment Methods, North-Holland, Amsterdam, 1983.
35. J. F. Ziegler, Ion Implantation (Academic Press, Inc., 1984) 51.
36. O. F. Goktepe, Mat. Sci. Eng. 69 (1985) 13.
37. M. L. Roush, F. Davarya, T. D. Andreadis and O. F. Goktepe, J. Vac. Sci. Technol. A1 (1983) 491.
38. W. Moller and W. Eckstein, Nucl. Instr. and Meth., B7/8 (1985) 645.
39. M. L. Roush, O. F. Goktepe, T. D. Andreadis and F. Davarya, Nucl. Inst. and Meth. 194 (1982) 611.
40. W. Moller and W. Eckstein, Nucl. Inst. and Meth., B2 (1984) 814.
41. M. L. Roush, T. D. Andreadis, F. Davarya and O. F. Goktepe, Appl. Surf. Sci. 11/12 (1982) 235.

42. M. L. Roush, T. D. Andreadis, F. Davarya and O. F. Goktepe, Nucl. Instr. and Meth. 191 (1981) 135.
43. W. Eckstein and W. Moller, Nucl. Inst. and Meth., B7/8 (1985) 727.
44. O. F. Goktepe, Ph. D. Thesis (1977) (unpublished).
45. M. L. Roush, T. D. Andreadis and O. F. Goktepe, Rad. Effects, 55 (1981) 119.
46. J. P. Biersack and L. G. Haggmark, Nucl. Instr. and Meth., 174 (1980) 257.

THE THEORY OF THE PREFERENTIAL SPUTTERING OF ALLOYS, INCLUDING THE ROLE OF GIBBSIAN SEGREGATION

Roger KELLY and Antonino OLIVA*

IBM Thomas J. Watson Research Center, Yorktown Heights, NY 10598, U.S.A.

ABSTRACT: Preferential sputtering of alloys leads, not to a profile showing a simple depletion of one component starting at the surface, but to a more complicated situation in which the outermost atom layer shows a composition spike of one component while a depleted region begins at atom layer two. This situation is manifested in experimental profiles, in conflicting results between experiments done (for example) with Ion Scattering and Auger Electron Spectroscopy, in measurements of the angular and energy distributions of sputtered atoms, and in examples such as Ag-Si where one component resists sputtering altogether. At steady state, the composition spike is governed by a conservation relation which equates the ratio of the sputtered fluxes (taking into account both mass and binding effects) to the ratio of bulk compositions. The steady-state second-atom-layer composition is determined by a relation similar to that for equilibrium Gibbsian segregation, but with the segregation ratio governed by a kinetic quantity (K) rather than a thermodynamic one ($\exp[-\Delta G^{seg}/kT]$). Nevertheless, K is similar numerically to the composition ratio observed in equilibrium Gibbsian segregation. Beyond the second atom layer there is a composition profile ideally describable (for bombardments at or below ambient temperature) by ion-beam mixing theory, but which can be shown also to admit a tractable diffusion approximation.

*Permanent address: Dipart. di Fisica, Università degli Studi della Calabria, Cosenza, Italy.

1. INTRODUCTION

When binary or ternary alloys are bombarded with ions, it is normally found, provided <u>subsurface</u> composition probes such as Auger Electron Spectroscopy (AES) are used, that one component is depleted. The component has, in effect, been preferentially sputtered and, because of the concurrent ion-beam mixing, a composition profile has been set up. This is an important effect in the domain of particle-surface interactions and has led to a decade and a half of major effort in experiment, simulation, and theory.

Until recently, almost all effort in trying to understand preferential sputtering and the attendent composition profiles, whether in explaining experiments, setting up a simulation, or forming the basis of theory, has centered on mass and chemical binding differences. These will be discussed in Sections 2 and 3, and shown to be of limited importance. Mass, for example, is most relevant in isotopic effects and in near-threshold events.

Very recently the failure (in most situations) of models based on mass and binding has been emphasized and the surprising conclusion that Gibbsian segregation (either intrinsic or impurity-driven) plays a dominant role in preferential sputtering has been reached. The argument was initially based on a trend analysis [1-7]: by attributing preferential sputtering to segregation it was possible to account, in virtually all cases, for <u>the</u> <u>sense</u> of the experimental data. Later, explicit composition profiles [8-26], angular distributions [e.g. 27-30], energy distributions [e.g. 31-33], resistance to sputtering [e.g. 33A-36], and conflicts between outer-layer spectroscopies (Ion Scattering Spectroscopy (ISS); Field Ion Microscopy (FIM); Secondary Ion Mass-Spectroscopy (SIMS)) and sub-surface probes (AES; X-ray Photoelectron Spectroscopy (XPS)), as in Ref. [11] vs. [37], were reported which strongly suggested a role for some form of segregation. We note also that Williams and Baker [35] correctly predicted the form of the composition profile, while the first comprehensive attempts to quantify the profile were by Swartzfager et al. [11], Lam and Wiedersich [11A], Itoh and Morita [5], and one of the present authors [6,38]. Segregation, especially that induced by bombardment, will be discussed in Section 4.

It is thus intended to summarize the theoretical basis of preferential sputtering. Some of the material is familiar [39,40] but the overall picture is here presented for the first time.

In view of the overriding importance of explicit composition profiles to the origin of preferential sputtering, the systems showing them are summarized in table 1 and examples of the four principle types are given in figs. 1-4. In these figures, $\alpha_i(x)$ stands for atom fraction of component i as a function of depth x. Figure 1 [11] is an ISS profile of Au-Pd, obtained by prebombarding at the indicated temperatures with 2 keV Ne$^+$ and then profiling at $-90°C$ with 2 keV Ne$^+$. It is perturbed noticeably by a continuing loss of Pd, i.e. the apparent bulk Au composition exceeds the real value. Figure 2

TABLE 1. Examples of composition-depth profiles interpretable in terms of bombardment-induced Gibbsian segregation.

System	Technique	Figure best representing the profile	Ref.
Ag-Au	conflict between ISS and AES	...	8
	low-E profile with AES	4	9,10
	low-E, low-T profile with ISS[a]	1	11
Ag-Pd	low-E profile with AES	4	9
Ag-Si	low-E profile with AES	4	12,12A
Au-Cu	low-E, low-T profile with AES[a]	2	13,14
Au-Pd	low-E, low-T profile with ISS[a]	1	11
Cr-Si	low-E profile with AES	4	15
Cu-Ni	conflict between SIMS or ISS and AES	...	8,16
	low-E, low-T profile with ISS[a]	1	11,18
	low-E, low-T profile with AES[a]	2	17
	field ion microscopy	...	19
	simulated	...	11A,18
Cu-Pt	low-E profile with AES	4	9
Fe-Si	AES involving many transitions	3	20
Mo-Si	low-E profile with AES	4	15,21
Ni-Mo	AES involving many transitions	3	22
Ni-Si	low-E, low-T profile with AES[a]	2	23
	low-E profile with AES	4	15,24
Ni-W	AES involving many transitions	3	20
Pt-Si	low-E profile with AES	4	15,24
Ta-O	low-E profile with AES	4	25
Zn-Cu	low-E profile with AES	4	26

[a]In [17,18,23] the profiling temperature was $\sim$25°C and it was therefore inferred, incorrectly, that there was little or no segregation during bombardments just above $\sim$25°C. In [11, 13,14], by contrast, the profiling temperature was -90 or -120°C and segregation was found even at -120°C [14].

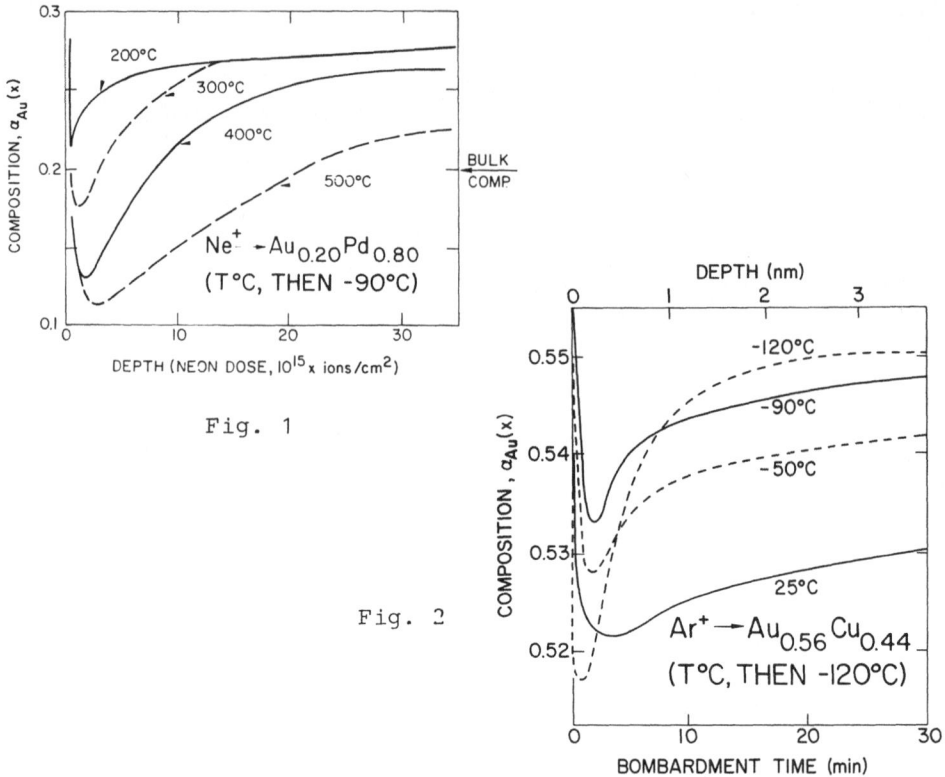

Fig. 1

Fig. 2

FIGURE 1. Composition profiles for bombarded $Au_{0.20}Pd_{0.80}$ as obtained by first bombarding to steady state with 2 keV Ne^+ at the indicated temperatures, then quenching to -90°C, and finally profiling with 2 keV Ne^+. Compositions were obtained with ISS using the same 2 keV Ne^+ as the probe ion. The reason that the bulk composition implied by the figure, $Au_{0.28}Pd_{0.72}$, disagrees with the stated composition, $Au_{0.20}Pd_{0.80}$, is that the system is subject to on-going Pd loss (Section 4.2). Due Swartzfager, Ziemecki, and Kelley [11].

FIGURE 2. Composition profiles for bombarded $Au_{0.56}Cu_{0.44}$ as obtained by first bombarding to steady state with a 40 $\mu A/cm^2$ beam of 2 keV Ar^+ at the indicated temperatures, then cooling to -120°C, and finally profiling with a 0.4 $\mu A/cm^2$ beam of 2 keV Ar^+. Compositions were obtained with low-energy AES, namely 69 eV for Au and 60 eV for Cu. It was shown that the bombard-ment-enhanced diffusion needed for creating the profiles during the initial bombardment occurs to a greater extent for high current density and, accordingly, that a 0.4 $\mu A/cm^2$ beam accomplishes a fairly good profiling. Nevertheless, ISS work on a similar target [41,42] shows that the surface composition spike was significantly truncated. Due to Li and Koshikawa [14].

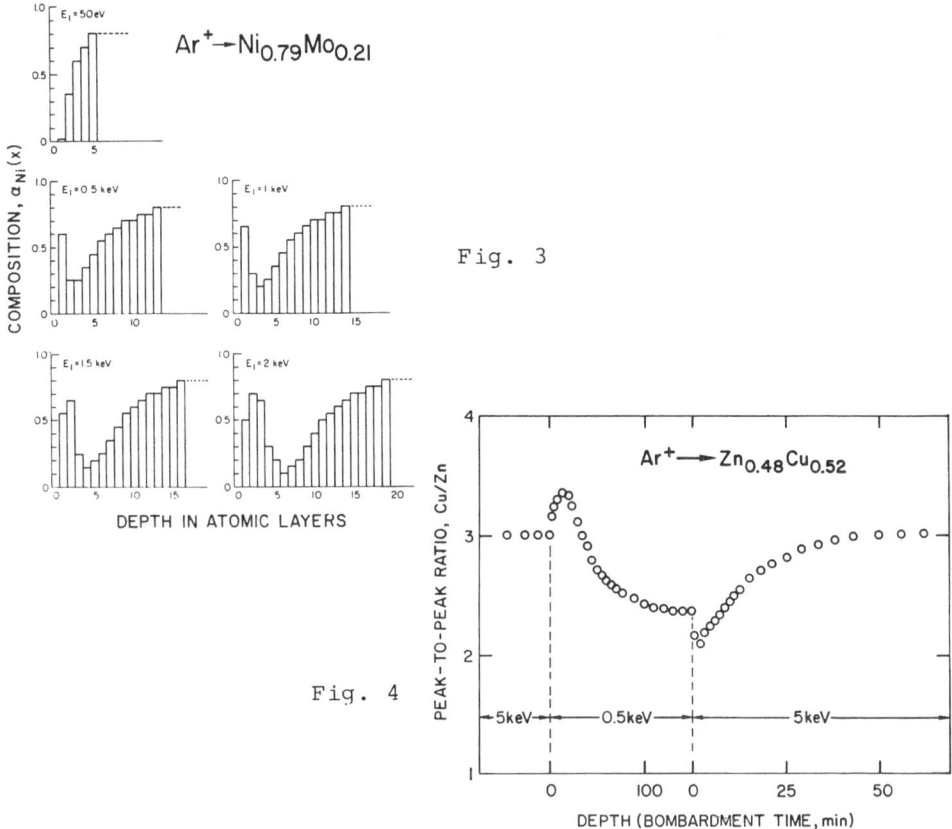

Fig. 3

Fig. 4

FIGURE 3. Composition profiles for bombarded $Ni_{0.79}Mo_{0.21}$ as obtained by first bombarding to steady state with 50 eV to 2 keV Ar^+, then terminating the bombardments altogether, and finally using AES. Compositions as a function of depth were obtained by deconvoluting the information from 10 Auger transitions of different energy, and should in principle be free of the perturbations which were noted in figs. 1 and 2. Due to Bartella and Oechsner [22].

FIGURE 4. Composition profile expressed as the ratio Cu/Zn for bombarded $Zn_{0.48}Cu_{0.52}$ as obtained by first bombarding to steady state with a 1.3 $\mu A/cm^2$ beam of 5 keV Ar^+ at room temperature, and then abruptly reducing the incident energy to 0.5 keV. Compositions were obtained with medium-energy AES, namely 994 eV for Zn and 920 eV for Cu. Noting that according to both its bond energy and size Zn should be the segregating species, the maximum shown by the ratio Cu/Zn can be regarded as equivalent to the minima in figs. 1-3. Nevertheless, as in fig. 2, the extent of the surface composition spike was probably underestimated. The fact that the steady-state signal for 5 keV Ar^+ exceeds that for 0.5 keV is easily understood in terms of the altered layer depths. Due to Ferron et al. [26].

[14] is a conventional AES profile of Au-Cu, obtained by pre-bombarding at the indicated temperatures with 40 $\mu A/cm^2$ of 2 keV Ar^+ and then profiling at $-120°C$ with 0.4 $\mu A/cm^2$ of 2 keV Ar^+. It is perturbed mainly by a truncation of the Au surface composition spike, which, according to ISS studies [41,42], should have begun at 0.70 ± 0.05. Figure 3 [22] is an AES profile of Ni-Mo obtained by prebombarding with 50 eV to 2 keV Ar^+ and then profiling without further bombardment by deconvoluting the information from 10 different Auger transitions. It is in principle free of significant perturbations. Similar profiles have also been obtained by simulation [11A,18].

The common feature in all such profiles is the existence of a surface composition spike, an explicit indication of Gibbsian segregation.

Figure 4 [26] is at first sight a further clear example of a segregation-related profile. Transients of this type were first encountered by bombarding [9,12A] (or simulating bombardment [43]) initially at an intermediate energy such as 5 keV and then at an abruptly reduced energy such as 0.5 keV. Alternatively they can be obtained by bombarding first normal to the surface and then abruptly changing to oblique incidence [44]. Without going into detail we note that such transients can be expected for preferential sputtering of any origin [9,44], though (at least with alloys) will be as prominent as in fig. 4 only if spike-type profiles are present. It follows that fig. 4 is equivalent to figs. 1-3.

2. MASS DIFFERENCES

The first concerted effort to understand preferential sputtering was based on mass differences. They enter most simply in recoil sputtering, i.e. direct beam-to-surface interactions (primary recoils). Recoil sputtering has two basic regimes, characterized by high and low incident energies. At high energies, the important aspect is that a lighter target species is moved further, whether deeper into the target or out of the target, as can be inferred by considering projected ranges. For example, 10 keV He^+ has a mean projected range in Si of 86 nm, whereas for 10 keV Ar^+ the range is only 12 nm [45]. A fully analytical theory is possible [46], but leads to the conclusion that, for normal incidence, the effect is unimportant owing to the concurrent cascade sputtering. This conclusion remains true also for grazing incidence provided the incident particles are heavy [47]. On the other hand, for light ions at grazing incidence, recoil sputtering can be quite important [44,48-50].

At lower energies recoil sputtering becomes more important relative to cascade effects [51], especially if the masses of the target constituents are sufficiently different [52]. Furthermore, if the incident mass is low, and the incident energy approaches the sputtering threshold, there is a tendency for the lighter target species to be the only one which is relocated and for the effect to be significant. An example [53] is the evolution of $Au_{0.03}Cu_{0.97}$ to $Au_{0.05}Cu_{0.95}$ due to bombardment with 150 eV $^3He^+$. The sputtering threshold E_1^{th} will be taken [51,54] as that appropriate to an incident ("1") light ion (energy E_1) which reflects once beneath the surface (whence retaining an energy $E_1[1 - \gamma_{12}]$) and then transfers $E_1(1 - \gamma_{12})\gamma_{12}$ to a surface or subsurface ("2") atom, the binding energy of which is U_2:

$$E_1^{th} = U_2/(1 - \gamma_{12})\gamma_{12},$$

where γ_{12} is the usual energy transfer factor $4M_1M_2(M_1 + M_2)^{-2}$, M_i is mass, U_{Cu} is 3.50 eV (from eq. (8)) and U_{Au} is 3.76 eV (from eq. (8)). It follows that E_1^{th} is 25 eV for Cu and 68 eV for Au, so that the incident energy, 150 eV, is in the one case 6 times above threshold and in the other only twice. In practice, also 1000 eV H_3^+ causes a preferential loss of Cu from **Au-Cu** [55]. Nevertheless, low-energy recoil sputtering can normally be neglected precisely because it is a low-energy effect.

Rather more subtle is the role of mass in cascade sputtering, i.e. in the more extended interactions involving beam-to-cascade-to-surface (all generations of recoils). A compact restatement of the result of Andersen and Sigmund [56] was given in [38], the essence being as follows.

For a bombarded binary alloy, A-B, let α_A be the atom fraction of A and α_B be the atom fraction of B. Also, let C_{AB} be the constant in the power-law scattering cross section, which is related to mass, to atomic number (Z_i), and to the power-law scattering parameter (m) by [57]

$$C_{AB} \propto M_A^m M_B^{-m} (Z_A Z_B)^{2m} (Z_A^{2/3} + Z_B^{2/3})^{-1+m}.$$

Then a simple argument based on the total energy stored in moving A atoms averaged over many cascades being a steady-state quantity gives the following for the flux ratio across any plane in the target:

$$\frac{flux_A}{flux_B} = \frac{\alpha_A C_{BA}}{\alpha_B C_{AB}} = \frac{\alpha_A M_B^{2m}}{\alpha_B M_A^{2m}}, \tag{1}$$

where m is $\leq 1/4$ for atom ejection [58]. The argument finishes by introducing the conservation relation for the steady-state ("∞") value of the surface composition:

$$\frac{surface\ flux_A}{surface\ flux_B} = \frac{\alpha_{A(2)}^\infty M_B^{2m}}{\alpha_{B(2)}^\infty M_A^{2m}} = \frac{\alpha_{A(3)}}{\alpha_{B(3)}}, \tag{2}$$

where "3" denotes the bulk. Eq. (2) presupposes that sputtered atoms originate solely from the outermost atom layer, a detail which is largely but not completely correct [29,59-62]. (This restriction will be relaxed in Section 4.2.)

Further consideration of the problem, based on [63] and discussed previously [39,40], suggests the use of the somewhat extended relation

$$\frac{surface\ flux_A}{surface\ flux_B} = \frac{\alpha_{A(2)}^\infty M_B^{2m}}{\alpha_{B(2)}^\infty M_A^{2m}} \cdot \frac{\alpha_{A(2)}^\infty + \alpha_{B(2)}^\infty \gamma_{AB}}{\alpha_{B(2)}^\infty + \alpha_{A(2)}^\infty \gamma_{AB}} = \frac{\alpha_{A(3)}}{\alpha_{B(3)}}. \tag{3}$$

This presupposes that eq. (1) describes the flux correctly from the bulk to the second atom layer, and the flux finally interacts with a static outer atom layer. Such a modification is reasonable but contributes nothing to the argument, as the additional factor is of order unity unless the masses are markedly different.*

Thus, we find that, from the point of view of cascade sputtering, lighter components should be lost preferentially. Although the underlying theory has a firm basis and cascade sputtering is itself a numerically important effect, the predicted result is disobeyed in about one-half of the studied systems. For example [4], with Ag-Ni, Au-Ni, Au-Pd, Gd-Co, Gd-Fe, In-Ga, Pb-In, Pb-Sn, and Pd-Ni it is the heavier species which shows an overall preferential loss, as is sensed when subsurface probes such as AES or XPS are used (e.g. Au-Pd, fig.1 of [38]).With Cu-Ni significant overall Cu loss is found in spite of the masses being nearly equal (fig. 5).

It must be emphasized that, for the 10 examples listed above, we have been concerned only with overall loss, as it is this loss that eqs. (1) and (2) fail to explain. It would be expected, on the other hand, that eqs. (1) and (2) give a reasonably good description of the outermost atom layer and, when ISS or SIMS is used, this has to some extent been shown to be true. Au-Cu, for example, shows a marked outer-surface loss of Cu (fig. 6), Au-Pd shows a similar loss of Pd (fig. 1), while Cu-Ni shows no outer-surface loss at all (fig. 5). A notable inconsistency is found with Ag-Au, however, where two ISS studies failed to reveal the expected loss of Ag (fig. 7). Au-Cu, Au-Pd, and Ag-Au will be reconsidered in Section 4.2 in terms of what is to be expected when significant emission occurs from beneath atom layer one.

We have seen that both recoil and cascade sputtering suggest that a lighter species is lost preferentially. A contrary conclusion was reached by Roush et al. [79,80]. In molecule dynamics simulations of relocation due to cascades, they found that the excessive transport occurred with the heavier species, and this result was used to rationalize changes found with silicides [15,21,24]. It is almost certain to be wrong in view of alternative simulations [81] as well as of the overall pattern of experimental results [4].

3. CHEMICAL BINDING DIFFERENCES

A more complete approach to understanding preferential sputtering is based on taking into account both mass and binding. We will not consider recoil sputtering in this respect, as it is a more or less ballistic process in which binding is unimportant. But cascade sputtering with pure metals shows a well-established [82] inverse dependence on the heat of atomization, thence, by inference, on the surface binding energy

*Equation (3) contains surface ("2") composition on the left side since composition-depth profiles, in the absence of segregation, can be inferred from figs. 1-4 to change sufficiently gradually. The forms given in [39,40] both had bulk ("3") compositions and are erroneous in this respect.

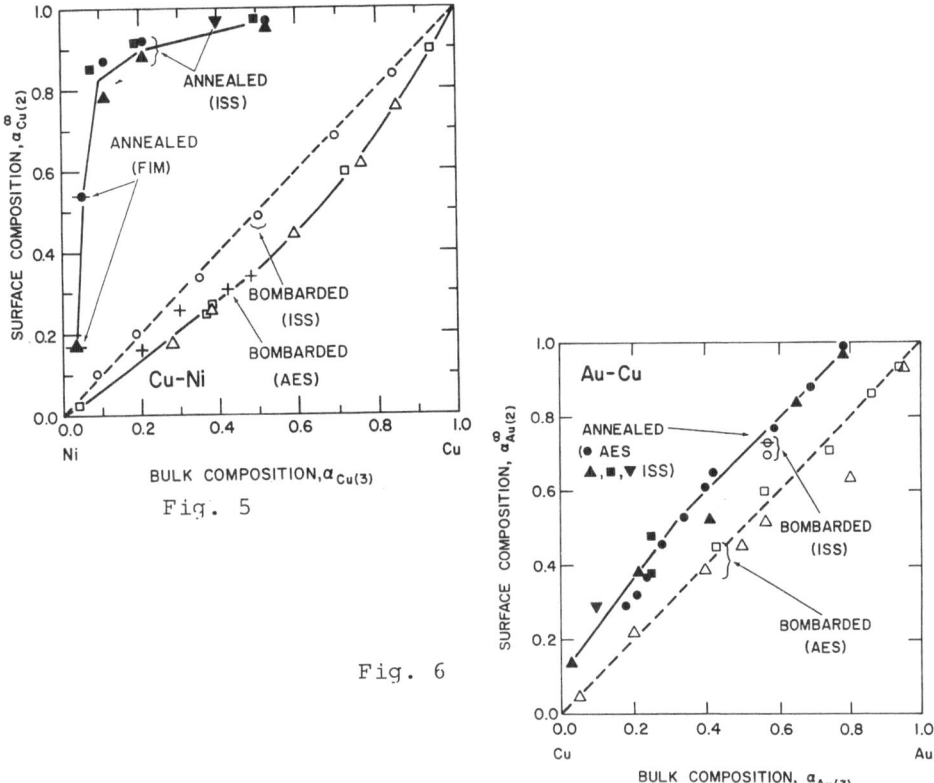

Fig. 5

Fig. 6

FIGURE 5. Comparison of surface and bulk compositions for Cu--Ni specimens which have been either annealed (●, 400°C then ISS [64]; ▲ and ▼, 500°C then ISS [64,18]; ■, 600°C then ISS [11]; ●-, 550°C then FIM [65]; ▲-, 650°C then FIM [65]) or bombarded (O, 2 keV Ne⁺ [11]; △ and Ⅱ, 0.5 keV Ar⁺ [66,67]; +, 2 keV Ar⁺ [68]). The diagonal line corresponds to no surface alteration. The bombarded surfaces would be described as showing bulk composition on the basis of the ISS study [11] but as being deficient in Cu on the basis of the AES studies [66-68]. The results for the bombarded surfaces are straightforward if viewed in terms of a composition spike as in figs. 1-4.

FIGURE 6. Comparison of surface and bulk compositions for Au-Cu specimens which have been either annealed (●, 300°C then AES [69]; ▼, 113 and 200°C then ISS [70]; ▲, 327°C then ISS [71]; ■, 550 and 830°C then ISS [72]) or bombarded (O and ⊖, 1.5-3 keV Ar⁺ [42,41]; △ and Ⅱ, 1 keV Ar⁺ [73,74]). The diagonal line corresponds to no surface alteration. The bombarded surfaces would be described as being deficient in Cu on the basis of the ISS studies [41,42] but as showing bulk composition on the basis of the AES studies [73,74]. As with fig. 5, the results are straightforward if viewed in terms of a composition spike. Evidently the AES measurements [73,74] show a compensation in which the surface spike and subsurface depletion cancel.

50

TABLE 2. Parameters used for evaluating eqs. (5) and (6). ΔH_i^a is the heat of atomization of component i and h_m is defined by eq. (6). Those values of h_m taken from [88] are averages for all compositions.

System, A-B	ΔH_A^a (eV)	ΔH_B^a (eV)	h_m (eV)
Ag-Au	2.94	3.82	- 0.193 [88]
Au-Cu	3.82	3.49	- 0.197 [88]
Au-Pd	3.82	3.90	- 0.320 [88]
Cu-Ni	3.49	4.46	+ 0.074 [88]
Ni-Mo	4.46	6.82	- 0.042 [87]
Ni-W	4.46	8.80	...

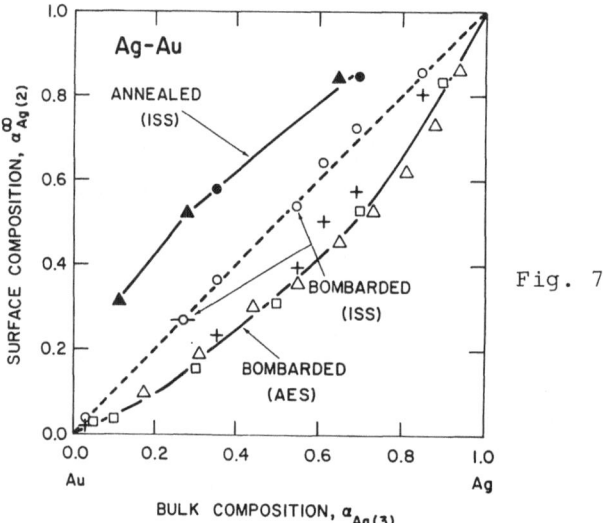

Fig. 7

FIGURE 7. Comparison of surface and bulk compositions for Ag-Au specimens which have been either annealed (●, 300 and 430°C then ISS [75]; ▲, 500°C then ISS [76]) or bombarded (O, $\ominus$, and +, 1.5-2 keV Ne$^+$ [77,11,10]; Δ and ▯, 0.7-1 keV Ar$^+$ [73,78]). The diagonal line corresponds to no surface alteration. The bombarded surfaces would be described as showing bulk composition on the basis of the ISS studies [11,77] but as being deficient in Ag on the basis of the AES studies [10,73,78]. The results for the bombarded surfaces are discussed in Section 4.2.

U_i, such that eq. (2) can be generalized to

$$\frac{\text{surface flux}_A}{\text{surface flux}_B} = \frac{\alpha^x_{A(2)} M^{2m}_B U_B}{\alpha^x_{B(2)} M^{2m}_A U_A} = \frac{\alpha_{A(3)}}{\alpha_{B(3)}}. \tag{4}$$

Eq. (4) can be justified by a dimensional analysis in the sense that, if α_A/M^{2m}_A is proportional to the total energy stored in moving A atoms at any depth, then the yield will scale as $\alpha^x_{A(2)}/M^{2m}_A U_A$. It is not correct to identify U_A and U_B with the corresponding quantities for pure A and B, as is sometimes done [e.g.11A,83,84], since the binding in an alloy is governed in all cases by the statistics of site occupancy (including ordering and segregation) and in some cases (as when species such as B or Si are involved) by changes in the character of the binding.

The problem of the statistics of site occupancy can be taken into account to first order by using the "quasichemical" variant of thermodynamics, which is commonly used to analyze phase diagrams, ordering, and surface properties with alloys [85,86]. The basis is to postulate the existence of nearest-neighbour binding energies, U_{AA}, U_{BB}, and U_{AB}. These are not formal bond energies, as with a covalent crystal, but rather negative quantities which when summed in pairwise fashion reproduce the bulk binding (cohesive) energy of the system. It is clear that, when only nearest neighbour interactions are considered, the summation will be correct if U_{AA} and U_{BB} are defined in terms of the heats of atomization of the pure substances, ΔH^a_A and ΔH^a_B, e.g.

$$\Delta H^a_A = -\frac{1}{2} Z_{A(3)} U_{AA}, \tag{5}$$

where $Z_{A(3)}$ is the bulk coordination number of substance A, and U_{AB} is defined in terms of the heat of mixing, ΔH_m:

$$\Delta H_m \equiv \alpha_{A(3)} \alpha_{B(3)} h_m = \alpha_{A(3)} \alpha_{B(3)} Z_3 (U_{AB} - \frac{1}{2}[U_{AA} + U_{BB}]). \tag{6}$$

In the latter expression, the bulk coordination number is assumed to be species independent, whence the notation Z_3, and site occupancy is assumed to be random. The average bulk binding (cohesive) energy to this approximation is

$$\langle U_3 \rangle = -\frac{1}{2} Z_3 [\alpha^2_{A(3)} U_{AA} + \alpha^2_{B(3)} U_{BB} + 2\alpha_{A(3)} \alpha_{B(3)} U_{AB}]. \tag{7}$$

The parameters needed for evaluating eqs. (5) and (6) for certain systems are summarized in table 2.

Vaporization is regarded as involving the emission of half-space atoms with coordination number $\frac{1}{2} Z_3$ [89] as in fig. 8, whereas sputtering more reasonably involves in-surface atoms with coordination number Z_2 [90] as in fig. 9. A more or less continuous supply of such atoms will exist since surface vacancies will tend to be filled by low-temperature relocation processes such as ion-beam mixing [91], picosecond-timescale recrystallization [92], interstitial diffusion [93], and adatom motion (fig. 9). The argument proceeds as follows:

52

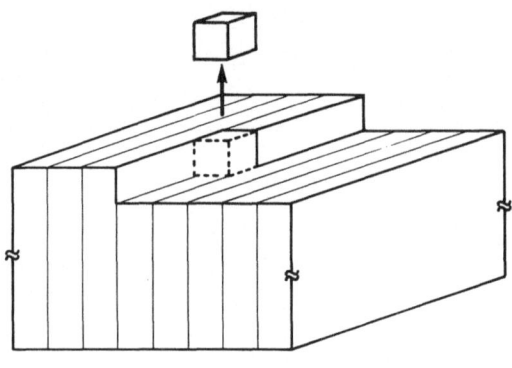

Fig. 8

FIGURE 8. Sketch of a half-space atom with coordination number $\frac{1}{2}Z_3$ which has been emitted by vaporization. There is no result-ing surface vacancy but rather the jog is displaced.

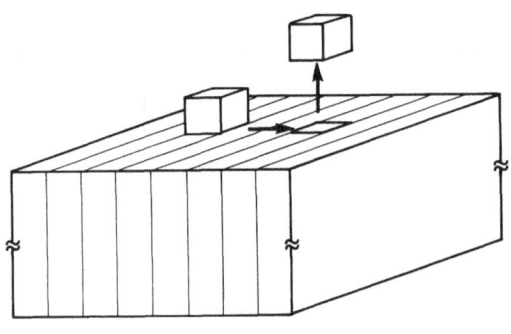

Fig. 9

FIGURE 9. Sketch of an in-surface atom with coordination number Z_2 which has been expelled by sputtering. The resulting surface vacancy is shown as being subsequently filled by a low-temperature relocation process such as ion-beam mixing [91] or, in the particular example, adatom motion.

(a) The sites in the alloy must be taken as being occupied with a particular statistics. For a bombarded system, random occupation is more usual than ordered, even if in particular cases bombardment is found to lead to ordering [e.g. 94-96]. Segregation, as will be seen, is also usual.

(b) Evaluate the surface binding energy, e.g. U_A, as the sum of nearest-neighbor binding energies. For a random system one obtains

$$U_A = -Z_2(\alpha_{A(2)}U_{AA} + \alpha_{B(2)}U_{AB})$$
$$= (Z_2/Z_3)\{(1 + \alpha_{A(2)})\Delta H_A^a + \alpha_{B(2)}(\Delta H_B^a - h_m)\}, \qquad (8)$$

where Z_2 has been assumed to be species-independent and the subsurface has been assumed to show the surface composition. (The latter point was, incorrectly, not realized in setting up eq. (9) of [39]). For a segregated system one obtains

$$U_A = -Z_\ell(\alpha_{A(2)}U_{AA} + \alpha_{B(2)}U_{AB}) - Z_v(\alpha_{A(2')}U_{AA} + \alpha_{B(2')}U_{AB}), \qquad (9)$$

where $\alpha_{i(2')}$ is the segregation-induced subsurface composition as given by eq. (15B) (to follow), Z_ℓ is the lateral coordination number, Z_v is the vertical coordination number, and $Z_2 = Z_\ell + Z_v$.

(c) Form U_B/U_A. For a random system one obtains

$$U_B/U_A = \frac{[1 + \alpha_{B(2)}]\Delta H_B^a + \alpha_{A(2)}[\Delta H_A^a - h_m]}{[1 + \alpha_{A(2)}]\Delta H_A^a + \alpha_{B(2)}[\Delta H_B^a - h_m]}, \qquad (10)$$

while for a segregated system the result is easily inferred.

(d) Solve eqs. (4) and (10) iteratively for $\alpha_{A(2)}^\infty$.

The results of this more complete approach are partially successful. If mass effects are overlooked, the species with the lower binding energy is predicted to be lost preferentially, i.e. $\alpha_{i(2)}^\infty < \alpha_{i(3)}$, and this trend is in rather good agreement with experiments in which subsurface probes were used [4]. For example, the AES measurements for Ag-Au (fig. 7), Au-Pd (fig. 1 of [38]), and Cu-Ni (fig. 5) are understandable in these terms. But the loss is in all cases predicted to be smaller   than is observed (columns 3 and 5 of table 3).

The reason for the failure of binding differences to lead to losses similar to those observed is brought out in columns 2 and 3 of table 4. The examples show that, for a given ratio of heats of atomization, the corresponding ratio U_B/U_A is always substantially nearer unity. On the other hand, when both mass and binding are taken into account, the situation becomes identical to that already noted in Section 2: instances where there is an overall loss of a heavier or equal-mass component become problematical (columns 3 and 6 of table 3).

The situation is not changed if segregation is allowed for by using eq. (9).

TABLE 3. Comparison of observed steady-state surface composi-
tions after bombardment, namely $\alpha_{A(2)}^{\infty}$, the <u>outer-surface</u> composition, and $\alpha_{A(2')}^{\infty}$, the <u>subsurface</u> composition, with $\alpha_{A(2)}^{\infty}$
from eq. (4). The entries in column 6 correspond to the use
of m = 1/4 for the power-law parameter.

System, A-B	Observed $\alpha_{A(2)}^{\infty}$	Observed $\alpha_{A(2')}^{\infty}$	Ref.	Calculated $\alpha_{A(2)}^{\infty}$ from eq. (4) without mass	Calculated $\alpha_{A(2)}^{\infty}$ from eq. (4) with mass
$Ag_{0.27}Au_{0.73}$	0.28	0.22 [a]	11	0.25	0.20
$Au_{0.56}Cu_{0.44}$	0.70 ± 0.05	0.52 [a]	41, 42,14	0.58	0.70
$Au_{0.20}Pd_{0.80}$	0.28	<0.21 [a,b]	11	0.202	0.26
$Cu_{0.50}Ni_{0.50}$	0.48	0.35 [a]	11	0.47	0.48
$Ni_{0.79}Mo_{0.21}$	0.62 [c]	0.18 [c]	22	0.75	0.70
$Ni_{0.82}W_{0.18}$	0.44 [c]	0.18 [c]	20	0.76	0.65

[a] $\alpha_{A(2')}^{\infty}$ for Ag-Au [11], Au-Cu [14], Au-Pd [11], and Cu-Ni
[18] depends on the bombardment temperature. For example, the
values for Au-Pd are 0.21 at 200°C, 0.18 at 300°C, 0.13 at
400°C, and 0.11 at 500°C [11].

[b] Because of the on-going loss of Pd, the profiles in fig. 7
of [11] or fig. 1 here are somewhat perturbed. The apparent
values both of $\alpha_{A(2')}^{\infty}$ (given in table 3) and of the bulk composition (seen in fig. 1) are too high.

[c] Average for incident Ar^+ energies ranging from 0.5 to 2
keV.

TABLE 4. Examples of surface—binding—energy ratios.

System, A-B	$\Delta H_B^a/\Delta H_A^a$	U_B/U_A [a]	α_A^∞ [2] corresponding to U_B/U_A
$Ag_{0.50}Au_{0.50}$	1.30	1.14	0.47
$Au_{0.50}Cu_{0.50}$	0.91	0.96	0.51
$Au_{0.50}Pd_{0.50}$	1.021	1.010	0.498
$Cu_{0.50}Ni_{0.50}$	1.28	1.13	0.47
$Ni_{0.50}Mo_{0.50}$	1.53	1.23	0.45
$Ni_{0.50}W_{0.50}$	1.97	1.38	0.42

[a]Deduced by iterative solution of eqs. (4) and (10), with mass neglected in the case of eq. (4).

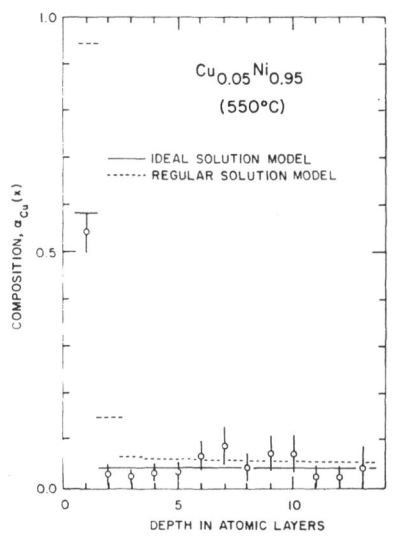

Fig. 10

FIGURE 10. Composition profile for thermally equilibrated $Cu_{0.05}Ni_{0.95}$ as obtained by first annealing at 550°C and then profiling by FIM. The ideal solution model of segregation predicts the composition spike to be confined to atom layer <u>one</u>, while the regular solution model suggests that there may <u>be</u> a slight composition excess in layers <u>two</u> and <u>three</u>. Other models of segregation suggest, in agreement <u>with</u> what <u>is</u> observed, that there may be a composition depletion in the near-surface layers, but do not change the basic idea that the main effect of segregation is confined to atom layer <u>one</u>. Due to Ng, Tsong, and McLane [99].

Nor is there any change if thermal sputtering is postulated to play a role. It is true that the surface flux ratio would become a strong function of ΔH_i^a [4]:

$$\frac{\text{surface flux}_A}{\text{surface flux}_B} = \frac{\gamma_A \alpha_{A(3)} M_B^{1/2} (\Delta H_B^a)^2}{\gamma_B \alpha_{B(3)} M_A^{1/2} (\Delta H_A^a)^2} \exp\left\{\frac{\Delta H_B^a - \Delta H_A^a}{k\hat{T}}\right\}, \quad (11)$$

where γ_i is the thermodynamic activity coefficient, defined most simply in terms of vapor pressures,

$$p_{i(alloy)} = p_{i(pure)} \gamma_i \alpha_{i(3)}, \quad (12)$$

and $\hat{T}$ is the "thermal-spike" temperature. However, there is no experimental evidence for thermal sputtering with metals [e.g. 97,98].

The use in eqs. (11) and (12) of $\alpha_{i(3)}$ rather than $\alpha_{i(2)}^\infty$ constitutes a departure from [4]. We believe it to be correct in that, although vaporizing atoms do indeed come from the surface ("2"), the γ_i are defined in terms of bulk compositions ("3") so automatically take into account the difference between $\alpha_{i(2)}^\infty$ and $\alpha_{i(3)}$.

It is clear that binding energy is not conserved during the sputtering of in-surface atoms for two reasons: Z_2 differs from $\frac{1}{2}Z_3$, and $\alpha_{i(2)}$ differs from $\alpha_{i(3)}$. The magnitude of the discrepancy is the difference between the average steady-state surface binding energy, which for stoichiometric loss is just

$$\langle U_2 \rangle = \alpha_{A(3)} U_A + \alpha_{B(3)} U_B,$$

and the average bulk binding energy as given by eq. (7). This difference is found to be ≤ 0.1 eV. Not only are such values numerically unimportant, but, since in real systems targets are maintained at fixed temperatures, they are irrelevant.

4. BOMBARDMENT-INDUCED GIBBSIAN SEGREGATION
4.1 General comments

When alloys are heated to a sufficient temperature that thermally activated diffusion can occur, it is found that one component or the other shows a composition spike in the outermost atom layer (e.g. fig.10[99]). This effect is termed equilibrium Gibbsian segregation if it occurs in the absence of bombardment. A standard argument based on minimizing the free energy of the system (surface plus bulk) suggests that the magnitude of the composition spike at equilibrium, $\alpha_{A(2)}^\infty$, is related to the bulk composition, $\alpha_{A(3)}$, by the relation:

$$\frac{\alpha_{A(2)}^\infty}{\alpha_{B(2)}^\infty} = \frac{\alpha_{A(3)}}{\alpha_{B(3)}} \exp\left\{\frac{-\Delta H^{seg}}{kT}\right\} \exp\left\{\frac{\Delta S^{seg}}{k}\right\}$$

$$= \frac{\alpha_{A(3)}}{\alpha_{B(3)}} \exp\left\{\frac{-\Delta G^{seg}}{kT}\right\}, \quad (13)$$

where ΔH^{seg} and ΔS^{seg}, the heat and entropy of segregation, are given in very general form by [100]. To lowest approximation ΔH^{seg} is the energy difference when an A atom in the bulk is exchanged with a B atom at the surface and ΔS^{seg} is zero [4,86]. The driving force, in the absence of chemical effects such as surface oxygen, is a combination based on binding (the species with the weaker binding, e.g. A if $|U_{AA}| < |U_{BB}|$, enriches) and, for a dilute component, oversize. Undersize appears to be irrelevant for a dilute component [101].

The temperature required for equilibrium Gibbsian segregation to develop is normally well above ambient temperature since self-diffusion is involved. Only with low-melting systems such as In-Ga [29] or combinations of Pb-In-Sn [102,103] does a low temperature suffice. Nevertheless, it must be anticipated that similar segregation would occur at or below ambient temperature for a bombarded target. One reason is bombardment-enhanced diffusion, as when point defects are formed collisionally and then migrate by activated jumps, especially interstitials [93] and adatoms. Another reason lies with effects where the atomic motion is wholly collisional. They include ion-beam mixing [104], known to persist without temperature dependence as the temperature is lowered from ~400 to 20 K (fig. 11 [91]), and picosecond-timescale recrystallization [92]. That processes exist which are able to bring about a segregated state without self-diffusion follows from experiments where segregation-related profiles (e.g. figs. 2,3) or angular distributions [28,105] are found in low-temperature bombardments.

If a bombarded system develops a composition spike of the type expected for Gibbsian (or similar) segregation, then (i) the steady-state profile is found to be not as in fig. 10, but rather as in figs. 1-4. (ii) Because of the existence of such profiles, the segregated species is preferentially removed from the target subsurface and the non-segregating species is present in the subsurface to excess. This leads to claims that the segregating species is preferentially sputtered (which is true in the sense that there is a subsurface depletion) and, occasionally, that the non-segregating species segregates (which is false except as a description of the subsurface [106]). (iii) Because of being present in the target subsurface to excess, the non-segregating species is emitted to an atypical extent from the subsurface and it therefore shows a narrowed angular distribution (e.g. fig. 12 [30]). It is possible that the non-segregating species also shows a narrowed energy distribution [106], though experiments on neutral partical emission rather than ions [31-33] are here needed. It is conceivable that the non-segregating species shows an enhanced degree of ionization [106]. (iv) In extreme cases, where the non-segregating species is largely excluded from the outer atom layer and, at the same time, shows bombardment-induced mobility at the ambient temperature, it resists sputtering altogether (e.g. fig. 13 [35]).

The most fundamental of these features, to be expected when segregation occurs in a bombarded system, concerns the change in the steady-state composition profile from that of fig. 10 (showing a surface spike alone) to that of figs. 1-4 (showing a surface **spike followed by** depletion). This will now be discussed.

58

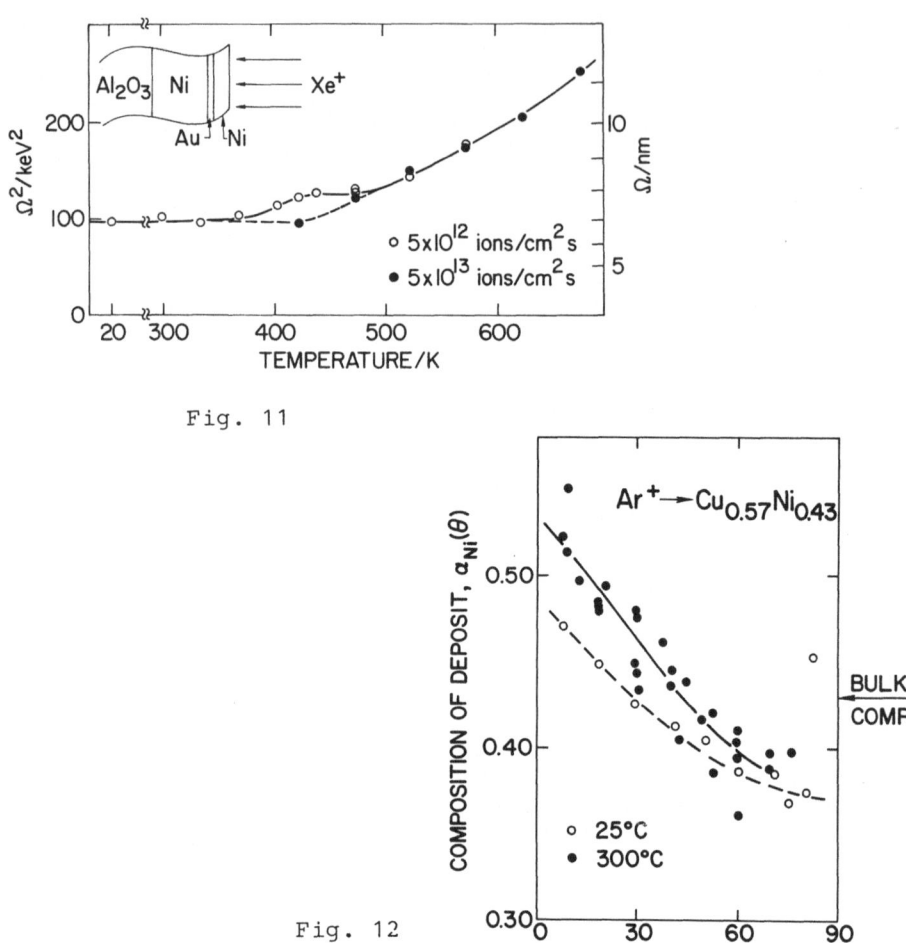

Fig. 11

Fig. 12

FIGURE 11. Temperature dependence of the mixing due to bombardment with 500 keV Xe[+] of thin (1.0 nm) Au markers located 40 nm
beneath a Ni surface. Each data point corresponds to a new
target bombarded to a constant dose of 5×10^{15} cm^{-2}. The mixing,
measured by Rutherford backscattering, was expressed in terms
of the variance (Ω^2) of a fitted Gaussian. The effective diffusion coefficient is $D_{eff} = \Omega^2/2t$. Due to Bøttiger, Nielsen,
and Thorsen [91].

FIGURE 12. Angular variation of the sputtered-atom yields for
3 keV Ar[+] normally incident on Cu$_{0.57}$Ni$_{0.43}$ as determined by
X-ray analysis of the collected sputter deposits. Bombardments
were carried out at 25 and 300°C, with the angular variation
significantly stronger at the higher temperature. The results
show that Ni, which is the non-segregating species, has a
narrowed angular distribution. Due to Ichimura et al. [30].

TABLE 5. Examples of $\alpha^{\infty}_{A(2)}$ and $\alpha^{\infty}_{A(2')}$ as governed <u>only</u> by segregation (eq. (15B)) and conservation of matter (eq. (14)). The assumptions are these: $\beta = 0.14$ or 0.28; $\alpha_{A(3)} = \alpha_{B(3)} = 0.5$; $M^{2m}_B U_B / M^{2m}_A U_A = 1$, i.e. no true preferential sputtering.

Segregation ratio, K	$\alpha^{\infty}_{A(2)}$ for $\beta = 0.14$	$\alpha^{\infty}_{A(2')}$ for $\beta = 0.14$	$\alpha^{\infty}_{A(2)}$ for $\beta = 0.28$	$\alpha^{\infty}_{A(2')}$ for $\beta = 0.28$
1	0.500	0.500	0.500	0.500
2.5	0.531	0.312	0.563	0.340
5	0.549	0.196	0.604	0.234
10	0.563	0.114	0.637	0.149
20	0.571	0.062	0.661	0.089
40	0.576	0.033	0.676	0.050

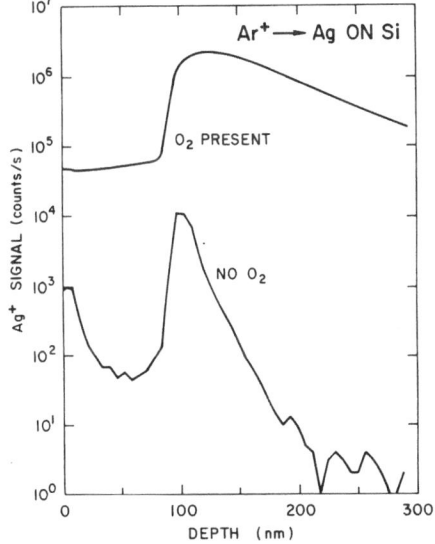

Fig. 13

FIGURE 13. Composition profile for a thin-film sandwich consisting of 100 nm of Si on a thin layer of Ag on an Si substrate. The lower profile was obtained with 8 keV Ar$^+$ in the <u>absence</u> of O_2 and shows a nominally broadened interface. The upper profile was obtained with 8 keV Ar$^+$ in the presence of O_2 flooding and shows a much greater broadening (factor of 5). The latter effect is understandable in terms of chemically-induced Gibbsian segregation of Si so that Ag is largely excluded from atom layer <u>one</u>. Due to Williams and Baker [35].

4.2 The outermost atom layer

Because sputtered atoms originate mainly from the outer-most atom layer [29,59-62], this layer must have a steady-state composition, $\alpha^{\infty}_{A(2)}$, governed to first approximation by the conservation relation eq. (4). More correctly, if it is supposed that a fraction $(1 - \beta)$ of the sputtered atoms comes from within layer one and a fraction β from beneath layer one, then instead of eq. (4) one has [62,106]

$$\frac{\alpha^{\infty}_{A(2)}(1 - \beta) + \alpha^{\infty}_{A(2')}\beta}{\alpha^{\infty}_{B(2)}(1 - \beta) + \alpha^{\infty}_{B(2')}\beta} \cdot \frac{M_B^{2m}U_B}{M_A^{2m}U_A} = \frac{\alpha_{A(3)}}{\alpha_{B(3)}}. \tag{14}$$

Equation (14) is most useful when segregation occurs and the relation between $\alpha^{\infty}_{A(2)}$ and $\alpha^{\infty}_{A(2')}$ is fixed (eq. (15B), to fol-low). The overall result is that the value of $\alpha^{\infty}_{A(2)}$ is con-trolled by 3 ill-defined parameters: the power-law parameter m, the subsurface sputtering fraction β, and the segregation ratio K (eq. (15B)). It is therefore probably unwise to make quantitative statements about $\alpha^{\infty}_{A(2)}$ except in one respect: the difference between eqs. (4) and (14) serves rather well to explain why $\alpha^{\infty}_{A(2)}$ for the segregating species of Ag-Au, Au-Cu, and Au-Pd is generally too high, i.e. column 2 of table 3 generally exceeds columns 5 or 6. The reason is that the following inequality **always** holds for the segregated species:

$$\alpha^{\infty}_{A(2)} \text{ (from eq. (14))} > \alpha^{\infty}_{A(2)} \text{ (from eq. (4))},$$

while the following **frequently** holds:

$$\alpha^{\infty}_{A(2)} \text{ (from eq. (14))} > \alpha_{A(3)}.$$

Examples are given in table 5. As would be expected, $\alpha^{\infty}_{A(2)}$ increases systematically as $\alpha^{\infty}_{A(2')}$ falls, an effect recently confirmed both experimentally and by simulation [18].

4.3 The second atom layer

The second atom layer, with steady-state composition $\alpha^{\infty}_{A(2')}$, bears a relation to the first governed by the kinetics of atom movement as treated first by Swartzfager et al. [11], Lam and Wiedersich [11A], and Itoh and Morita [5]. We here restate the argument of [5] in a form which differs by requir-ing that lattice sites be conserved though which retains the simplification that only atom layer one contributes to sput-tering ($\beta = 0$). Sites are conserved in a bombardment-induced relocation process if the lattice relaxes appropriately fol-lowing each elementary jump [60,104,107]. For example, let A tend to segregate. Then a transfer of A from (2') to (2), with a rate per second $k_{+}\alpha_{A(2')}$, triggers a relaxation in which a converse transfer occurs. For homogeneous (i.e. stoichiome-tric) relaxation, the converse transfer could involve either A or B so that a net change in the system occurs only at a reduced rate $k_{+}\alpha_{A(2')}\alpha_{B(2)}$. Concurrent with the radiation-enhanced jumps, A is lost from (2) at a rate $(v_A/\lambda)\alpha_{A(2)}\alpha_{B(2')}$, where v_A is the velocity of surface recession at an A site and the factor $\alpha_{B(2')}$ recognizes that a net change occurs in (2) only if a B is exposed in (2').* This leads to a rate equa-

*The use of the factor $\alpha_{B(2')}$ agrees with [5,11A] but not [11].

tion,

$$d\alpha_{A(2)}/dt = (k_+ + v_B/\lambda)\alpha_{A(2')}\alpha_{B(2)} - (k_- + v_A/\lambda)\alpha_{A(2)}\alpha_{B(2')},$$

so that at steady state one obtains

$$\frac{\alpha_{A(2)}^\infty}{\alpha_{B(2)}^\infty} = \frac{\alpha_{A(2')}^\infty}{\alpha_{B(2')}^\infty} \cdot \frac{k_+ + v_B/\lambda}{k_- + v_A/\lambda} = \frac{\alpha_{A(2')}^\infty}{\alpha_{B(2')}^\infty} \cdot \frac{D_+/v_A\lambda + v_B/v_A}{D_+/v_A\lambda K + 1} \qquad (15A)$$

$$\approx \frac{\alpha_{A(2')}^\infty}{\alpha_{B(2')}^\infty} \cdot K, \qquad (15B)$$

where we have used k_- for the rate per second of a jump unfavorable to segregation, $k_+\lambda^2 \equiv D_+$ (a sort of diffusion coefficient), $K \equiv k_+/k_-$ (the segregation ratio), and, as will be justified in Section 4.4,

$$D_+/v_A\lambda \approx D/v\lambda \gg 1.$$

The ratio v_B/v_A is identical with $M_A^{2m}U_A/M_B^{2m}U_B$ as appears in eqs. (4) and (14).

The expressions for bombardment-induced segregation, eqs. (15A) and (15B), are thus of similar form to that for equilibrium segregation, eq. (13). This result differs from both [11] (their eq. (14)) and [5] (their eq. (34)), where the proposed expressions can be written, in the present notation,

$$\alpha_{A(2)}^\infty \approx \alpha_{A(2')}^\infty K. \qquad (16)$$

This is a significant difference, as the form of eqs. (15A) and (15B) restricts compositions to the interval $0 \leq \alpha_i \leq 1$ whereas that of eq. (16) does not.*

It is unclear how to evaluate K from first principles, as it depends on the quantities, mobilities, and configurations of a wide variety of defects which are being relocated in a system of potential wells perturbed by segregation. Nevertheless, K can be evaluated empirically from profiles as in figs. 1 to 4. As seen in table 6 here plus table 3 of [38], the following holds:

(a) K is an increasing function of temperature starting at roughly $0.5\exp(-\Delta G^{seg}/kT)$ at or below room temperature, where $\exp(-\Delta G^{seg}/kT)$ should be assigned a value appropriate to the temperatures where **equilibrium** Gibbsian segregation is normally studied, and approaching $\exp(-\Delta G^{seg}/kT)$ as the temperature increases.

(b) Cu-Ni is deviant in that we find K $\ll \exp(-\Delta G^{seg}/kT)$.

(c) Judging from the ion-beam mixing work of [91] (fig. 11), it is likely that K will remain greater than unity even at absolute zero.

*It is true that [5] discusses a relation identical to the present eq. (15), their eq. (33). They obtain it empirically, however, and propose that it is **not** valid for a collisional process.

TABLE 6. Comparison of the bombardment-induced segregation ratio, K, as in eq. (15B) with the equilibrium segregation ratio, $\exp(-\Delta G^{seg}/kT)$, as in eq. (13). Corresponding information on Au-Cu and Au-Pd is given in [38].

System	Temperature (°C)	K	$\exp(-\Delta G^{seg}/kT)$	Technique and ref.
Ag-Au	- 163	1.2		AES [10]
	200	1.3		ISS [11]
	250	1.5		"
	300	2.2		"
	335	2.7		"
	300-500		2.8 ± 0.3	ISS [75,76]
Cu-Ni	25	1.8		SIMS + AES [8]
	100	1.9		"
	200	3.0		"
	200	1.7		ISS [11]
	300	5.3		SIMS + AES [8]
	400	7.3		"
	400		49 ± 6; 103	ISS [64,18]
	500		28 ± 1; 41	"
	650		12 ± 6; 15	FIM [65], ISS [18]
Ni-Mo	(25)	4.5-15 [a]		AES [22]
Ni-W	(25)	1.0-14 [a]		AES [20]

[a] For incident Ar^{+} energies ranging from 0.5 to 2 keV.

There is therefore a severe depletion in the second layer of the species which segregates, even for K as low as 2.5 (table 5).

It can be objected that a composition spike in the first atom layer would not be stable with respect to ion-beam mixing but we suggest that it develops in the vicinity of each impacting ion subsequent to the maximum development of the collision cascade, i.e. after $\sim 10^{-12}$s. Relevant transport processes include bombardment-enhanced diffusion (i.e. the delayed motion of residual point defects) and post-cascade collisional processes such as picosecond-timescale recrystallization [92].

4.4 Beyond the second atom layer

The altered second atom layer acts as a boundary condition and induces a composition profile to develop extending to a depth comparable to that to which the ions penetrate. The transport in question can be assumed to be of the same type which enables the segregation to occur. At sufficient depths, the bulk composition is normally resumed unless (trivially) the system is at a sufficiently high temperature that self-diffusion can contribute. An approach to this situation is seen in figs. 1 and 2.

The simplest treatment of the region depleted by preferential sputtering near an alloy surface is to assume the existence of diffusion-like motion with a species-independent diffusion coefficient D, to assume surface recession at velocity v, and either to terminate the diffusion-like motion abruptly at depth L or to use a spatially decreasing diffusion coefficient. The diffusion equation appropriate to the former case (diffusional freezing at depth L) is

$$\partial\alpha_i(x,\phi)/\partial\phi = D(\partial^2\alpha_i(x,\phi)/\partial x^2) + v(\partial\alpha_i(x,\phi)/\partial x), \quad (17)$$

where ϕ is fluence (dose) and x is depth beneath the surface. The steady-state solution is

$$\alpha_i(x) = \alpha_{i(3)} - (\alpha_{i(3)} - \alpha^x_{i(2')})\exp(-vx/D) \quad \text{for } 0 < x < L;$$

$$\alpha_i(x) = \alpha_{i(3)} \quad\quad\quad\quad\quad\quad\quad\quad \text{for } x > L, \quad (18)$$

where $\alpha^x_{i(2')}$ has been written rather than $\alpha^x_{i(2)}$, as is appropriate to profiles as in figs. 1-4. An example of the latter case (a spatially decreasing D) is when $D(\partial^2\alpha_i(x,\phi)/\partial x^2)$ is written [108]

$$(\partial/\partial x)(D^*e^{-kx}\partial\alpha_i(x,\phi)/\partial x).$$

Except for the lack of a surface composition spike, eq. (18) has the general shape of figs. 1-4. A similar shape was obtained also in the simulations of [81] but not (incorrectly) in those of [79,80].

For the same reason that a theory of K is not possible, i.e. extreme complexity, it is not possible to develop a theory of D. Equation (18) can, however, be applied to experimental composition-depth profiles as in figs. 1-4, and values of D/v as in table 7 obtained. It follows that $D/v\lambda$ is consistently >>1 and this is why eq. (15B) was approximated as it was.

TABLE 7. Examples of D/v and of D as deduced with eq. (18) from experimental composition profiles as in figs. 1-4. Since $\lambda \simeq 0.25$ nm, it follows that $D/v\lambda$ is consistently >>1.

System	Ion and energy	Temperature (°C)	D/v (nm)	D ($10^{-16} \times$ cm^2/s)	Ref.
$Ag_{0.27}Au_{0.73}$	2 keV Ne$^+$	200-300	8.5-15	1.2-3	11
$Au_{0.56}Cu_{0.44}$	2 keV Ar$^+$	-120-25	0.6-7.6	10-90	14
$Au_{0.20}Pd_{0.80}$	2 keV Ne$^+$	200-400	1.0-3.7	0.6-?	11
$Cu_{0.50}Ni_{0.50}$	2 keV Ne$^+$	200-400	4.8-11.0	3-6	11
$Cu_{0.40}Ni_{0.60}$	3 keV Ne$^+$	100-400	1.6-10.2	1.5-10	18
$Cu_{0.60}Ni_{0.40}$	5 keV Ar$^+$	400, 500	24-63	...	109

64

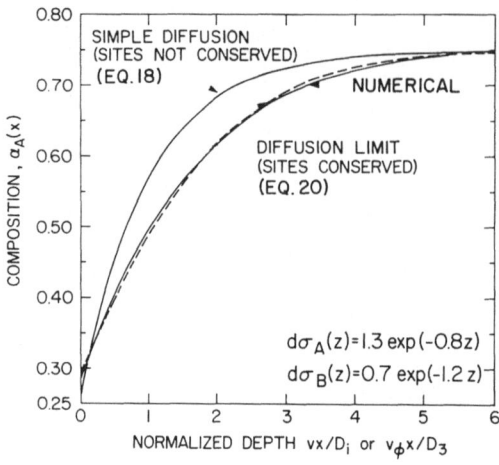

FIGURE 14. Comparison of eq. (18), eq. (20), and a numerical solution for the composition profile of a binary system. The system is described initially by compositions $\alpha_{A(3)} = 0.75$ and $\alpha_{B(3)} = 0.25$, and is then bombarded with particles which cause atom relocations to proceed with the indicated cross sections [107]. The numerical solution is from eqs. (9) and (10) of [107]. The simple-diffusion and "diffusion limit" solutions are from eqs. (18) and (20), with D_i, D_j, $\alpha_{i(2)}^\infty$, and v evaluated as discussed in [107]. There is seen to be reasonable agreement between eq. (20) and the numerical solution, but eq. (18) deviates somewhat. Due to Oliva, Kelly, and Falcone [107]

Equation (17) conserves lattice sites only because D was assumed to be species independent. It otherwise does not and, if conservation is important, an alternative formalism must be used [107]. The diffusion coefficient of each species, D_i, is now taken to be distinct and the final result is the following diffusion equation:

$$\partial\alpha_i(x,\phi)/\partial\phi = D_i(\partial^2\alpha_i(x,\phi)/\partial x^2) + v(\partial\alpha_i(x,\phi)/\partial x) -$$

$$- (\partial/\partial x)(\alpha_i(x,\phi)\sum_i D_i\partial\alpha_i(x,\phi)/\partial x). \qquad (19)$$

If eq. (19) is summed over all atom types, the first and third terms become zero while the second and fourth cancel each other. This shows that the equation conserves lattice sites no matter how much the individual values of D_i differ. The equation is non-linear but is still easily solved under steady-state conditions. For two components, i and j, one obtains:

$$[\alpha_{i(3)} - \alpha_i(x)] \exp[-(D_i-D_j)(\alpha_i(x)-\alpha_{i(2')}^\infty)/D_3] =$$

$$= [\alpha_{i(3)} - \alpha_{i(2')}^\infty] \exp(-vx/D_3). \qquad (20)$$

Here D_3, given by

$$D_3 \equiv D_i \alpha_{j(3)} + D_j \alpha_{i(3)},$$

is a sort of interdiffusion coefficient analogous to that encountered in conventional diffusion work.

Figure 14 compares eq. (18), eq. (20), and a numerical solution for the composition profile of a bombarded binary system. The method of obtaining D_i, D_j, $\alpha^\infty_{i(2')}$, and v is discussed elsewhere [107] and it is sufficient for present purposes to note that the formally more correct solution, eq. (20), agrees closely with the numerical solution. The simplified solution, eq. (18), is at fault mainly in predicting too little subsurface depletion.

5. CONCLUSIONS

Bombardment-induced Gibbsian segregation in alloys leads typically to composition profiles as in figs. 1-4, thence to preferential sputtering of the segregating species.

The steady-state <u>outer-surface</u> composition, $\alpha^\infty_{A(2)}$, is given by the conservation relations, eq. (4) or (14). The more correct relation, eq. (14), predicts significantly higher values of $\alpha^\infty_{A(2)}$ than the simpler relation, eq. (4), so that the following inequality often holds: $\alpha^\infty_{A(2)} > \alpha_{A(3)}$. $\alpha^\infty_{A(2)}$ is measurable only with truly surface-sensitive probes such as ISS (figs. 5-7) or SIMS. The steady-state <u>subsurface</u> composition, $\alpha^\infty_{A(2')}$, is determined by the relation for bombardment-induced segregation, eq. (15B). It is similar in form to eq. (13), describing equilibrium segregation, except that $\exp(-\Delta G^{seg}/kT)$ is replaced with the essentially unevaluable rate constant ratio K. The difficulty in understanding K is that it depends on the quantities, mobilities, and configurations of a wide variety of bombardment-induced defects. The difference between $\alpha^\infty_{A(2)}$ and $\alpha^\infty_{A(2')}$ is sensed most easily when subsurface probes such as AES (figs. 5-7) are used.

The altered subsurface acts as a boundary condition and induces a composition profile to develop extending (at least for low enough bombardment temperatures) to a depth comparable to that to which the ions penetrate. It is in principle due to the same bombardment-induced defects that lead to the segregation. A low-order treatment of the diffusion profile, eqs. (17) and (18), does not conserve lattice sites. Conservation of sites is, however, easily taken into account as in eqs. (19) and (20).

ACKNOWLEDGEMENTS

The authors are grateful to Professor Giovanni Falcone, Università degli Studi della Calabria, for stimulating discussions on ion-beam mixing.

REFERENCES

1. R. Kelly, in: Proc. Symp. on Sputtering, eds. P. Varga et al. (Inst. für Allgemeine Physik, Technische Univ., Vienna, Austria, 1980) p. 390.
2. H. Nakamura, K. Morita, and N. Itoh, Nucl. Instr. Meth. 191 (1981) 119.
3. P.H. Holloway and R.S. Bhattacharya, J. Vac. Sci. Technol. 20 (1982) 444.
4. R. Kelly, in: Chemistry and Physics of Solid Surfaces, Vol. V, eds. R. Vanselow and R. Howe (Springer-Verlag, Berlin, 1984) p. 159.
5. N. Itoh and K. Morita, Rad. Effects 80 (1984) 163.
6. R. Kelly, Surf. and Interface Anal. 7 (1985) 1.
7. M.P. Thomas, R.J. Oldman, and B. Ralph, Thin Sol. Films (in press).
8. M. Yabumoto, H. Kakibayashi, M. Mohri, K. Watanabe, and T. Yamashina, Thin Sol. Films 63 (1979) 263.
9. G. Betz, M. Opitz, and P. Braun, Nucl. Instr. Meth. 182/183 (1981) 63.
10. P.H. Holloway and S.K. Hofmeister, Surf. and Interface Anal. 4 (1982) 181.
11. D.G. Swartzfager, S.B. Ziemecki, and M.J. Kelley, J. Vac. Sci. Technol. 19 (1981) 185.
11A.N.Q. Lam and H. Wiedersich, J. Nucl. Mat. 103/104 (1981) 433.
12. P. Braun, Vakuum-Technik 28 (1979) 76.
12A.P. Braun, W. Färber, G. Betz, and F.P. Viehböck, Vacuum 27 (1977) 103.
13. R.-S. Li, T. Koshikawa, and K. Goto, Surface Sci. 121 (1982) L561.
14. R.-S. Li and T. Koshikawa, Surface Sci. 151 (1985) 459.
15. T. Wirth, V. Atzrodt, and H. Lange, Phys. Stat. Sol. a82 (1984) 459.
16. T. Okutani, M. Shikata, and R. Shimizu, Surface Sci. 99 (1980) L410.
17. L.E. Rehn and H. Wiedersich, Thin Sol. Films 73 (1980) 139.
18. N.Q. Lam, H.A. Hoff, H. Wiedersich, and L.E. Rehn, Surface Sci. 149 (1985) 517.
19. M.P. Thomas and B. Ralph, J. Vac. Sci. Technol. 21 (1982) 986.
20. J. Bartella and H. Oechsner, unpublished.
21. V. Atzrodt, W. Titel, T. Wirth, and H. Lange, Phys. Stat. Sol. a75 (1983) K15.
22. J. Bartella and H. Oechsner, Surface Sci. 126 (1983) 581.
23. L.E. Rehn, V.T. Boccio, and H. Wiedersich, Surface Sci. 128 (1983) 37.
24. T. Wirth, V. Atzrodt, and H. Lange, Phys. Stat. Sol. a72 (1982) K89.
25. P. Varga and E. Taglauer, Nucl. Instr. Meth. B2 (1984) 800.
26. J. Ferron, L.S. de Bernardez, E.C. Goldberg, and R.H. Buitrago, Appl. Surface Sci. 17 (1983) 241.
27. R.R. Olson and G.K. Wehner, J. Vac. Sci. Technol. 14 (1977) 319.
28. H.H. Andersen, B. Stenum, T. Sørensen, and H.J. Whitlow,

Nucl. Instr. Meth. 209/210 (1983) 487.

29. M.F. Dumke, T.A. Tombrello, R.A. Weller, R.M. Housley, and E.H. Cirlin, Surface Sci. 124 (1983) 407.
30. S. Ichimura, H. Shimizu, H. Murakami, and Y. Ishida, J. Nucl. Mat. 128/129 (1984) 601.
31. M.L. Yu and W. Reuter, J. Appl. Phys. 52 (1981) 1478.
32. M.L. Yu and W. Reuter, J. Appl. Phys. 52 (1981) 1489.
33. M. Riedel and H. Düsterhöft, in: Sec. Ion Mass Spectrometry SIMS IV, eds. A. Benninghoven et al. (Springer-Verlag, Berlin, 1984) p. 73.
33A. H.H. Andersen, Rad. Eff. 19 (1973) 257.
34. R.R. Hart, H.L. Dunlap, and O.J. Marsh, J. Appl. Phys. 46 (1975) 1947.
35. P. Williams and J.E. Baker, Nucl. Instr. Meth. 182/183 (1981) 15.
36. V. Deline, W. Reuter, and R. Kelly, in: Sec. Ion Mass Spectrometry SIMS V (in press).
37. A. Jablonski, S.H. Overbury, and G.S. Somorjai, Surface Sci. 65 (1977) 578.
38. R. Kelly and D.E. Harrison, Mat. Sci. and Eng. 69 (1985) 449.
39. R. Kelly, Nucl. Instr. Meth. 149 (1978) 553.
40. R. Kelly, Surface Sci. 100 (1980) 85.
41. H.J. Kang, R. Shimizu, and T. Okutani, Surface Sci. 116 (1982) L173.
42. H.J. Kang, E. Kawatoh, and R. Shimizu, Surface Sci. 144 (1984) 541.
43. M.L. Roush, T.D. Andreadis, F. Davarya, and O.F. Goktepe, Nucl. Instr. Meth. 191 (1981) 135.
44. B. Baretzky and E. Taglauer, in: Erosion and Growth of Solids Stimulated by Atom and Ion Beams (this volume).
45. J.F. Gibbons, W.S. Johnson, and S.W. Mylroie, Projected Range Statistics, Semiconductors and Related Materials, 2nd ed. (Dowden, Hutchinson, and Ross, Stroudsberg, PA, U.S.A., 1975).
46. S. Dzioba and R. Kelly, J. Nucl. Mat. 76 (1978) 175.
47. I. Reid, B.W. Farmery, and M.W. Thompson, Nucl. Instr. Meth. 132 (1976) 317.
48. H.L. Bay, J. Bohdansky, W.O. Hofer, and J. Roth, Appl. Phys. 21 (1980) 327.
49. P.-J. Schneider, W. Eckstein, and H. Verbeek, Nucl. Instr. Meth. B2 (1984) 655.
50. B. Baretzky and E. Taglauer, Surface Sci. (in press).
51. R. Behrisch, G. Maderlechner, B.M.U. Scherzer, and M.T. Robinson, Appl. Phys. 18 (1979) 391.
52. R.P. Webb and D.E. Harrison, J. Appl. Phys. 53 (1982) 5243.
53. G.C. Nelson and R. Bastasz, J. Vac. Sci. Technol. 20 (1982) 498.
54. J. Bohdansky, J. Roth, and H.L. Bay, J. Appl. Phys. 51 (1980) 2861.
55. R. Bastasz and J. Bohdansky, in: Proc. Symp. on Sputtering, eds. P. Varga et al. (Inst. für Allgemeine Physik, Technische Univ., Vienna, Austria, 1980) p. 430.
56. N. Andersen and P. Sigmund, Kgl. Danske Vid. Selsk. Mat. Fys. Medd. 39 (1974) No. 3.

57. K.B. Winterbon, P. Sigmund, and J.B. Sanders, Kgl. Danske
    Vid. Selsk. Mat. Fys. Medd. 37 (1970) No. 14.
58. M. Jakas and D.E. Harrison (private communication, 1984).
59. G. Falcone and P. Sigmund, Appl. Phys. 25 (1981) 307.
60. P. Sigmund, A. Oliva, and G. Falcone, Nucl. Instr. Meth.
    194 (1982) 541.
61. M.H. Shapiro, P.K. Haff, T.A. Tombrello, D.E. Harrison,
    and R.P. Webb, Rad. Effects 89 (1985) 243.
62. R. Kelly and A. Oliva, Nucl. Instr. Meth. B (in press).
63. H.F. Winters and P. Sigmund, J. Appl. Phys. 45 (1974) 4760.
64. H.H. Brongersma, M.J. Sparnaay, and T.M. Buck, Surface Sci.
    Sci. 71 (1978) 657.
65. Y.S. Ng, S.B. McLane, and T.T. Tsong, J. Vac. Sci. Technol.
    17 (1980) 154.
66. H. Shimizu, M. Ono, and K. Nakayama, J. Appl. Phys. 46
    (1975) 460.
67. P.S. Ho, J.E. Lewis, H.S. Wildman, and J.K. Howard, Surface
    Sci. 57 (1976) 393.
68. M.A. Burke and J.J. Schreurs, Surf. and Interface Anal. 5
    (1983) 155.
69. J.M. McDavid and S.C. Fain, Surface Sci. 52 (1975) 161.
70. M.J. Sparnaay and G.E. Thomas, Surface Sci. 135 (1983) 184.
71. G.C. Nelson, J. Vac. Sci. Technol. Al (1983) 1037.
72. T.M. Buck, G.H. Wheatley, and L. Marchut, Phys. Rev. Lett.
    51 (1983) 43.
73. W. Färber, G. Betz, and P. Braun, Nucl. Instr. Meth. 132
    (1976) 351.
74. H.G. Tompkins, J. Vac. Sci. Technol. 16 (1979) 778.
75. G.C. Nelson, Surface Sci. 59 (1976) 310.
76. M.J. Kelley, D.G. Swartzfager, and V.S. Sundaram, J. Vac.
    Sci. Technol. 16 (1979) 664.
77. G.C. Nelson, J. Vac. Sci. Technol. 13 (1976) 974.
78. M. Yabumoto, K. Watanabe, and T. Yamashina, Surface Sci.
    77 (1978) 615.
79. M.L. Roush, O.F. Goktepe, T.D. Andreadis, and F. Davarya,
    Nucl. Instr. Meth. 194 (1982) 611.
80. M.L. Roush, F. Davarya, O.F. Goktepe, and T.D. Andreadis,
    Nucl. Instr. Meth. 209/210 (1983) 67.
81. W. Möller and W. Eckstein, Nucl. Instr. Meth. B2 (1984)
    814.
82. D. Rosenberg and G.K. Wehner, J. Appl. Phys. 33 (1962)
    1842.
83. G. Betz, Surface Sci. 92 (1980) 283.
84. H. Gnaser, J. Marton, F.G. Rüdenauer, and W. Steiger,
    Spectrochim. Acta 37B (1982) 797.
85. R.A. Swalin, Thermodynamics of Solids, 2nd ed. (Wiley, New
    York, 1972) p. 141.
86. R.A. van Santen and M.A.M. Boersma, J. Catal. 34 (1974) 13.
87. P.J. Spencer and F.H. Putland, J. Chem. Thermodyn. 7
    (1975) 531.
88. R. Hultgren, P.D. Desai, D.T. Hawkins, M. Gleiser, and
    K.K. Kelley, Selected Values of the Thermodynamic Propert-
    ies of Binary Alloys (Am. Soc. for Metals, Metals Park, OH
    44073, U.S.A., 1973).
89. W. Hirschwald and I.N. Stranski, in: Condensation and Evap-

oration of Solids, eds. E. Rutner et al. (Gordon and Breach, New York, 1964) p. 59.

90. D.P. Jackson, Rad. Effects 18 (1973) 185.
91. J. Bøttiger, S.K. Nielsen, and P.T. Thorsen, Proc. Mat. Res. Soc. (Europe) (Strasbourg, France, 1984) p. 111.
92. D.E. Harrison and R.P. Webb, Nucl. Instr. Meth. 218 (1983) 727.
93. P. Sigmund, Appl. Phys. A30 (1983) 43.
94. H.M. Naguib and R. Kelly, Rad. Effects 25 (1975) 1.
95. G. van Tendeloo, J. van Landuyt, and S. Amelinckx, Rad. Effects 41 (1979) 179.
96. R.P.W. Lawson, W.A. Grant, and P.J. Grundy, Nucl. Instr. Meth. 209/210 (1983) 243.
97. K. Besocke, S. Berger, W.O. Hofer, and U. Littmark, Rad. Effects 66 (1982) 35.
98. W.O. Hofer, K. Besocke, and B. Stritzker, Appl. Phys. A30 (1983) 83.
99. Y.S. Ng, T.T. Tsong, and S.B. McLane, Phys. Rev. Lett. 42 (1979) 588.
100. P. Wynblatt and R.C. Ku, Surface Sci. 65 (1977) 511.
101. N.H. Tsai, G.M. Pound, and F.F. Abraham, J. Catal. 50 (1977) 200.
102. R.P. Frankenthal and D.J. Siconolfi, Surface Sci. 119 (1982) 331.
103. R.P. Frankenthal and D.J. Siconolfi, Surface Sci. 125 (1983) 595.
104. U. Littmark, Nucl. Instr. Meth. B7/8 (1985) 684, and preceding articles.
105. H.H. Andersen, B. Stenum, T. Sørensen, and H.J. Whitlow, Nucl. Instr. Meth. B2 (1984) 623.
106. R. Kelly, Nucl. Instr. Meth. B (in press).
107. A. Oliva, R. Kelly, and G. Falcone, Surface Sci. (in press).
108. A.H. Eltoukhy and J.E. Greene, J. Appl. Phys. 51 (1980) 4444.
109. L.E. Rehn, N.Q. Lam, and H. Wiedersich, Nucl. Instr. Meth. B7/8 (1985) 764.

THEORY OF SURFACE EROSION AND GROWTH

G CARTER
University of Salford, Dept. of Electronic and Electrical Engineering,
U.K.

ABSTRACT
   The general theory of wavefront propagation is summarised and it is shown
that surface growth and erosion stimulated by atom and ion beams are a
special case of this formalism. Expressions are deduced to show how both
continuous surfaces and surface discontinuities evolve during growth and
erosion and it is demonstrated how these can be used to advise geometrical
and numerical methods of surface evolution prediction.

1. THE SURFACE AS A PROPAGATING WAVEFRONT
   A surface may be regarded as the discontinuity between a solid (or
liquid) and its environment – the gas, vacuum or liquid phase. If, for any
reason, this discontinuity moves in space or time then its progress is
quite analogous to that of a moving wavefront which separates regions of
different local behaviour. The spatial paths or trajectories which connect
successive surfaces after infinitesimal time increments carry the
information about the regional discontinuity forward with time and are
known as the characteristics (1). In geometrical optics,which traces
optical (or electromagnetic) wavefront motion,the characteristics are the
optical rays. The above statements are valid whatever the mechanism by
which the surface (or wavefront) is caused to move and can include local
and extensive erosion and growth and externally impressed motion. It is
for this reason that, for the last ten years or so, the erosion and growth
of solids stimulated by atom and ion beams has come to be regarded as a
moving wavefront and theoretical treatments of such processes have
increasingly made appeal to various aspects of the theory of wavefront
motion (2-13). This had already been applied to other erosional and growth
processes including chemical dissolution and crystal growth (14-16) and
geomorphological development (17,18). This generality of approach will be
discussed in the present contribution which sets out a theoretical framework
for determining and predicting the evolution of surface contours under a
variety of ion and atom beam stimulated erosion and accretion situations.
It will be shown that although theory can be used to guide evaluation and
prediction, graphical and/or numerical computation techniques are necessary
to reveal the actual temporal and spatial evolution of surfaces for
comparison with experimental studies. These techniques will be discussed
fully in the complementary contributions to these proceedings by Nobes (19)
and Smith (20).
   Although we will follow this rather powerful general rationale it should
be noted that, historically, other theoretical treatments had been
developed earlier (21-24) but which have, more recently (10), been shown to
be special cases of the wavefront propagation method.

2. THEORY OF WAVEFRONT MOTION - CONTINUOUS SURFACES

In Figure 1 we consider a general surface, S, which is modified (the wavefront propagates) in space and time and which obeys a general describing equation:

$$S(\underline{x},t) = 0 \tag{1}$$

where $\underline{x}$ are position coordinates (x, y, z) of any surface point at time t. If an infinitesimal surface element moves along its normal direction with a speed C, then in order that neighbouring points at $(\underline{x},t)$ and $(\underline{x}+\delta\underline{x}, t+\delta t)$ should lie on the surface at successive times, expansion to first order of $S(\underline{x}+\delta\underline{x}, t+\delta t) = 0$ and subtraction of equation (1) reveals that:

$$S_t \pm (\underline{C}.\nabla)S = 0 \tag{2}$$

or

$$(S_t)^2 = C^2 (S_n)^2 \tag{3}$$

where subscripts indicate partial derivatives with respect to that variable (and n is the normal to lines of constant S). Equation (3) is known as the Eikonal equation for the wavefront and is the fundamental relationship which will be used throughout the following discussion in which detailed attention will be paid to the nature and magnitude of the speed C. The behaviour of the surface (wavefront) is more easily determined if equation (1) is rewritten in the alternative form:

$$S(\underline{x},t) = \sigma(\underline{x}) - t = 0 \tag{4}$$

When this equation (4) is substituted into equation (3) the result is:

$$\sum_{i=1}^{3} \sigma_{x_i}^2 = 1/C^2 \tag{5}$$

where the x_i represents the Cartesian coordinates x,y,z.

An alternative formulation to equation (4) can sometimes be helpful. This is (4):

$$x_i - \eta(x_j,x_k,t) = 0 \tag{6}$$

Equations (3) and (5) are, for many conditions of the speed C, first order non-linear or hyperbolic equations and the behaviour of $\sigma(\underline{x})$ as a function of time is optimally determined by defining a Hamiltonian operator

$$H = \tfrac{1}{2}C\, p_i^2 - \tfrac{1}{2}C^{-1}. = 0 \tag{7}$$

where $p_i = \sigma_{x_i}$.

If the surface is described in modified spherical polar coordinates, $(\underline{x},\theta,\phi)$, then useful geometrical relationships are:

$$\cos\theta = \frac{-\sigma_z}{(\sigma_x^2 + \sigma_y^2 + \sigma_z^2)^{\frac{1}{2}}} \tag{8a}$$

and

$$\tan\phi = \frac{\sigma_y}{\sigma_x} \tag{8b}$$

Moreover if the surface $\sigma(x)$ has a normal n, then

$$n^2 = \sigma_x^{\,2} + \sigma_y^{\,2} + \sigma_z^{\,2} \tag{8c}$$

and the plane angle, γ, between $\underline{n}$ and any unit vector $\underline{L}$ in the direction defined by its unit coordinate vectors, L_1, L_2 and L_3 is given by:

$$-\cos\gamma = \frac{L_1\sigma_x + L_2\sigma_y + L_3\sigma_z}{(\sigma_x^{\,2} + \sigma_y^{\,2} + \sigma_z^{\,2})^{\frac{1}{2}}} \tag{8d}$$

Equations (5) and (8a) may be combined to give

$$\frac{C^2\sigma_z^{\,2}}{\cos^2\theta} - 1 = 0 \tag{9}$$

The aim of constructing solutions to equation (3) and its equivalents of equations (7) or (9) is to determine those trajectories in space or characteristics which link positions on the surface at successive infinitesimally small time increments. Whitham (1) has described how such solution is accomplished as follows. Consider a function $\sigma(x_1 \ldots x_n)$ which satisfies the differential equation

$$H(\underline{p},\sigma,\underline{x}) = 0 \tag{10}$$

where $p_i = \dfrac{\partial\sigma}{\partial x_i}$

Next define a curve C in a space written in parametric form $\underline{x} = \underline{x}(\lambda)$. The total derivative of σ along the curve C is given by

$$\frac{d\sigma}{d\lambda} = \frac{\partial\sigma}{\partial x_j} \cdot \frac{dx_j}{d\lambda} = p_j \frac{dx_j}{d\lambda} \tag{11}$$

where repeated use of the subscript (j) indicates summation over all $j = 1 \ldots n$.

In addition the total derivative of p_i along the curve C is given by

$$\frac{dp_i}{d\lambda} = \frac{d}{d\lambda}\left(\frac{d\sigma}{dx_i}\right) = \frac{\partial^2\sigma}{\partial x_i \cdot \partial x_j}\frac{dx_j}{d\lambda} \tag{12}$$

Whilst the x_i derivative of equation (10) yields:

$$\frac{\partial^2\sigma}{\partial x_i \cdot \partial x_j} \cdot \frac{\partial H}{\partial p_i} + \frac{\partial H}{\partial\sigma} \cdot \frac{\partial\sigma}{\partial x_i} + \frac{\partial H}{\partial x_i} = 0 \tag{13}$$

Comparing equations (12) and (13) reveals that if the curve C is chosen such that

$$\frac{dx_j}{d\lambda} = \frac{\partial H}{\partial p_i} \tag{14}$$

then

$$\frac{dp_i}{d\lambda} = - p_i \frac{\partial H}{\partial \sigma} - \frac{\partial H}{\partial x_i} \tag{15}$$

Equations (14) and (15) together with equation (1) written as

$$\frac{d\sigma}{d\lambda} = \sum_j p_j \frac{\partial H}{\partial p_j} \tag{16}$$

now form a set of (2n+1) <u>ordinary</u> differential equations for determining a 'characteristic curve' and the values of σ and its spatial derivatives p_i along the curve. Each curve connects successive points on the developing surface and in principle the values of σ and p_i can be determined at successive intervals by integration along the characteristics. As we will demonstrate shortly, equation (14) determines speeds along characteristic directions whilst equation (15) determines the rate of change of the direction cosines of surface points along characteristics.

If in equation (9) the substitution g = C/cosθ is made, where g is the effective wavefront velocity of a point in the Oz direction, then we may define a Hamiltonian equivalent to equation (7) as

$$H \equiv 1 - g^2 \sigma_z^2 \equiv 0 \tag{17}$$

In this equation g may be a function of all or some of the x_i and all or some of the σ_{x_i}. In this context the x_i may also represent a time t dimension.

If, now, we apply the results of equation (16), together with equations (8a) and (8b) to equation (17), then after some mathematical manipulation one deduces (10):

$$\frac{d\sigma}{d\lambda} = \frac{dt}{d\lambda} = -2 \tag{18}$$

Using this result, again with equations (8a) and (8b) and applying equations (14) and (15) to equation (17) one deduces the complete set of characteristic equations for a three-dimensional system as:

$$\frac{dx}{dt} = - g_\theta \cos^2\theta \cos\phi - g_\phi \cot\theta \sin\phi \tag{19a}$$

$$\frac{dy}{dt} = - g_\theta \cos^2\theta \sin\phi + g_\phi \cot\theta \cos\phi \tag{19b}$$

$$\frac{dz}{dt} = - g + g_\theta \sin\theta \cos\theta \tag{19c}$$

and

$$\frac{d\sigma_x}{dt} = \frac{1}{g} \{\sigma_x g_t + g_x\} \tag{20a}$$

$$\frac{d\sigma_y}{dt} = \frac{1}{g} \{\sigma_y g_t + g_y\} \tag{20b}$$

$$\frac{d\sigma_z}{dt} = \frac{1}{g} \{\sigma_z g_t + g_z\} \tag{20c}$$

Equations (20) clearly represent the variations of surface orientation with time and equations (19) clearly represent the speeds of surface points which follow these variations as indicated earlier.

If the velocity components given by equations (19) are projected along the direction of the surface normal (with orientations θ and ϕ) it is found that their sum is simply equal to $- g \cos\theta$. This indicates that the projection of the total velocity:

$$\nu = \{(\frac{dx}{dt})^2 + (\frac{dy}{dt})^2 + (\frac{dz}{dt})^2\}^{\frac{1}{2}}$$

along the characteristic onto the direction of the surface normal is precisely equal to the speed of the wavefront, $C = g\cos\theta$, along the normal.

Without being specific about the processes which lead to surface motion some quite general results can be deduced from equations (19) and (20). Firstly it may be noted that if g is a function of x, y or z, the wavefront propagation is appropriate to that in an inhomogeneous system, whilst if g is a function of θ and ϕ (and consequently of σ_{x_i}) the system is anisotropic.

If g is a function of time then the wavefront propagation is in a time dependent system.

Considering next the speed equations (19) it is noted that if g_θ and g_ϕ are identically zero (i.e. local maxima, minima or saddle points in the $g(\theta,\phi)$ function) then

$$\frac{dx}{dt} = \frac{dy}{dt} = 0 \quad \text{and} \quad \frac{dz}{dt} = -g,$$

i.e. the velocity along characteristics has only a component in the $-Oz$ direction. This conclusion is also true for $\cos\theta = 0$ ($\theta = \pm\pi/2$), which implies the lack of motion of planes parallel to the $-Oz$ direction except in the $-Oz$ direction.

It is also readily demonstrated that, except for the trivial case $g = 0$, the components $dx/(dt)$, $dt/(dt)$ and $dz/(dt)$ cannot be identically reduced to zero. If, however, $\theta = \pi/2$, $\phi = 0$ (planes parallel to xOz) then $dx/(dt) = 0$; for $\theta = \pi/2$, $\phi = \pi/2$ (planes parallel to yOz) then $dy/(dt) = 0$, whilst for $\theta = 0$ all ϕ (planes parallel to xOy) $dz/(dt) = -g$.

Finally, if surface elements of different azimuthal orientation, ϕ, but fixed polar orientation, θ, are to possess equal but differently directed speeds, i.e. symmetric speed conditions, then both dz/dt and the radial speed $dr/(dt) = \{(dx/(dt))^2 + (dy/(dt))^2\}^{\frac{1}{2}}$ must be equivalent for different ϕ values. It is readily shown, from equations (19a) and (19b), that

$$(\frac{dr}{dt})^2 = \cot^2\theta \{(g_\theta \sin\theta \cos\theta)^2 + g_\phi^2\} \tag{19d}$$

From this equation (19d) and equation (19c), for the speeds to be azimuthally equivalent g. g_θ and g_ϕ must possess identical values for given θ and ϕ. Such conditions again require extrema in the general $g(\theta,\phi)$ function but further demand (because of constant g requirement) that these extrema be identical in all respects.

Considering finally the orientation equations (20) it is noted that if all the $d\sigma_{x_i}/(dt)$ are identically zero, then the σ_{x_i} are constants which, from equation (8) implies that the characteristics are those of constant orientations θ and ϕ. The $d\sigma_{x_i}/(dt)$ may be rendered equal to zero, individually or collectively in a number of ways. Firstly if g is neither a function of space coordinates, x_i, nor of time t, then $d\sigma_{x_i}/(dt) = 0$ for all x_i and t. Secondly, if, for all x_i, $\sigma_{x_i} g_t + g_{x_i} = 0$, then $d\sigma_{x_i}/(dt) = 0$.

This suggests that for each surface coordinate point and associated direction cosines or orientations locally variable control of g_{x_i} and g_t could be exercised to ensure constancy of all σ_{x_i}. A specific example would be one where the spatio-temporal x_i variation of g was such that the temporal variation was spatially uniform then g would become a separable function $g(x_i)$ $g(t)$. Use of equations (8) and (20) then reveals that, in order for θ and ϕ to be constants, the demand is that $g_x = k_1 g_y = k_2 g_z$ where k_1 and k_2 are constants.
If g is temporally invariant ($g_t = 0$), then g_{x_i} is zero when either g = constant, independent of all x_i, or when g exhibits an extremum (maxima, minimum or saddle point) identically for all x_i, and again σ_{x_i} = constant.

On the other hand, if g is spatially invariant ($g_{x_i} = 0$), then, for any g_t, from equation (20)

$$\frac{d\sigma_{x_i}}{dt} = \frac{\sigma_{x_i}}{g} \cdot \frac{\partial g}{\partial t} ,$$ which solves to give $\sigma_{x_i} = k_i g$, where k_i

are constants, and so, from equations (8) θ and ϕ are again constants on the characteristics, although from equations (19) the speeds vary along the characteristics since these depend upon instantaneous values of g. In the specific case of an initially planar surface geometry the surface propagates as a plane.
Finally, if g is temporally invariant ($g_t = 0$) then if any of the g_{x_i} are individually zero, the characteristics correspond to σ_{x_i} = constant in the i'th direction for which g_{x_i} is zero.
In particular if $g_z = 0$, then equation (19c) indicates that σ_z = constant, which from equation (17) identifies the characteristics as those of constant g. It may, in some circumstances, therefore be easier for particular systems which exhibit $g_x = 0$ in an arbitrary direction x_a to effect an axial rotation to render x_a parallel to the -Oz direction so that characteristics are equivalent to those of constant g(a).
It should be noted that when the σ_{x_i} are constants and the characteristics are those of constant orientations θ and ϕ, then these characteristics are linear since the speed components of equation (19) are, for each θ and ϕ, fixed values. In such cases, since θ and ϕ, remain constant along the characteristics, then only values of θ and ϕ between lower (θ_L and ϕ_L) and upper (θ_u and ϕ_u) bounds present on the initial surface are present at some later time. In the case where the σ_{x_i} vary with time this restriction is no longer valid and new surface orientations are generated. The preceding restrictions on the bounded values of θ and ϕ were first noted explicitly by Frank (14) in discussions of chemical erosion of solids.
In considering surface or wavefront evolution generally no, specification of the origin or nature of the normal velocity has been made. This velocity could arise from one or both of two sources. Firstly processes

acting within the system (defined as both spatial regions divided by the surface) can locally modify the surface and secondly externally impressed processes can either locally or extensively move the surface.

In the case of a rigid solid: vacuum system, translation and/or rotation of the solid provides uniform extensive motion. In all cases the velocity C represents the local vector sum of all processes which modify the surface velocity at that local point. It is sometimes more convenient to effectively remove the extensive motion component so that surface modification resulting from the variable local processes can be identified. In this case, Carter, Cruz and Nobes (9) have demonstrated a generalisation of Whitham's (1) one-dimensional result that all of the equations developed earlier can be rewritten identically with $\underline{C_e}$

replacing C, where $\underline{C_e}$ is the effective local normal velocity and is equal

to the vector difference between the total normal velocity $\underline{C}$ and the normal velocity C_m resulting from externally impressed motion. It will be demonstrated $\underline{}$ later, and particularly in the accompanying contributions by Nobes (19) and Smith (20), that such a strategy is particularly useful in determining the evolution of solid surfaces which are being eroded in spatio-temporally variable erosion conditions or where the solid is being externally manipulated simultaneous to erosion.

The preceding analyses are general and valid for continuous surfaces or wavefronts in which σ and the σ_{x_i} are continuous and single valued for

all x_i. Special considerations are necessary when any of the σ_{x_i}

(surface orientations) are discontinuous and multivalued at any surface coordinate set and these special circumstances will be analysed shortly. Before doing this, however, we note that equations (19) and (20) enable, for a continuous surface, precise evaluation of surface (wavefront) evolution if the values of g, g_θ, g_ϕ, g_{x_i} and g_t are known or prescribed

for a given surface evolution process. The detailed methods of evaluation will be described by Nobes (19) and Smith (20) later but we note here that a general strategy is as follows.

At a given time t, coordinate values (x,y,z) for a set of points on the surface are selected and the corresponding values of θ and ϕ (and hence σ_{x_i}) together with the local values of g and its derivatives are

determined. A time step δt is then considered during which the σ_{x_i} change

to new values given by equation (20), i.e. $$\delta\sigma_{x_i} = \frac{\delta t}{3}\left\{\sigma_{x_i}g_t + g_x\right\}$$

and the coordinates change to new values

$x_i + \delta x_i = x_i + \frac{dx_i}{dt}\,\delta t$, where the values of $\frac{dx_i}{dt}$ are given by equations

(19). These new coordinates and orientation values then allow reconstruction of the modified surface at the time $t + \delta t$. This process is commenced at $t = 0$ and iterated over a sequence of whatever optimum time steps δt are chosen. Essentially the method is equivalent to that of ray tracing in geometrical optics and is illustrated schematically in Figure 2.

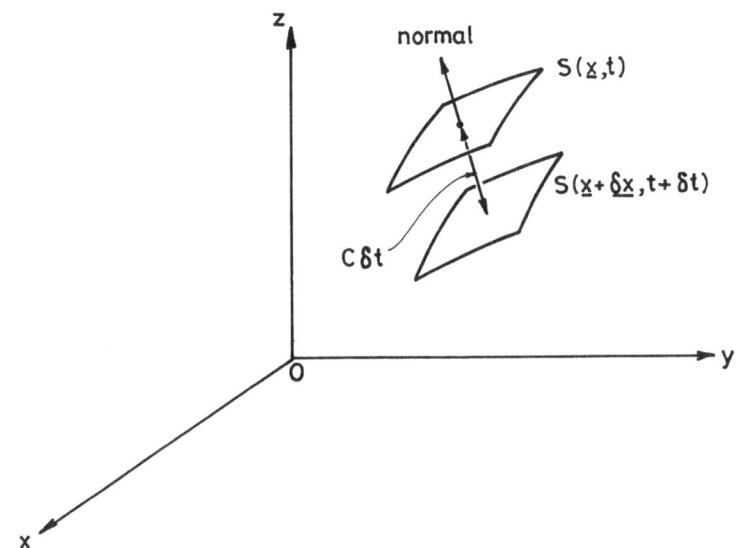

FIGURE 1. The evolution of a surface $S(\underline{x},t)$ which propagates along its normal with a speed C.

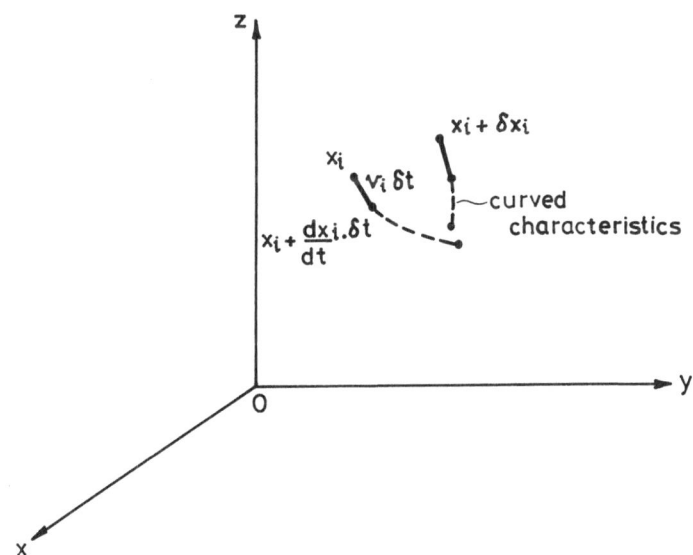

FIGURE 2. The progress of surface points (x_i), $(x_i + \delta x_i)$ along curved characteristics.

3. THEORY OF WAVEFRONT MOTION - DISCONTINUOUS SURFACES

When rays intersect in this latter system a focus is formed and
information from different regions of an object is coalesced. In the
case of a propagating surface the separate information which coalesces
when two characteristics intersect are the local surface orientations (θ
and ϕ from the σ_{x_i} (15)) appropriate to the intersecting characteristics.
Consequently, the surface, σ, remains continuous when characteristics
intersect but orientation or slope discontinuities occur at this intersection. Once formed this point of orientation discontinuity or 'edge'
propagates in space and time in a manner constrained by the orientations
which bound the edge and which may themselves change spatio-temporally.
Since the characteristics originate from individual points on the initial
($t = 0$) surface, their subsequent paths or trajectories and their points
of intersection and subsequent evolution are intimately dependent upon
the initial surface contour. The general process of edge formation is
represented schematically for a two-dimensional section of a surface in
Figure 3. The prediction of characteristics intersection and edge
formation and motion is therefore an initial value problem (1). The time
t at which edge formation first occurs from an initially everywhere
continuous surface can be deduced in a formal manner from equations (19)
as follows. If two points $(x_o, 0)$, $(x_o + \delta x_o, 0)$ define a line element
δx_o on the initial surface at $t = 0$, then at a later time t, these points
transpose to new points, $(\underline{x}_f, t)$, $(\underline{x}_f + \underline{\delta x}_f, t)$, defining a line element $\underline{\delta x}_f$
on the new surface. The spatial coordinates of these new points are

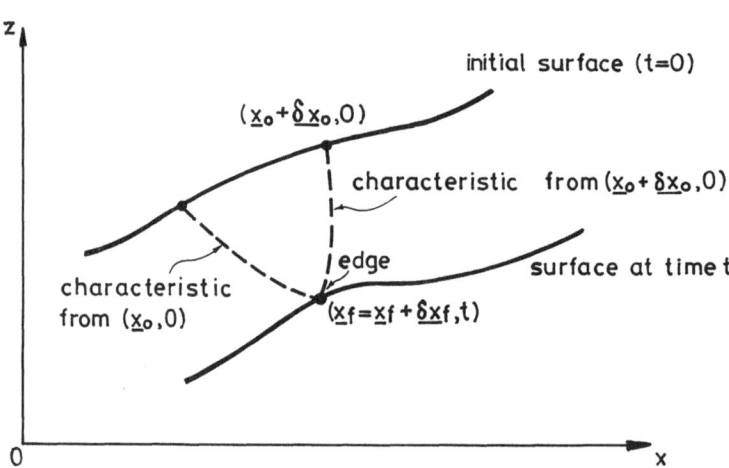

FIGURE 3. The formation of a surface discontinuity (edge) in a two-
dimensional surface section at $(\underline{x}_f = \underline{x}_f + \underline{\delta x}_f, t)$ by the intersection of
curved characteristics originating from initial surface positions
$(x_o, 0)$, $(x_o + \delta x_o, 0)$.

determined by adding to the initial values the incremental distances
travelled along trajectories, projected on the coordinate axes, in the
time interval t. If the instantaneous characteristic velocity components
at some intermediate time t' between O and t relevant to position vectors
at intermediate positions $\underline{x}'$ and $\underline{x}'+\delta\underline{x}'$ are written as

$$\nu_i(\underline{x}',t') = \frac{dx_i}{dt}(x',t'),\text{ which are given from equations (19), then the}$$

values of the coordinate components of $\underline{x}_f$ and $\underline{x}_f+\delta\underline{x}_f$, are respectively:

$$x_{io} + \int_o^t dt'\ \nu_i(\underline{x}',t') \text{ and } x_{io} + \delta x_{io} + \int_o^t dt'\ \nu_i(\underline{x}' + \delta\underline{x}',\ t') \qquad (21)$$

where the different instantaneous values of the ν_i's reflect the fact that
the characteristics both originate from points of differing initial
orientation and propagate through different space and time positions in the
system. In order that the characteristics should intersect and an edge be
formed it is required that $\delta\underline{x}_f$ should be zero, or, equivalently, that the
component values of $\underline{x}_f$ and $\underline{x}_f+\delta\underline{x}_f$ should be identically equal. From
equation (21) this demands that the following set of equations be satisfied
i.e.

$$\delta x_{io} = \int_o^t dt'\ \nu_i(\underline{x}',t') - \int_o^t dt'\ \nu_i(\underline{x}'+\delta\underline{x}',\ t') \qquad\qquad \text{or}$$

$$\delta x_{io} = \int_o^t dt'\ \{\nu_i(\underline{x}',t') - \nu_i(\underline{x}'+\delta\underline{x}',\ t')\} \qquad (22a)$$

This equation can be expanded to first order in the integrand to give:

$$\delta x_{io} = -\int_o^t dt'\ \frac{d\nu_i}{d\underline{x}}(\underline{x}',t')\ \delta\underline{x}_o(t') \qquad (22b)$$

This set of equations can be solved, in principle, to determine the time t
for edge initiation for specified forms of the δx_{io}, $d\nu_i/(d\underline{x})$ and $\delta\underline{x}(t')$.
It should be noted that in determining $\nu_i(\underline{x}',t')$, and hence $d\nu_i/(d\underline{x})$ that
this is not only a function of $\underline{x}'$ and t' but of the local orientations θ
and ϕ also. These orientations, however, depend upon initial conditions
(θ_o,ϕ_o) and the changes in θ and ϕ which result from the changes in the
δx_i's along the characteristics.
Unfortunately even if the δx_{io} (from the initial surface) and $d\nu_i/(d\underline{x})$
(from equations (19) and (20) with defined functions for g and its
derivatives) can be specified analytically the $\delta x_i(t')$, the spatial
coordinate separation of the two trajectories at a specific time, cannot be
so specified analytically. It is necessary to determine the character-
istics appropriate to $\underline{x}_o$ and $\underline{x}_o+\delta\underline{x}_o$ from equations (19) and (20) as
described earlier and evaluate the individual coordinate separations at
each intermediate time. More directly once these characteristics have
been determined for a whole set of $(\underline{x}_o,o)$ values it becomes graphically
obvious where, in both space and time, characteristics first intersect to
form edges.
There is a special circumstance, however, in which equation (22) is
solvable analytically, which occurs when the characteristics are linear.
In this condition the distances $\delta x_i(t)$ are readily determinable from
geometric analysis of progress along individual characteristics even if

the characteristic velocity components are time dependent but <u>not</u> spatially dependent. In these conditions

$$v_i(t') = v_i(o) + \int_o^t dt'' \frac{\partial v_i(t'')}{\partial t} \tag{23}$$

Indeed, as indicated earlier, the essential requirement for linear characteristics is that the system is homogeneous, i.e. g is independent of spatial position and further simplification is possible if the system is considered time independent also. In these circumstances equation (22a) is replaced by its finite difference form.

$$\delta x_{io} = t\{v_i(\underline{x}_o,o) - v_i(\underline{x}_o + \delta \underline{x}_o,o)\} \tag{24a}$$

or

$$\delta x_{io} = -\delta \underline{x}_o \, t \, \frac{dv_i}{dx}(x_o) \tag{24b}$$

The set of equations given by equation (24b) was derived in alternative form by Carter et al. (23) and Ducommum et al. (25) for the sputtering erosion, in a spatio-temporally uniform ion flux, of the two-dimensional section of a solid, and for the complete three-dimensional case by Smith et al (4). The result for a one-dimensional situation given by just one of the set of equation (24b), is also identical to that given by Whitham (1). In the case of a spatially homogeneous, time-independent system, the function g and its orientation derivatives are spatially independent and so that variation in v_i with $\underline{x}_o$ arises from the initial conditions of the surface contour. If from knowledge of the g function, and consequently the v_i values from equations (19), and of the initial surface contour, a generalised map of the variations of v_i with x_i for all three coordinate components is made then equation (24b) indicates that edge formation occurs when:

$$t = -\{\frac{dv(x)}{dx}\}^{-1} = -\{\frac{dv(y)}{dy}\}^{-1} = -\{\frac{dv(z)}{dz}\}^{-1} \tag{25}$$

The time t for edge formation can then be deduced from the v_i / x_i map by determining the x_{io} coordinates for which the identity of equation (25) is satisfied. The space coordinates at which edge formation occurs are then readily deduced by adding $v_{io}.t$ to the initial values x_{io}. It should also be noted that equation (25) indicates that edges first form when t is a minimum or when

$$\frac{dv(x)}{dx} = \frac{dv(y)}{dy} = \frac{dv(z)}{dz}$$ are identical maxima.

Again the general v_i / x_i map can be used to determine the points on the initially continuous surface which first coalesce to form an orientation discontinuity and the time and spatial coordinates at which this first occurs.

Once an edge has formed it will continue to propagate and its direction and speed are readily determined as described by Frank (15). If three planar elements, with unit normal vectors n_1, n_2, n_3 intersect at a vertex point p_{123} and advance with speeds, measured normally to themselves, of C_1, C_2 and C_3 then provided that no coordinate discontinuity occurs at p_{123}, the displacement of the vertex in time δt is $U_{123}\delta t$, which must satisfy the set of three equations:

$$\underline{U}_{123} \cdot \underline{n}_1 = C_1$$

$$\underline{U}_{123} \cdot \underline{n}_2 = C_2 \qquad (26)$$

$$\underline{U}_{123} \cdot \underline{n}_3 = C_3$$

This set of equations merely reflects the fact that the effective velocities of the planar motions along the direction of the common motion of their intersection point p_{123} must be equal to maintain no coordinate discontinuity.

For a given set of planar orientations and prescribed values of the normal speeds, equation (26) enables evaluation of the velocity (both the speed and direction of motion) of the point of intersection. Only three equations are needed for such an evaluation so if more than three planes meet at a singular point this only propagates as a singular point provided that the orientations $(\underline{n})$ and the speeds (C) also satisfy equation (26). If the velocity $\underline{U}_{123}$ lies in a direction (θ', ϕ') with unit vector components L_1, L_2 and L_3 along the x, y, z, coordinates, and subtends planar angles γ_1, γ_2 and γ_3 to the three planar elements considered, then using equations (8d) remembering that $g = C/\cos\theta$, equation (26) can be rewritten:

$$\underline{U}_{123} = \frac{g_1 \cos\theta_1}{\cos\gamma_1} = \frac{g_2 \cos\theta_2}{\cos\gamma_2} = \frac{g_3 \cos\theta_3}{\cos\gamma_3} \qquad (27a)$$

or alternatively

$$\underline{U}_{123} = \frac{-g_1 \sigma_{z_1}}{L_1 \sigma_{x_1} + L_2 \sigma_{y_1} + L_3 \sigma_{z_1}} = \frac{-g_2 \sigma_{z_2}}{L_1 \sigma_{x_2} + L_2 \sigma_{y_2} + L_3 \sigma_{z_2}} = \frac{-g_3 \sigma_{z_3}}{L_1 \sigma_{x_3} + L_2 \sigma_{y_3} + L_3 \sigma_{z_3}} \qquad (27b)$$

Finally, remembering the earlier statement in the discussion of equations (19) that the total velocity along the trajectory projected onto the surface normal is equal to the normal speed, then equation (27a) can also be rewritten

$$\underline{U}_{123} = \frac{\nu_1 \cos\alpha_1}{\cos\gamma_1} = \frac{\nu_2 \cos\alpha_2}{\cos\gamma_2} = \frac{\nu_3 \cos\alpha_3}{\cos\gamma_3} \qquad (27c)$$

where the ν's are the characteristic velocities of each plane and the angles α are the corresponding angles between the direction of each characteristic and the normals to those planes.

Any of these equations allow complete evaluation of the velocity $\underline{U}_{123}$ and its direction. For example, if three intersecting planes, characterised by g values of g_1, g_2, g_3 and orientations (θ_1, ϕ_1), (θ_2, ϕ_2) and (θ_3, ϕ_3) are considered it is readily shown that the component of the edge velocity in the Ox direction, $U(x)$, is given by

$$U(x) = \frac{A}{B} \qquad (27d)$$

where $A = g_1(\tan\theta_2 \sin\phi_2 - \tan\theta_3 \sin\phi_3) - g_2(\tan\theta_1 \sin\phi_1 - \tan\theta_3 \sin\phi_3) + g_3(\tan\theta_1 \sin\phi_1 - \tan\theta_2 \sin\phi_2)$

$$B = \tan\theta_2\tan\theta_3\sin(\phi_3-\phi_2)+\tan\theta_1\tan\theta_3\sin(\phi_1-\phi_3)+\tan\theta_1\tan\theta_2\sin(\phi_2-\phi_1)$$

Similar geometric relationships are deduced for the Oy and Oz components of $\underline{U_{123}}$ and these may be used to deduce both the magnitude of $\underline{U_{123}}$ and the direction (θ_e,ϕ_e). When the $U(x_i)$'s are so evaluated and if three neighbouring planes $(\theta,\phi+\delta\phi)$, (θ,ϕ) and $(\theta+\delta\theta, \phi)$ are specifically considered, the resulting values become identical to those of equations (19) as they should since the surface has then been relaxed to a continuous form.

This may be easily illustrated by considering the further simple example of a two-dimensional section of a surface contour (with no variation in the Oy direction) and let the direction of $\underline{U_{13}}$ make an angle θ_e with the Oy direction.

Consequently $\pi/2 - \gamma_1 = \theta_e - \theta_1$, $\pi/2 - \gamma_2 = \theta_2 - \theta_e$ and equation (27a) adopts the form:

$$\underline{U_{12}} = \frac{g_1\cos\theta_1}{\sin(\theta_e-\theta_1)} = -\frac{g_2\cos\theta_2}{\sin(\theta_2-\theta_e)} \tag{27e}$$

This equation is readily solved to show that:

$$\tan\alpha = \frac{g_1\tan\theta_2-g_2\tan\theta_1}{g_1-g_2} \tag{28a}$$

and that the components of $\underline{U_{13}}$ in the Oz and Ox directions are

$$U(z) = \frac{g_1\cot\theta_1-g_2\cot\theta_2}{\cot\theta_1-\cot\theta_2} \tag{28b}$$

and

$$U(x) = \frac{g_1-g_2}{\tan\theta_2-\tan\theta_1} \tag{28c}$$

If in the limit of a continuous surface $\theta_2 \to \theta_1 + \delta\theta$, then expansion of equations (28b) and (28c) show that they relax, respectively to

$$U(z) = -g - g_\theta \sin\theta \cos\theta \tag{29a}$$

and

$$U(x) = -g_\theta \cos^2\theta \tag{29b}$$

These expressions are equivalent to equations (19c) and (19a) respectively for an azimuthally invariant system. Equations (28b) and (28c) are therefore the finite difference forms of equations (29a) and (29b) and can therefore be written, more generally, as

$$U(z) = -\frac{[g\cot\theta]}{[\cot\theta]} \tag{30a}$$

and

$$U(x) = -\frac{[g]}{[\tan\theta]} \tag{30b}$$

where the square brackets indicate a discontinuous jump in the appropriate quantity. This formulation is typical of the defining equations for shock

processes (1) and indicates that the propagation of a surface orientation
discontinuity (an edge) is equivalent to that of a shock wavefront.
Mathematically the intersection of characteristics is equivalent to
formation of a shock.

In an entirely intersecting planar system the edge propagates linearly
in space with the speed and direction derived, in general, by solution of
equation (27). If, however, the edge forms part of a continuous surface
then the motion of the edge changes in speed and direction as further
characteristics from the initial surface intersect the edge characteristic.
As a result, the edge characteristic is generally curved in space and the
velocity along the characteristic is variable.

An important interpretation of equations (27) is that they define
Snell's law in optical wavefront propagation such that the rays
(characteristics) refract at the interface between systems for which the
wavefront velocities are different. This occurs via constancy of the
effective speed of propagation parallel to the plane of discontinuity. In
the case of a spatio-temporally varying surface which intersects several
media the shock velocity components derived from equations (27) and in
equation (30) for a two-dimensional case, indicates that the speed of
propagation of the discontinuous surface edge along the interface between
the media must be such that its projection parallel to the normals to each
of the surface components which intersect at the common media interface
must be equal to the speed of each surface component along its normal.
This is a general boundary condition (1) which will be found to be useful
and necessary in some of the examples of surface evolution considered in
the contribution by Nobes (19).

Although the prediction of initial edge formation is, in principle,
straightforward from equation (22b) and in practice equally straightforward
for homogeneous systems with linear characteristics (equations (24b) and
(25)), whilst the velocity and direction of edge motion can be found
directly from equations (27) and (30), description of the spatio-time
evolution of the characteristics appropriate to edges is more difficult and
in inhomogeneous systems can only be easily undertaken by iterative
raytracing techniques. The reason for this difficulty is that the
requirement is to determine, as a function of increasing time, which
characteristics from the initially continuous surface intersect the
trajectories of the edge (or edges) which have formed and, when they do so
intersect, what new surface orientations are coalesced into the edge. In
the case of homogeneous systems the problem is less severe since the
characteristics of continuous parts of the surface are linear and of
constant orientation whilst neither of these properties are valid in non-
homogeneous systems. Initial discussion on edge motion is therefore
maintained specific to homogeneous time-independent systems where
$g_{x_i} = g_t = 0$.

For any general initial surface an edge will first form at a time t_e
from a region of the surface with orientations (θ, ϕ) for which equation
(25) is locally satisfied, providing that for this region the time t_e is
positive. This edge forms, initially, as a surface discontinuity for
which the bounding angles are θ and ϕ. With increasing time the
characteristics from neighbouring points on the initial surface, with
different initial values of (θ, ϕ), may intersect at a different spatial
location than that of first edge formation and when this occurs the edge
propagates along the locus of intersecting characteristics and the bounding
orientations to this edge vary. The position of an edge formed at a time t
by intersection of the linear characteristics from two initial surface

points (x_1, y_1, z_1) and (x_2, y_2, z_2) with orientations (θ_1, ϕ_1) and (θ_2, ϕ_2) and corresponding velocities along the characteristics of ν_1 and ν_2 may be deduced by application of equations (24a), i.e.

$$x_1 - x_2 = \{\nu_1(x_1) - \nu_2(x_2)\} t \qquad (31)$$

$$y_1 - y_2 = \{\nu_1(y_1) - \nu_2(y_2)\} t \qquad \text{and}$$

$$z_1 - z_2 = \{\nu_1(z_1) - \nu_2(z_2)\} t$$

Eliminating t from these equations results in the identities:

$$\frac{z_1 - z_2}{x_1 - x_2} = \frac{\nu_1(z_1) - \nu_2(z_2)}{\nu_1(x_1) - \nu_2(x_2)} \qquad \text{and} \qquad (32a)$$

$$\frac{z_1 - z_2}{y_1 - y_2} = \frac{\nu_1(z_1) - \nu_2(z_2)}{\nu_1(y_1) - \nu_2(y_2)} \qquad (32b)$$

These equations indicate that if a general plot of the characteristic velocity components $\nu(x)$, $\nu(y)$, $\nu(z)$ is made for all surface orientations then an edge can be found (or continue to be propagated) if the slope of the chord joining two points (x_{i1}), (x_{i2}) on the initial surface is equal to the slope of the chord joining the vectors ν_1 and ν_2 on the characteristic velocity component plot which correspond to the initial surface points. As will be demonstrated in the subsequent contribution by Nobes (19) this component velocity plot has a number of valuable features which can be used in the evaluation of surface evolution. For the present discussion of edge motion we note one of these features. Mutual differentiation of equations (19) reveals that the slope of a point on the surface defined by the ν_i's is identical to those of the point(s) on the initial surface contour which possess the relevant ν_i values. If tangent planes are now constructed to different points (i.e. different ν values) on this component velocity plot, then if they intersect at a point and the vector from the origin of the component velocity plot to this point is constructed, it is ready shown geometrically that the vector OE is equivalent to U_{123}

and satisfies equation (27a). This means that the vector OE completely represents the velocity and direction of motion of the edge formed from intersecting surface element on a discontinuous surface if, and only if, such edges form according to the criteria of equations (32). Since the edge vector is now defined, then its locus can be determined geometrically as follows. The initial surface and the component velocity plot are compared to determine if and when an edge first forms for a specific surface point of defined orientations (θ, ϕ). Incremental orientations surrounding this point are then selected and the chord slope equality criterion of equations (32) are examined. If this criterion is satisfied the edge vector appropriate to these orientations is determined as described above. The process is then iterated for increasing incremental orientation changes from the initial value and the variation in the edge velocity and direction is determined. This allows direct translation into a locus of the edge as a function of time since equations (31) determine the time at which an edge possesses a given velocity. Moreover, since each edge vector represents the intersection of quite specific surface orientations, the evolution of the orientations bounding the edge as it propagates is equally readily determined. This is effectively equivalent to determining, geometrically, the evolution of the edge discontinuity bounding angles with time, which is described, analytically, for a continuous surface by equations (20). To-date no similar analytic expressions have been derived for discontinuous surfaces but a comparable set of equations may be determined as follows.

Equations (31) may be solved to yield the time, t_e, at which an edge forms from initial surface points (x_1, y_1, z_1) and (x_2, y_2, z_2) and the spatial location of this edge is then given by the set of equations $x_{1e} = x_1 + v_1(x_1)t_e$ etc. This edge will then propagate with velocity components $U(x_1)$ given by equations such as (27) (28) and (30) appropriate to the initial surface points (x_{1i}) (x_{2i}). If neighbouring initial surface points at $(x_{1i} + \delta x_{1i})$, $(x_{2i} + \delta x_{2i})$ then propagate to form a new edge position at time $t_e + \delta t$, then the coordinate of this new position must not only be given by the requirement for characteristic intersection from the initial surface points but also by the coordinates of the edge formed at t_e with the incremental change of $U(x_i)\delta t$. The coordinates of the edge at t_e are given by use of equations (30) as:

$$x_{ie} = \frac{v(x_{2i}) \cdot (x_{1i}) - v(x_{1i}) \cdot (x_{2i})}{v(x_{2i}) - v(x_{1i})} \qquad (33a)$$

and the change in these coordinates, due to intersection of characteristics from neighbouring surface points at $t_e + \delta t$ must be equal to $U(x_i)\delta t$. Consequently:

$$\delta \left\{ \frac{v(x_{2i}) \cdot (x_{1i}) - v(x_{1i}) \cdot (x_{2i})}{v(x_{2i}) - v(x_{1i})} \right\} = U(x_i)\delta t$$

which may be written as

$$\frac{d}{dt} \frac{[\frac{x_i}{v(x_i)}]}{[\frac{1}{v(x_i)}]} = U(x_i) \qquad (33b)$$

which are the discontinuity analogues of equations (20) and where, as before, the square braces indicate jumps in the relevant quantities.

At the instant of edge initiation, t_e, it is readily demonstrated that equation (33b) relaxes to the form:

$$\frac{d}{dx_i} \frac{v(x_i)}{\frac{dv_i(x_i)}{d(x_i)}} = 0 \qquad (33c)$$

This equation may also be derived by differentiation of equation (25) and is a more precise condition for edge initiation.

In order to trace the locus of the edge, however, it is necessary to use this set of equations and numerical iteration or to employ geometric methods. Although not exact equivalents to equations (20), since equation (33b) defines the changing initial surface coordinates which continue to form the propagating edge, it does from specification of the initial surface describe the temporal change in the orientations which bound the edge.

The technique employed above can also be used to deduce edge motion in terms of variation of the instantaneous surface contour rather than in terms of the initial surface contour, i.e. the x_i in equation (33b) refer to initial surface coordinates. If the edge is located at $\underline{x}_o$ at time t and points on the surface contour at this time, located at $\underline{x}_o - \delta x_1$ and $\underline{x}_o + \delta x_2$, subsequently collapse to the new edge at t + δt, then use of equivalent expressions to equation (33a) rapidly reveals that the equivalent expression to equation (33b) is:

$$\frac{\left[\dfrac{\dfrac{dx_{if}}{dt}}{\nu(x_{if})}\right]}{\left[\dfrac{1}{\nu(x_{if})}\right]} = U(x_{if}) \tag{33d}$$

where the x_{if} refer to coordinates on the surface contour at the specific time considered, and, again, the square braces indicate jumps in the bracketed parameters.

Equation (33d) may also be derived by application of the time derivative operator d/(dt) in equation (33b) to the x_i only, which is valid for homogeneous systems since the $\nu(x_i)$ remain spatio-temporally constant. Equation (33d) is, however, fully valid for heterogeneous systems since, over the small time increments considered in its derivation, point motion along a characteristic is essentially linear. This set of equations (33d) is therefore the appropriate analogue of equations (20) for heterogeneous systems.

It may be noted, finally, that the right hand side $U(x_i)$'s of equations (33b) and (33d) may also be replaced by jump function equivalents by use of equations (19) and (28) which lead to relationships for the $U(x_i)$'s in terms of discontinuities of the g values and surface orientations.

Although such relationships may be stated formally, their major use is in advising the evolution of edge characteristics by geometric and numerical methods and it is for this reason that we have, in earlier discussion, preempted some of the subsequent contribution by Nobes (19) which will describe geometrical techniques in some detail.

Some further progress can be made analytically if constraints are placed upon edge motion. One such constraint has been examined by Carter and Nobes (11) where it was demanded that edges should propagate linearly. It was demonstrated that, for prescribed g functions, this constraint places specific requirements on the initial surface geometry. Generally speaking (11), only planar segmented surfaces or surfaces which are symmetric in θ and φ, and then only over a limited orientation range, give rise to linear edge motion.

Finally it is noted that there may be special circumstances in which motion is artificially constrained. One such area would be a system of finite extent where certain regions of a surface are constrained not to evolve or where the evolution is prescribed by, for example, time modulated exposure of regions of the surface to processes which modify the surface. In these conditions, again discussed in more detail by Nobes (19), the edge development is determined by equations (27) appropriately applied at the boundary between evolving and constrained surface regions.

In evaluating surface evolution by the method of iteration along the characteristics described earlier the analysis described for continuous regions of the surface is used and the motion of the discontinuities is, again using the methods outlined for these edges, fitted into the solution for the surrounding. Specifically, constrained motion at media boundaries is handled in the same way.

4. GROWTH AND EROSION

The preceding discussion about wavefront or surface evolution has been made deliberately general without specifying the processes which lead to evolution since, once the framework has been established, the more specific areas of interest can be easily treated within the general formalism. The present Institute is concerned with the erosion and growth of solids stimulated by atom and ion beams and so it is to these topics that major attention will be devoted. However, such incident species represent only part of the general classes of erosion and growth phenomena and so reference to other processes will also be made to preserve generality.

Whatever the processes generating surface evolution, it is evident from the discussion of Sections 2 and 3 that complete description of surface evolution is possible if the local normal velocity C, and its directed equivalent, $g = C/(\cos\theta)$ is known or prescribed for every point on a surface at all times. These velocities may be spatially, temporally, surface orientation and higher derivatives of the surface contour dependent and only very rarely and in special circumstances will it be possible to deduce solutions for the evolving surface in closed analytic form. Much more frequently the analysis presented provides the necessary guidance to construct numerical solutions using iterative techniques and the geometric and computational methods to be described by Nobes (19) and Smith (20). In the succeeding discussion growth and erosion processes will be discussed separately but in many circumstances both may operate simultaneously. Such circumstances are readily incorporated into the general formalisms by noting that the fundamental parameter dictating surface evolution is the local net surface normal velocity. If C_e is the local erosion velocity, C_a, the local growth (or accretion) velocity and C_m is the local normal velocity resulting from external or internal forces, the net velocity C_n is given by

$$C_n = C_e - C_a \pm C_m \tag{34}$$

It is this net normal velocity C_n and its directed equivalent $g_n = C_n/(\cos\theta)$ which is then employed in place of C and g in the preceding analysis.

A cautionary note should be added, however, in that the preceding analysis has considered surface evolution as a continuous process in which incrementally small differences in behaviour are independent. In many growth and erosion processes, however, effects occur on a discrete atomic scale and local atomic behaviour may influence neighbouring behaviour. In some circumstances this 'interference' may be accommodated into the general formalisms presented by suitable approximation techniques but this is not always possible and the preceding methods are unsuitable for the description of surface evolution. One such area is that of atomic occlusion in the bulk of solids, which leads to surface blistering and exfoliation as described later in these proceedings by Van Veen (26) and Scherzer (27). This and similar areas will therefore be excluded from further consideration here.

4.1. Growth processes

In addition to processes such as expansion and inflation (swelling) which increase the dimensions of a solid the most common growth techniques are those which involve deposition from the liquid or vapour phases. In liquid phase deposition the processes are chemical or electrochemical and, generally speaking, the flux of depositing atoms or molecules onto any surface point is both spatially and surface orientation independent. The growth process, however, will generally occur by sequential atomic deposition and the rate will be dictated by requirements of the energy to transfer species from the liquid to the surface to be minimised. This will lead to growth rates which are crystal orientation dependent for crystalline solids and in which growth may be enhanced at crystal imperfections such as dislocations. For a defect free surface the crystallographic dependence of growth rate implies, at the macroscopic level, a normal velocity, C, which is orientation dependent. This case has been considered in some detail by Frank (14,15) using a geometrical aid (the erosion slowness plot to be discussed later by Nobes (19) to determining characteristic speeds and directions). It was shown that, as growth proceeds, those crystal orientations for which the normal velocity C is both largest and symmetric dominate and the crystal shape which evolves is that associated with maximum growth-rate planes.

In the case of growth from the vapour phase, a number of techniques are available including direct evaporation or sputtering from a source (or distributed sources) on to a substrate, deposition of one or more species from a plasma or directed ion (or other radiation sources) impinge and deposit onto a substrate. The incident flux in such cases depends upon the nature of the system geometry and process, and can range from random incidence (chemical vapour deposition) to some or all directed incidence. This deposition from an extended vapour source will include incidence onto specific surface points integrated over all solid angles from the point to the source whilst in plasma and directed incidence techniques the flux may be more monodirectional. In plasma systems where plasma boundaries adjust to surface contours for electrostatic reasons, flux incidence may be everywhere normal to the local surface. Where energetic species also form a part of the total incident flux, e.g. in plasma deposition, their influence may be to enhance atomic mobility of surfaces (see for example, the subsequent contribution by Rossnagel (28)),which may tend to equalise growth rates onto different crystallographic faces and/or to enhance growth at specific nucleation sites such as crystal defects or impurity complexes. In order to treat such deposition processes by characteristic methods the individual fluxes of all species, their spatial, temporal and energy distributions at each surface point, must be known and integrated over all possible fluxes. In some cases, e.g. radiation enhanced atomic migration into and across surfaces, the differential fluxes may be synergistic in their effect rather than linearly additive and integrable. The existence and result of such processes must be known, e.g. the contributions by Rossnagel (28), Kay (29) and Auciello (30).

If fluxes are additive in prescribing growth then the total normal growth velocity of a point on a surface of infinitesimal area dS resulting from condensation of atomic fluxes derived from an extended source is given by:

$$C_a = \frac{1}{N} \oint dE \ dJ \ \frac{dS}{r^2} \cos\varepsilon.f \tag{35a}$$

In this equation N is the atomic density of the condensate, dE is an infinitesimal area of the emitter source, dJ is the flux per unit solid angle from this infinitesimal area and may be spatially, temporally species and energetically distributed, r is the radius vector from source to receiving surface, ε is the angle between the surface normal to dS and r̲, and f is the condensation coefficient on the surface. The integral is performed over the whole emitter region E. If the flux is moderated between source and receiver by, for example, gas phase collision the flux dJ must be appropriately modified.

In the case of deposition from the vapour phase and a condensation coefficient independent of incidence flux direction but dependent upon surface crystallographic orientation equation,(35a) is replaced by its kinetic theory of molecular impingement equivalent:

$$C_a = \frac{n\bar{c}}{4N} \cdot f(hkl) \tag{35b}$$

where n is the vapour phase molecular density and $\bar{c}$ the molecular mean velocity and f(hkl) indicates the crystallographic dependence of condensation coefficient. If f is crystal orientation independent, the growth velocity C_a is uniform for all points on the surface and the characteristics are those of constant C_a and normal to each surface point. The evolution of the surface is thus determined by extending the surface everywhere by constant increments $C_a \delta t$ for each time step δt. This will give rise to temporary edge formation but these will slowly wash out and the surface will tend towards one of plane geometry. If, however, f is crystal orientation dependent, then growth will be most rapid in planes of highest condensation coefficient and although edges will form and propagate along the intersecting regions of growth habit on neighbouring planes, the evolution will tend towards dominance of those planes of maximum growth rate as in the case of liquid phase deposition.

In the case of directed flux sources the evolution will depend upon both the geometry of the initial surface and of the extended source since some fluxes may be shadowed from some surface regions by interposed surface protuberances. In such cases detailed knowledge of initial surface and source geometries are required before surface evolution can be predicted. In the case of energy-stimulated deposition (e.g. from plasma sources such as in ion plating or with directed ion beams) there is some evidence to suggest (31) that the energy deposition generates substantial atomic motion of the condensing flux which, artificially, renders the growth rate constant in the macroscopic direction of energy incidence, i.e. g not C is constant. In such conditions the characteristics are those of constant g and growth proceeds uniformly in the Oz direction. This results in the initial surface contour being constantly reproduced parallel to the initial surface and any initial microfeatures are preserved.

Finally we note an area of deposition and growth relevant to and arising from the sputtering erosion processes discussed in detail in the next subsection. When incident ions sputter a surface the ejected sputtered atoms are distributed spatially and energetically. If some component of this sputtered flux is intercepted by neighbouring regions of the surface,atomic redeposition can occur simultaneous with any sputtering of the intercepting region. The effective normal velocity of this region is then given by equation (34) with the growth velocity component C_a derived from equation (35). The integration in equation (35) extends over all regions of the surface from which sputtered flux can be transported to the region at which C_n is to be determined. In general this leads to time (and spatial) dependence of C_n and the effects of these combined sputtering and redeposition processes have been considered in simplified (non time

dependent) form by Wilson and Belson (32) and Wilson (33) and in more
detail by Makh, Smith and Walls (12). Detailed attention will also be
given to this problem in the subsequent contribution by Smith (20) to these
proceedings, since it is often a limitation in the application of ion
bombardment sputter etching for microlithographic pattern delineation on
semiconductor device surfaces.

4.2. Erosion processes

Just as with growth processes, the erosion of solids can be effected by
both liquid-phase and vapour-phase attack, including processes such as
dissolution and corrosion. The formalisms to describe erosion or etching
are quite identical to those described previously for deposition and indeed
in the case of crystalline substrates, Frank's (14,15) and Cabrera's (16)
studies apply with equal validity to their complementary observations on
growth processes. The main difference in the end result of determining
crystal surface evolution by characteristic methods under etching
conditions is that the surface facets which tend to dominate the form of
the crystal are those of minimum normal erosion velocity. Such results
are equally applicable to vapour phase corrosion and are of some significance in reactive gas etching of semiconductors in situations where the
etch rate with a particular vapour may be strongly crystallographically
orientation dependent as demonstrated by Donnelly (34) and referred to by
Dzioba (35) in these proceedings. The erosion of solids may also occur as
a result of ablative action between solids in contact between which there
is relative motion. Examples of such processes are machining, grinding and
cutting whilst particulate impact generally leads to erosion also.
Hydrodynamic and aeolian interaction with solids frequently shapes
surfaces and characteristic methods have been used by Luke (17,18) to
predict geomorphological feature development. The present author and his
colleagues (36) have also investigated geomorphlogical erosion processes and
the surface morphological evolution of sand blasted solids. In all these
cases the local normal erosion rate is a function of local surface
coordinates, orientations and sometimes higher surface derivatives (39),
frequently with a not dissimilar functional behaviour to that associated
with the atomic scale erosion or sputtering of solids by energetic ion
impact. The detailed treatment of such processes is therefore largely
contained within the framework of sputtering induced erosion, a main topic
of this Institute, to which we now turn attention. Two subsets of
sputtering phenomena are interesting to present discussion. These are,
respectively, physical or ballistic sputtering, the fundamental principles
and measurements of which are discussed in these proceedings by Kelly (40)
and Hofer (41) and bombardment induced or enhanced chemically reactive
sputtering which are discussed by Heslop (42) and Dzioba (35).

In the following we discuss physical sputtering first, outlining both the
general formalism and indicating specific areas which will be elaborated
in the contributions by Nobes (19) and Smith (20).

As described in the contributions by Kelly (40) and Hofer (41), sputtering
is essentially a statistically distributed process so that any one ion
incident onto a substrate may eject a variable number of atoms which, in
turn, may be emitted from different surface positions relative to the
point of ion impact. Consequently until many cascades produced by
individual ions have overlapped spatially the number of atoms ejected per
incident ion, the sputtering yield Y, may be variable and an atomic scale
surface topography may develop because of uneven surface distribution of
atomic ejection. As ion fluence increases it becomes feasible to consider

a mean sputtering yield which, as discussed by Kelly (40) and Hofer (41), will be a function of incident ion energy and mass, substrate mass and temperature, the crystallographic orientation of the substrate and the polar and azimuthal angles of incidence of ions with respect to the normal to the surface and a fixed direction in the crystal. It has also been argued by Sigmund (43) that because sputtering results from a series of atomic collisions in the substrate, then if the collision cascade cannot fully develop because of inadequate volume locally (i.e. atomic scale microtopographic features) the macroscopic sputtering yield is an inadequate descriptor. Carter et al.(44) have shown that this perturbation may be described, approximately, by a local surface radius of curvature dependence of sputtering yield, which allows the characteristics approach, which is now elaborated, to be employed.

For a given set of ion mass and energy, substrate mass, temperature and orientation parameters the sputtering yield becomes a function $Y(\theta,\phi)$ of the angles of incidence relative to specified crystal directions. As a first approximation for a crystalline substrate, and fully descriptive for an amorphous substrate, $Y(\theta)$ tends to increase from a minimum for normal ion incidence ($\theta = 0$) rise to a maximum at some angle $\pm\theta_p$ and then decline towards zero as grazing incidence ($\theta \rightarrow \pm \pi/2$) is approached. For an ion flux density dJ incident at angles θ, ϕ, to the normal to a surface element, the sputtering rate is merely the product of the flux density projected onto the surface plane and the sputtering yield. For a substrate of atomic density N, the erosion velocity normal to the surface is simply

$$C_e = \frac{dJ\cos\theta}{N} \; Y/(\theta,\phi) \tag{36}$$

If the ion flux is spatially distributed, equation (36) is integrated (or summed for several discrete fluxes) over all incidence directions. Such summed distribution has been discussed by Makh et al. (45) and will be further considered in the contribution by Smith (20). If the flux is mono-directional but spatially and temporally variable, equation (36) relaxes to

$$C_e = \frac{J(x_i,t).Y(\theta,\phi,x_i,t)\cos\theta}{N(x_i,t)} \tag{37}$$

where both the sputtering yield and atomic density have been generalised to potential space and time variation which will occur in inhomogeneous substrates and time variable ion species and energies. For simplification we write equation (37) as

$$C_e = \frac{JY\cos\theta}{N} \tag{38a}$$

for which

$$g_e = \frac{JY}{N} \tag{38b}$$

If equation (38b) is substituted into the defining characteristics equations deduced in Section 2 for continuous surface and in Section 3 for discontinuous surface, the results obtained by many earlier authors (2-25, 36-38,44,45) using a variety of methods and initial assumptions about the spatio-temporal and orientation dependent behaviour of J, Y and N are readily recovered. Specific applications to spatio-time varying J, Y and N will be considered in detail in the contributions by Nobes (19) and Smith (20) and so we only make two general observations at this stage and do not reproduce the special forms of the characteristics which will be deduced and employed by these authors. Firstly we note that during sputtering some variable fractions of the incident flux may be scattered and the sputtered atom flux may be ejected to other regions of the surface. In addition to generating growth by deposition both these fluxes may contribute to modified local erosion rates. In principle all that is required to fit such processes into the general scheme is to accommodate their influence into the integrated form of equation (36). This, however, requires knowledge of the scattering and sputtering flux distributions from all surface regions and this is rarely known. Consequently approximate theoretical estimates are sometimes made of such processes and these form the basis of surface evolution treatments which will be discussed by Smith (20). An experimental observation of the importance of such fluxes is the production of pedal depressions adjacent to high elevation angle (large θ) boundaries on solids which result from additional fluxes from the elevated boundaries into the surrounding surface as noted in the contributions by Whitton (46), Heslop (42), Dzioba (35) and Auciello (30). Such pedal features are often unwelcome artefacts in many, particularly lithographic, technical applications and minimisation may be achieved by substrate or ion flux rocking and/or rotation.

The second general result is the important role played by regions in which the spatial divergence of characteristics are at extrema. Such regions exhibit, at least temporarily stability during surface evolution and it is for this reason that features such as cones and pyramids are frequent outcomes of sputtering with orientations for which sputtering yields are maxima and, in the case of pyramids, symmetrically equivalent. It should be noted however that unless the system is heterogeneous (i.e. spatially variable substrate composition or ion flux distribution) surface orientations associated with such extrema in characteristics behaviour will emerge in the evolving surface only if they were included within the range of orientations present in the initial surface. Examples of heterogeneous systems are discussed by Nobes (19) and exemplified by Heslop (42), Dzioba (35) and Whitton (46) where surface or bulk impurities and defects perturb either local ion flux or sputtering yield so that new orientations can develop. Thus etch pits develop in regions of enhanced sputtering yield at dislocations and intersecting etch pits produce elevated boundaries which lead to ridge and pyramid evolution. Accidental surface contamination and deliberate lithographic masking lead to plateaux and mesa structures and non-parallel side wall etch channels respectively.

We turn next to radiation enhanced or activated chemical erosion which will be discussed in detail by Heslop (42) and Dzioba (35). Whereas with physical or ballistic sputtering the general mechanisms are reasonably well understood (40,41) and extensive data and theoretical modelling exists the situation is much less clear with 'chemical' sputtering. A number of different physical and chemical processes (34,37) are believed to operate and dominate according to experimental circumstances. Consequently chemical sputtering is usually substrate and etchant species specific to a much greater extent than in ballistic sputtering, i.e. there is increased substrate 'selectivity' in chemical sputtering. In the various manifesta-tions of chemical sputtering such as sputtering with chemically reactive ions, sputtering by chemically inactive ions in the presence of a reactive gas vapour or flux and sputtering in reactive gas plasmas a variety of species generally strikes the substrate. These include parent molecular, dissociated fragment and atomic ions (positively and negatively charged) neutral molecular, atomic and free radical species, electron and photons, together with species which may be sputtered from other surfaces onto the substrate (e.g. the contribution to these proceedings by Kay (29)). The flux and energy distributions of these species will often be complex and may change with time as specific features are etched and flux access is modified to different regions of the surface. Whilst it is of fundamental interest to understand the physical and chemical processes involved in erosion this is not vital to prediction of morphological evolution. What is important to know is the spatio-time distribution of all of the relevant fluxes and how these interest synergistically or by linear superposition to produce a measurable surface normal erosion velocity. The important requirement, therefore, is to be able to specify C_e in terms of the parameters of the erosion system and, when this is available, the techniques of evolution prediction described in Sections 2 and 3 may be used directly. A first attempt at such prediction has been made for ion beam enhanced reactive gas etching by Carter and Nobes (48) assuming a simple etching model based upon radiation induced bond dissociation in the substrate and associated enhanced etching. Whether the details of the model are correct is relatively unimportant since the major requirement of the model is that etching is only enhanced over a substrate volume local to the points of radiation or energy (directed ion beam) incidence. Given this requirement the 'anisotropy' commonly associated with chemical sputtering is an immediate result since the lateral spread of the enhanced reaction relative to the direction of energy (ion) incidence will generally be only of the order of tens of atomic spacings. Anisotropy can also result, of course, from blocking processes where reaction products deposit upon surface regions exposed to little or no incident energy flux so that they are not continuously scavenged and etching is minimised. Such processes, when parametrically described are readily accommodated into the characteristics' approach since the implications of anisotropy are merely that C_e is small or zero for specific regions of surface contours. If, for example, enhanced erosion is stimulated by a defined (e.g. by a litho-graphic mask) ion flux then lateral erosion of channel walls can only occur over a very limited region since the transverse dimensions of displacement and dissociation cascades will be only tens of atomic distances and unless there is scattered ion flux only the direct line of sight to the ion flux region of the substrate surface will erode rapidly and uniformly. The characteristics are normal to the initial unmasked surface so that parallel side wall channels develop. If, however, as erosion proceeds and gas phase and surface scattering processes randomise the incident fluxes, then the etching will be more isotropic and over etched or mask undercut channel geometries can ensue. Such processes have been discussed, in general terms, by Carter, Nobes and Cruz (8) who demonstrate that, once the etch process is appropriately parameterised, the characteristic methods are fully applicable to chemical sputtering situations. It should be noted that, whereas the techniques will be identical to those for physical sputtering, the execution may be far more demanding because of the variety of incident flux parameters at work and their potential space and time variation which may be difficult to predict or measure. Consequently although characteristic methods are applicable to chemical sputtering processes the effort expended in determining even the parametric dependences of the system variables may be better spent in empirical observations of morphological evolution when system variables are controllably adjusted. Nevertheless this process may be guided by the preceding approximate erosion analysis if the probable magnitudes and effects of system variations are reasonably understood and characterised.

4.3. <u>Growth and erosion processes</u>

In the preceding discussion growth and erosion have largely been discussed as separately identifiable processes although examples of simultaneous operation of both have been noted. In general terms the surface evolution is fully described by insertion of the appropriate form of equation (34) into the characteristics defining equations of Sections 2 and 3. Both processes may act simultaneously in technical applications such as ion beam or energy deposition enhanced thin film growth (29,31,49) and during sputtering in the presence of externally controlled impurity deposition (28). These combined processes can admit to evaluation by characteristics methods provided, again, that either the processes can be fully understood or properly parameterised in terms of system variables.

In the case of impurity deposition during simultaneous sputtering (28) there is added complexity since the system is essentially two phase and processes occurring with the substrate and with the deposit must be treated separately, particularly when deposit coverage of the substrate surface is incomplete. A similar difficulty arises with sputtering only of a non-elemental solid in which thermodynamic forces can also act to modify surface composition (40) and potentially the morphology. Such processes may also act in elemental solids but these can be treated more readily in the characteristics approach. Thermodynamic and other influences generally result in atomic transport both parallel and normal to surfaces because of the existence of driving gradients such as mechanical stress, electrostatic, thermal and chemical potential differences. These give rise to atomic flux gradients which, in turn, lead to local surface accretion (growth) or erosion. A number of such processes have been considered by Mullins (50,51) and surface morphological evolution predicted by other than characteristics methods. Carter (52) and Carter et al.(53), however, have considered two of these transport processes, surface and volume diffusion, and demonstrated how characteristics methods may be applied to surface evolution. The surface free energy generally varies with crystal orientation (54) and this variation, together with local radius of curvature variations gives rise to gradients in chemical potential which drive atomic and defect fluxes. Carter (52) and Carter et al.(53) have shown that in the simple case of a two-dimensional surface section the resulting effective normal velocity of a surface element is given, for surface and volume diffusion, respectively by:

$$c_T = - \frac{D_s N_o \Omega^2}{kT} \gamma \frac{\partial^3 \theta}{\partial s^3} \tag{39a}$$

$$c_T = - \frac{D_v \Omega}{kT} \gamma \tan\theta \frac{\partial^2 \theta}{\partial n^2} \tag{39b}$$

In these equations D_s and D_v are surface and bulk diffusion coefficients, N_o is the surface atomic density and Ω the atomic volume, T is temperature and k is Boltzmann's constant. s and n are distances measured along and normal to the surface respectively. These equations assume no crystallographic orientation dependence of γ but are readily modified to include such dependence (50) and for three-dimensional contours whilst the variations with respect to s and n are easily transformed to Cartesian axial equivalents. It is already clear from equations (39) that the surface normal velocity will be a function of higher order surface derivatives than for growth or sputtering erosion processes and so, although characteristics methods can still be used (52,53) the characteristics will be complex, non-linear and not descriptive of constant surface orientations. If sputtering is also active and linearly additive to these transport processes then the net surface normal velocity locally is given by $C = C_e + C_T$ and again characteristics methods may be used.

In general, however, the irradiation causing sputtering will also modify the near surface defect densities and, as a consequence, the diffusion coefficients D_s and D_v will adopt values appropriate to radiation enhanced conditions. The precise forms to be used for the diffusion coefficients will depend upon any model assumptions as to how the instantaneous enhanced defect density is related to incident radiation flux, but it might be anticipated (52) that the overall surface normal velocity would eventually contain three terms. One term, describing sputtering, would be linearly incident flux dependent, but relatively substrate temperature independent. A second term describing atomic transport under no irradiation conditions, with thermodynamic defect populations, would be temperature dependent but incidence flux independent. The third term, describing additional transport resulting from radiation enhanced defect populations, would be both incident flux and temperature independent. The morphological evolution would therefore be expected to be a complex function of both incidence flux and substrate temperature conditions but, as yet, there have been no comprehensive experimental studies to evaluate such dependences.

Frequently, however, surface atomic transport tends to minimise discontinuities in elemental solids and smoother surfaces may be an expected result during sputtering at elevated temperatures. Some studies by Whitton et al.(55) with crystalline Cu substrates and Vasiliu et al. (56) with Fe substrates suggest that this general prediction is valid. The investigations reported in these Proceedings by Rossnagel (28) for continuously contaminated surfaces also illustrate the even more complex interplay of thermal and irradiation conditions in modifying surface evolution. It is suggested that further experimental work for both elemental and compound substrates and for continuously contaminated well defined crystal substrates would aid understanding in this interesting area.

5. CONCLUSIONS

In this contribution a general theoretical formalism has been developed which describes a spatio-temporally evolving surface as a propagating wavefront. The evolution of the surface morphology was shown to be describable by following progress along characteristics and the form of these characteristics was explored for spatially and temporally inhomogeneous and anisotropic systems. Such systems were shown to include many situations of both growth and erosion of surfaces by atom and ion fluxes, and the general results of a number of such discrete or combined effects were examined. Attention was also given to other processes which modify surface morphology including atomic transport by diffusion processes.

Although it is relatively straightforward to formulate both the general and system specific theory for surface evolution of continuous surfaces, surface discontinuities and media interfaces, it is only rarely that the theory gives rise to closed solutions which specify the spatio-temporal evolution of the surface morphology. The theory is, however, necessary to advise optimum use of geometric and numerical iterative techniques which are used to elucidate the details of the evolution. This contribution should therefore be seen as supporting and explaining the rationale of the following discussions by Nobes (19) and Smith (20) particularly and of relevance to the contributions on the experimental studies of surface

96

feature development by Whitton (46), Rossnagel (28), Heslop (42), Dzioba (35), Kay (29) and Auciello (30).

REFERENCES

1. Whitham GB: Linear and Non-Linear Waves, John Wiley, New York, (1974).
2. Carter G, Colligon JS and Nobes MJ: J Mat Sci $\underline{8}$, 1473, (1973).
3. Smith R and Walls JM: Phil Mag $\underline{A42}$, 235 (1980).
4. Smith R, Valkering TP and Walls J M: Phil Mag $\underline{A44}$, 879 (1981).
5. Carter G, Nobes MJ, Lewis GW and Cox J: Vacuum $\underline{34}$, 263 (1984).
6. Nobes MJ, Carter G, Lewis GW and Whitton JL: Vacuum$\underline{33}$, 381 (1983).
7. Carter G and Nobes MJ: Nucl Instrum & Meth in Phys Res $\underline{230B2}$, 635 (1984).
8. Carter G, Nobes MJ, Cruz SA and Lewis GW: Nucl Instrum & Meth in Phys Res. To be published (1985).
9. Carter G, Nobes MJ and Cruz SA: J Mat Sci Letters $\underline{3}$, 523 (1984).
10. Carter G and Nobes MJ: Ion Bombardment Modification of Surfaces, Fundamental and Applications (eds O Auciello and R Kelly), Chapter 5, Elsevier Book Publ Co, Amsterdam (1984).
11. Carter G and Nobes MJ: Phil Mag. To be published (1985).
12. Makh SS, Smith R and Walls JM: J Mat Sci $\underline{17}$, 1689 (1982).
13. Tagg MA, Smith R and Walls JM: J Mat Sci. To be published (1985).
14. Frank F C, Growth and Perfection of Crystals, p411, John Wiley, New York (1958).
15. Frank FC: Zeits Phys Chem Neue Folge $\underline{77}$, 84 (1972).
16. Cabrera N and Vermilyea DA: Growth and Perfection of Crystals, p420, John Wiley, New York (1958).
17. Luke JC, J Geophys Res $\underline{77}$, 2460 (1972).
18. Luke JC, Zeits Geomorph Supp $\underline{25}$, 114 (1976).
19. Nobes MJ: These Proceedings.
20. Smith R: These Proceedings.
21. Stewart ADG and Thompson MW: J Mat Sci $\underline{4}$, 56 (1969).
22. Nobes MJ, Colligon JS and Carter G: J Mat Sci $\underline{4}$, 730 (1969) and Carter G, Colligon JS and Nobes MJ: J Mat Sci $\underline{6}$, 115 (1971).
23. Carter G, Colligon JS and Nobes MJ: Rad Effects $\underline{31}$, 65 (1977).
24. Barber DJ, Frank FC, Moss M, Steeds JW and Tsong IST: J Mat Sci $\underline{8}$, 1030 (1973).
25. Ducommum JP, Cantagrel M and Marechal M: J Mat Sci $\underline{9}$, 725 (1974) and Ducommum JP, Cantagrel M and Moulin M: J Mat Sci $\underline{10}$, 52 (1975).
26. Van Veen A: These Proceedings.
27. Scherzer BMV: These Proceedings.
28. Rossnagel SM: These Proceedings.
29. Kay E: These Proceedings.
30. Auciello O: These Proceedings.
31. El-Sherbiny M, Halling J andTeer DG: Advance in Surface Coating Technology, Proc Int Conf, London (Welding Institute), p147 (1978).
32. Belson J and Wilson IH: Rad Effects $\underline{51}$, 27 (1980) and Belson J and Wilson IH: Nucl Instrum & Meth $\underline{182/183}$, 267 (1981).
33. Wilson IH, Belson J and Auciello O: Ion Bombardment Modification of Surfaces, Fundamentals and Applications, (eds O Auciello and R Kelly), Chapter 6, Elsevier Book Publ Co, Amsterdam (1984).
34. Donnelly VM, Ibbotson DE and Flamm DL: Ion Bombardment Modification of Surfaces, Fundamentals and Applications, (eds O Auciello and R Kelly), Chapter 8, Elsevier Book Publ Co, Amsterdam (1984).
35. Dzioba S: These Proceedings.
36. Carter G and Nobes MJ: Earth Surface Processes $\underline{5}$, 131 (1980).

37. Carter G, Nobes MJ and Arshak KI, Wear 53, 245 (1979).
38. Carter G and Nobes MJ: Wear 96, 227 (1984).
39. Finnie I and Kabil Y H: Wear 8, 60 (1965).
40. Kelly R: These Proceedings.
41. Hofer W O: These Proceedings.
42. Heslop C J: These Proceedings.
43. Sigmund P: J Mat Sci 8, 1545 (1973).
44. Carter G, Nobes MJ and Webb RP: J Mat Sci 16, 2091 (1981).
45. Makh SS, Smith R and Walls JM: Low Energy Ion Beams (Inst Phys Conf Series 54) p246 (1980) and Smith R, Makh SS and Walls JM: Phil Mag A47, 453 (1983).
46. Whitton JL: These Proceedings.
47. Winters HE: Nucl Instrum & Meth in Phys Res. To be published (1985).
48. Carter G and Nobes MJ: Vacuum 32, 593 (1982).
49. Harper JME, Cuomo JJ, Gambino RJ and Kaufman HR: Ion Bombardment Modification of Surfaces, Fundamentals and Applications (eds O Auciello and R Kelly) Chapter 4, Elsevier Book Publ Co, Amsterdam (1984).
50. Mullins WW: J Appl Phys 28, 333 (1957).
51. Mullins WW: Metal Surfaces (eds N A Gjostein and W D Robertson), American Society for Metals, Chapter 2, (1963).
52. Carter G: J Mat Sci 11, 1091 (1976).
53. Carter G, Colligon JS and Nobes MJ: Proc 8th Symp on Physics of Ionised Gases (ed B Navinsek, Inst J Stefan, Ljubljana) (1976).
54. Herring C: Structure and Properties of Solid Surfaces: (eds R Gomer and C S Smith) Univ of Chicago Press, Chapter 1 (1953).
55. Whitton JL and Carter G: Symposium on Sputtering, Vienna 1980 (eds P Varga, G betz and F P Viehbock, Inst f Allge, Phys Tech Univ Wien) p552 (1980).
56. Vasiliu F, Teodorescu A and Flodeanu F: J Mat Sci 10, 399 (1975).

ON THE COMPOSITION OF THE SPUTTERED FLUX FROM METAL TARGETS AND
THEIR COMPOUNDS

G. BETZ and W. HUSINSKY
Institut f. Allgemeine Physik, Technische Universität Wien,
Karlsplatz 13, A-1040 Wien, Austria

1. INTRODUCTION

Sputtering of clean metal targets with rare gas ions generally
results in a sputtered flux consisting predominantely of
neutral ground state metal atoms. This is no longer true for
oxides as well as for targets with adsorbed oxygen on the
surface. In addition, excited and ionized particles as well as
molecules are sputtered under such conditions and their
contribution to the sputtered flux can increase by orders of
magnitude(1,2). This effect is routinely used in surface
analytical techniques, like Secondary Ion Mass Spectrometry
(SIMS) to increase the signal strength (2). On the other hand,
relatively little is known about the quantitative composition
of the sputtered flux under such conditions. Such knowledge is
not only important for quantification of SIMS but will also be
of importance for more recent analytical techniques like
Secondary Neutral Mass Spectrometry (SNMS) (3) and Multi Photon
Resonance Ionisation (MPRI) (4). For this reason we have tried
to investigate the composition of the sputtered flux for metal
targets and some of their compounds under clean conditions as
well as exposed to a number of gases (O_2, N_2, NH_3, CH_4).

2. EXPERIMENTAL SET-UP

The measurements have been performed under UHV conditions
(residual gas pressure less than 10^{-9} mbar). Ion bombardment
was performed with 15 keV Ar^+ and current densities of the
order of a few $\mu A/cm^2$. The experimental techniques used in
situ were Laser Induced Fluorescence Spectroscopy (LIF) or
Doppler Shift Laser Fluorescence Spectroscopy (DSLFS),
Bombardment Induced Light Emission (BLE) and Quartz
Microbalance measurements. The experimental set-up is shown in
Figure 1. The emitted light from short lived excited sputtered
atoms or ions has been analyzed in front of the target surface
with a grating monochromator in the case of BLE. Sputtered
ground state and metastable atoms were analyzed using a cw-ring
dye laser pumped by a 20 W Argon ion laser in the case of LIF.
For the case of Quartz Microbalance measurements sputtered
particles emitted under an angle of 45° to the surface normal
were collected on a quartz oscillator. Sputtering rates have
been evaluated under the assumption of a cosine distribution
for the emitted material. The obtained yields however are not
very sensitive to this assumption. Even for the extreme
assumption of a cos^2 distribution the calculated yields will be
only 6% lower. For measurements under a high oxygen partial

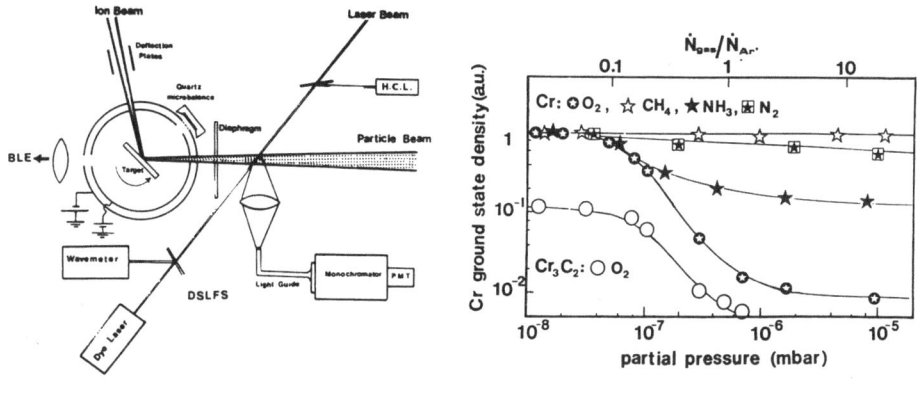

FIGURE 1   FIGURE 2

FIGURE 1 Experimental set-up. In addition to an angle of 45° between laser and particle beam (shown in the figure) as used in DSLFS also an angle of 90° was possible as used in LIF.
FIGURE 2 Dependence of the neutral Cr ground state yield for Cr and Cr_3C_2 targets under 15 keV Ar^+ bombardment as a function of the partial pressure (oxygen to Ar^+ flux ratio N_{gas}/N_{Ar}) of O_2, CH_4, N_2, and NH_3.

pressure we have assumed the formation of the metal oxide on the quartz, e.g. Cr_2O_3 formation for sputtering a Cr target, and corrected the measured frequency changes for the incorporated oxygen. In addition the target was surrounded by two concentric cylinders to which positive and negative voltages could be applied. Thus either all sputtered particles or only neutral particles were collected on the quartz, i.e. total sputtering yields as well as the contribution of charged particles to the sputtered flux could be determined. A more detailed description of the experimental set-up is given elsewhere (5).
Pure Cr and Zr targets as well as Cr_2O_3 and Cr_3C_2 targets were analyzed. The Cr compound targets were mounted together with a Cr reference target to achieve exactly identical conditions. The measurements were performed in the dynamic mode, i.e. the partial gas pressure for the different gases introduced into the system (O_2, N_2, CH_4 and NH_3) was increased between consecutive measurements up to 10^{-5} mbar.

3. RESULTS AND DISCUSSION
3.1 LIF results
We have measured the relative partial sputtering yield of Cr ground state atoms from pure Cr targets under increasing partial pressure of O_2, N_2, NH_3 and CH_4 (Fig.2). The influence for the different gases is very pronounced for O_2, much weaker for NH_3 and nonexistent for N_2 and CH_4, which can be explained by the low sticking coefficients of N_2 and CH_4 on metals at room temperature (6). Results were also obtained for Cr_2O_3 targets which were very similar to heavily oxygen covered Cr targets. On the other hand the results for Cr_3C_2 targets are

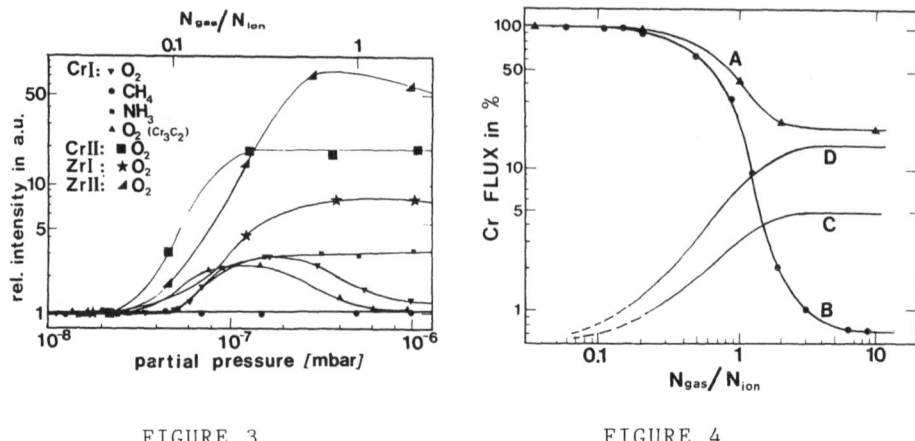

FIGURE 3   FIGURE 4

FIGURE 3 Dependence of the yield of excited Cr and Zr atoms and
ions for Cr, Zr and Cr_3C_2 targets under 15 keV Ar^+ bombardment
as a function of the partial pressure (oxygen to Ar^+ flux ratio
N_{gas}/N_{ion}) of O_2, CH_4 and NH_3. The following lines were
measured: CrI..425,4nm; ZrI..468.7nm; CrII..283.5nm; ZrII..
343.8nm.
FIGURE 4 Composition of the sputtered flux as a function of the
ratio of the oxygen to Ar^+ flux densities for 15 keV Ar^+
bombardment of Cr.
Curve A: Total Cr flux
Curve B: Contribution of sputtered ground state Cr atoms
Curve C: Contribution of Cr sputtered in ionized particles
Curve D: Contribution Cr sputtered in neutral molecules

very similar to the results for metallic Cr (Fig 2). The
reduced sputtering yield of the Cr ground state by a factor of
~ 8 can be reasonably explained by the lower Cr concentration
at the surface and an overall lower sputtering yield of Cr_3C_2.
Otherwise C does not influence the sputtering of Cr, quite
contrary to additionally added oxygen. For oxidation Cr_3C_2
behaves quite similar to Cr. Similar experiments (only oxygen
exposure) were also performed for a metallic Zr target (7).
Again a strong decrease of the Zr ground state signal with
increasing O_2 surface coverage of the target was observed. In
addition in the case of Zr a fine structure splitting of the
ground level exists. Besides measuring the partial yield for
the lowest ground state fine structure level, a^3F_2, we also
measured the yield for the first excited level, a^3F_3, (0.07 eV
above the ground level) and observed that the population of
that level decreased with oxygen surface coverage in the same
manner as the ground state.

3.2 BLE measurements
Short lived excited states of Cr and Zr were found to behave
differently from the metastable Zr levels of the ground state
fine structure, but similar to high lying metastable states
measured for Ca (8) and Ba (9). For oxygen exposure we found

an increase in the yield by up to 2 orders of magnitude for the dominant lines of the excited neutrals as well as ions.
For Cr and Ti the yield of excited atoms increased by a factor of 3 or 7, respectively, in the range above 10^{-7} mbar O_2 partial pressure and decreased again for Cr at a higher oxygen coverage (Fig. 3). In addition for Cr we observed no intensity changes for N_2 and CH_4 exposure in accordance with zero sticking probability, while a signal increase slightly larger than for O_2 was observed with NH_3, indicating that NH_3 sticks to the surface, although no pronounced changes were observed for the emission of sputtered ground state atoms (Fig. 2). The influence of oxygen on Cr_2C_3 was very similar to a metallic Cr target (Fig. 3). For metallic Cr and Zr targets much stronger increases with oxygen were found for the dominant excited ion lines, amounting to a factor of 20 for Cr and 80 for Zr .

3.3 Microbalance measurements

The strong decrease in the sputtering yield of ground state atoms with increasing oxygen surface coverage, as observed with LIF, does not correspond to a similar decrease in the total sputtering yield, as was found with a quartz microbalance (7). Under 15 keV Ar^+ bombardment only a decrease from the clean metal yield for Cr and Zr of 4.1 or 3.1 to 0.8 or 1.4, respectively, can be observed for an oxygen partial pressure of $1x10^{-5}$ mbar (flux of impinging oxygen molecules larger than flux of incident ions). This means that other sputtered components, like excited atoms, ions or molecules must dominate the sputtered flux at a high oxygen partial pressure. Furthermore it was found that with increasing oxygen coverage the contribution of sputtered ions increases from 1% (not measurable) to 25 or 35 % of the total flux for Cr or Zr, respectively.

3.4 Composition of the sputtered flux with increasing oxygen surface coverage

From the results obtained a quantitative picture of the composition of the sputtered flux as a function of the oxygen coverage can be obtained (Fig.4). The total sputtered flux decreases by about a factor of 4 with increasing oxygen surface coverage (Curve A in Fig. 4) in agreement with lower sputtering yields for oxides as compared to pure metals (10). Sputtered ground state atoms give the dominant contribution to the sputtered flux for clean targets. Their contribution however decreases drastically with increasing oxygen surface coverage and contributes less than 1% to the flux for an oxygen to ion flux ratio larger than one (Curve B in Fig. 4).
In our experimental set-up the distance between the target and the interaction of the sputtered particles with the laser beam is 5 cm. Thus in the LIF experiments most of the atoms sputtered in short lived excited states have decayed to the ground state before reaching the interaction volume and are measured as "ground state" atoms. In addition it was found that the excited flux increases strongly with oxygen coverage (Fig.3).Thus the measured value of $8x10^{-3}$ sputtered ground state atoms per incoming ion can be taken as an upper limit for

the yield of all short lived excited atoms, indicating that they represent only a minor component to the total flux. Furthermore the quartz microbalance measurements revealed that Cr sputtered as charged particles(ions or ionized molecules) amounts to about 25% of the total sputtered flux at 10^{-6} mbar O_2 partial pressure (Curve C in Fig. 4). Thus for oxygen to Ar^+ flux ratios larger than 1 more than 70 % of the sputtered flux have not been accounted for and we conclude that this major component consists of neutral molecules or clusters (Curve D in Fig. 4).

A similar dependence of the composition of the sputtered flux on oxygen coverage has sofar been also observed for Zr (7), Ti (7,11) and probably Fe (12).

This work has been supported by the Austrian "Fonds zur Förderung der wissenschaftlichen Forschung" under Proj. Nr. 4547 and 5577.

REFERENCES

1. Thomas E.W.: Vacuum 34 (1984) 1031
2. Williams P.: Applic. Surf. Sc. 13 (1982) 241
3. Gnaser H., Fleischhauer J., Hofer W.O.: Appl. Phys. A37 (1985) 211
4. Pellin M.J., Young C.E., Calaway W.F., Gruen D.M.: Surface Science 144 (1984) 619
5. Husinsky W., Betz G., Girgis I., Viehböck F.P., Bay H.L.: J. Nucl. Mater. 128/129 (1984) 577
6. King D.A. and Woodruff D.F.(ed): "The chemical physics of solid surfaces and heterogeneous catalysis" Vol.3 (1982) and Vol.4 (1984), Elsevier, Amsterdam, Oxford, New York
7. Betz G., Husinsky W.: Nucl. Instr. Meth. Phys. Res. B, in print
8. Husinsky W., Betz G., Girgis I.: Phys. Rev. Lett. 50 (1983) 1689
9. Grischkowsky D., Yu M., Balant A.C.: Surface Science 127 (1983) 315
10. Kelly R., Lam N.Q.: Radiat. Eff. 19 (1973) 39
11. Dullni E.: Nucl. Instr. Meth. Phys. Res. B 2 (1984) 610
12. Behrisch R., Roth J., Bohdansky J., Martinelli A.P., Schweer B., Rusbüldt D., Hintz E.: J. Nucl. Mater. 93/94 (1980) 645

GEOMETRIC METHODS OF ANALYSIS

M.J. NOBES
University of Salford, Department of Electronic and Engineering, U.K.

1. INTRODUCTION

This paper reviews the methods of geometric analysis which may be widely
applied to the prediction of surface morphology changes under conditions of
either accretion or erosion, but which have been utilised specifically in
connection with the topographical modification of crystalline and amorphous
solids subjected to energetic atom bombardment. Both the methods, and
where appropriate, the mathematical theory are treated in chronological
order, even though some advanced aspects of each have been used previously
by investigators in unrelated subjects (1-5). It must be stressed that the
parameters which determine the ablation rate of a surface may depend upon
both spatial and time coordinates even when mono-energetic atoms of a
single species are incident upon a specific surface at constant temperature.
For such a general case, no geometric analysis is possible and solution is
only achievable by iterative computation. As the number of variables which
influence a parameter is reduced, or assume an invariance, so the degree of
geometric analysis may be extended, but only for the simplest of cases can
this analysis accurately determine the long-term evolution and tendency
towards equilibrium - if attainable by the nature of the process.

In consequence, the early work in this field, in which basic parameters
were assumed to be invariant with respect to time and position coordinates,
yielded the greatest potential for geometric methods and established the
vital importance of characteristic methods of analysis (6,7). Even in more
complex situations, these same basic characteristics and plotting cursors
derived from their properties are shown to be essential to the geometric
estimation of surface development. When it became apparent that a
sputtered surface behaved as a wavefront in kinematic or non-linear wave
theory (8-10), then although extensive geometric treatment was still a
difficult problem, it was possible through certain simplifying assumptions
to apply characteristic methods such that the general nature of long-term
evolution in the region of grain boundaries and extended dislocations
could be assessed (10,11). It is acknowledged that during the sputtering
of a surface there may be accompanying processes operating which would have
to be allowed for in a general theory. These processes may be considered
to be intrinsic and extrinsic to the basic sputtering mechanism. The
former would include the redeposition of sputtered material and the enhancement of local ion or atom flux due to scattering and reflection from
asperities to lower regions. In the latter category one might include
surface and bulk diffusion and thermal migration of target atoms and
defects. Such processes have been included in theoretical treatments (12-
15) and their effects determined by iteration, but as yet no geometric means
has been found of exhibiting other than the instantaneous motion of surfaces
under such conditions. Since the object of geometric analysis is to avoid
repetitive construction to the extent that the form of a surface after

prolonged erosion should be determined from a single construction, such processes will henceforth be ignored.

As a simple introduction to the geometric determination of morphology resulting from an erosion process, consider the following hypothetical example. Let experiment indicate that a planar surface element of a bulk material erodes at a rate which is a function only of the elevation of that element above a reference plane within the bulk, that rate being measured in a direction which is parallel to the reference plane and lies in the plane defined by the normals to the reference plane and the element itself. Then one might initially construct a set of planes parallel to the reference plane and in each plane, measure inwards from the surface intersection with that plane a distance proportional to the elevation dependent function appropriate to it. By joining the contours at each level, the eroded surface shape would be revealed. Repetition of this process would exhibit the time-varying shape of the surface. This method of construction would quite probably display cusps at certain positions on certain planes. These would indicate that the real surface has generated edges or sharp angular discontinuities, the actual cross-section of surface at any level being determined by ignoring these cusps. It is considered inadvisable at this stage to dwell unduly on this latter point, but it should be noted that although many other examples could be quoted, the essence of the solution to them all is the seeking out of a particular direction for which some property of the local surface element is preserved, or, at least depends upon a minimum number of variables. In the above example, that particular property was the invariance of erosion rate, and in general, the path along which the constancy is preserved is known as a characteristic path.

In the case of physical sputtering it has been established by experiment that the erosion rate of a planar element is dependent upon the erosion flux density J, the atomic density N of the material and its sputtering yield Y. These three parameters may in absolute generality be dependent upon spatial coordinates (x,y,z) and time. In this communication, however, the time variable will be disregarded. Furthermore because of their inherent anisotropy, the sputtering yields of crystalline materials are dependent not only upon the inclination of the surface normal to the beam direction (θ) but also upon the azimuthal displacement (ϕ) of that same element from a reference plane containing the flux direction. The erosion rate in the surface normal direction may then be expressed with reference to a rectangular reference frame as

$$C = \frac{J(x,y,z)}{N(x,y,z)} \, Y(x,y,z,\theta,\phi) \cos\theta$$

Due to the limitation of early mathematical theory (16,17) the relevance of characteristic paths was for some years unknown. Consequently the investigations summarised in the next section revealed scant information concerning the instantaneous tendency of surfaces to alter their shape. The advent of two-dimensional characteristic theory opened the field for simple geometric techniques as shown in section 3 and realisation that a sputtered surface may be fully described by three-dimensional wave theory allowed for the introduction (section 4) of azimuthal dependence of the sputtering yield in addition to orientation dependence. At this stage, in addition to the vector methods used by theoreticians, some insight into the precise way in which azimuth and orientation change with time along a characteristic path was revealed by analytical geometry. Consequently the equivalent of a basic law of geometric optics is shown in section 5 to hold at all times along characteristic paths. Finally the full complexity of general surface evolution is outlined, and examples illustrate (in two dimensions for simplicity) the effects due to striated or stratified media.

2. INITIAL TRENDS IN ANALYSIS

Throughout this and the following section it is assumed that the sputtering yield Y is dependent only upon angle of incidence θ of the ion beam onto a surface element and that the Y-θ function is of a form typical of most amorphous or polycrystalline media subject to bombardment from monoenergetic ions of a single species at a defined temperature as shown in Figure 1.

It is further assumed that a plane containing the ion beam direction may be found which contains the normals to the surface elements defined by the intersection of the surface with that plane i.e. all the surface elements at a given section are defined by the same azimuthal plane which in this section will be the plane of illustration. Also J and N will be assumed to be spatially invariant.

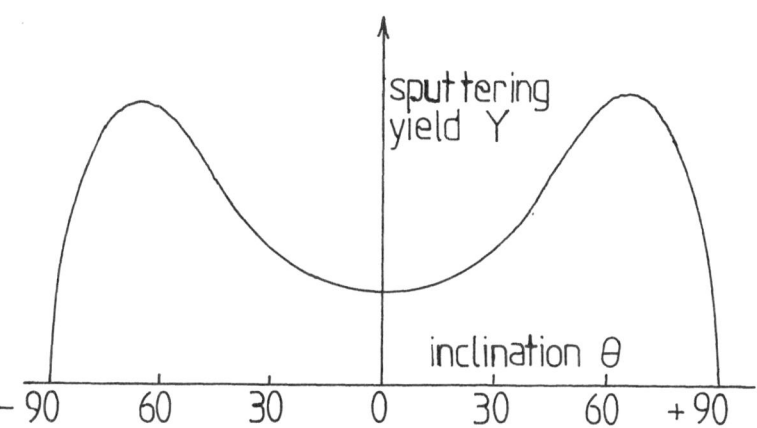

FIGURE 1. A typical variation of sputtering yield with incidence angle.

The work of Stewart and Thompson (16) which provoked a wide interest in sputtered topography, considered a continuous surface cross-section to be composed of a large number of planelets at different orientations. By allowing these elements to regress in their respective normal directions by amounts $J/N.Y(\theta)\cos\theta.\delta t$ in the time interval δt, they showed that certain elements were progressively obliterated whilst others, depending upon the initial profile, expanded. Although their treatment assumed that the fine edges established on the initial profile were preserved by the sputtering mechanism which will be shown later to be generally incorrect, their broad conclusions agreed with those derived from more sophisticated methods. Considering the junction of any two adjacent planelets which might form either a ridge or a valley, they showed that: (a) the crest of a ridge would move towards the plane for which $Y(\theta)$ is least, and (b) the foot of a valley would move towards the plane for which $Y(\theta)$ is greatest. Although it was not specifically noted in their work, the elegance of this approach lay in the fact that an element of orientation θ could be reproduced (at a normal displacement of $J/N.Y(\theta)\cos\theta.\delta t$) outside the rectilinear bounds of the initial element.

Catana et al (17) attempted to reconstruct a surface profile by noting the orientation at a large number of x positions on the initial profile and then allowing each to recede in the ion flux direction by an amount J/N.S(θ).δt. It was assumed that linking all these points would produce a first-stage time increment profile from which the new orientations at the same x position could be determined. For various reasons, however, this technique failed to give the ultimate profiles which reasonably might have been expected. Ishitani (18) later used a modified method which by repeated iteration revealed a more promising estimation of final morphology. Progressing from this approach, Nobes et al (19) considered the differential regression of adjacent surface elements in the ion flux direction. They obtained the expression:

$$\left|\frac{\partial \theta}{\delta t}\right|_x = -\frac{J}{N} \cos^2\theta \frac{dY}{d\theta} \left|\frac{\partial \theta}{\partial x}\right|_t \qquad (1)$$

with similar expressions for other regression directions. This related the rate of change of tangential angle of orientation (θ) at constant x coordinate to its rate of variation with x at specified time (i.e. the local radius of curvature). This approach allowed for formal specification of the surface approach to an equilibrium configuration. Barber et al (6) in a semi-quantitative analysis used the solutions of an analogous problem previously evaluated by Frank (20,21) which concerned the dissolution and growth of crystal which was based on kinematic wave theory described by Lighthill and Whitham (3,4). This revealed that a surface orientation (unless it is removed by the formation of edges during erosion) is always reproduced at successive stages of erosion in a precise direction and at a certain speed, both of which are functions only of that orientation. Similar conclusions were drawn by Ducommum et al (22,23) who considered an initial profile to be enveloped by a family of straight lines which were transposed by sputtering into a related set. This latter set then enveloped the sputtered surface. Both these methods produced the first single-step means of describing finite-time evolution - the essence of both being that a surface orientation follows a specific path during erosion which is the characteristic path for that orientation. Therefore the erosion slowness curve - the first milestone towards comprehensive geometric analysis - will be considered in the first part of the next section.

Three basic construction aids will be described in the next section, including the prementioned "erosion slowness curve" and what this author and colleagues have chosen to call the "characteristic velocity component plot". The latter was developed by Nobes et al (7,10,24,25) as a consequence of investigation of equation (1) and related expressions. If the total change of orientation from any point of an initial profile to any point of the sputtered surface is $\delta\theta$, then

$$\delta\theta = \left|\frac{\partial \theta}{\delta t}\right|_x \delta t + \left|\frac{\partial \theta}{\partial x}\right|_t \delta x$$

Then, if $\delta\theta = 0$, i.e. we follow the path of constant orientation,

$$\left|\frac{dx}{dt}\right|_\theta = -\left|\frac{\partial \theta}{\delta t}\right|_x / \left|\frac{\partial \theta}{\partial x}\right|_t$$

which, from expression (1) $= J/N.Y_\theta \cos^2\theta = vx$ \qquad (2)

Consideration of recession in the x direction yields a similar equation to (1), i.e.

$$\left.\frac{\partial\theta}{\partial t}\right|_y = -\frac{J}{N}(\sin\theta\cos\theta\, Y_\theta - Y)\left.\frac{\partial\theta}{\partial y}\right|_t$$

from which

$$\left.\frac{\partial y}{\partial t}\right|_\theta = \frac{J}{N}(\sin\theta\cos\theta\, Y_\theta - Y) = vy \qquad (3)$$

where $vx^2 + vy^2 = v\theta^2$.

The expressions (2) and (3) then represent the velocity components of an element of slope θ which is constantly reproduced during erosion. The relative merits of these "cursors" as they may now be called will be discussed at the end of the next section.

3. BASIC GEOMETRIC TECHNIQUE

Consider a Cartesian plot of $1/Y$ vs $\tan\theta/Y$ which is simultaneously a polar plot of $1/Y\cos\theta$ vs θ. The gradient of the normal from the origin to a tangent at any point on the curve is $-vy/vx = \tan\beta$ (Figure 2) which thus gives the direction of the characteristic, whilst the perpendicular distance from the coordinate origin to the tangent at θ is $1/v\theta$. This plot, or cursor, is known as the "erosion slowness curve" (6).

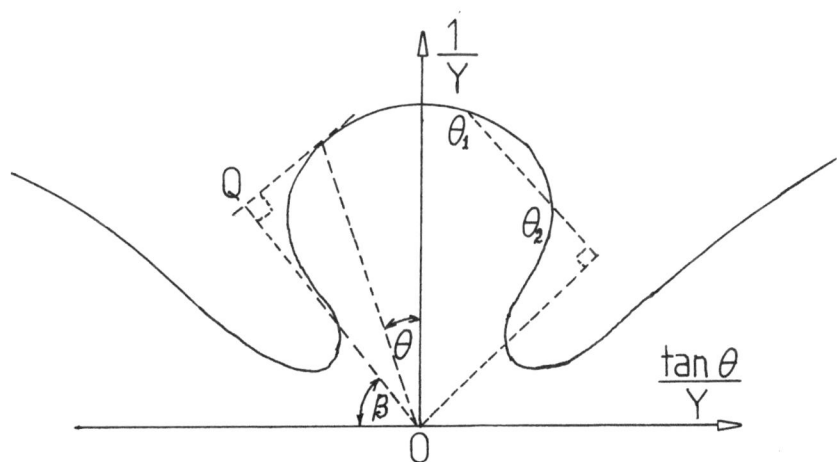

FIGURE 2. The erosion slowness curve showing construction for characteristic and edge motion.

Geometric construction would then be performed as follows: The origin could be placed at a point of known orientation on a profile with the $1/Y$ axis maintained parallel to the ion flux direction, and if the erosion slowness curve were graduated for different θ values the direction of recession OQ of the element at that point would be drawn and marked off at

intervals proportional to the inverse of OQ. Similar construction at
other points, followed by joining of the marked intervals would then
reveal a progressive series of erosion profiles. If the surface under
examination contains angular discontinuities, the influence of all the
included angular range of potential orientations must not be neglected.
In this case therefore, the point O would be maintained at the edge
position whilst the above process is repeated for several orientations
within the bounds of the adjacent continuous elements. At any stage of
evaluation, if cusps appear at the evolved surface, they indicate that
ranges of orientations have been absorbed into edges. These cusps should
be ignored in order to display the true cross-section of the eroded
surface, and the instantaneous velocity and direction of the edge may be
determined from the erosion slowness curve by drawing a line through the
points on the curve corresponding to the orientations θ_1 and θ_2 which meet
at the edge (Figure 2). Then, as for characteristics the direction of the
normal from the origin to this line would correspond to the direction of
edge motion, whilst the inverse of its length would be proportional to the
edge velocity.

A second cursor, known as the characteristic velocity component plot
(Figure 3) may be formed. Consider again, in a Cartesian system, a plot

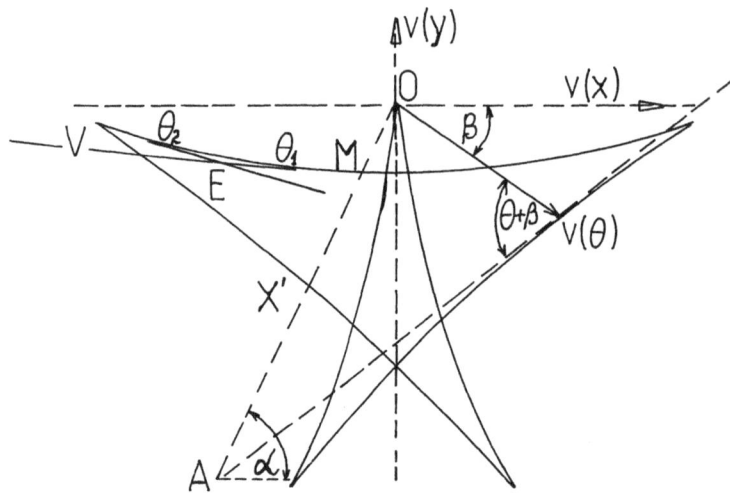

FIGURE 3. The characteristic velocity component plot illustrating a stable
edge at E.

of vy against vx where these quantities, as already shown, are the
velocity components of the orientation which is unaltered during sputtering. As the plot is constructed, let selected orientation points be marked.
Since $v\theta = \{ v_y^2 + v_x^2 \}^{\frac{1}{2}}$ the vector from the origin O to any point of
the plot represents the constant orientation velocity, and the inclination
at that point of the diagram must equal the value attributed to that point.
The recession angle β of that orientation is then given by $\tan\beta = v_y/v_x$
and it is easily shown that the tangent at this point intercepts the v_x
and v_y axes at values $J/N.Y(\theta)\cot\theta$ and $J/N.Y(\theta)$ respectively. It is then
apparent that the plot could be constructed by the following procedure:

(a) Select an orientation θ and determine from the Y(θ)-θ function the corresponding value of Y(θ).

(b) In the negative y axis on a rectangular system of coordinates mark a point at a distance from the origin proportional to Y(θ).

(c) Through this point draw a line inclined at the angle θ to the x axis.

(d) Repeat the process for other orientations in the range $-\pi/2$ to $+\pi/2$.

The family of straight lines so produced, form an envelope to the required curve.

Finally, although it is of no importance to the mapping of erosion contours, but has uses concerning the area of sputtered surfaces, it must be noted that the radius of curvature at any orientation θ on the cursor is proportional to the rate of change of radius of curvature along the characteristic θ path during erosion (7).

A sputtered profile might then be constructed using this cursor in several obvious ways. Perhaps the easiest method would be to make the cursor tangential to a specific point of the contour under examination, keeping the -y axis coincident with the incident sputtering flux direction. The line joining the origin to this point, if extended to the other side of the common tangent, would then coincide in direction to the constant-orientation characteristic, and the distance from the origin to the point would be proportional to the characteristic velocity. It must now be apparent that construction of erosion profiles need not involve a cursor at all. By marking off points below the test contour at distances proportional to Y(θ) and then through these points drawing extended lines at inclination θ it would be found that the complete line set envelopes a profile which would then accurately represent the eroded surface. Under normal sputtering experiments it is usual to erode a target to a degree which is small in comparison with its original dimensions. It is worth noting, however, that if a substrate is bounded by orientations which lie in the ion flux direction, i.e. $\theta = \pm\pi/2$ (pinning planes) and extensive ablation is permitted, then erosion distances become so large that in the final surface all information regarding the initial surface is lost. The ultimate surface would then comprise, depending upon its initial concave or convex nature, certain shapes identical in form to portions of the cursor derived from the constant velocity component plot (1,2,25). As for the erosion slowness curve, cusps displayed via geometric construction must be disregarded.

The instantaneous velocity and direction of an edge bounded by orientations θ_1 and θ_2 are then represented by the vector OE from the origin O to the junction E of tangents drawn at θ_1 and θ_2 to the cursor. Such an edge might be stable as in Figure 3 or unstable. In the latter case, certain ranges of orientations within the bounds of the initial edge instantly break into the original edge. The injected range or ranges will then correspond to the shortest route, along the cursor, which links the tangents at θ_1 and θ_2. Any angular discontinuities apparent between these ranges represent stable edges. Thus, Figure 4 shows how the edge AOB instantly assumes the form A'CDB'. The injected range CD is tangential to the displaced plane AO at C and the edge at D is stable.

From the above construction aid it can be seen that if β represents the recession angle of orientation θ, then the angle between the characteristic direction and the θ plane is θ+β and

$$\sin(\theta+\beta) = \frac{Y\cos\theta}{v\theta} \tag{4}$$

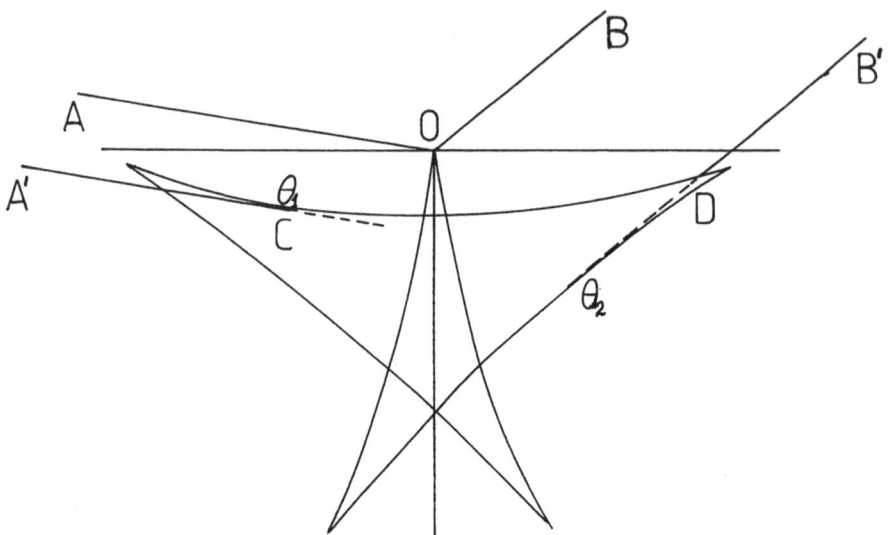

FIGURE 4. The evolution of an initially unstable edge.

Also it may be shown that $v\theta^2 = \{Y\cos\theta\}^2 + \{d/d\theta(Y\cos\theta)\}^2$ and this leads to yet another useful cursor design. Figure 5 displays a plot of $d/d\theta(Y\cos\theta)$ vs $Y\cos\theta$ based on a $Y-\theta$ function of the form shown in Figure 1. During

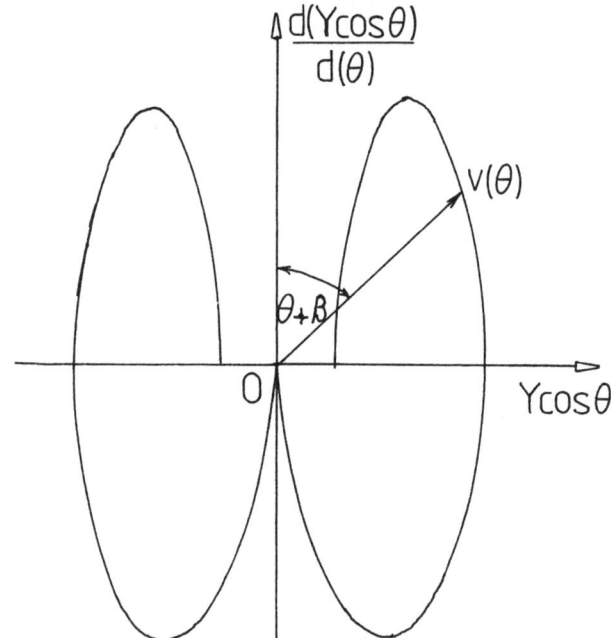

FIGURE 5. Another example of a plotting cursor.

construction it is necessary to mark off θ values on the curve. From (4) it is apparent that the vector joining the origin to any part on the cursor makes an angle with the d/dθ(Ycosθ) axis equal to θ+β and is proportional to the constant orientation velocity vθ '. During use such a cursor would be placed such that the d/dθ(Ycosθ) axis is tangential to a point of known θ value on the surface profile at the origin. As before, the line joining 0 to that θ value on the cursor would lie in the direction of the characteristic, and its magnitude would be proportional to vθ . Again, as in all previous cases, repetition for various θ, followed by linking of the construction points, exhibits the erosion contour.

There are several reasons why the above-mentioned cursor and the erosion slowness curve are somewhat inferior to the characteristic velocity-component plot. The erosion slowness curve at its extremities extends to infinity along both axes, that is, in the region of θ = ±π/2, thereby rendering its use rather impractical. Furthermore, in common with the latter cursor, it necessitates graduation of the derived curve. The elegance of the characteristic velocity-component plot lies in its ability to reveal instantly evolution in the region of surface angular discontinuities. Since at such discontinuities a range of orientations may be considered to exist at one point, that range must generally evolve into a shape corresponding to part of the characteristic plot. In addition, this cursor instantly reveals which edges on an initial profile are stable against changing form.

4. THREE-DIMENSIONAL SURFACES

The disadvantage of surfaces encountered in experimental work, from the aspect of geometric morphology construction is that surface section produced by intersection with any plane does not comprise elements of constant azimuth. In any plane of illustration therefore, it is possible to reconstruct only those elements which lie in that azimuthal plane. Even if the surface under consideration were axially symmetric, any dependence of sputter yield upon azimuthal plane ϕ would take some planar elements out of the plane of illustration, because in such an event as was shown in the contribution by Carter (30) that in addition to component velocities in the azimuthal plane of an element of surface, a third component directed normal to this plane would be introduced of magnitude

$$\frac{J}{N} \cdot \left| \frac{\partial Y(\theta,\phi)}{\partial \phi} \right|_{\theta} \cot \theta$$

The characteristic velocity component plot would then be a three-dimensional surface obtained by mutual plotting of vx , vy , and vz for all θ and ϕ values. For an axially symmetric yield function (no ϕ dependence) any θ,ϕ element of surface illustrated in its azimuthal plane would be reproduced in that plane with the velocity components noted earlier. Provided that J, N and Y were still spatially independent parameters, straight line characteristics would still exist such that any specified element on the initial surface characterised by θ and ϕ would be reproduced somewhere on any eroded state of that surface. In principle a reconstructed surface would be achieved by constructing planes displaced from each surface element by J/N.Y(θϕ)cosθ.T in the local surface normal direction. They would then form an envelope to the eroded surface at time T. Therefore, although no iteration is involved in determining long-term surface morphology, the limitations of planar illustration are obvious, and three-dimensional illustration by a computer is advisable.

5. SPATIALLY VARIABLE PARAMETERS

In order to allow for the dependence of J, N and Y upon spatial parameters and to appreciate the consequent influence upon characteristic paths,
velocity components and ultimately surface modification, the condition
which must be maintained along a characteristic will now be deduced. For
ease of manipulation it is convenient to imagine that the general sputtering yield may be expressed in the variables-separable form:

$$Y(x,y,z,\theta,\phi) \;=\; S(x,y,z)\, Y(\theta,\phi)$$

and that the sputtered medium is divided by a planar boundary such that
there is a step-function change in the value $J(x,y,z).S(x,y,z)/N(x,y,z)$
across the boundary. Then in the usual nomenclature of analytical geometry
in Cartesian coordinates, let the planar boundary be represented by
$Lx + My + Nz = P$, where L, M and N are the direction cosines of the normal
to the plane, and P is the perpendicular distance of the plane from the
origin. An incident planar element may similarly be expressed as:

$$\ell x + my + nz = p_1 \text{ where } \frac{dp_1}{dt} = (\frac{JS}{N})_1 \, Y(\theta_1,\phi_1)\cos\theta_1$$

In order to preserve surface continuity at the boundary, the emergent
planar element must then have the form:

$$(\ell+\lambda L)x + (m+\lambda M)y + (n+\lambda N)z - (p_1+\lambda P) = 0,$$

or, in terms of actual direction cosines:

$$(\frac{\ell+\lambda L}{K})x + (\frac{m+\lambda M}{K})y + (\frac{n+\lambda N}{K})z = \frac{p_1+\lambda P}{K}$$

where

$$d(\frac{p_1}{K})/dt = (\frac{JS}{N})_2 \, Y(\theta_2,\phi_2)\cos\theta_2$$

and

$$K^2 = (\ell+\lambda L)^2 + (m+\lambda M)^2 + (\dot{n}+\lambda N)^2 = 1+\lambda^2+2\lambda(\ell L+mM+nN) \qquad (5)$$

since $\ell^2 + m^2 + n^2 = L^2 + M^2 + N^2 = 1$

It then follows that $K = (\frac{JS}{N})_1 \, Y(\theta_1,\phi_1)\cos\theta_1 / (\frac{JS}{N})_2 Y(\theta_2,\phi_2)\cos\theta_2$

Let β_1 and β_2 represent the angles between the boundary and incident and
emergent planes respectively. Then

$$\cos\beta_1 = \ell L + mM + nN$$

and

$$\cos\beta_2 = (\frac{\ell+\lambda L}{K})L + (\frac{m+\lambda M}{K})M + (\frac{n+\lambda N}{K})N = \frac{\cos\beta_1+\lambda}{K}$$

and from (5),

$$K^2 = 1+\lambda^2+2\lambda\cos\beta_1 = (\lambda+\cos\beta_1)^2 + (1-\cos^2\beta_1) \qquad (6)$$

Also, $\sin^2\beta_1 = 1-\cos^2\beta_1$

and $\sin^2\beta_2 = 1-(\dfrac{\cos\beta_1+\lambda_2}{K})^2$

which yield from (6),

$$= \frac{1 - \cos^2\beta_1}{K^2}$$

Thus,

$$\frac{\sin^2\beta_1}{\sin^2\beta_2} = K^2 \quad \text{or} \quad \frac{\sin\beta_1}{\sin\beta_2} = K = \frac{(\frac{JS}{N})_1 Y(\theta_1\phi_1)\cos\theta_1}{(\frac{JS}{N})_2 Y(\theta_2\phi_2)\cos\theta_2} \tag{7}$$

(7) is simply Snell's Law relating the normal recession velocities on each side of the boundary to the change in wavefront inclination at the boundary. Viewed in the local reference frame, i.e. in the plane containing the three surface normals there appears to be a change of θ only on passage through the interface. It must be noted however that in order to establish this illustration plane, the interface should first be rotated about the z axis until it is normal to the xOz plane, then the incident(and interface) planes rotated about the interface normal until both planes are normal to the xOz plane. Therefore, referred to the initial coordinates, there is generally a change in both θ and ϕ at the interface. It is difficult to express the change in θ between the incident and emergent planelets, but it is clear that the emergent azimuth coincides with that of the plane which is perpendicular to both the incident plane and the interface. Thus ϕ_2 may be deduced as a function of ℓ, m, n, L, M and N and it follows that θ_2 (and hence $Y(\theta_2\phi_2)$) adjust to suit the requirement of Snell's Law. If this analytical approach is pursued , it is useful to note that $\cos\theta$ and $\tan\phi$ are respectively equivalent to n and m/ℓ. Obviously, no single-step construction is possible here, and even computer calculation would be tedious because surface contours of invariant JS/N would have to be identified and stored.

Since, through the convention adopted here, an azimuthal plane contains the particle flux direction, if an incident plane and an interface lie in the same azimuth, then the condition:

$$\frac{\frac{J}{N} Y\cos\theta}{\sin\beta} = \text{constant along a characteristic trajectory}$$

may be recast into a more usable form. Let Figure 6 represent an element of surface (orientation θ) approaching an interface of elevation α in a medium for which J and N are position independent. Then, $\sin\beta$ may be replaced by $\sin(\alpha-\theta)$ and if the interface were broadened to finite width such that all contours of invariant S were parallel and inclined at angle α (as above), then θ would constantly adjust to keep the ratio

$$\frac{JS}{N} \cdot \frac{Y\cos\theta}{\sin(\alpha-\theta)} \quad \text{invariant.}$$

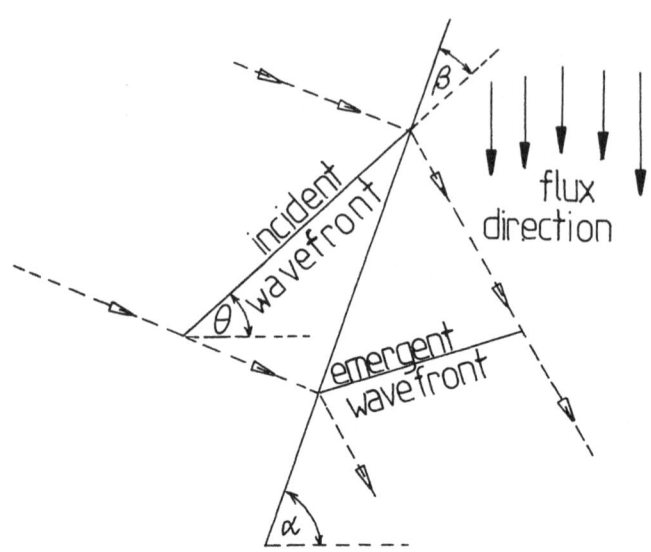

FIGURE 6.   Refraction at an interface.

With reference to Figure 3 the orientation θ would continuously alter so as to maintain constancy of the product of

$$\frac{J}{N} \ Y(\theta\phi)\cos\theta$$

and S, i.e. OA.S = constant. This form of Snell's Law, which is appropriate to plane and interface in common azimuth, will be used extensively in the next section to convey the nature of surface change in the neighbourhood of striated media.

Finally, it is worth reiterating the basic meaning of a characteristic path and the condition which must be fulfilled along its length. The direction and velocity of the path at any time must correspond to those in which the azimuth and orientation of a surface element existent on the path at that time would constantly be reproduced under spatially invariant parameters J, Y and N, existent at that time upon the surface element. At all times a surface element upon that path progressively adjusts its azimuth and orientation such that in relation to surface contours of spatially invariant J, Y and N, Snell's Law is obeyed.

6.   BASIC MATRIX INHOMOGENEITIES

When all, or indeed any of the fundamental sputtering parameters are complex functions of spatial coordinates, any form of geometric construction without iteration is impossible, and rendered even more so if no set of characteristics from part of the initial surface lie in one plane. Computer estimation of eroded surface shape is then the only possible solution. The necessary iteration should then be executed along the characteristic paths emanating from each element of the initial surface. Although such a procedure is time-consuming, its appeal lies in the fact that if the spatial variation of parameters is known, the planelet which ultimately lies on a particular characteristic depends only upon the one which existed at the start, and not upon the nature of the initial surface surrounding that element. However, if spatial variation is of a well-defined form as might reasonably be assumed in a model of a matrix grain boundary or planar dislocation, the evolutionary trend in the vicinity of such bulk irregularities may be forecast. The following example should clarify this:

Let Figure 7 represent the section of a grain boundary viewed in the azimuthal plane of the initial surface, in which the line of intersection of surface and boundary is normal to the erosion flux direction. Let the ion flux J and substrate atomic density N be spatially invariant whilst the spatially dependent component of the sputtering yield varies continuously as indicated between the two grains, both of which are initially at orientation α. This would assume that within the varying yield region the boundary of which is represented by two parallel lines of inclination   the yield may be expressed as:

$$Y(\theta X) = Y(\theta)_{X=0} \ f(X)$$

Since we are now dealing with a crystalline matrix it is necessary to imagine a $Y(\theta\phi)$ surface produced by rotating Figure 1 about the ion beam direction axis and perturbing this symmetric surface at specific regions. This would simulate the influence of yield variation due to the sputtering of various crystal faces by an ion beam incident in a fixed crystal direction. The corresponding characteristic velocity component plot would then be a complex three-dimensional figure, but the cross-section appropriate to the azimuthal plane under consideration could be used, as before, to

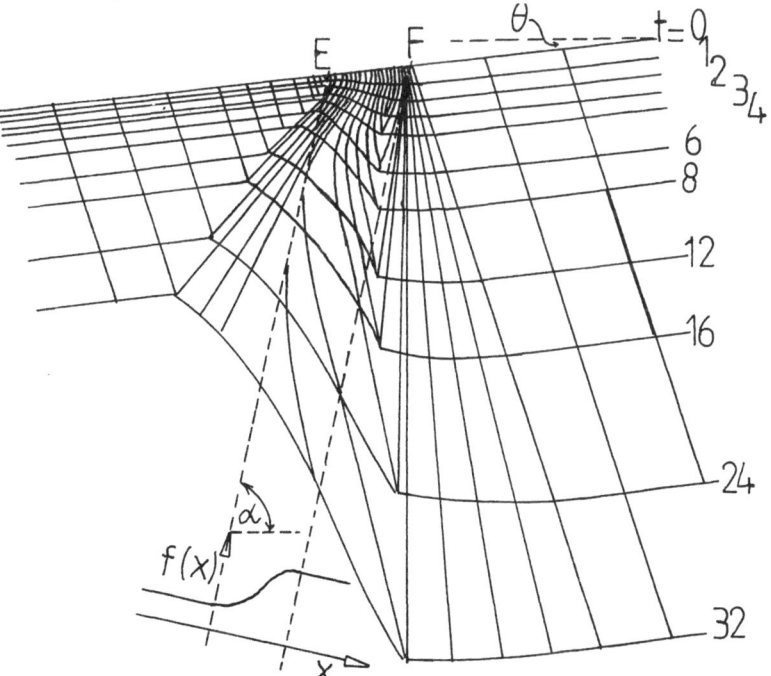

FIGURE 7.   Progression of a sputtered surface at a grain boundary by the construction of characteristic trajectories.

plot characteristic trajectories. In the varying-yield region, these would not be straight lines. However, in order to illustrate the power of this method, rather than accurately plot characteristics and resulting profiles at this stage, the random matrix form of the component plot shown in Figure 3 will be used in this example. As noted at the end of the last section, since all contours of invariant spatial yield component are parallel, the quantity

$$\frac{J}{N} \, Y(\theta)_{X=0} \, \frac{\cos\theta}{\sin(\alpha-\theta)} \cdot f(X) \tag{8}$$

must remain constant along a characteristic. Therefore, using Figures 7 and 3 in conjunction, and noting that OA.f(X) must remain constant along a characteristic path, it is possible to sense the way in which local surface inclination varies with its spatial position. All characteristics originating within the boundary move into a region of higher yield. Thus OA must decrease so that orientations will decrease until they reach the value at M unless they have already passed into the high-yield grain. Differentiation of expression (8) = constant, with respect to time, gives the rate of change of orientation along a characteristic as

$$\frac{d\theta}{dt} = \frac{J}{N} \, Y(\theta)\cos\theta \, \sin(\phi-\theta).f_X$$

Since $f_X = 0$ only at the boundaries of the varying-yield region, a characteristic on which $\theta = \theta_M$ at a point which does not lie on a boundary will progress towards the left, the orientation θ moving through the cusp at V and thence towards X'. If by this stage the characteristics have not moved into the low-yield grain they re-traverse the boundary region, eventually emerging into the other grain.

The exact form of the evolving surface depends on the ratio of the values of f(X) in each grain, and it will fall into one of three categories: low, medium and high ratio. The illustrated example is of the latter class. In the low-yield case, all characteristics from the low-yield grain and the region EF would move into the high-yield grain. The original surface inclination of the entire left-hand grain would be preserved and in the highest ratio limit, the characteristic emerging into the right-hand grain would have inclination θ_M (Figure 3). For medium-yield the inclination θ_M would remain, bordering the high-yield region, whilst the surface in the low-yield grain would be modified near the varying-yield region. Ultimately the orientation θ_X' would emerge into the low yield area. For higher ratios, this same orientation would remain at the left of the boundary, whilst higher ones develop on the other side.

By similar reasoning, the form of evolution of a surface in the vicinity of a planar dislocation may be deduced. With similar assumptions the effects due to ion beams scanned at low frequency can be assessed and this in turn reveals the possibility of predicting how masks should be scanned across bombarded substrates in order to deliberately produce a required surface morphology.

If the direction of invariance of J/N f(X) coincides with that of the particle flux, the constant appropriate to a characteristic path in two dimensions reduces to $J/N \, f(X) \, Y(\theta) = J/N \, Y(X,\theta)$. If J, instead of Y were spatially non-uniform, then the condition becomes $J(X) \, Y(\theta) =$ constant, which has been considered (26,27) in the literature and extended to embrace three-dimensional contours (8,9). Further developments of basic theory, which are not amenable to geometric interpretation, have prompted

publications concerning moving substrates (28) and yield modifications due
to substrate local curvature (29).

As a further example of this technique, the effect of slowly scanning
a particle flux across a surface may be fully evaluated. This is equivalent to the gradual exposure of substrate to incident flux as a beam
defining mask is eroded or moved across a target, and this is depicted in
Figure 8.

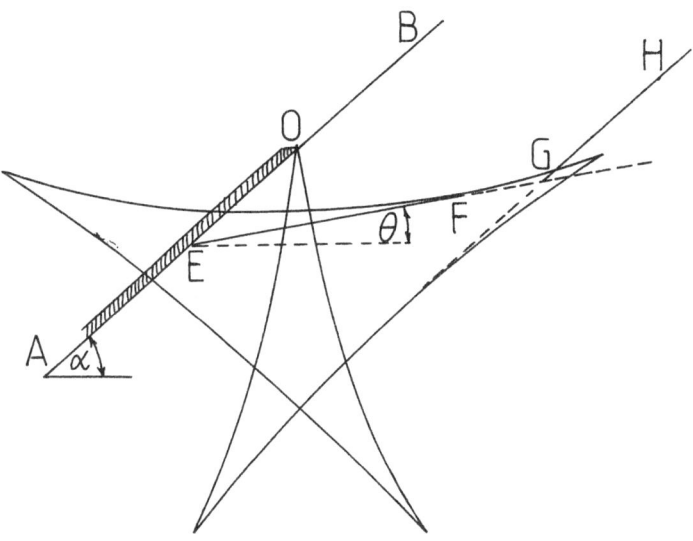

FIGURE 8. Surface development by increasing exposure to erosive flux
limited by a moving mask.

Let the line AB, of inclination α represent a planar substrate which is
half masked over the region AO. Let the mask edge at O move towards A
with velocity v at the onset of bombardment. The velocity v is analogous
to the velocity of intersection of an element of sputtered surface with
a surface contour of constant $Jf(X)/N$ which has been described earlier.
In that case an emerging surface of orientation θ was formed which is
equivalent here to the line EF. Therefore at a time at which the mask
edge has receded from O to E, the sputtered surface has acquired the
shape indicated by EFGH. The orientation θ, which is constantly generated
by the receding mask edge may be expressed, as before, by the relationship

$$v = \frac{J}{N} \, Y(\theta)\cos\theta/\sin(\alpha-\theta)$$

and this may be re-cast as $\tan\alpha - \tan\theta = \dfrac{JY(\theta)}{N} \cdot \dfrac{1}{v\cos\alpha}$

However, $v\cos\alpha$ is simply the mask velocity V_n normal to the flux
direction. Hence $\tan\alpha - \tan\theta = JY(\theta)/NV_n$. For a mask advancing over
the same surface, a similar expression may be obtained, i.e.

$$\tan\theta - \tan\alpha = \frac{JY(\alpha)}{NV_n}$$

in the absence of an analytic relationship between Y and θ, it is possible
to deduce the value of the generated orientation θ in the former of these
equations, either from the characteristic velocity component plot as
previously shown, or from the erosion slowness curve, as follows. Since

$$\tan\alpha - \tan\theta = \frac{JY(\theta)}{NV_n} \; , \; 1/Y(\theta) = \cot\alpha\tan\theta/Y(\theta) + J\cot\alpha/NV_n$$

and as the curve itself is a plot of $1/Y(\theta)$ vs $\tan\theta/Y(\theta)$, a straight line
constructed through the point $y = J\cot\alpha/NV_n$ or $x = J/NV_n$ making an angle
α with the y axis would intersect the curve at the required value of θ.

In view of the potential of the procedures described above, their
application to prediction of evolution in more complex processes is conceivable. Such a process could be the well-established method of simultaneous sputtering and chemical vapour etching of materials as illustrated in
Figure 9, in which, for simplicity, the etching component could be
represented by an additional term, which is orientation independent, in the
expression for the normal ablation rate. This rate R_N would then become
$R_N = Y\cos\theta+A$ which would effectively produce a sputtering yield equal to

$Y+A\sec\theta$. This would introduce added terms of $A\sin\theta$ and $A\cos\theta$ respectively
to the components vx and vy of the characteristic velocity component plot,
i.e. the addition of the constant vector A normal to the diagram at all
points would produce the appropriate cursor for the estimation of surface
evolution under the combined processes.

It would appear that no generally applicable expressions for erosion rate
are derivable, since, depending on spatial position, either one or both
processes are operative on any element. Furthermore the processes may be
mutually interactive, but it is believed that through correct formal
specification of local erosion parameters, quasi-solution may be achievable
by geometric analysis.

7. CONCLUSION

In the experimental study of general erosion and growth processes, it is
often necessary to compare observed surface forms with the estimations of
accurate theoretical modelling. It is hoped that the methods reviewed in
this communication, since they are based on characteristic theory will
assist in such comparisons and also be used as a check against improper
computer coding. Furthermore, to the researcher who has little knowledge
of, or access to, computer facilities, these methods should provide useful
forecasts of surface behaviour.

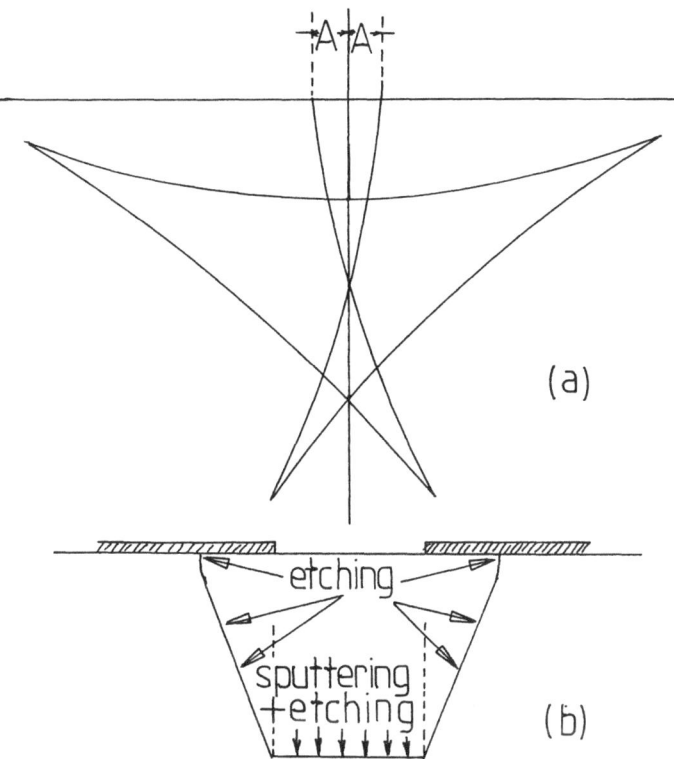

FIGURE 9. (a) Modified diagram incorporating the effect of chemical etching (b) The consequent undercutting of a masked substrate.

REFERENCES

1. Luke JC: J Geophys Res 77 (1972) 2460.
2. Luke JC: Zeits Geomorph Supp 25 (1976) 114.
3. Lighthill MJ and Whitham GB: Proc Roy Soc 229A (1955) 281.
4. Lighthill MJ and Whitham GB: Proc Roy Soc 229A (1955) 317.
5. Whitham GB; Linear and Non-Linear Waves (John Wiley, New York) (1974).
6. Barber DJ, Frank FC, Moss M, Steeds JW and Tsong IST: J Mat Sci 8 (1973) 1030.
7. Carter G, Colligon JS and Nobes MJ: Rad Effects 31 (1977) 65.
8. Smith R and Walls JM: Phil Mag A42 (1980) 235.
9. Smith R, Valkering TP and Walls JM: Phil Mag A44 (1981) 879.
10. Carter G and Nobes MJ: Nucl Inst & Meth B2 (1984) 635.
11. Carter G and Nobes MJ: in Ion Bombardment Modification of Surfaces, eds O Auciello and R Kelly (Elsevier, Amsterdam) (1984).
12. Wilson IH: Rad Effects 18 (1972) 95.
13. Belson J and Wilson IH: Symposium on Sputtering, Vienna 1980 (eds P Varga G Betz and F P Viehbock, Inst F Allgemeine Phys, Tech Univ Wien) (1980) 574.

120

14. Belson J and Wilson IH: Nucl Instr & Meth 182/183 (1980) 275.
15. Makh SS, Smith R and Walls JM: Low Energy Ion Beams (Inst Phys Conf Series) 54 (1980) 246.
16. Stewart ADG and Thompson MW: J Mat Sci 4 (1969) 56.
17. Catana C, Colligon JS and Carter G: J Mat Sci 7 (1972) 467.
18. Ishitani T, Kato M and Shimizu R: J Mat Sci 9 (1974) 505.
19. Nobes MJ, Colligon JS and Carter G: J Mat Sci 4 (1969) 730.
20. Frank FC; Growth and Perfection of Crystals (John Wiley, New York) (1958) 411.
21. Frank FC; Zeits Phys Chem Neue Folge (1972) 84.
22. Ducommun JP, Cantagrel M and Marchal M: J Mat Sci 9 (1974) 725.
23. Ducommun JP, Cantagrel M and Moulin M: J Mat Sci 10 (1975) 52.
24. Nobes MJ; PhD Thesis, University of Salford (1976).
25. Lewis GW, Carter G, Nobes MJ and Cruz SA: Rad Effects Lett 58 (1981) 119.
26. Carter G, Nobes MJ, Arshak KI, Webb RP, Evanson D, Eghawary BDL and Williamson JS: J Mat Sci 14 (1979) 728.
27. Nobes MJ, Webb RP, Carter G and Whitton JL: Rad Effects Lett 50 (1980) 133.
28. Carter G, Nobes MJ and Cruz SA: J Mat Sci Lett 3 (1984) 523.
29. Carter G, Nobes MJ and Webb RP: J Mat Sci 16 (1981) 2091.
30. Carter G: These Proceedings.

SURFACE TOPOGRAPHICAL EVOLUTION: COMPUTATIONAL METHODS OF ANALYSIS

R. SMITH and M.A. TAGG
Maths Department
Loughborough University of Technology
Loughborough LE11 3TU, U.K.

1. INTRODUCTION

The earlier paper by Carter[1] described in a very general way how an eroding or growing surface can be described in terms of non-linear wave theory and some geometrical examples of developing surfaces have been described by Nobes[2]. For many practical applications however, where the characteristics are not straight lines, only computational models allow detailed quantitative calculations to be made. In three dimensions even when the characteristics are linear, the geometrical nature of the surface is usually such that computer graphics have to be used in order to visualise the physical behaviour of the eroding or growing morphology.

A fundamental parameter in both the computational and geometric methods of surface erosion and growth is the sputtering yield, or rate of growth of the surface. The applications described here will be entirely concentrated in the area of erosion and therefore a means of determining the sputtering yield itself, of a material under atom and ion bombardment is required. The sputtering yield depends on many parameters, the composition of the beam and of the target material, the beam energy, angle of incidence, particle flux etc. and can be determined either by direct measurement or by using physical models. There exists in the literature a vast amount of data for the sputtering yields of many different materials as a function of beam energy[3]. There appears to be good agreement between the various theoretical approaches to the problem of predicting sputtering yields as a function of beam energy[4,5]. However, the growth of surface morphology on an ion eroded surface, with the discontinuities in surface gradients which occur as a result of such bombardments (cones, etch pits, facets etc.) can be described in terms of the variation of sputtering yield with angle of incidence and this data is less fully documented. Some theoretical calculations have been carried out, but existing transport theory models[4] are not appropriate for beams of near grazing incidence, and it has been the Monte Carlo binary[5] collision approach which has been used with most noted success to predict the experimentally observed shapes. Some experimental results have been documented[6], see figure 1a, for amorphous materials but for crystals, due to channelling and lattice effects, the relationship is more complex[7] see figure 1b. For crystals, the sputtering yield depends not solely on the angle of incidence of the beam but also on the particular crystal plane under ion bombardment. It would be a major experimental task to carry out such measurements for even one crystalline substance with a particular ion beam species of given energy, to include all the important crystal planes and ion incidence angles. Even computationally, this would require a considerable amount of computing time. Only fairly general aspects of the erosion of crystals will therefore be discussed here. The binary collision code used to determine the sputtering yield variations as a function of ion

122

incidence angle is based on the TRIM[5] model, but with some significant differences, in particular in the tracking of the recoil atoms in the solid. It is assumed that each collison is binary and that atoms are set in motion when they receive an energy greater than a threshold energy E_B. The trajectory of each atom set in motion by an incident ion, is followed through the target until the atom is sputtered or until its energy falls below a cut-off value E_C. A recursive algorithm follows each recoil and the recoils generated by recoils until sputtering occurs or until the atom comes to rest. In calculating the sputtering yield, the isotropic surface binding energy model is used, i.e. an atom upon reaching the target surface will escape if its energy exceeds a set value E_S.

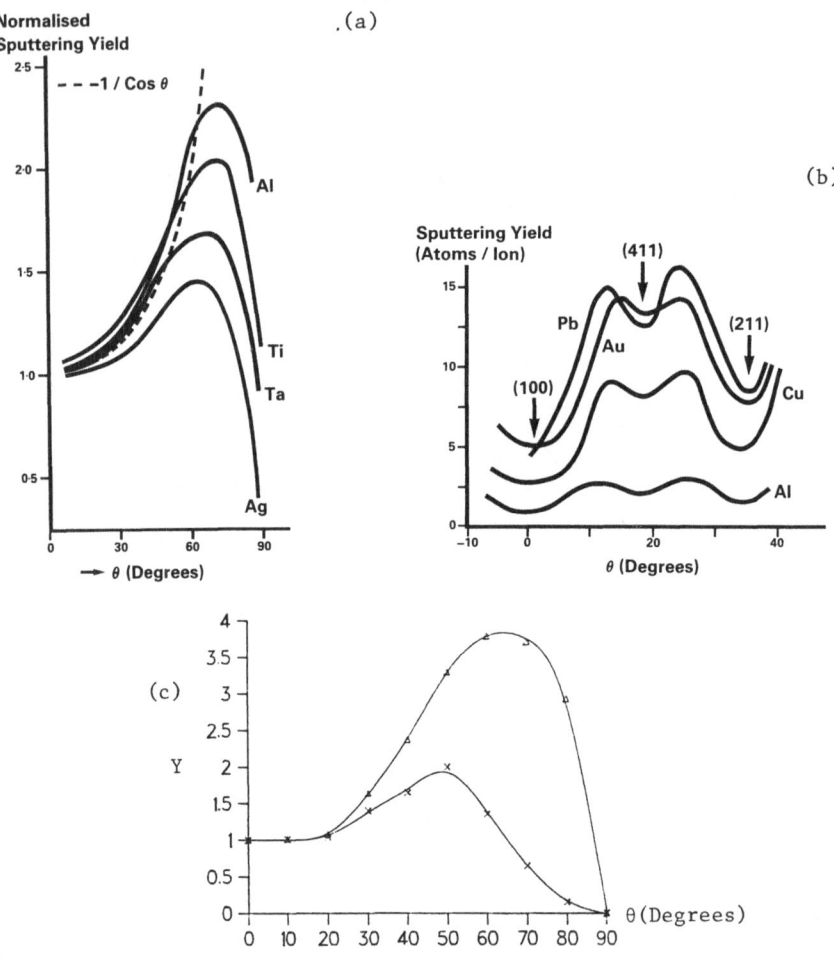

FIGURE 1. Sputtering yield dependence as a function of angle of incidence a) Experimental data for different polycrystalline metals bombarded by Ar[+] ions of 1.05eV.   b) Experimental data for 20keV Ar[+] ions on the (100) face if various crystals.   The rotation axis is the (011)   axis.   c) Normalised sputtering yield curves for amorphous Si under bombardment by Ar[+] ions, △ 1keV, × 500eV, calculated using a binary collision model.

Figure 1c shows the results for the angular dependence of low energy Ar$^+$ ions incident on amorphous Si. The results for ions of 1 keV energy are in excellent agreement with experimentally obtained data[13]. The results for ions of lower energy indicate a zero sputtering yield at an angle slightly less than 90° incidence, probably due to the reflection of ions from the surface. Nevertheless, these preliminary results indicate the success that simple binary collision models have in determining the sputtering yield dependence on angle of incidence for low energy inert gas ion bombardment.

Continuum models of primary ion beam erosion cannot in themselves describe all the surface artefacts which are possible on eroding surfaces, and secondary effects such as ion reflection causing secondary sputtering and redeposition are also responsible for surface topographical change. Again in order to model these correctly, it is necessary to know the spatial and energy distributions of the reflected and sputtered ions. Generally, the reflected ions will have sufficient energy to cause secondary sputtering, whilst the sputtered ions being of much lower energy, will tend to redeposit particularly along the sides of high-walled structures. Once these distributions are known, they can be incorporated into computer programmes which describe the general wavefront motion of the surface. Specific examples will be presented which illustrate computationally and graphically these secondary effects.

2. SURFACE MORPHOLOGY - PRIMARY BEAM EFFECTS

The basic equations which govern the progression of a surface under primary ion or atom beam bombardment have been described by Carter[1]. The progression of this surface can be determined by the solution of a set of differential equations for the characteristic trajectories along which the surface moves and by integration of a set of invariant equations which hold along these characteristics[8,9]. These characteristics have a direct physical meaning in many non-linear wave motions, being equivalent to the 'rays' of geometrical optics and simple waves in acoustics. If it is required to obtain analytical information about the surface, then the characteristic approach gives important information about the development of edges, facets and other surface gradient discontinuities.

Computationally, the characteristic method is fairly straightforward to programme. Its drawback, however, is that it can produce a large set of unwanted data after the characteristic intersections have indicated the formation of edges. Once this has occurred, all subsequent information obtained by integration along the characteristics must be discarded, since the uniqueness of the solution breaks down. If such points were included, the surface would appear to consist of cusps and folds but physically such points have no meaning and must be deleted from the final result. Although it is computationally possible to include pieces of computer code in the programmes to deal with these spurious points, it has been found to be more convenient to delete them manually from the data sets after the eroded surfaces have been calculated. However, a simpler computational method has been developed and programmed in two dimensions which overcomes the problem of intersecting characteristics. This is described in detail later.

Just as edges can form from an initially smooth surface, so can smooth surfaces form from initial gradient discontinuties in a surface. This is computationally more difficult to deal with, but in practise this situation does not arise if a smooth initial surface is to be eroded. It can arise, for example in pattern delineation where an initial angular profile is subject to further erosion by ion beams. The motion of these angular points requires a special consideration in the computation[10].

2.1 Development of points and edges

The erosion of a surface, modelled as a non linear wave, shows that edges and vertices can develop from an initally smooth surface due to characteristic intersections. Carter[1] has shown how the motion of these surface shape discontinuities can be tracked in terms of the intersecting planes which define that discontinuity. It is however, possible to predict the time and spatial co-ordinates of the formation of these discontinuities from a knowledge only of the initial surface shape and sputtering yields. From equation (24a) of Carter's paper we obtain

$$\delta x_{io} = - \sum_{j=1}^{3} \delta x_{jo} \ t \ \frac{\partial v_i}{\partial x_j} \qquad i = 1,2,3$$

i.e.

$$\sum_{j=1}^{3} \left[\delta_{ij} + t \ \frac{\partial v_i}{\partial x_j} \right] \delta x_{jo} = 0 \qquad i = 1,2,3$$

$$(2.1)$$

where δ_{ij} is the Kronecker delta.

In terms of the angles θ and φ defined by Carter[1], his equation (19a) becomes

$$\delta x = t (G_1 \delta\theta + G_2 \delta\varphi)$$

$$\delta y = t (G_3 \delta\theta + G_4 \delta\varphi)$$

where $\delta x = \delta x_{10}$; $\delta y = \delta x_{20}$.

$$G_1 = -g_{\theta\theta}\cos^2\theta\cos\varphi + 2g_\theta\sin\theta\cos\theta\cos\varphi + g_\varphi\cosec^2\theta\sin\varphi - g_{\theta\varphi}\cot\theta\sin\varphi$$

$$G_2 = g_\theta\cos^2\theta\sin\varphi - g_{\theta\varphi}\sin\varphi\cos^2\theta - g_\varphi\cot\theta\cos\varphi - g_{\varphi\varphi}\cot\theta\sin\varphi$$

$$G_3 = -g_{\theta\theta}\cos^2\theta\sin\varphi + 2g_\theta\cos\theta\sin\theta\sin\varphi - g_\varphi\cosec^2\theta\cos\varphi + g_{\theta\varphi}\cot\theta\cos\varphi$$

$$G_4 = -g_\theta\cos^2\theta\cos\varphi - g_\varphi\cot\theta\sin\varphi - g_{\theta\varphi}\cos^2\theta\sin\varphi + g_{\varphi\varphi}\cot\theta\cos\varphi$$

Hence

$$\begin{bmatrix} \delta x \\ \delta y \end{bmatrix} = t \begin{bmatrix} G_1 & G_2 \\ G_3 & G_4 \end{bmatrix} \begin{bmatrix} \theta_x & \theta_y \\ \varphi_x & \varphi_y \end{bmatrix} \begin{bmatrix} \delta x \\ \delta y \end{bmatrix} \qquad (2.2)$$

and so t is given by

$$\det \left[\begin{bmatrix} 1 & 0 \\ 0 & 1 \end{bmatrix} - t \begin{bmatrix} G_1 & G_2 \\ G_3 & G_4 \end{bmatrix} \begin{bmatrix} \theta_x & \theta_y \\ \varphi_x & \varphi_y \end{bmatrix} \right] = 0 \qquad (2.3)$$

This equation gives not only t, but also the orientations of the line elements on the original surface which collapse to elements of zero length.

We have in fact

$$\begin{bmatrix} I & - & GHt \end{bmatrix} \begin{bmatrix} \delta x \\ \delta y \end{bmatrix} = 0 \qquad (2.4)$$

where I is the identity matrix

$$G = \begin{bmatrix} G_1 & G_2 \\ G_3 & G_4 \end{bmatrix} \quad ; \qquad H = \begin{bmatrix} \theta_x & \theta_y \\ \varphi_x & \varphi_y \end{bmatrix}$$

Letting the matrix $I - GHt$ have components given by

$$I - GHt = \begin{bmatrix} A_1 & A_2 \\ A_3 & A_4 \end{bmatrix} \qquad (2.5)$$

then $A_1 A_4 - A_2 A_3 = 0$.

There are in fact two seperate cases to consider illustrated by figure 2. In case (a) the initial surface forms an edge e.g. the intersection of the side walls of a pyramid. The elements of the matrix $I - GHt$ are not degenerate and only line elements of a specific orientation given by

$$\frac{dy}{dx} = - \frac{A_1}{A_2} \qquad (2.6)$$

on the original surface collapse in to form the edge. In case (b) when the elements in the matrix are all zero, a point forms, such as at the top of the pyramid. In this case line elements of all orientations collapse to form the point.

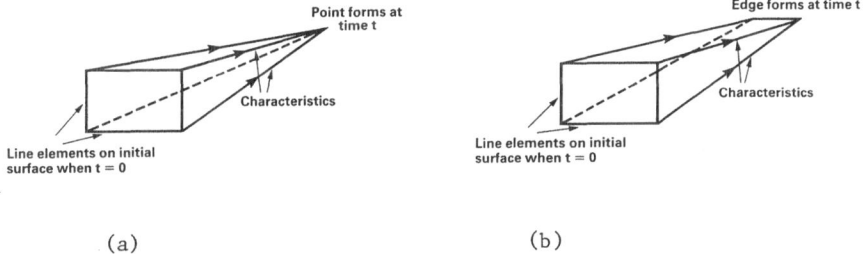

(a)   (b)

FIGURE 2. A schematic diagram, illustrating the formation of a surface gradient discontinuity after erosion. a) a point forms, b) an edge forms.

Equation 2.3 gives the time at which a surface element at at given point P, collapses to an element of zero area. The edge on the eroded surface first forms at the minimum value of t, as a function of the initial point P. However, as erosion continues after the initial formation of the edge, adjacent surface elements collapse in and so the edge trajectory is given by the locus of the intersection of characteristics whose spatia coordinates (x, y, z) are achieved at the same time.

This is perhaps more easily seen in two dimensions, when the characteristic velocity (vx,vy) is a function only of θ. In this case two characteristics emanating from the points. (x_1,y_1) whose angle of incidence is θ_1. and (x_2,y_2), θ_2 intersect at a time t at a point (x,y) where

$$x = x_1(\theta_1) + vx(\theta_1)t = x_2(\theta_2) + vx(\theta_2)t$$

$$y = y_1(\theta_1) + vy(\theta_1)t = y_2(\theta_2) + vy(\theta_2)t \qquad (2.7)$$

i.e. $-t = \dfrac{x_1(\theta_1) - x_2(\theta_2)}{vx(\theta_1) - vx(\theta_2)} = \dfrac{y_1(\theta_1) - y_2(\theta_2)}{vx(\theta_1) - vx(\theta_2)} \qquad (2.8)$

For a given θ_1, equation (2.8) can be solved to give θ_2, the other angle bounding the edge. The edge coordinates (x,y) are then given by direct solution of equations (2.7) and the edge, has a discontinuity in the value of θ given by $(\theta_2-\theta_1)$

A very detailed discussion of edge motion in two dimensions is given by Carter and Nobes[11].

2.2 **Erosion of crystalline materials**.

Because the sputtering yield $Y(\theta,\varphi)$ is a difficult quantity to measure or calculate for all θ and φ, only qualitative conclusions can be drawn from an analysis of crystalline materials.

For a crystalline material of atomic density N with sputtering yield Y subjected to an ion bombardment of uniform incidence flux Φ (ions cm^{-2}s^{-1}) the function g defined by Carter's[1] equation (17) becomes

$$g = \frac{\Phi}{N} Y .$$

It has been found to be more convenient to use angles θ and ψ defined in figure 3 rather than θ and φ used by Carter.

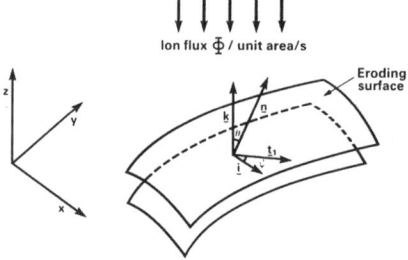

Ion flux Φ / unit area/s

Eroding surface

FIGURE 3. A schematic diagram illustrating surface erosion in three dimensions.

In this case for a time and spatially independent flux the characteristics have equations (velocities)

$$\frac{dx}{dt} = -\frac{\Phi}{N} \left\{ \cos^2\theta \sin\psi Y_\theta + \cos\psi \cot\theta Y_\psi \right\} \qquad (2.9a)$$

$$\frac{dy}{dt} = +\frac{\Phi}{N} \left\{ \cos^2\theta \cos\psi Y_\theta - \sin\psi \cot\theta Y_\psi \right\} \qquad (2.9b)$$

$$\frac{dz}{dt} = \frac{\Phi}{N} \left\{ \cos\theta\sin\theta Y_\theta - Y \right\}$$

(2.9c)

The characteristics are lines of constant θ and ψ and are straight lines. In order to compute an eroded surface after time t, it is necessary to start from a set of initial surface points (x_i, y_i, z_i) and produce an envelope of points (x_i', y_i', z_i') corresponding to the subsequently eroded surface at time t. The relationship between these points is

$$x_i' = x_i + vx.t \; ; \qquad y_i' = y_i + vy.t \; ; \qquad z_i' = z_i + vz.t$$

where the characteristic velocity $\underline{v} = (vx, vy, vz)$ and all points after the intersection of the characteristics have been deleted. Figure 4 represents the variation of sputtering yield with angle of incidence for two different materials. The dependence for an amorphous material generally rises to a maximum at about 60° incidence before falling to zero at 90°. Because of crystal lattice effects the sputtering yield for a crystalline material has generally many more maxima and minima as a function of incidence angle.

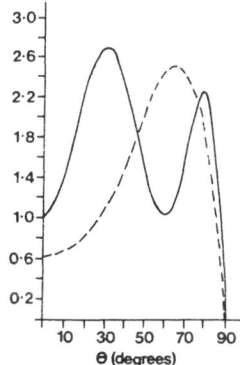

FIGURE 4. The variation of sputtering yield with ion incidence angle. Dashed line: amorphous material; solid line: crystalline material.

To illustrate the difference in behaviour between the different types of material, the erosion of a two dimensional cylinder has been computed. Calculations for the erosion of such contours have been carried out by other authors[12-14]. The two sets of results are shown in figure (5). For the amorphous materials, an edge first forms near to the top part of the cylinder and this converges to a point at a later time. A similar phenomenon occurs for the crystalline material but in addition an edge also forms further down the side of the cylinder. Note that near the maximum values of $Y(\theta)$, the characteristics diverge and since these are lines of constant θ, a much wider area of the sputtered surface now subtends angles near to θmax; i.e. a facet has formed corresponding to the direction θmax.

Near to a minimum of θ, the characteristics converge and an edge forms. If the surface had been concave towards to the beam, rather than convex, it is easily seen geometrically that the reverse situation would have occurred, with edges forming in the direction of θmax. The computer simulations show this directly but important qualitative information can be obtained from analysis.

128

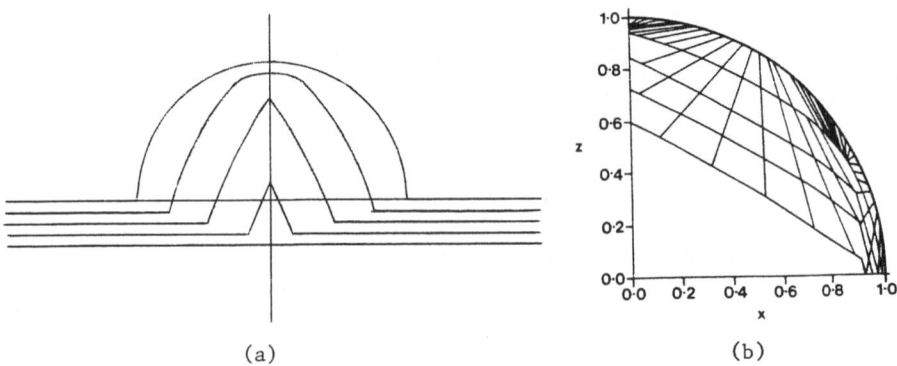

(a)   (b)

FIGURE 5. The erosion of a circular contour in two dimensions. a) Amorphous material, b) crystalline material. The straight lines are the characteristic lines.

If the characteristic velocity is recast in terms of components parallel to unit vectors $(\hat{n}, \underline{t}_1, \underline{k})$ see figure 3, then $\underline{v}$ becomes

$$\underline{v} = \frac{\Phi}{N}\left\{ -Y_\theta \cos^2\theta \hat{\underline{n}} - Y_\psi \cot\theta \underline{t}_1 + (\sin\theta\cos\theta Y_\theta - Y)\underline{k} \right\} \qquad (2.10)$$

The simulations in figure 5 are for the case $Y_\psi = 0$ in which case from equations (3.9) the characteristics move parallel to the z direction when $Y_\theta = 0$, and diverge near $Y_\theta = 0$ provided Y is a maximum with respect to θ.

The characteristics converge near $\theta = 0$, where Y is a minimum. If α is defined by

$$\tan \alpha = \frac{Y_\theta \cos^2\theta}{Y - Y_\theta\sin\theta\cos\theta} . \qquad (2.11)$$

then the characteristics begin to diverge as θ increases from 0, and an edge forms when α is stationary with respect to θ. i.e. the value of θ corresponding to

$$Y_{\theta\theta}\cos\theta - 2\sin\theta Y_\theta = 0 \qquad (2.12)$$

To illustrate the effect of non zero Y_ψ consider a cylindrically symmetric initial surface. The intersection of a horizontal plane with the surface is a circle around which θ is constant. The characteristics emanating from such a contour now subtend on angle γ with the inward $\hat{n}$ direction where

$$\tan\gamma = \frac{Y_\psi}{Y_\theta\sin\theta\cos\theta} \qquad (2.13)$$

This is shown in figure 6. From equation (2.10) it also follows that the characteristic projections move inward if $Y_\theta > 0$ and outward if $Y_\theta < 0$.

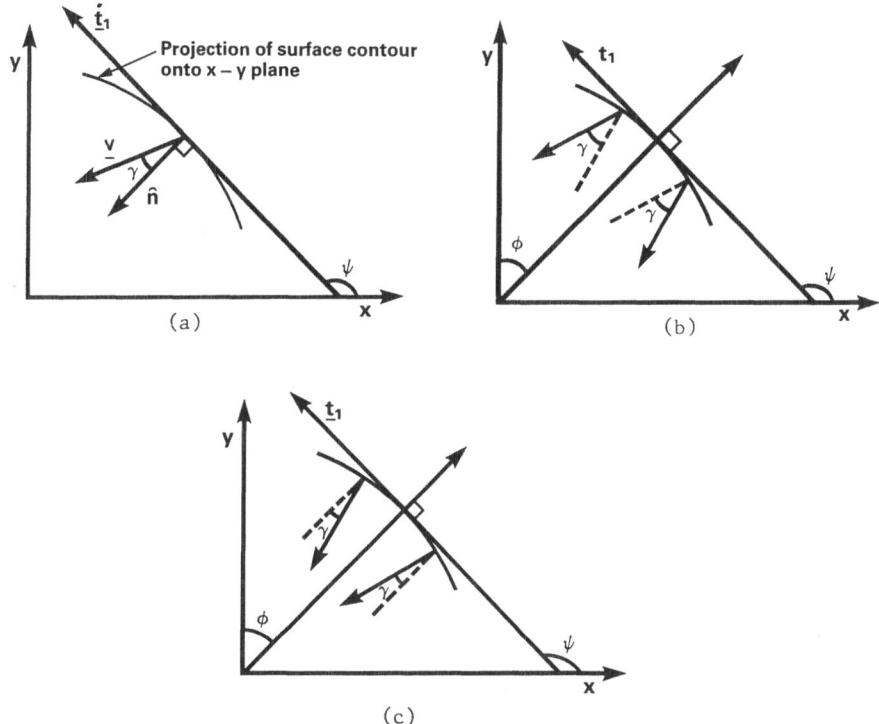

FIGURE 6. A schematic diagram illustrating the characteristic projections for a cylindrically symmetric convex surface. a) Illustrating the relationship between $\hat{n}$, a unit vector in the direction of projected surface normal $\underline{v}$, the projected characteristic vector and $\underline{t}_1$ the unit tangent parallel to the x-y plane. Diagrams b) and c) illustrate the characteristic directions near $Y_\psi = 0$. b) $Y_\theta > 0$. c) $Y_\theta < 0$.

If it is assumed that the orientation of the crystal is such that the z-axis is an axis of n fold symmetry then Y must be invariant for the corresponding symmetry operation

$$Y(\psi+\gamma_r) = Y(\psi) \qquad \gamma_r = \frac{2\pi r}{n} \quad r = 1,\ldots,n \ . \qquad (2.14)$$

For a simple cubic crystal with one of its main axes in the z direction, n = 4.

The edge or facet formation can be demonstrated by considering an initially convex surface with axial symmetry near $Y_\psi = 0$; with $Y_\theta > 0$. The directions of the characteristic projections are shown in figures 6b and 6c. In case (b) Y has a minimum with respect to ψ and in case (c) a maximum. A facet eventually forms in case (b) and an edge in case (c). The situation is reversed if the surface shape is concave. This dual behaviour has also been observed in the etch pits. The facets of the side wall of the etch pit (concave) subtend a different angle to the ion beam compared to the facets at the crater bottom (convex)[15].

2.3 Three Dimensional Amorphous Surfaces.

In this section the characteristic equations for an amorphous or polycrystalline material will be solved for a number of important cases. Suppose the direction of the ion beam is specified by the unit vector $\underline{\ell} = \underline{\ell}(t)$, where t is time. The rate of erosion of the target in the direction of the surface normal $\underline{n}$ is

$$\frac{\Phi}{N} Y(\theta) \underline{\ell}.\underline{n}$$

If the equation of the surface at time t (as given by equation (4) of reference 1) is

$$\sigma(\underline{r}) - t = 0 \qquad (2.15)$$

The the Hamiltonian for the system is

$$H = 1 - \frac{\Phi}{N} Y(\theta) \cos\theta (\sigma_x^2 + \sigma_y^2 + \sigma_y^2)^{1/2} = 0 \qquad (2.16)$$

and

$$\cos\theta = - \frac{(\ell_1 \sigma_x + \ell_2 \sigma_y + \ell_3 \sigma_z)}{(\sigma_x^2 + \sigma_y^2 + \sigma_z^2)^{1/2}} \qquad (2.17)$$

where $\underline{\ell} = (\ell_1, \ell_2, \ell_3)$. And so

$$H = 1 + \frac{\Phi}{N} Y(\theta) \left\{ \ell_1 \sigma_x + \ell_2 \sigma_y + \ell_3 \sigma_z \right\} = 0 \qquad (2.18)$$

The characteristic equations then become

$$\frac{dx}{dt} = \frac{\Phi}{N} \left[A\left[\ell_1 + B\sigma_x\right] - Y(\theta)\ell_1 \right] \qquad (2.19a)$$

$$\frac{dy}{dt} = \frac{\Phi}{N} \left[A\left[\ell_2 + B\sigma_y\right] - Y(\theta)\ell_2 \right] \qquad (2.19b)$$

$$\frac{dz}{dt} = \frac{\Phi}{N} \left[A\left[\ell_3 + B\sigma_z\right] - Y(\theta)\ell_3 \right] \qquad (2.19c)$$

If the ion flux Φ depends only on the spatial co-ordinates x and y then the invariant equations holding along the characteristics are

$$\frac{d\sigma_x}{dt} = C\sigma_x + \frac{\Phi_x}{\Phi} \qquad (2.20a)$$

$$\frac{d\sigma_y}{dt} = C\sigma_y + \frac{\Phi_y}{\Phi} \qquad (2.20b)$$

$$\frac{d\sigma_z}{dt} = C\sigma_z \qquad (2.20c)$$

where $\qquad A = Y'(\theta)\cot\theta \qquad ; \qquad B = \dfrac{\cos\theta}{(\sigma_x^2 + \sigma_y^2 + \sigma_z^2)^{1/2}}$

$$C = \left[\dot{\ell}_1 \sigma_x + \dot{\ell}_2 \sigma_y + \dot{\ell}_3 \sigma_z \right] \left[Y(\theta) - A \right]$$

$$Y'(\theta) = \frac{dY}{d\theta}, \qquad \dot{\ell}_i = \frac{d\ell_i}{dt} \qquad i = 1,2,3 .$$

The essential difference between the above equations and those used to simulate static, uniform beams is the apppearance of the terms on the right hand side of equations (2.20). This means that σ_x, σ_y, σ_z are not constant along the characteristics and the solution to the problem is determined by solving the six simultaneous ordinary differential equations (2.19) and (2.20).

2.4 Spatially uniform flux, time independent beam.

In this section, surface erosion due to the application of a uniform ion flux is considered. These conditions prevail in industrial fabrication processes where wide ion beams or plasma sources are employed. In this simplified case, the right hand sides of equations (2.20) are all zero and the characteristics defined by equations (2.19) are straight lines. This allows a direct calculation of the eroding surface without recourse to integrating a set of ordinary differential equations. The results of a typical sequence of simulations are shown in figure 7. In this example an elliptical hummock is used as a three dimensional illustration of the technique. The hummock is described by the equation

$$z = \exp - (x^2 + 0.4y^2)$$

The particular form of the $Y(\theta)$ curve used, corresponds to argon bombardment of silicon. It can be seen from figure 7 that the overall shape of the hummock changes with increasing ion dose.

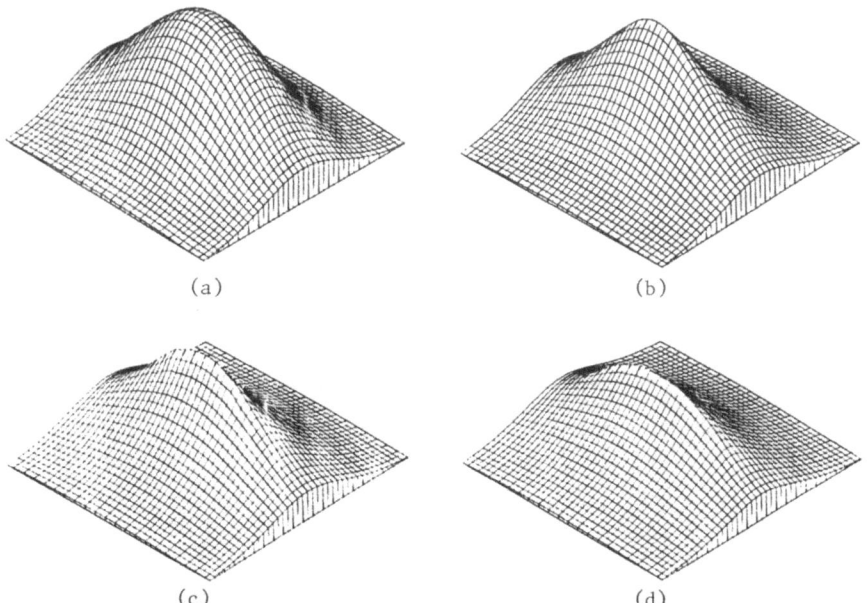

(a)

(b)

(c)

(d)

FIGURE 7. The erosion of the surfaces $z = \exp(-x^2 - 0.4y^2)$ after equal doses of ions

Note that starting from a non-axially symmetric protrusion first produces a finite edge at the top of the hummock. As erosion continues, the sides of the structure sputter at a faster rate than the plane beneath. The ultimate equilibrium shape is a flat plane.

2.5 <u>Time Dependent Beam Orientation - Sample Rocking and Rotation</u>.

Sample rocking and rotation is a technique that is used to minimise the growth of unwanted topography when using ion beams to mill, or thin samples. One particular application is the use of ion beams to thin quartz crystal resonators where it is imperative to obtain as smooth and flat a final surface as possible. Sample rocking and rotation can smooth out "initial surface defects" which would, with a static beam, produce cone and pyramids.

The computational technique involves setting up a digital representation of the surface (x_i, y_i, z_i) and calculating the unit normal $\underline{n}_i$ at these points. The components $(\sigma_x, \sigma_y, \sigma_z)$ are initialized by noting that at times $t = 0$ this is in the same direction as $\underline{n}_i$ and then using equation (2.16). Originally, the digital representation of surface was carried out using 7000 data points, which was found to be a suitable number which would allow reasonable graphical simulations of the eroded surfaces. Such a number requires the solution of 7000 sets of the six ordinary differential equations with the consequence that the standard computer packages could not be used without incurring a large CPU time. Subsequently a Fortran code based on a Runge Kutta Merson fifth order method was written specifically to solve these equations. The codes uses the specialized nature of the equations and also their symmetry in order to minimise the function evaluations. The advantages in using the Merson method is that error estimates are available. These were checked to ensure errors were within a suitable tolerance. A typical run for 7000 points is about 40 minutes on the Honeywell Multics system, although this time does depend to a large extent on the rocking and rotation rates and also the timestep. Two dimensional simulations were also calculated, and essentially the method of solution is identical with the one just described, the system now reducing to a set of four differential equations by omitting one spatial co-ordinate.

In both the two and three dimensional simulations it is possible to obtain physically unrealistic solutions caused by the characteristic lines crossing. The physical explanation of this is that the surface has formed an edge (discontinuity in gradient) and the characteristic method produces a multivariate solution and cannot be applied after such nonlinear wave interactions. It is essential to remove these spurious points since they are unrealistic, and graphical techniques are employed to achieve this. Finally Gino F, Ginosurf and Nag plotting routines are used to produce the final graphical simulations.

2.6 <u>Three dimensional simulations - beam or sample rotation</u>

The same initial surface as described earlier has been chosen to illustrate the effect of rotation. To illustrate the effect of rotating the beam, the direction cosines were chosen to vary according to

$$\ell_1(t) = \sin\alpha\sin(2\pi\beta t)$$

$$\ell_2(t) = \sin\alpha\cos(2\pi\beta t) \qquad\qquad (2.21)$$

$$\ell_3(t) = \cos\alpha .$$

With these parameters, the ion beam describes cones of fixed half angle α

around the z axis with one rotation per $1/\beta$ secs. The simulations shown in figure 8 were all carried out using equal timesteps. For the case $\alpha = 0$, $\beta = 0$, and a normal incidence beam, an axially symmetric structure stays symmetric. Consider now the hummock defined by the equation

$$z = e^{-(x^2+y^2)} .$$

The development of a sharp cone from an initially smooth surface is clearly shown. Figures (a–d) are simulations with $\alpha = \pi/9$, $\beta = 1$. The main feature to observe is that the conical feature which develops, tends to follow the beam causing bent cones. For a fast rotation rate the development of sharp cones is suppressed and the eroded profiles seem to retain similar shapes to the original surface. A fuller description of this is given elsewhere[16].

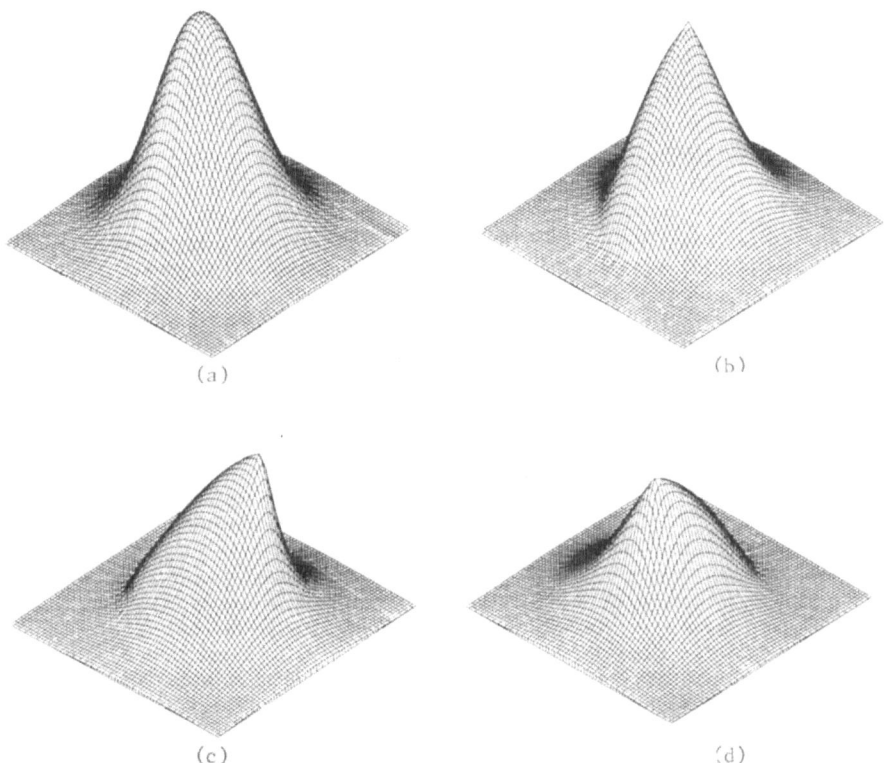

(a)   (b)

(c)   (d)

FIGURE 8. The erosion of the surface $z = \exp(-(x^2+y^2))$ after successively equal doses: illustrating the effect of beam rotation for a non normally incident beam at an angle of $\pi/9$ to the z-axis. The rotation frequency is $1/(2\pi)$ secs and the z scale is magnified to highlight the effects.

2.7 <u>Non Uniform Ion Flux</u>

In practice, ion etching is often carried out using an ion gun which produces a non uniform ion current, distribution in space. This is generally the case in surface analysis where to a first approximation the beam profile can be considered to be Gaussian.

To illustrate the effect of such a non uniform flux, two dimensional examples are chosen for simplicity of calculation. In two dimensions the characteristic lines become

$$\frac{dx}{dt} = \frac{\Phi(x)}{N} \cos^2\theta Y_\theta(\theta) \qquad\qquad (2.22a)$$

$$\frac{dz}{dt} = \frac{\Phi(x)}{N} (\sin\theta\cos\theta Y_\theta(\theta) - Y(\theta)) \qquad\qquad (2.22b)$$

and the invariant equations holding along the characteristics simplify to

$$\frac{d\theta}{dt} = -\frac{\Phi_x(x)Y(\theta)}{N} \cos^2\theta . \qquad\qquad (2.22d)$$

Although the invariant equation integrates to give

$$\frac{\Phi(x)Y(\theta)}{N} = \text{constant}$$

along the characteristics it is in practice easier to integrate equations (2.22) together rather than use this condition, in the calculations.

The two dimensional simulations illustrate the erosion of part of the circle $x^2 + z^2 = 1$; $x, z > 0$. The results shown in figure (9) are for a material with the same sputtering yield as that chosen to illustrate the three dimensional theory. The figures compare then effects of varying the effective width of the Gaussian distribution,

$$\Phi(x) = \Phi_0 e^{-x^2/\rho^2} .$$

Figure 5 illustrates how a circle is eroded by a uniform beam into a sharp wedge-like structure. This and subsequent diagrams show the surface contour at various stages of erosion. In figure 5a the characteristics are straight lines. For wide Gaussian beams, figures 9a, 9b, the wedge structure is still present but the wedge angle has increased. As the beam narrows, the edge takes longer to form and eventually a sharp rimmed crater develops, figure 9c, due to the larger ion flux despite a higher rate of erosion elsewhere due to angle of incidence effects. The characteristics for these narrow beams, have large curvature.

2.8 Two Ion Beams.

Rocking and rotation of the sample can be an effective way of reducing the growth of unwanted surface topography during ion beam etching but this may not be a practical possibility in many systems. In surface analysis there is some evidence to suggest that[17] the growth of unwanted topography, which can effect the accuracy of composition/depth profiles, can be minimised by the use of two static ion beams. The calculations for the erosion of a surface by two ion beams in two dimensions can either be deduced directly from characteristic calculations or by a line element method which will be described in more detail in section 2.9.

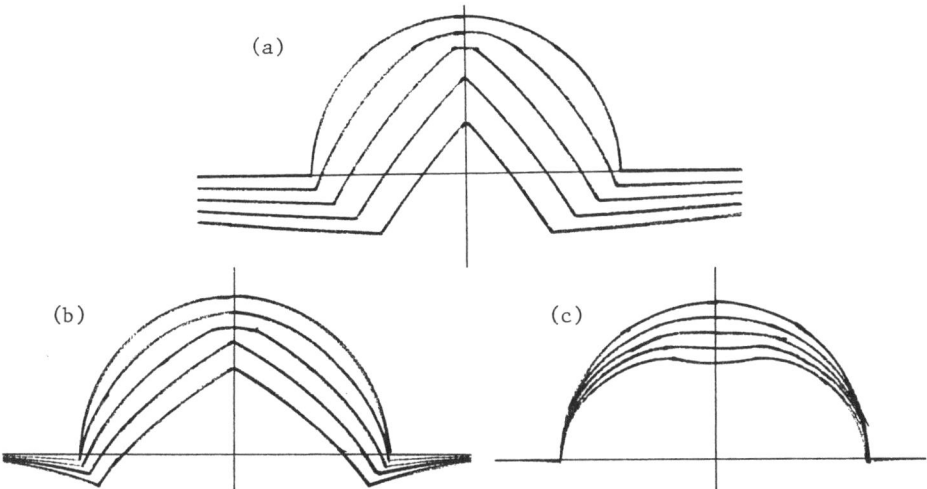

FIGURE 9. Successive erosion contours of a circular hummock on a flat plane subjected to erosion by an ion beam with a Gaussian current distribution $\Phi = \Phi_0 \exp(-x^2/\rho^2)$. a) $\rho = 10$, b) $\rho = 1$, c) $\rho = 1/\sqrt{3}$.

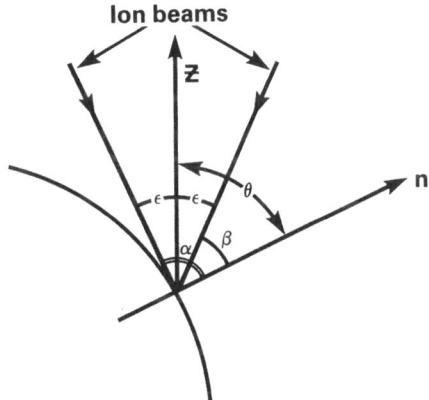

FIGURE 10. A schematic diagram illustrating the angles involved when a two dimensional surface is subject to erosion by two ion beams.

The geometry of surface erosion by two ion beams is shown in figure 10. If the fluxes of the two beams are Φ_1 and Φ_2 and sputtering yields Y_1 and Y_2, then with the angles β, ϵ, θ defined as in figure 10, the equations for the characteristics become[18]

$$\frac{dx}{dt} = -\frac{1}{N}\left[\cos\theta\left\{\Phi_1 Y_1' \cos\alpha + \Phi_2 Y_2' \cos\beta\right\} + \left\{\Phi_2 Y_2 - \Phi_1 Y_1\right\}\sin\epsilon\right]$$

$$\frac{dz}{dt} = \frac{1}{N}\left[\sin\theta\left\{\Phi_1 Y_1' \cos\alpha + \Phi_2 Y_2' \cos\beta\right\} - \left\{\Phi_1 Y_1 + \Phi_2 Y_2\right\}\cos\epsilon\right] \tag{2.23}$$

136

where primes denote differentiation with respect to θ.

The characteristics are again straight lines if the flux is uniform and the beam time independent. Some simulations for the erosion of a cylinder are shown in figure 11. For the case of two beam erosion (or even erosion with one beam at non-normal incidence) surface protrusions can shield other parts of the surface from the beams. In the case of the erosion of a circle, some parts of the circular contour are exposed to both beams but others are eroded only by one beam. This means that the side walls of the circle are eroded less than its top. The figures illustrate the flatter topped structures that result with two beam erosion, when the beams are symmetrically placed about the planar surface normal.

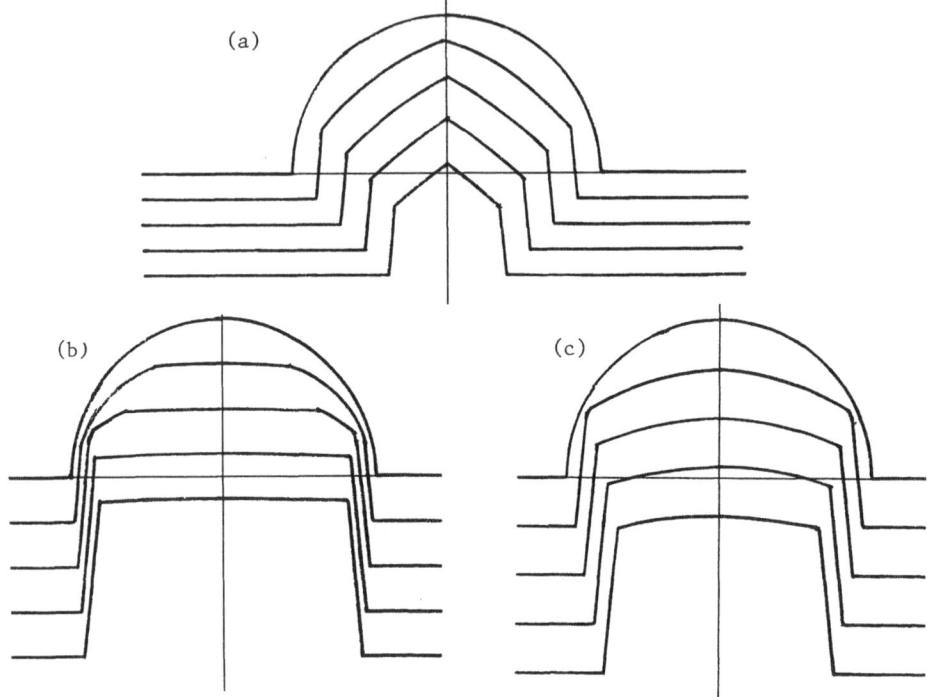

FIGURE 11. The erosion of a circular hummock on a flat plane by two ion beams of the same uniform flux distribution. a) $\epsilon = 30°$ b) $\epsilon = 50°$ c) $\epsilon = 70°$.

2.9 Two Dimensional Segment Motion Model.

Although the characteristic method of calculation is extremely useful, particularly in obtaining analytical and descriptive information regarding the formation of points, edges and facets, it does have the drawback of being computationally unwieldly after the formation of surface shape discontinuities This has resulted in a new method which is computationally easier to programme, based on the advance of line segments in two dimensions.[19, 20]

This algorithm is particularly useful when a smooth contour is to be eroded and when parts of that contour collapse to edges as the erosion proceeds. It requires modification in the case of a surface with initial angular points which open out to form a part of a smooth curve.

In the simulation the initial curvilinear profile at time t_o is approximated by a number of line segments. The new profile after a time $t_o + \Delta t$ is obtained by tracking the points of intersection of the line in the direction of the normal to these line segments. The points of intersection of the new line segments then define the surface at the later time. The technique is illustrated in figure 12a.

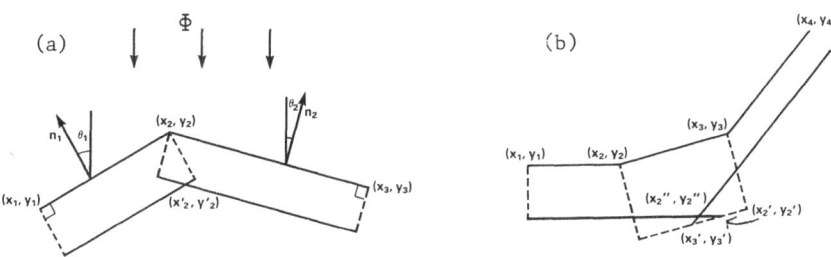

FIGURE 12. a) A schematic diagram illustrating the erosion of two line segments $(x_1, y_1), (x_2, y_2)$ and $(x_2, y_2), (x_3, y_3)$. The point (x_2, y_2) defines the intersection point of the new line segments representing the eroded contour at time $t_o + \Delta t$. b) Illustrating the collapse of a line segment initially between (x_2, y_2) and (x_3, y_3) into an edge, located at time $t_o + \Delta t$ at the point (x_2, y_2).

From the definition of the quantities in figure (12), the points (x_1', y_1') (x_3', y_3') on the new profile are given by

$$x_1' = x_1 + \frac{\Phi}{N} Y(\theta_1) \sin\theta_1 \ \Delta t$$

$$y_1' = y_1 - \frac{\Phi}{N} Y(\theta_1) \cos\theta_1 \ \Delta t$$

(2.24)

$$x_3' = x_3 + \frac{\Phi}{N} Y(\theta_2) \sin\theta_2 \ \Delta t$$

$$y_3' = y_3 - \frac{\Phi}{N} Y(\theta_2) \cos\theta_2 \ \Delta t$$

(2.25)

The point (x_2', y_2') is then obtained by finding the intersection point of the two planar fronts passing through (x_1, y_1) and (x_3, y_3) and for a uniform ion flux is given by

$$x_2' = \frac{y_1' - y_3' - \tan\theta_1 x_1' + \tan\theta_3 x_3'}{\tan\theta_3 - \tan\theta_1}$$

(2.26a)

$$y_2' = y_1' + \tan\theta_1 (x_2' - x_1')$$

(2.26b)

In the case when $\theta_1 = \theta_3$, the intersection point is

$$x_2' = x_2 + \frac{\Phi}{N} Y(\theta_2) \sin\theta_2 \Delta t \qquad (2.27a)$$

$$y_2' = y_2 - \frac{\Phi}{N} Y(\theta_2) \cos\theta_2 \Delta t \qquad (2.27b)$$

The case of a beam with a non uniform flux can be easily incorporated and the algorithm also allows easily for oblique ion incidence, the use of more than one ion gun and beam shadowing by higher surface features. One particular advantage of the line segment motion algorithm compared to the characteristic method is the relative ease with which the motion of corners and edges can be tracked. This is accomplished by a piece of computer code which eliminates line segments which shrink to zero length. This means that adjacent line segments on the eroded profile were originally seperated by intermediate segments on the initial profile. Thus on the eroded profile a surface gradient discontinuity forms. The computational procedure for eliminating these segments is shown in figure (12b). Here the line joining $(x_2.y_2)$ and $(x_2.y_2)$ and the line joining $(x_3.y_3)$ and $(x_3.y_3)$ have crossed. When this occurs the line segment (2) is deleted and the new point $(x_2.y_2)$ defines the point of intersection of the line segments (1) and (3) on the eroded surface.

A second advantage of the method is the ability to incorporate secondary effects such as redeposition and ion reflection, and the ease with which additional points on a contour can be added when line elements have enlarged. For primary beam calculations this does not matter since facet formation is indicated but for secondary effects it is important to represent the surface contour by an accurate representation of small line elements. In most of these calculations a check in the programme maintains approximately the same number of elements as the erosion proceeds with time.

3. REDEPOSITION

In the sputtering process a secondary mechanism which can occur is the redeposition of sputtered material. This can be important along the edges of steep surface features and can be observed along the edges of microcircuit photoresist patterns where ridges of material remain when the resist pattern is removed (figure 13).

This redeposition, due to emitted secondary particles of very low energy, is deleterious to the performance of devices produced by ion etching. In order to model this phenomena, it is first necessary to calculate the secondary flux at a point where redeposition is occuring.

Particles can arrive and redeposit which originate at many points on the bombardment surface. It is first necessary therefore to evaluate flux integrals in order to sum this net contribution. A number of authors[21,22] have evaluated these flux integrals for a number of different surface shapes and spatial emitted particle distributions. It is then necessary to incorporate these flux integrals into the continuum equation which describes the build-up of redeposited material. Only in simple cases is it possible to evaluate these integrals analytically and in general a complicated integro-differential equation results. However, to illustrate the method a particularly important two-dimensional case will be illustrated. This is the situation which models the growth of material along the sides of groove patterns.

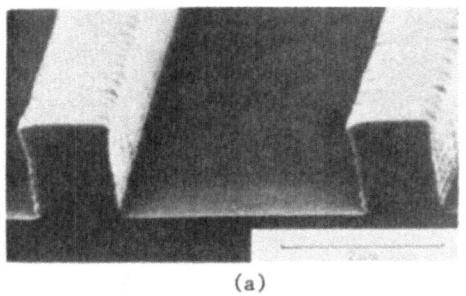

(a)

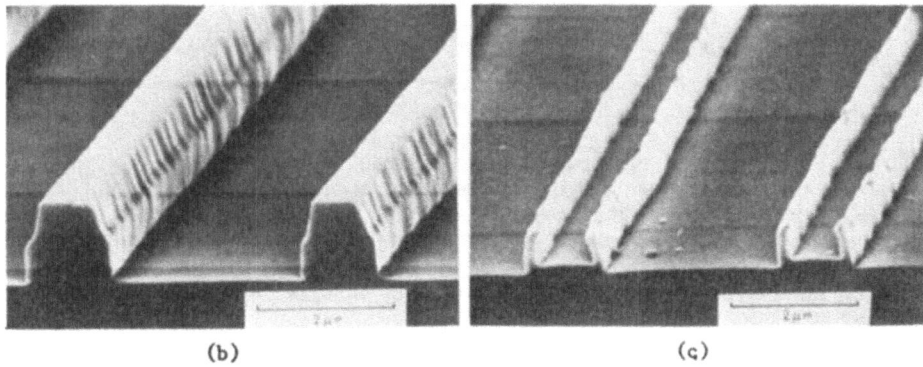

(b)   (c)

FIGURE 13.   Scanning electron micrographs illustrating phenomena of faceting and redeposition during ion etching.   AZ1350 photoresist mask on silicon substrate.   a) before ion beam etching,   b) after ion beam etching to a depth of 1300Å, and   c) the redeposited material left after dissolution of the photoresist in a solvent.   (Courtesy of H.I. Smith).

In order to simplify the algebra it will be assumed that the flux φ_1 of secondary particle emission from the surface at a point, is proportional to the angle between the direction of emission and the surface normal Ω, i.e.

$$\varphi_1 = \varphi_0 \cos\Omega \qquad (3.1)$$

where φ_0 is the peak emitted flux density.   This is the so-called cosine distribution and is an adequate model for many applications although the precise distribution depends both on θ and the ion energy (see figure 14). Consider the two surfaces in two-dimensions as shown in figure 15.   The surface OX" is being bombarded by a uniform beam and the point A on the surface OX', is receiving a flux from the surface elements at B.   The flux density of redeposited material at A from the surface element at B is

$$\delta\varphi_2 = \frac{\Phi Y \cos\theta \, dy \, |ds_1|}{\pi N |AB|^4} \, (\underline{AB} \cdot \underline{n}_1)(\underline{AB} \cdot \underline{n}_2). \qquad (3.2)$$

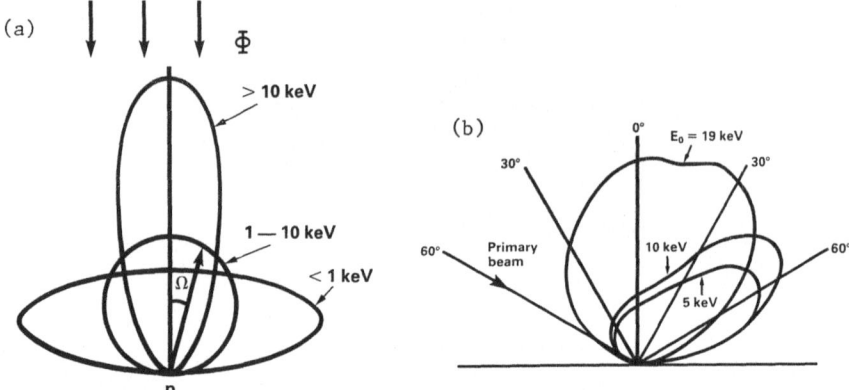

FIGURE 14. Angular distribution of sputtered material. a) The general shape of the distribution curves for material sputtered from a point P by normally incident ions at different energies. The amount of material sputtered in a given direction is proportional to the length of the vector from P to the distribution curve. b) Angular distribution of sputtered material for polycrystalline W bombarded with oblique incidence Kr^+ ions of different energies.

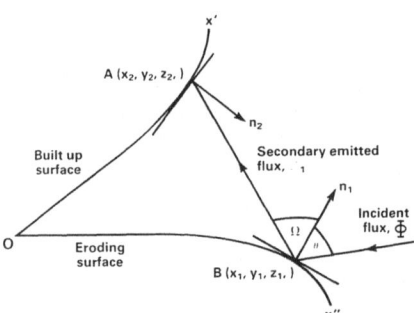

FIGURE 15. A schematic diagram illustrating the redeposition of sputtered material from a point B to a point A.

where ds_1 is the arc element of the surface at B and all the other quantities are defined in figure 15. The total flux density at A from a sputtered two dimensional surface between the lines $s_1 = a$ and $s_1 = b$ is

$$\varphi_2 = \frac{\Phi}{\pi N} \int_a^b \int_{-\infty}^{\infty} \frac{Y\cos\theta(\underline{AB} \cdot \underline{n_1})(\underline{AB} \cdot \underline{n_2})}{|\underline{AB}|^4} \, dy_1 ds_1 \qquad (3.3)$$

The analysis simplifies for the practical case when the surface OX" is a horizontal plane. This situation covers the important class of problems in microelectronics where the redepositing material originates from the bottom of grooves and gives an equation for the build-up of material at A in the normal direction r_n as

$$\frac{\partial r_n}{\partial t} = \frac{\mu \Phi Y}{2N} \left\{ 1 + \left[\frac{dz_2}{dx} \right]^2 \right\}^{1/2} \left[\frac{dz_2}{dx_2} \left\{ \cos\beta_2 - \cos\beta_1 \right\} + \sin\beta_1 - \sin\beta_2 \right] \quad (3.4)$$

where μ is the sticking coefficient and

$$\beta_1 = \arctan \frac{x_0 - x_2}{z_1 - z_2}$$

$$\beta_2 = \arctan \frac{d - x_0 + x_2}{z_1 - z_2} \ .$$

Here d represents the initial groove width and x_0 is the lower limit for the s_1 integration in equation (3.3).

If the erosion between the patterned structures is not that of a flat plane, then the integral (3.3) cannot be evaluated analytically and a complex integro-differential equation results. Otherwise equation (3.4) is solved by the method of characteristics and numerical integration. The characteristic equations for equation (3.4) become

$$\frac{dx_2}{dt} = \frac{\mu \Phi Y \cos\theta}{2N} \left\{ \cos\beta_2 - \cos\beta_1 \right\} \quad (3.5a)$$

$$\frac{dz_2}{dt} = \frac{\mu \Phi Y \cos\theta}{2N} \left\{ \sin\beta_2 - \sin\beta_1 \right\} \quad (3.5b)$$

Some computed results for different groove widths are presented in figure 16

The theory discussed here does not take into account simultaneous erosion and redeposition. However, the calculations show that when redeposition occurs, the groove walls remain close to perpendicular. The important parameters which govern this redeposition are the groove widths and the depth of ion etched material. The angles of the side walls lie close to 90° for the larger groove widths but become smaller with diminishing groove width. It is only for such steep walls that redeposition dominates the erosion process, since the sputtering yield is small near grazing incidence. However, approximate calculations show that for a groove width of $0.5\mu m$ in Si, with side walls at an angle of 82°, then erosion dominates the redeposition process under 1keV Ar^+ ion bombardment. It is likely that a thin crust of redeposition material builds up at such an angle

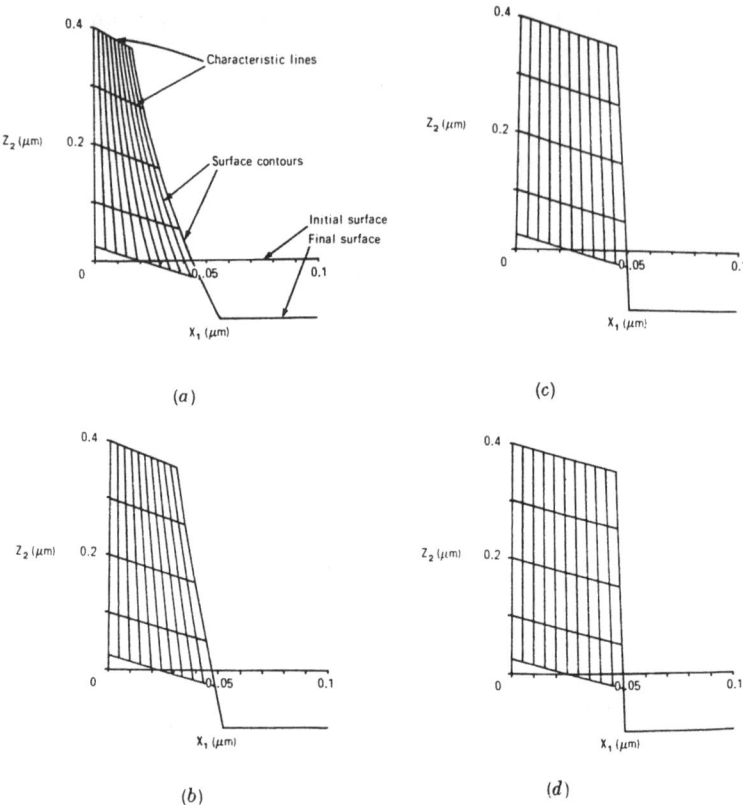

FIGURE 16. Theoretically calculated redeposition profiles for rectangular grooves of different widths and for different depths of erosion using the cosine distribution for the sputtered particles. The depth of erosion corresponding to each surface contour is 10nm and the groove height is $0.4\mu m$. Groove widths are: a) $0.5\mu m$, b) $1.0\mu m$, c) $3.0\mu m$ and d) $6.0\mu m$.

that the redeposition and erosion balance. It would not be possible in practise for this effect of redeposition to continue without check. Since if it did the surface would subtend angles to the primary beam which would imply a high sputtering yield. Near the tops of the grooves a facet forms at the 90° angle of the side wall with the top of the pattern. If this facet corresponds to angles close to the maximum sputtering yield angle and if the erosion effect continues to dominate a triangular wedge structure can develop from the initially rectangular profile after erosion for a sufficient length of time.

4. Computational Model for the Study of Ion Reflection

Ions reflected from the sides of steeply inclined structures generate an enhanced erosion flux usually directed towards underlying sections of the profile. This flux creates an additional erosion to be imparted to various sections of the profile, and thus alters the subsequent contour development. The model described here assumes that ions are reflected specularly for angled structures greater than a fixed initial angle θ_m. although any distribution can in practice be modelled. A spatially distributed reflection coefficient is calculated for each time step, and this is used to update the etch rates of the appropriate points on the profile within the influence of the secondary flux. The angle θ_m is assumed to be the value of θ corresponding to the maximum sputtering yield $Y(\theta)$, since the subsequent decrease in yield as θ increases to $\pi/2$ is due, at least in part, to the increased probability of ions being reflected from the surface with little energy loss. The reflected ion flux is assumed to be a linear function of angle of incidence given by:

$$\Phi_R(\theta) = \begin{cases} 0 & 0 \leqslant \theta < \theta_m \\ \dfrac{\Phi}{N}\left[\dfrac{\theta-\theta_m}{\pi/2-\theta_m}\right] & \theta_m \leqslant \theta \leqslant \pi/2 \end{cases}$$

Work is presently underway to determine a more exact distribution by using a binary collision model to determine Φ_R.

The initial profile is divided into a number of line segments, each with a particular orientation and etch rate. For each line segment with orientation greater than the fixed critical angle θ_m, the intersection of the specularly reflected ion flux Φ_R, with the underlying profile is computed. Assuming that the ith segment with end points (x_i, y_i) and (x_{i+1}, y_{i+1}) has orientation θ_i, with $\theta_i > \theta_m$, the gradient of the reflected ion flux emanating from this segment is $2\theta_i - \pi/2$. The intersection $(\bar{x}, \bar{y})$ of this line with the jth segment is calculated from

$$\bar{x} = (y_i - y_j - \tan(2\theta_i-\pi/2)x_i + \tan(\theta_j)x_j/[\tan(\theta_j) - \tan(2\theta_i-\pi/2)]$$

Where θ_j in the orientation of the jth segment.

For this intersection point to lie on the part of the contour under investigation it must satisfy the criteria $x_i \leqslant \bar{x} \leqslant x_{i+1}$. Segments of the profile are then sampled until the appropriate intersection segment i say, is found. Having obtained this point, the magnitude of the enhanced flux Φ' can be found by

$$\Phi' = \frac{\Phi_R(\theta_i)}{N} \ Y(\pi-2\theta_i-\theta_i') \ \cos(\pi-2\theta_i-\theta_i')$$

This process is repeated for all segments with orientations greater than θ_m. A cubic polynomial is then fitted to obtain a spatially distributed reflection coefficient. The modified values of the etch rates for all segments lying inside the region affected by the enhanced ion flux are then updated by summing the additional ion flux calculated at the midpoint of the segment. In the case when two or more distinct sections of the profile generate an enhanced flux, seperate reflection coefficients are calculated for each section. The new profile at time $t_o + \Delta t$ is then obtained by the method described in section 2.9.

The new profile at $t_o + \Delta t$ is checked to ensure that the segments sizes were less than a suitable fixed length, specifiable in the programme but usually taken as the initial segment length. Any larger segments were subdivided into a suitable number of smaller segments. The reason for this was to minimise the error induced into the algorithm since the secondary flux can vary substantially over a long segment. Since the reflection coefficient is not only spatially distributed but also time dependent, small timesteps were also taken in order to reduce the errors.

The whole process is then repeated using the newly created profile as the initial profile, to obtain a set of contours showing the time evolution of the sputtered topography.

A test for the model was to investigate the behaviour of a semi-circular profile on a flat plane since this model has been used by previous authors to illustrate first order erosion theory. The sputtering yield dependence on ion incidence angle was taken as for amorphous Si under 1keV Ar$^+$ ion bombardment. Figure 5a shows the evolution of this profile using the line segment algorithm for constant timesteps taken at $\Phi t/N$ = 0.15, where ion reflection is not incorporated. The development of a pointed structure from an initially smooth profile is clearly shown together with its subsequent decay. The equilibrium situation as t → ∞ is a flat plane. The influence of ion reflection is shown in figure 17. Initially a shallow depression is formed at the sides of the cone, this deepens with time due to the secodary ion flux and also spreads due to erosion effects. A pit gradually forms as the cone eventually shrinks into the base and the ultimate shape is a pit with shallow sides. The sides expand laterally with time but the depth of the pit will remain costant, relative to the surrounding flat surface. For the semicircular structure, the walls at the point, where the protrusion joins the flat plane are at grazing incidence to a normally incident ion beam, and in this region ion reflection is of primary importance. The ultimate effect of secondary sputtering which arises from this relatively small part of the initial surface contour is to produce a shallow pit and a difference in levels between the flat plane and the crater bottom equal to 30% of the height of the original protrusion.

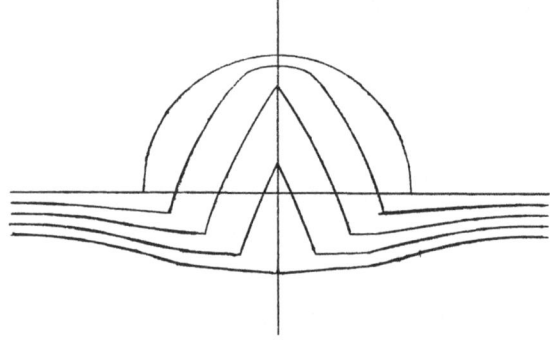

FIGURE 17. Surface contours illustrating the effect of secondary sputtering by ion reflection from the side walls of an initially circular contour (cf figure 5a).

In using ion beams for pattern delineation, steep sided rectangular profiles are often eroded. The calculations above indicate the importance that ion reflection can have in deepening the grooves seperating such steep sided surface patterns.

REFERENCES

1   Carter G: These proceedings.
2   Nobes MJ: These proceedings.
3   Andersen HH and Day HL: Chapter 4 in Sputtering by Particle Bombardment I (ed. Behrisch, R.) Springer Verlag, Berlin 1983.
4   Sigmund P: Chapter 2 in Sputtering by Particle Bombardment I (ed. Behrisch, R.) Springer Verlag, Berlin, 1983.
5   Biersak JP and Haggmark LG: Nucl. Inst. and Meth. 174, 257, 1980.
6   Oechsner H: Appl. Phys. 8, 185, 1975.
7   Onderdelinden D: Appl. Phys. Lett. 8, 189, 1966.
8   Whitham GB: Linear and Nonlinear Waves: Wiley: New York, 1974.
9   Smith R and Walls JM: Phil. Mag. A 42, 235, 1980.
10 Smith R, Tagg MA, Carter G and Nobes MJ: J. Mater. Sci. Lett. In Press, 1985.
11 Carter G and Nobes MJ: Phil. Mag. A 51, 745, 1985.
12 Barber DJ, Frank FC, Moss M, Steeds JW and Tsong ITS: J. Mater. Sci. 8 1030, 1973.
13 Ducommun JP, Cantagrel M and Moulin M: J. Mater. Sci. 10, 52, 1975.
14 Carter G, Colligon JS and Nobes MJ: J. Mater. Sci. 6, 115, 1973.
15 Smith R, Valkering TP and Walls JM: Phil. Mag. A 44, 879, 1981.
16 Smith R, Tagg, MA and Walls JM: J. Mater. Sci. In Press 1985.
17 Sykes DE, Hall DD, Thurstans RE and Walls JM: Appls. of Surf. Sci. 5, 103, 1980.
18 Makh SS, Smith R and Walls JM: J. Mater. Sci. 17, 1689, 1982.
19 Neureuther AR, Liu CY and Ling CH: J. Vac. Sci. Technol. 16, 1772, 1979.
20 Tassopoulos N, Hagouel P and Kriezis E: 'Microcircuit Engineering '83' (Eds. Ahmed H, Cleaver JRA and Jones GAC) Academic Press, New York, 421, 1983.
21 Wilson IH, Belson J and Auciello O: Chapter 6 in Ion Bombardment Modification of Surfaces (eds. Auciello, O. and Kelly, R.) Elsevier, Amsterdam, 1984.
22 Smith R, Makh SS and Walls JM: Phil. Mag. A. 47, 453, 1983.

PREFERENTIAL SPUTTERING OF TANTALUM OXIDE:
REEMISSION OF HELIUM AND TRANSIENT EFFECTS IN THE ALTERED LAYER

B. BARETZKY and E. TAGLAUER
Max-Planck-Institut für Plasmaphysik, EURATOM-Association,
D-8046 Garching/München, FRG

Samples of Ta_2O_5 were bombarded with 1.5 keV and 2.0 keV He^+ ions at different angles of incidence until equilibrium surface concentration was reached as measured by Auger Electron Spectroscopy. The reemission of the implanted helium during sputtering with 1 keV Ar^+ ions was measured by the change of the helium partial pressure. The correlation between the depth distribution of the sputtering particles and the concentration depth profile supports the assumption that the altered layer corresponds to the range profile of the bombarding ions. Changes of the angle of incidence result in characteristic transient surface concentration changes which can be explained by applying existing models for the development of the altered layer.

1. INTRODUCTION

Ion bombardment of multicomponent systems generally results in a change of the surface composition due to preferential sputtering. This effect occurs whenever the ratio of the sputtering yields of the different components differs from that of the surface composition /1/. For compounds consisting of components with large mass differences being sputtered with light ions (He^+ and H^+) at low energy (0.3 - 3 keV) depletion of the lighter constituent has been observed /2, 3/. In a previous paper it was shown that this effect also depends on the angle of incidence, more grazing incidence leading to less preferential sputtering /4/. The explanation for these experimental observations is that in this mass and energy range, i.e. near the threshold energy of the heavier component, the mass dependence of the energy transfer between projectile and different target atoms leads to a sputtering selectivity. During preferential sputtering an altered layer with a concentration profile into a certain depth is established.

In this paper we report on simultaneous helium partial pressure and concentration depth profile measurements which confirm the assumption that the depth of the altered layer is correlated with the range of the impinging ions /5/.Furthermore, changes of the angle of incidence lead to transient effects in the surface concentration ratio which can be explained by applying the model of Betz et al. /6/ for the development of the altered layer.

2. EXPERIMENTAL

Anodically oxidized Ta_2O_5 samples with an oxide layer of about 3500 Å thickness were bombarded with 1.5 keV He^+ or 2 keV He^+ ions under UHV conditions and at different angles of incidence (for details see ref. 4). For the calculation of the sputtered material and the corresponding depth we assumed a dependence of the sputtering yield on the angle of incidence ψ relative to the target surface like

$$Y(\psi) = Y(90^\circ)/\sin \psi$$

which was found for TaC by Roth et al. /7/. The sputtered depth was determined by using a sputtering yield of $Y(90^\circ) = 9.3 \times 10^{-3}$ Ta_2O_5 molecules/ion for 1.5 keV He^+ /7 / and 0.25 Ta_2O_5 molecules/ion for 1 keV Ar^+ and the density for amorphous Ta_2O_5 of 5×10^{14} Ta_2O_5 /cm^2 with a thickness of 4.5 Å /5/. The depth profiles of the altered layers were measured by using 1 keV Ar^+ at an angle of incidence of $\psi = 30^\circ$ (argon shows a much smaller preferential sputtering effect than helium /2/). The helium partial pressure was measured with a quadrupole mass spectrometer which was calibrated by the Bayard-Alpert gauge reading.

3. RESULTS

Figure 1 shows the reemission of the implanted helium during sputtering with Ar^+ ions. The surface concentration ratio Ta/O and the helium partial pressure decrease in a similar way with increasing sputtering depth. We also see that the depth of the distribution increases with increasing ion energy.

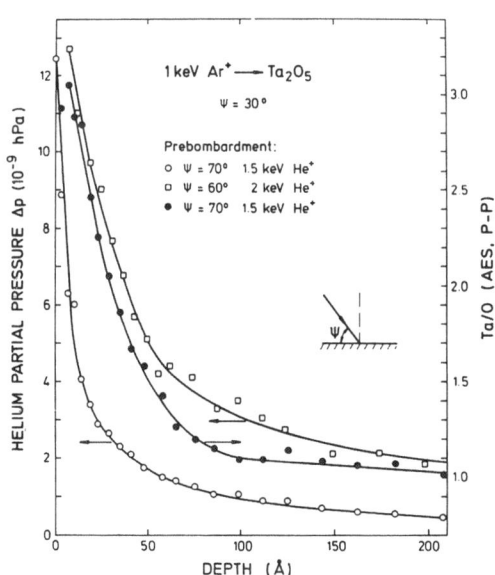

FIGURE 1. Change of the helium partial pressure and the composition depth profile for Ar^+ sputtering of Ta_2O_5 after prebombardment with 1.5 or 2.0 keV He^+ at $\psi = 70^\circ$ and $\psi = 60^\circ$.

148

By changing the angle of incidence from normal to grazing incidence, i.e.
from a higher Ta/O ratio to a lower ratio we observe a significant
overshoot, Fig. 2. The opposite case - the change to a higher angle of
incidence - is plotted in Fig. 3. The characteristic fluence F_0 to reach
equilibrium surface concentration /4/ apparently depends on the starting
conditions.

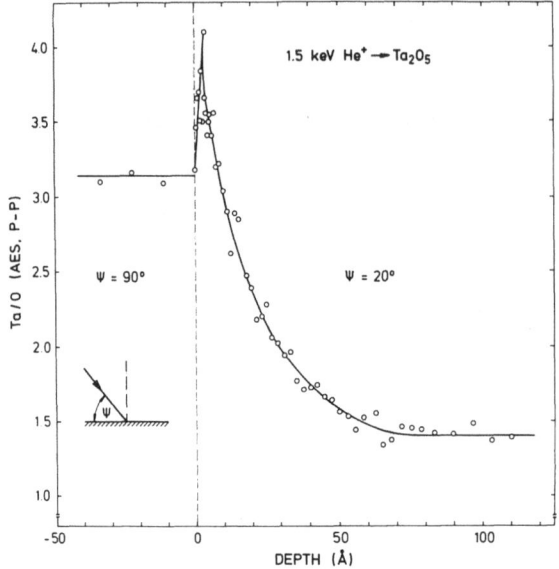

FIGURE 2. Change of the
surface composition of
Ta_2O_5 for the transition
to a smaller angle of
incidence (from $\psi = 90^0$
to $\psi = 20^0$) as a function
of the sputtered depth.

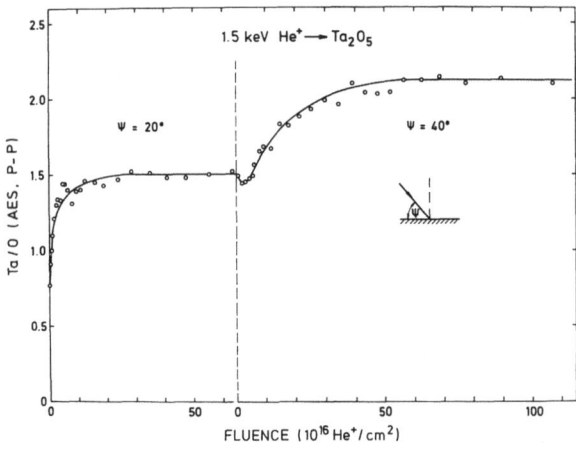

FIGURE 3. Change of the
surface composition of
Ta_2O_5 for the transition
to a larger angle of
incidence (from $\psi = 20^0$
to $\psi = 40^0$) as a function
of the fluence.

4. DISCUSSION

The helium reemission in Fig. 1 shows that the depth distribution of the implanted helium is correlated with the concentration depth profile of Ta/O. The observed helium distribution could in principle be different from the initial implantation profile if diffusion is significant. A comparison between the He$^+$ion fluence and the total reemitted He gas during depth profiling indicates that diffusion losses do not play a major role. Varga et al. have shown that the depth of the altered layer increases with increasing primary energy /5/. The difference in impact angle between ψ = 60^0 and ψ = 70^0 is negligible relative to the effect of the energy change /4/. The assumption that the depth of the altered layer increases because of the increasing range of the He$^+$ ions in the target is verified by these helium reemission measurements.

We can also see by comparing Fig. 2 and Fig. 3 that the characteristic fluence to reach equilibrium surface concentration depends on the initial surface composition: the characteristic fluence F_0 for going from ψ = 90^0 (Ta/O high) to ψ = 20^0 is F_0 = 8.6 x 10^{16} He$^+$/cm^2 , whereas for the transition from a virgin surface (Ta/O low) to the ψ = 20^0 equilibrium value yields F_0 = 2.2 x 10^{16} He$^+$/cm^2. For an unbombarded surface F_0 is determined by two factors, namely the depth of the depleted zone and the degree of depletion in this zone. Thus F_0 is correlated to the total amount of preferentially sputtered oxygen. If by prebombardment a higher degree of depletion has been established, this layer has to be sputtered off before a new equilibrium layer with lower depletion can develop.

The transient behaviour shown in Figs. 2 and 3 for the variation of the angle of incidence is similar to the observation of Betz et al. /6/ with alloys for the variation of the ion energy. Therefore the model proposed by these authors is used for explaining our results. The correspondence with the Betz model is shown in Fig. 4 for the change of the angle of incidence from ψ = 90^0 to ψ = 20^0. Changing to smaller ψ and thus smaller steady-state concentration ratio Ta/O the resulting concentration profile in the transient region is governed as follows: a pre-existing decreasing composition profile (dashed) is superimposed on a new rapidly increasing profile (dotted) and as a result there is an overshoot.

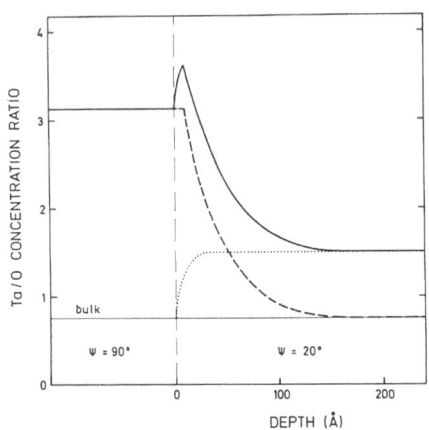

FIGURE 4. Schematic representation of the transient behaviour by changing the angle of incidence from ψ = 90^0 to ψ = 20^0. The resulting concentration ratio Ta/O (full curve) is a superposition of the deep depth profile of the altered layer from the preceding bombardment at ψ = 90^0 (dashed curve) with the rapidly completed exponential increase of the Ta/O ratio appropriate to bombardment at an angle of incidence ψ = 20^0 (dotted curve).

Changing to higher ψ and thus higher concentration ratio Ta/O, the resulting concentration profile in the transient region is dominated by the pre-existing shallow concentration depth profile of the prebombarded target, resulting in an undershoot.

In this model the observed transients do not give information on the possible existence of a surface composition spike due to Gibbsian segregation which was discussed for preferential sputtering of binary alloys /8/.

5. CONCLUSIONS

Helium reemission shows that the depth distribution of implanted helium is correlated with the concentration depth profile. This further confirms that the altered layer corresponds to the range profile of the sputtering particles. This conclusion is tentative in so far as possible diffusion effects are not explicitly taken into account, but they appear to be not very strong.

The transient behaviour of the surface composition by changing the angle of incidence can be explained by the model of Betz et al. /6 / which predicts the observed overshoot for the transition to a smaller angle of incidence and the undershoot for changing to a larger angle of incidence.

REFERENCES

1. For recent reviews, see e.g.:
   G.Betz and G.K. Wehner, in: Sputtering by Particle Bombardment II,
   Ed. R. Behrisch (Springer, Berlin, 1983) p. 11;
   H.H.Andersen, in: Ion Implantation and Beam Processing,
   Eds. J.S. Williams and J.M. Poate (Academic Press, Sydney, 1984)
   p. 128;
   R. Kelly, in: Chemistry and Physics of Solid Surface V,
   Eds. R. Vanselow and R. Howe (Springer, Berlin, 1984) p. 159.
2. E. Taglauer, Appl. Surf. Sci. (1982) 80.
3. E. Taglauer and W. Heiland, Appl. Phys. Lett. 33 (1978) 950.
4. B. Baretzky and E. Taglauer, Surf. Sci. (1985, in press).
5. P. Varga and E. Taglauer, Nucl. Instr. Methods B2 (1984) 800.
6. G. Betz, M. Opitz and P. Braun, Nucl. Instr. Meth. 182/183 (1981) 63.
7. J. Roth, J. Bohdansky and W. Ottenberger, IPP-Report 9/26, 1979,
   Max-Planck-Institut für Plasmaphysik, Garching.
8. R. Kelly, Surf.& Interf. Analysis, 7 (1985) 1.

EXPERIMENTAL STUDIES OF MORPHOLOGY DEVELOPMENT

J.L. WHITTON
Queen's University, Physics Department, Canada.

This contribution to the Advanced Study Institute Programme on the "Erosion and Growth of Solids Stimulated by Atom and Ion Beams" will be a resume' of the approximately eight years of experimental investigation of ion beam-induced modification of metal surfaces made by myself and my many collaborators. A list of these collaborators and the many institutes involved is given in the Acknowledgements section. However, I would like to preface this contribution by stating, clearly and unequivocally, that without the enthusiasm of, and contributions from, my collaborators, the study would have been far less effective and rewarding.

The aim, from the beginning, was to make a detailed series of experiments with well defined controllable parameters in an attempt to establish the mechanism responsible for the production of the topographical features observed so frequently on ion-bombarded surfaces of metals.

These features, in particular cones, were first reported by Günterschulze and Tollmien (1) in 1942 as appearing on the surfaces of bombarded Mg, Zn, Cd, Al, Sn, Pb, Bi, Cu and Ag so this obviously was a general effect. They proposed that surface contaminants, having lower sputtering yield than the substrate, would result in the formation of surface irregularities which would later develop into cones due, as later shown by Wehner (2), to the fact that the sputtering yield is a function of the incident angle between the ion beam and the specimen surface. Typically, for the system we began with, 40 keV argon ions directed on to copper, the sputtering yield is fairly constant from normal angle of incidence to about 30°, rising fairly smoothly to a maximum at 80°, then dropping rapidly to zero at around 82° (3). Much experimental and theoretical work has clearly shown this to be a general trend in the incident energy range of 1 to 100 keV (4,5).

Our interests were first sparked in our early experiments on observing a very strong grain orientation effect. When bombarded with 40 keV argon ions to a fluence of 1×10^{19} ions.cm^{-2} at normal incidence, a carefully polished specimen of polycrystalline copper showed some grains covered with conical-like features at a much lower elevation than adjacent grains. A good (and very beautiful) example of this is shown in Figure 1 (6). Before bombardment the surface was flat and featureless at magnifications up to 10,000x in the scanning electron microscope. The features shown in Figure 1 illustrate the cliff at the left being the grain boundary with an almost featureless plateau while the right-hand grain has receded (sputtered) at a much faster rate suggesting, therefore, an orientation dependent sputtering yield. This was not an isolated effect, many adjacent grains had this marked difference in both elevation and features. At lower doses, orientation effects were also observed in the different shapes of sputter-etch pits on adjacent grains. Figure 2 (7).

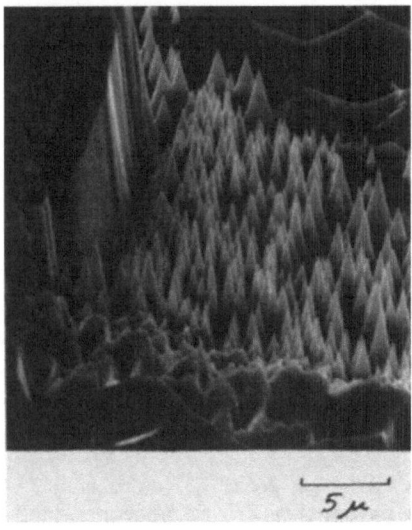

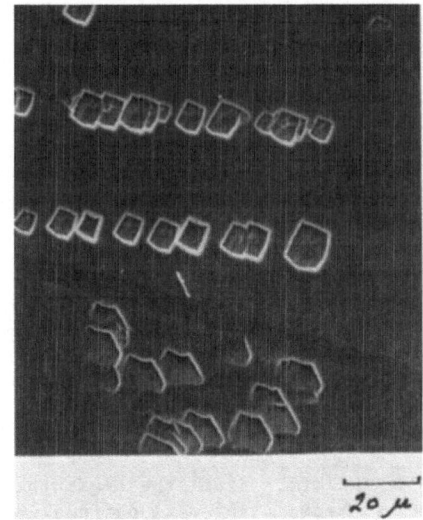

Fig.1 40 keV argon ion bombarded polycrystalline 99.999% copper.

Fig.2 40 keV argon ion bombarded polycrystalline 99.999% copper showing different etch pit habits in adjacent grains.

A very noticeable general effect in these early studies was that the conical features observed were never higher than adjacent grains. This is in direct contrast to the earlier-reported impurity induced cones which, as a consequence of the mechanism involved, always stand proud of their surroundings.

We decided, therefore, to adjust our experimental conditions in order to eliminate the possibility of residual impurity or scratches on the specimen surface and to avoid impurity deposition during bombardment. The sample preparation was made as follows: first a slice was cut from a 99.999% pure copper ingot by a very slow spark erosion technique followed by the sequence of polishing:

1) Grind (abrade) on 60 grit silicon carbide followed by the successively finer grades of 120, 220, 400 and 600

2) Polish on 6μm diamond impregnated cloths, followed by 3, 1 and 0.25 μm.

3) Vibratory polish for 16 hours in an aqueous slurry of 0.05 μm Al_2O_3

4) Chemically polish for 10 seconds in a solution of

55 parts	H$_3$PO$_4$	
25 parts	HA	at 65 – 75°C
20 parts	HNO$_3$	

Rinse with distilled H$_2$O then with C$_2$H$_5$OH and blow-dry. The resultant surface is flat, structureless at magnifications up to 10,000 x and, when applied to single crystals, excellent channeling characteristics are obtained. Impurities are not detectable by Rutherford backscattering of 1 MeV He$^+$ nor by EDAX in the scanning electron microscope. A detailed account of specimen preparation is given elsewhere (8).

Similar precautions were taken with the conditions of bombardment. Only mass analyzed beams were used and defining apertures were kept at least 10 cm from the specimen surface. Typical pressure in the target chamber was 10^{-6} Torr. Occasional bombardments at 10^{-9} Torr gave identical effects.

As mentioned earlier, initial observations of 40 keV argon ion bombarded polycrystalline 99.999% pure copper showed clearly that some grains eroded faster than others, sputter-etch pits were formed on all the grains but only on some grains were cones observed. These individual grains were carefully X-rayed to produce Laue patterns to see if indeed the grain orientation was important. It was. Analysis of the Laue patterns showed that all grains containing cones had an orientation of ± a few degrees from the < 11 3 1> direction. This direction lies between the low index planes {001} and {011} and is 15° from the <100> direction. Figure 3.

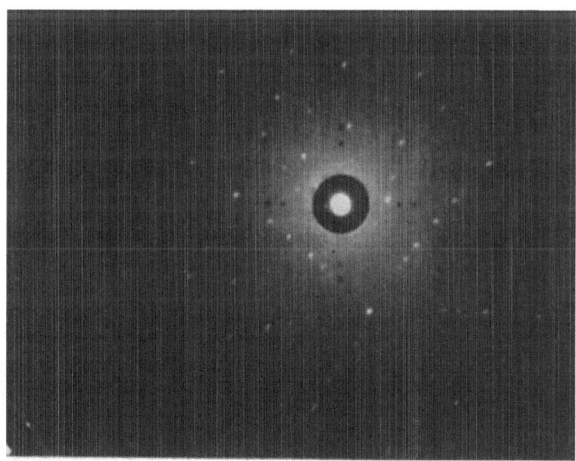

Fig.3 X-ray back reflection Laue pattern of a copper < 11 3 1> surface. The major intersection is the <001>.

Simultaneously, a more careful examination of the conical features showed them to be pyramids, as illustrated in Figures 4 and 5. These figures show the importance of knowing exactly at which angle the specimen is tilted to the electron beam in the scanning microscope since only small changes can result in significant differences in appearance of particular features.

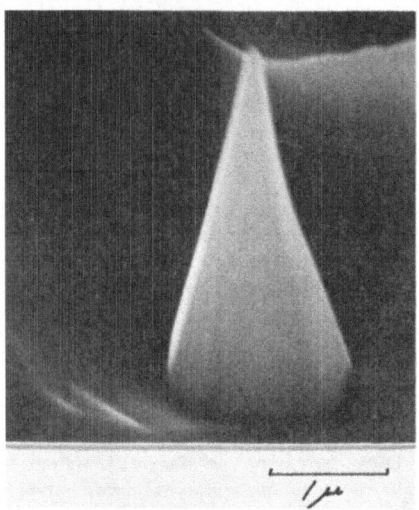

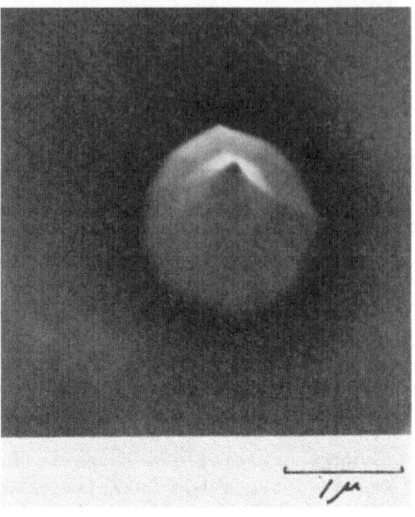

Fig.4 Single pyramid on 40keV argon ion bombarded copper. Viewing angle in SEM is 40°.

Fig.5 The same pyramid as shown in Fig.4. Viewing angle is 0°.

The importance of the <11 3 1> orientation was established beyond doubt when a single crystal of high purity copper was cut to that orientation and bombarded with 40 keV argon to a fluence of 1×10^{19} ions. cm^{-2}. The bombarded area was entirely covered with pyramids of such regular size and spacing that only red light was reflected in daylight. Figure 6. This type of total coverage has never been observed on polycrystalline copper nor have we seen any significant production of pyramids on surfaces other than <11 3 1> ± a few degrees.

We can, however, produce (reproducibly) pyramid covered surfaces on <11 3 1> surfaces within the following ranges of conditions (9).

Fig.6 40 keV argon ion bombarded <11 3 1> surface of copper showing complete pyramid coverage.

Incident ion energy : from 20 to 41 keV. Above this energy only few etch pits, pyramids are seen.

Ion dose : total coverage from 1×10^{19} ions.cm^{-2} to 5×10^{21} ions cm^{-2}. We have gone no higher in total dose.

Target temperatures : from 100K to 550K, above this temperature the pits lose their characteristic crystallographic shape, becoming rounded and only few pyramids form.

Ion beam current : must be greater than 100 microamps.cm^{-2} otherwise only few pits and pyramids form.

Working pressure : from 5×10^{-5} to 5×10^{-9} Torr; therefore excluding the presence of impurity, e.g. oxide.

In the earlier work using polycrystals (6), we invoked the presence of grain boundaries, etch pits and any other departure from flatness to explain the formation of these pyramidal features. We have never found it necessary, (as others have done (9-11)) except in an occasional experiment using impure copper, to invoke the presence of impurity as a necessary condition for the development of surface topography. Now, in the use of single crystals where grain boundaries are not present, we find it necessary to invoke only the development of sputter etch pits to provide the necessary edges for the later cutting out of pyramids. The habits of sputter etch pits are, like chemical or thermal etch pits, completely dependent on the grain orientation in polycrystals and, in single crystals on the crystal orientation. Figure 7, from an argon ion bombarded grain of a polycrystal is shown to illustrate the beautiful symmetry of sputter etch pits to show how the shape of the pits remains unchanged as the pits develop. Shape changes do occur when pits overlap and intersect. A similar illustration for a single crystal having orientation near the <11 3 1> is given in Figure 8 a-f where a sequence of micrographs is shown from the edge of a bombarded area to the centre thus showing the sequential development of pits, pyramids and overlapping effects (12) (Similar beautifully facetted effects may be seen on thermally and on chemically etched material).

Fig.7 40 keV argon ion bombarded polycrystalline copper showing the constancy of etch pit form with development.

For suitably oriented substrates of copper these etch pits are the fore-runners of the pyramids. We never, on single crystals, observe

156

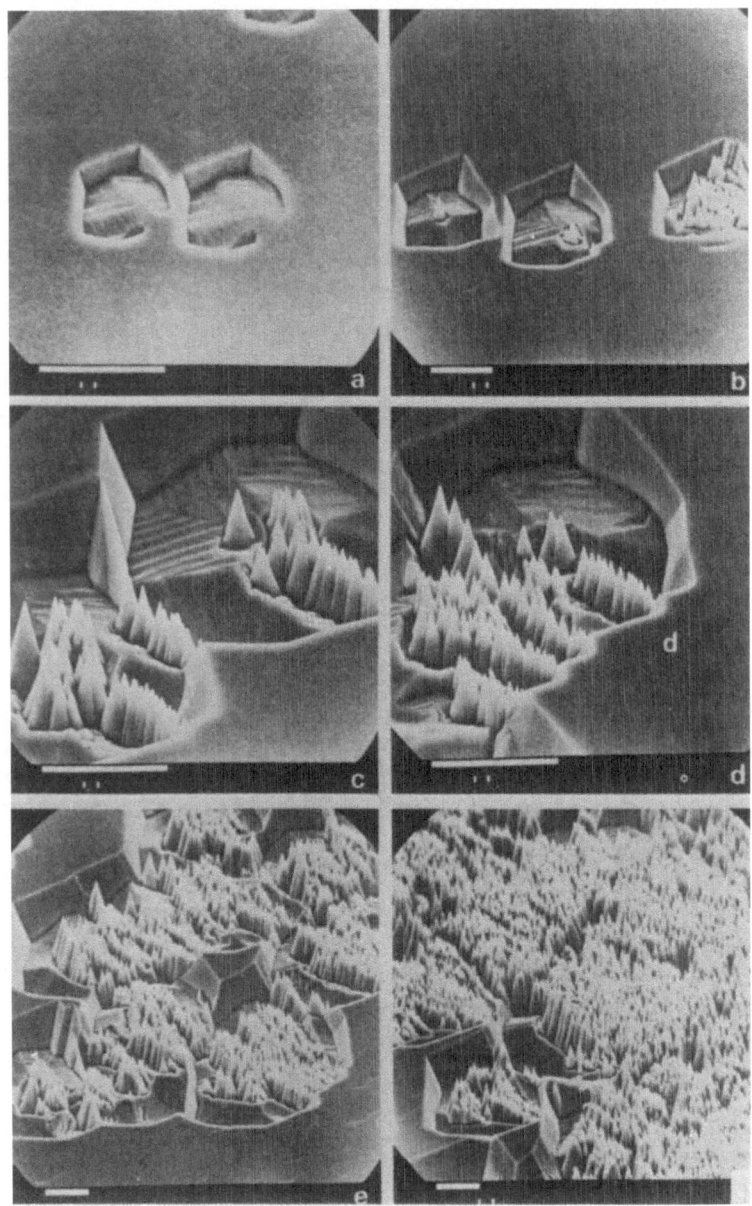

Fig.8. (a-f) Series of micrographs from 30 keV argon ion bombarded
copper showing development of pits and pyramids as a
function of increasing dose and dose rate.
Magnifications as shown. Marker = 10 μm.

pyramids without first seeing pits, as shown on Figures 8a–f. These figures illustrate how the pits first appear, pyramids then begin to be cut from the edges and bases and when the edges retreat due to further sputtering, the pyramids exist independently to be followed by others cut in the same crystallographic form and direction. Overlapping pits also result in rows of pyramids and the end result is a surface covered with these strongly crystallographic features, quite different to the irregularly spaced impurity-induced cones. Note that quite generally we NEVER observe (with two exceptions to be mentioned later) pyramids with apices higher than the flat surroundings. We emphasise here that it is only on crystallographic material that we can produce pyramids. Similar bombardments of amorphous metal (Metglas) results only in a few isolated CONES, having circular cross-section.

Our proposed mechanism(s) of pit and pyramid formation –based on our many observations follow(s), but first it is worth mentioning the basis of the most frequently proposed mechanism. That is, following protuberance initiation as a result of local perturbation in sputtering yield (either from impurity protection or defect structures) the change of sputtering yield with angle of ion beam to the changing surface (Sθ) assumes dominance and a quasi-stable situation is reached for the Sθ maximum (or minimum) condition. Finally, a cone is formed with an apex angle dependent on ion beam energy. We propose alternatively that crystallography is the most important parameter. The <11 3 1> surface is necessary for many reasons: it is necessary to have a high index direction where the argon ion range is short, therefore the energy deposition rate and near-surface defect production rate is high. It seems there is a competition between rate of defect production and annealing (note the absence of dense pyramid covered surfaces above 41 keV, above target temperatures of 550K and the necessity of beam currents higher than 100 microamps.cm^{-2}). The more detailed crystallographic requirements are as follows: Figure 9 – a model of a single pyramid cut from an <11 3 1> surface is used to explain the process.

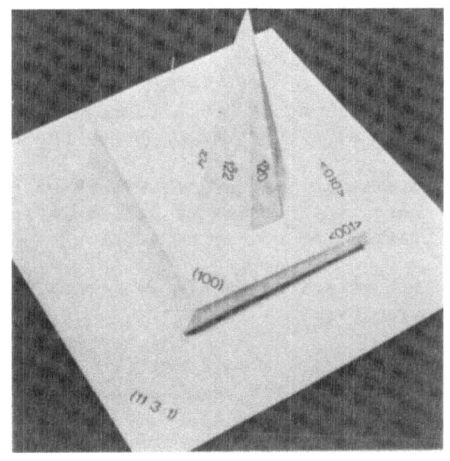

Fig.9 Model of a pyramid similar to that shown in Fig.4. The crystallographic nature of the pyramid is shown as is its accommodation to the crystal lattice (Courtesy of Uffe Littmark)

The <11 3 1> surface is not energetically stable against erosion—induced facetting so that when sputtered, an attempt is made to adjust locally to a more stable configuration, in this case the <100>, some 15° from the <11 3 1>.

The pyramid cut out from the original surface, has of necessity its axis parallel to the ion beam and must therefore tilt its base to accommodate to the new (1 0 0) base plane. It now sits in an etch pit, two of whose sides are low index directions, <0 0 1> and <0 1 0> with the diagonal <0 1 1>.

The octagonal bases of the pyramids have directions of the type <001> and <011> and the eight facets are planes lying in these zones, specifically (010) (011) (104) (122) (120) (233) (103) (144).

The influence of the change in sputtering yield with $S\theta$ is probably of importance only in the initiation of the pyramids. The final form is dictated by crystallographic habit so it is not of great interest to try to relate the apex angle to incident ion energy, as has so often been done in the past. Our studies (14) show only a very weak dependence of apex angle on incident ion energy in the interval 20–80 keV. Tanovic (15) sees a much stronger dependence of apex angle on crystal orientation of (100), (111), (110) and (11 3 1) surfaces with 40 keV Ar$^+$ bombardment. However, as stated earlier, the appearance of pyramids on low index surfaces is seldom. Wilson and Kidd (16) have mentioned the possibility of influence of orientation to account for their observation of adjacent pyramids havng quite different apex angles. We conclude from our detailed studies of argon bombarded copper that:

1) impurities are not necessary for the initiation of cones/pyramids, otherwise it would be difficult to explain the complete coverage of pyramids on some grains but not on others.

2) etch pits are, on single crystals, the necessary fore—runners of pyramids.

3) pyramids are always cut from edges or corners of pits and the over—lapping of pits results in further rows of pyramids.

4) pyramids exist independently in rows as the sides of etch pits retreat due to further sputtering.

5) since the pyramids remain at the same height as the surface from which they are cut, they must sputter at the same rate. This suggests a finite size plateau on top of the pyramids of (11 3 1) orientation.

6) the bases of the pits rather quickly assume a more stable surface than the original (11 3 1). This is the (0 0 1) to which the bases of the pyramids accommodate while still retaining their axes parallel to the direction of the beam. The octagonal base of the pyramids has direction <100> and <110>.

7) the pyramid semi—vertical angles are only weakly, if at all, energy

dependent, the pyramid facets preferring to stabilize to relatively low index planes which lie in the 100 and 110 zones.

8) The combination of a high index surface (11 3 1) close to a low index direction <001> is appropriate in providing the right conditions for high near-surface energy deposition (displacement rate) and for allowing the octagonal based pyramids to accommodate to a low index plane.

Now that our results, proposed mechanisms and conclusions are summarized, we list the most preferred other proposed mechanisms, necessary conditions, for the elaboration of conical features and comment accordingly.

1) Impurity induced departure from surface flatness followed by erosion because of change in Sθ (the most popular proposed mechanism).

2) Pre-existing asperities supplying the departure from flatness, followed by the Sθ effects

3) Growth

4) Redeposition

5) Surface migration

6) Erosion

These "mechanisms" have been proposed over the years either alone or in various combinations. We consider them in order.

1) The need for impurity, introduced either during bombardment or internal, can play a role in the seeding of cones but, as we have shown, when clean targets having crystallographic nature are used, it is not a necessary condition. Witness the formation of pyramids on one grain but not on adjacent, under similar bombardment conditions. Figure 1. Electron induced X-ray analysis of the pyramid covered region in Figure 1 showed only argon as impurity. Oxide impurity can also be ruled out as we observe pyramid covered surfaces after bombardments at pressures down to 10^{-9} Torr. With beam currents of > 100 μA.cm^{-2}. sec^{-1} the surface atoms are removed by sputtering at a rate faster than oxygen can settle. In a similar way, apart from the initial cutting out of cones, we are inclined to relegate the importance of the Sθ effect to a very minor level so far as the final form of morphology is concerned. Our reasons for this statement lie in the observed preference of pyramids to adopt facets in fairly low index planes and in the observation of the dependence of apex angle on target orientation (15).

2) Pre-existing asperities, due to mechanical or chemical treatment, inclusions, surface impurities are all invoked by Kelly and Auciello (17) for the starting conditions for cone erosion. This is undoubtedly so, but this type of uncontrolled surface leads to many complications in the later interpretation of developed features and is certainly not a

necessary condition. The ion beam itself generates the necessary
defects and discontinuities for eventual evolution of pyramids.

3) Growth is listed by many authors as a necessary condition of formation
of cones. We find, with our experimental conditions that growth is
conspicuous by its almost complete absence. We do observe a growth
at the tops of pyramids but generally only under the influence of
electron bombardment in the SEM. (Figure 10a and b). We have often
observed well facetted pyramids in the SEM and when re-examining the
identical features some time (weeks) later we find them to have changed
their form by growth at the tips. Most proponents of growth are also
proponents of impurity induced cone formation. This type of formation
can often result in an isolated cone on a plateau. A further bombardment can cause the cone to increase in size, therefore "proof of
growth". But this growth can also be achieved by recession of the
plateau and so the cone "grows", but only by removal of material
around its base. To unequivocally decide if a cone grows or the
plateau on which it sits retreats, some point of reference as e.g.
a sputter etch pit is necessary.

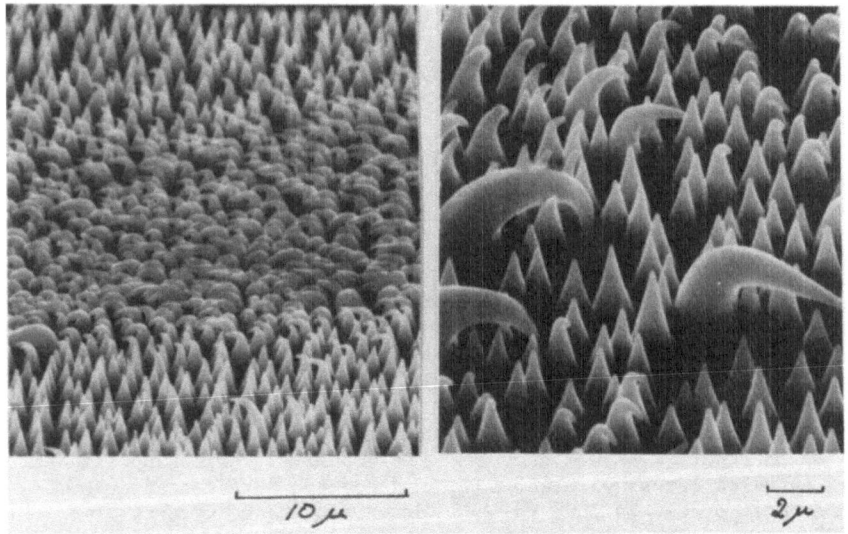

Fig.10a and b - Examples of bent and twisted pyramids on

a) argon ion bombarded copper   b) copper ion bombarded copper

4) Re-deposition, is claimed by Hauffe (18) to be an essential part of
the development of surface features, along with growth following foreign
atom initiation. The published evidence is difficult to assess and,
since re-deposition is an effect on the atomic scale, the possibility
of confirming it as a real effect, with present day SEMs is low. From
all the published results from Hauffe and others, we believe it to play
only a small role, if any, in the development of surface topography.

5) The role of surface migration, again occuring on atomic scale is as difficult to assess as redeposition - for similar reasons. The only indication we have of surface migration is that of rounding of pit edges when copper targets are deliberately heated above 550K. Obviously, if high current ion beams are used, gross heating with subsequent ease of surface migration can take place. More will be heard on this topic by Rossnagel (these proceedings).

6) All agree that some erosion mechanism is essential in the formation of cones/pyramids.

To summarize the argon bombardment of copper results, we believe that on the crystalline material a) it is not necessary to invoke the presence of impurity for the inititation of cone/pyramid development, b) the $S\theta$ effect is of importance only at the start of erosion, c) that growth is not a necessary mechanism of pyramid formation and that if redeposition and/or surface migration do play a role, they appear to be only minor. Confirmation of our statement that crystallography is all-important can be seen in the micrographs of other authors. Wilson and Kidd[16] e.g. noted that some of their cones were of hexagonal cross-section, Gvosdover et al. (10) show clearly facetted pyramids, produced, interestingly enough, when their bombardments made on single crystal copper have beam incidence $15°$ from [001] in the (100) plane. this is only a few degrees from our <11 3 1>. We also can produce pyramids on an (001) surface of copper when bombarded along the <11 3 1> but never, even up to doses of 5×10^{21} 40 keV $Ar^+.cm^{-2}$ when bombardment is normal to the low index surface. Erlenwein (19) showed many beautifully facetted pyramids on Ar^+ bombarded Au, Zn and Pb. Further confirmation of our results is given by Shimizu (20), who, using electrolytically polished copper found that the formation of facetted cones came primarily from eroded ridges between pits and from overlapping pits/craters. From the appearance of his micrographs, he has used large-grained material, essentially single crystal.

The question of stability of cones/pyramids under prolonged bombardment has been raised (21-23). Our observations show, conclusively, that a pyramid-covered surface remains pyramid-covered under prolonged bombardment (to 200 times the dose required to initially cover the surface with pyramids) but this is because of continuous change. We see pyramids being eroded while others are in the process of development so, a quasi-stable pyramid surface can be maintained (24).

However, the system argon into copper is quite insular so we decided to see if one would observe similar effects by, first, bombarding <11 3 1> copper with different ions and, second, bombarding <11 3 1> surfaces of different face centred cubic metals with energetic argon ions.

In the first case an <11 3 1> copper specimen was bombarded in different areas with 40keV neon, argon, krypton, xenon, nitrogen and copper (25) each to a fluence of $\approx 1 \times 10^{19}$ ions.cm^{-2} at normal incidence. The results may be summarized as follows:

162

(1) All ion species lead to characteristic morphological surface changes including etch pit, and frequently, pyramid generation.

(2) The lighter ion species tend to produce lower pit densities and less well-defined pit habits and fewer, sometimes no pyramids.

(3) The heavier ions (including Cu^+) produce denser arrays of all features and better geometrically defined features. (Figure 11)

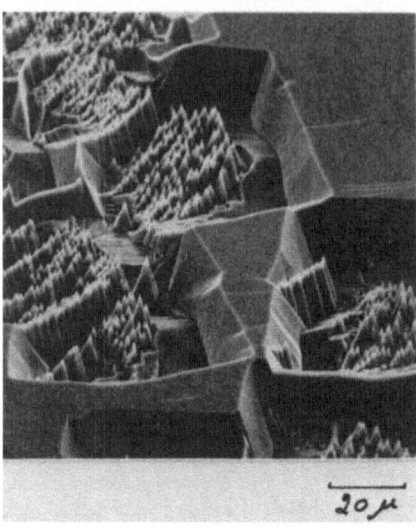

Fig.11 40 keV copper ion bombarded <11 3 1> single crystal copper showing well-defined geometric arrays of pyramids in etch pits.

(4) N^+, Ne^+ and Ar^+ ions produce a zone of surface blisters outside but surrounding the irradiated area but none within the irradiated area. (Figure 12)

(5) Some signs of pyramid bending occur from one to a subsequent examination but there are no clear correlations with ion species and surface zone.

(6) The as-formed pyramids are in dense crystallographic array after Cu, Kr and Xe irradiation. (Figure 13)

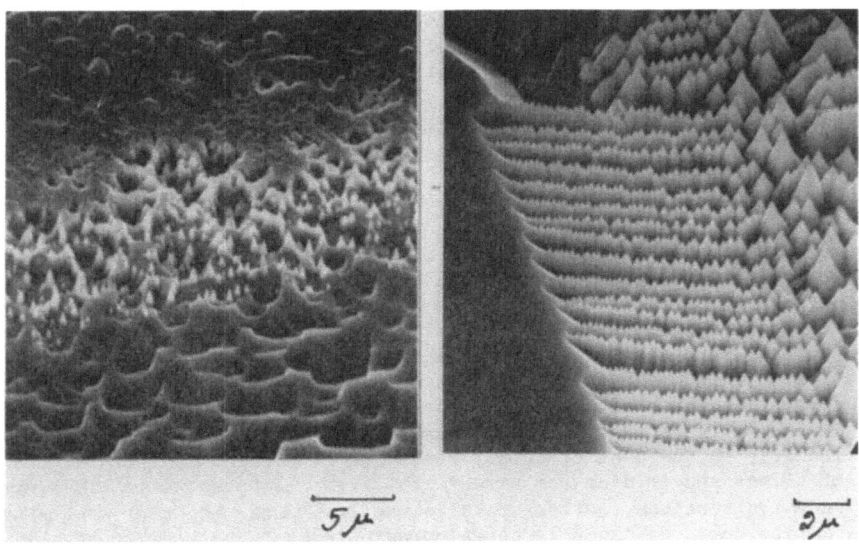

Fig.12   40 keV neon ion bombarded
         <11 3 1> single crystal copper
         in the region of the ion flux
         boundary. The upper part of
         the figure is in the low flux
         region and blisters are visible
         The lower part is in higher
         flux and shows pits and pyramids.

Fig.13   40 keV krypton ion bomb-
         arded <11 3 1> single
         crystal copper showing
         colinear pyramid geo-
         metries showing their
         connection to etch pit
         jogs.

These results are explained in terms of preferential sputtering of native and irradiation induced defect structures and differential atomic mobility of different ion species in the Cu. The Cu$^+$ irradiation results exhibit the lack of necessity of occluded gas for feature development.

We conclude that comparative studies of constant fluence, energy and incidence direction but different species ion bombardment of (11 3 1) Cu indicate that gas occlusion is not a necessary condition for morphological feature initiation. Blistering does occur with lighter ion irradiation but not in the zone of highest flux, a result it is believed of radial migration of gas to beneath a surface region which is lightly sputtered.

Pit initiation is believed to occur in the neighbourhood of extended defects and is surface area specific. Pits enlarge and adopt geometric habit determined by extrema in sputtering yields of atomic planes and are somewhat ion species dependent. Pyramids develop in direct association with pits and at discontinuities or jogs in pit facets or inter pit

164

ridges. In general, pits and pyramids are both results of differential local erosion processes. N[+] irradiation, although leading to etch pitting of similar magnitude to other ion species, results in somewhat less recognizably crystallographic pits, a result perhaps of chemical modification to the substrate. In the case of Cu[+] irradiation there is evidence of a feature growth process. This will be discussed later.

In the second case, we have bombarded <11 3 1> surfaces of the face centred cubic metals copper, nickel, palladium, silver, iridium and gold with 25, 30 or 40 keV argon ions to a fluence of 1×10^{19} ions.cm^{-2}. As in our earlier studies of argon bombarded <11 3 1> copper, we observe the production of pyramids follows the formation of sputter-etch pits of well-defined crystallography. Examples are given of all these specimens in Figures 14-19, showing first on lightly bombarded areas where the etch pits form, followed by heavily bombarded areas when the pyramids appear. The crystallographic nature of the pits and the pyramids is quite evident as is the alignment of rows of pyramids. These rows of pyramids, particularly in the case of copper bombarded with argon and with krypton, and of silver and iridium bombarded with argon are shown to align with low index directions, often parallel to low index edges of etch pits. Only in the case of the gold cones/pyramids does there seem to be a lack of facetting. This is not, as yet, understood. (26)

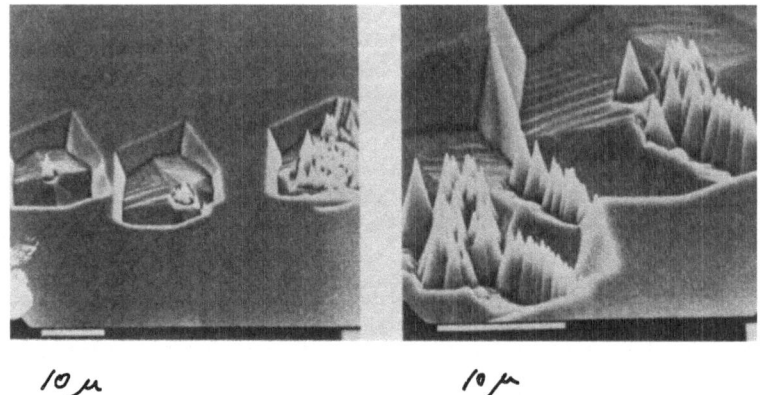

10 μ   *10 μ*

Figs. 14 a and b - 30 keV argon ion bombarded <11 3 1> copper

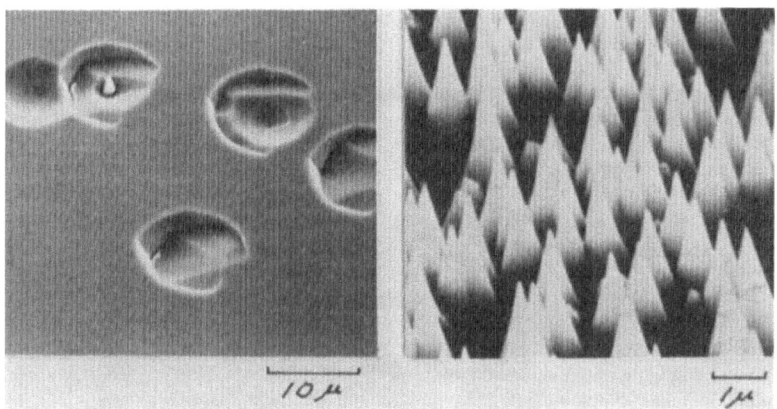

Figs. 15 a and b — 25 keV argon ion bombarded <11 3 1> nickel

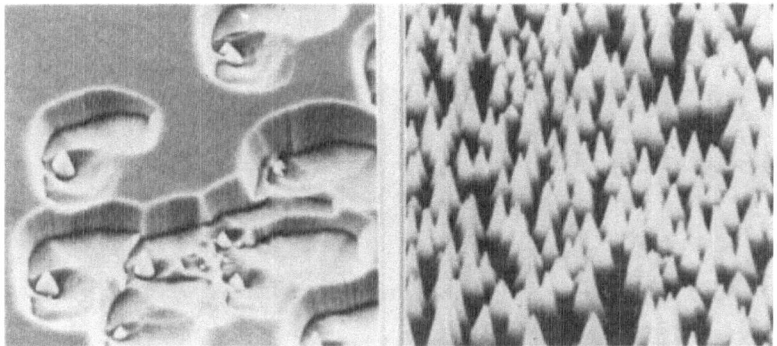

Figs. 16 a and b — 25 keV argon ion bombarded <11 3 1> palladium

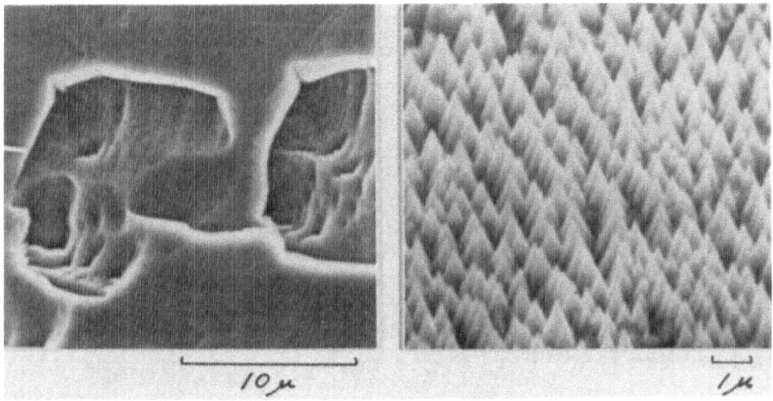

Figs. 17 a and b — 25 keV argon ion bombarded <11 3 1> silver

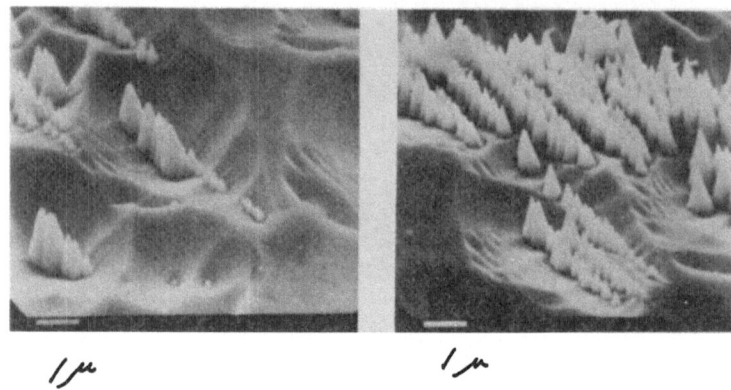

Figs. 18 a and b - 25 keV argon ion bombarded <11 3 1> iridium

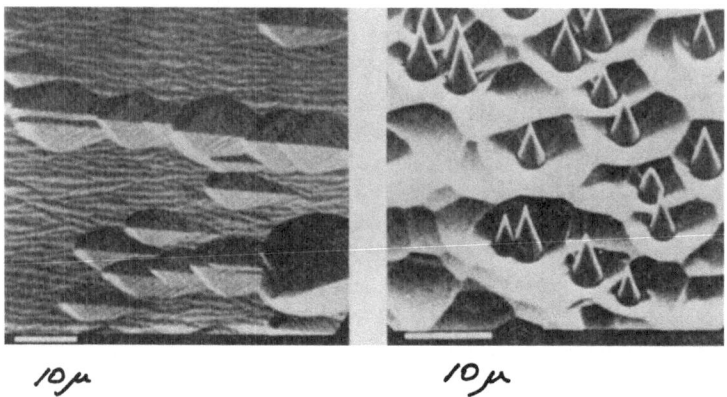

Figs. 19 a and b - 25 keV argon ion bombarded <11 3 1> gold

An unusual feature on the <11 3 1> copper bombarded with the self-ion copper is that of two distinct. types of pyramid, the smaller being of the size expected from the surrounding etch pit edges, the larger being above the surrounding pit edges and exhibiting apparent impurity on their summits. An explanation was kindly afforded us by Drs. G. Sørensen and H. H. Andersen who pointed out the possibility of $C_2H_4Cl^{35}$ and C_2H_4 Cl^{37} impurity being mass analyzed in the same way as Cu^{63} and Cu^{65}, arising from the use of CCl_4 for the production of Cu ions in ion source. This appeared initially as a growth mechanism but is, in fact, an impurity initiated feature. Figure 20.

Fig.20   40 keV argon ion bombarded <11 3 1> copper showing the
two distinct types of pyramid.

The experimental evidence presented so far makes clear that, given
the correct choice of parameters, sputter-etch pits of well-defined
crystallographic shape can be produced when <11 3 1> surfaces of copper
are bombarded with energetic neon, argon, krypton, xenon or copper ions
and act as fore-runners to the development of pyramids. Further, argon
ion bombardment of the face centred cubic metals copper, nickel,
palladium, silver, iridium and gold exhibit the same features of pit and
pyramid development. The evidence is clear, extensive,
well-substantiated and the bombardment induced features are readily
reproducible.

Nonetheless, a recent review article of ours (27) which goes into
more detail than in this chapter has been openly criticized, somewhat
vituperatively, by a proponent of the "impurity-induced" cones school
(28). This is most unfortunate since comparison of the mechanisms
involved in the two types of cone or pyramid production is meaningless.
A brief example will suffice to illustrate the point. Bear in mind that
in the systems we describe, sputter-etch pits are always fore-runners of
pyramid production; pyramids are never seen on exposed flat surfaces.

Recent impurity-induced cone production (29) where contamination was
introduced to a Pb (111) surface in a scanning electron microscope and
subsequently bombarded by argon ions shows the reverse effect i.e. a
protuberance eventually developing into a cone is seen to gradually
enlarge as sputtering proceeds and the base gradually develops a
surrounding trench which on single crystals develops into a
crystallographic pit. A schematic of the process is shown in Figure 21.
This impurity induced cone formation thus proceeds in the opposite way to
our defect-induced pyramid formation. This process is almost routinely
observed with impurity-induced cones but never in clean systems.

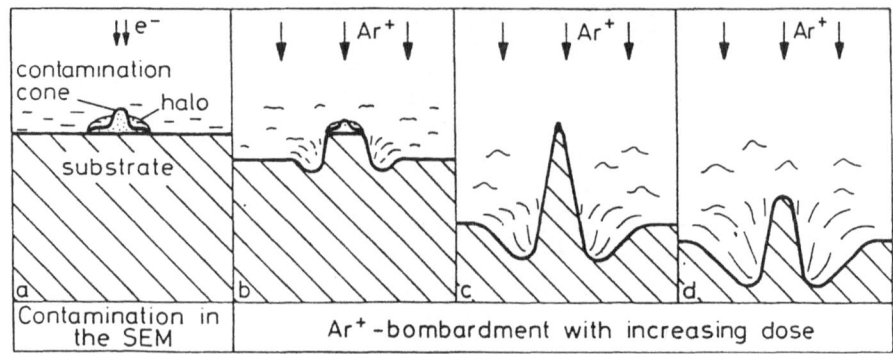

Fig.21   Schematic diagram showing impurity-induced cone and subsequent
         pit formation (from ref. 29).

Angular distribution of sputtered particles from flat and from pyramid covered surfaces

The ability to routinely produce these pyramid covered surfaces has
given us the opportunity to investigate their influence on the sputtering
yield and the angular distribution of sputtering yield and to compare
these to the yield from flat surfaces having the same <11 3 1>
orientation (30).   This measurement is of interest since it has been
claimed that a deliberately roughened surface made e.g.   by sputtering
can increase optical absorption (31,32) and secondary electron emission
(33).   Others have indicated (34,35) that "proper texturing" of a surface
could be beneficial in lowering the surface sputtering (erosion) rate by
energetic ions impinging on the walls of a controlled fusion reactor.

For our sputtering-yield experiment the (11 3 1) surface was
prepared from a large-grain 99.999% pure polycrystalline sample.   Part of
a 10-mm long (11 3 1) grain was bombarded at normal incidence to a total
dose of 1 x 10^{19} 40 keV argon ions.cm^{-2} in order to produce a completely
pyramid-covered surface.   From this surface the angular distribution was
measured by bombarding, again at normal incidence, with a further dose of
5 x 10^{18} ions cm^{-2} and collecting the sputtered copper atoms on a
semicircular Al foil in the plane (811) which contains both directions
<11 3 1> and <101>.   The angular distribution from the flat surface was
made in the same way, but from the part of the same (11 3 1) grain which
had not been pre-bombarded.   Scanning electron microscopy of the surfaces
before and after the collections showed the pyramid-covered surface to be
still pyramid covered, while the initially flat surface had changed only
slightly by developing a few sputter etch pits.   The pyramid-covered part
of the grain is shown in Figure 22.

The collector foils of 99.999% pure Al were of semicylindrical form of radius 5 cm and were liquid-nitrogen cooled to guarantee a sticking factor independent of the flux density of impinging particles (36). Subsequent analysis of the deposited copper was made by 1.8 MeV-He Rutherford backscattering analysis. With this energy the He backscattered from the Cu atoms is clearly separated from the Al substrate. Measurements were made at 5° (in places 2.5°) intervals and the results, expressed in terms of Cu atoms x 10^{16} cm^{-2}, plotted as a function of angle. Stopping-power values were taken from Ziegler (37).

The angular distributions thus obtained from both flat and pyramid covered surfaces are shown in Figure 23. Two effects are immediately obvious: First, the total sputtering yield from the pyramid-covered surface is higher than from the flat surface. This observation is in qualitative agreement with calculations of Littmark and Hofer(38), the quantitative conclusions of which pertain to faceted surfaces but which can allow qualitative inferences for pyramid-covered surfaces. A simple geometrical explanation for the 25–30% increase in yield from the pyramid covered surface is suggested and shown in Figure 24. If we consider the forward-biased energy/disorder distribution of a 40 keV-Ar-ion cascade in Cu and assume that Cu particles can be ejected only from some few Angstroms from the surface, then the available volume is higher from the side of a pyramid, (having a half angle of 15°) than from the surface normal by a factor 2.4. This trend is in reasonable agreement with our experimental results.

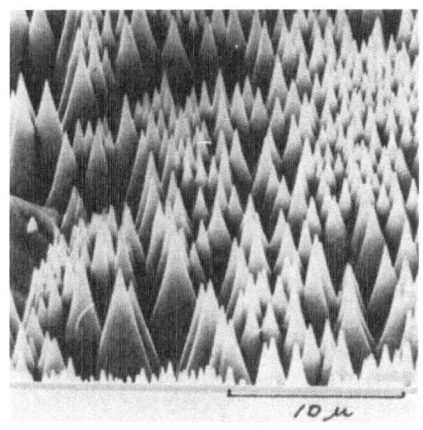

Fig.22 The pyramid-covered surface used for the angular distribution measurements.

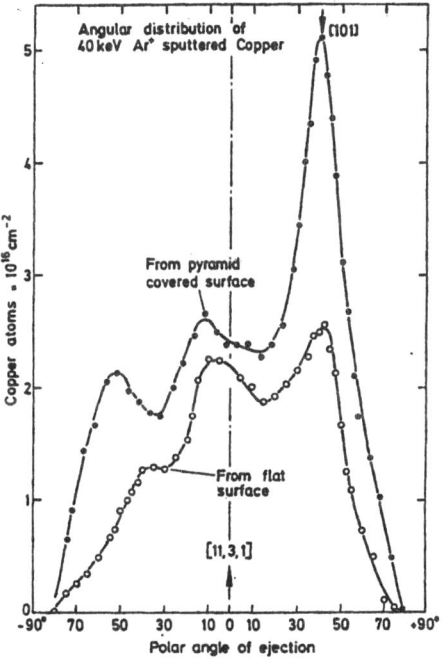

Fig.23 Angular distribution of 40 keV argon ion-sputtered copper from flat and from pyramid-covered surfaces.

However, the more striking effect seen in Figure 23 is the enhancement of preferential ejection from the pyramid-covered surface. Preferential ejection along low-index directions is a well-known phenomenon in the mechanism of sputtering, but this large enhancement due to the pyramid-covered surface is remarkable.

Although the mechanism responsible is undoubtedly more complicated than that proposed in Figure 24, we can again use a simple geometrical argument to account at least in part for this much-enhanced preferred ejection. We consider again the geometry of the beam (a) normal to the flat surface and (b) parallel to the axis of a pyramid. The <101> direction is 42° from the <11 3 1> (see Figure 23); thus the possibility of a focused collision sequence leaving the (11 3 1) surface is much reduced from that of leaving the side of a pyramid. This is clear if only by virtue of the increased length a focussed collision sequence must have in the former case. Simple trigonometry gives a ratio of about 4. This proposed mechanism is, in a way, similar to that proposed by Yurasova (39) for terrace formation on argon-bombarded copper crystals.

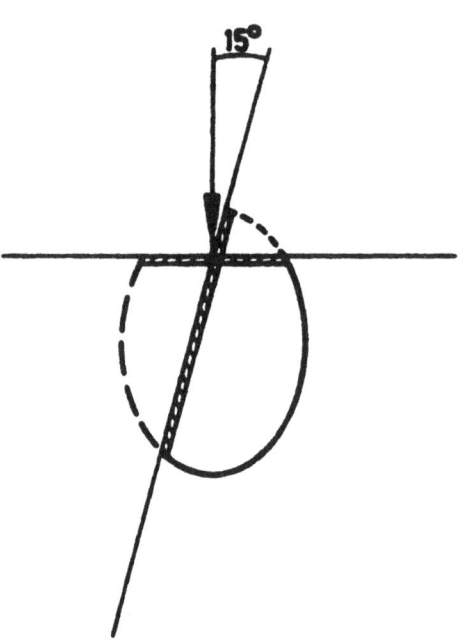

Since only one <110> direction is operative in this geometry, the enhanced yield can come from only ≈ 3 of the 8 pyramid facets. Those facets contributing to the left-hand side of the angular distribution (Figure 23) present too thick a barrier to allow focused collisions to be observed on the right-hand side of the angular distribution. Therefore the real effect of this enhanced ejection can be much greater than observed here.

Fig.24 Illustrating the near-surface contribution to sputtering for a 40 keV argon ion a) normal to the surface and b) parallel with pyramid axis.

Although the simple geometrical arguments we put forward are perhaps naive, we found the effect sufficiently remarkable that we extended the measurements to include the use of precisely oriented Cu single crystals and found similar effects (40).

Conclusions

We have shown that given the correct choice of ion bombardment parameters, pyramids of well-defined geometry can be developed on the ⟨11 3 1⟩ surfaces of copper bombarded with either neon, argon, copper, krypton or xenon ions. Conversely, argon ion bombardment of ⟨11 3 1⟩ surfaces of copper, nickel, palladium, silver, iridium and gold results in a development of the same well-defined geometric pyramids. At no time do we require seeding by low sputtering yield or high melting point impurities to initiate pyramid formation.

The production of these pyramid-covered surfaces has allowed a comparison to be made of the angular distribution of argon ion sputtered copper from flat and from pyramid-covered ⟨11 3 1⟩ surfaces. The overall sputtering yield is higher from pyramid-covered surfaces and much enhanced ejection is seen along low index – in particular the ⟨110⟩ – directions.

Acknowledgements

The experimental results reported here have been obtained consistently, no matter in which laboratory the ion bombardments took place. These laboratories include the H.C. Ørsted Institute, the Niels Bohr Institute and Aarhus University in Denmark, the Institute for Atomic Physics, Stockholm, Chalk River Nuclear Laboratories, Canada and, last but certainly not least, the University of Salford, England.

My collaborators in this venture are: G. Carter, J.S. Colligon, W.A. Grant, G.W. Lewis, M.J. Nobes and F. Paton of Salford, S. Kofod, K. Persson, J. Jensen and O. Holck of Copenhagen, V. Toft and J. Bjorglund of Aarhus, B. Emmoth and M. Braun of Stockholm, L. Tanovic of Sarajevo, G. Kiriakidis of Crete, U. Littmark and W.O. Hofer of Julich, O. Westcott of Chalk River, Chen Hao-Ming of Peking and again, last but not least, J.S. Williams of Melbourne.

The orientation, cutting and preparation of the various specimens was very effectively done in Copenhagen by Mogens Jaegher and Bendt Vaarby and the excellent photographic work by Inger Jensen.

The much-needed and much appreciated financial support has been provided by the Danish Natural Sciences Research Council, Philips Fund of 1958, Nordita, the U.K. Science and Engineering Council and by a NATO Travel Research Grant.

Final production of this camera-ready paper is due solely to the excellent efforts of Chris Bustard, Queen's University, who successfully manipulated the word processor to accommodate the many necessary juxtapositions.

My grateful thanks are extended to all who have been involved.

29 August 1985
Queen's University at Kingston
Ontario, Canada

REFERENCES

1. Günterschulze, A. and Tollmien, W. Z. Phys., 119 685, 1942.
2. Wehner, G.K. J. Appl. Phys. 30 1762 1959.
3. Cheney, K.B. and Pitkin, E.T. J. Appl. Phys. 36 3542, 1965.
4. Witcomb, M.J. Rad. Effects. 27 223, 1976.
5. Yamamusa, Y., Itikawa, Y and Itoh, N. Report IPPJ-AM-26, Nagoya University, Nagoya, Japan, 1983.
6. Whitton, J.L., Carter, G., Nobes, M.J. and Williams, J.S. Rad. Effects. 32 129, 1977.
7. Carter, G., Nobes, M.J., and Whitton, J.L. Applied Physics (in press) 1985.
8. Whitton, J.L. Proc. Roy. Soc. A311 63 1969.
9. Wehner, G.K. and Hajicek, D.J. J. Appl. Physics. 42, 1145, 1971.
10. Gvosdover, R.S., Efremenkova, V.M., Shelyakin, L.B. and Yurasova, V.E. Rad. Effects. 27 237, 1976.
11. Stewart, A.D.G. and Thompson, M.W. J. Mater. Sci. 4, 56, 1969.
12. Whitton, J.L., Holck, D., Carter, G. and Nobes, M.J. Nucl. Instr. and Methods 170, 371, 1980.
13. Cope, J.O. J. Appl. Cryst. 15 396, 1982.
14. Tanovic, L., Whitton, J.L. and Kofod, S. Symposium on the Physics of Ionized Gases Dubrovnik. Ed B. Navinsek, 1978.
15. Tanovic' L. private communication.
16. Wilson, I.H. and Kidd, M.W. J. Mater. Sci. 6, 1362, 1971.
17. Auciello, O and Kelly, R. Rad. Effects 66 195, 1982.
18. Hauffe, W. Rad Effects.
19. Erlenwein, P. Ph.D. Thesis, Technical Univeristy of Berlin, 1977.
20. Shimizu, R. Jap. J. Appl. Phys. 13, 228, 1974.
21. Lewis, G.W., Colligon, J.S., Paton, F., Nobes, M.J., Carter, G., and Whitton, J.L. Rad. Effects Lett. 43, 49 (1979).
22. Auciello, O. and Kelly, R. ibid. 43, 37 (1979) also 43, 117 (1979).
23. Chadderton, L.T. ibid. 43, 91 (1979).
24. Hao-Ming, Chen, Jensen, J. and Whitton, J.L. to be published.
25. Carter, G., Nobes, M.J., Lewis, G.W., Whitton, J.L. and Kiriakidis, G. Vacuum 34 Nos. 1 and 2, 167, 1984.
26. Whitton, J.L., Kiriakidis, G., Carter, G., Lewis, G.W. and Nobes, M.J. Nucl. Instr. and Methods in Phys. Research B2 640, 1984.
27. Carter, G., Navinsek, B and Whitton, J.L. Topics in Applied Physics. Ed. R. Behrisch (Springer-Verlag, Berlin) 52, Chapter 6, 1983.
28. Wehner, G.K. J. Vac. Sci. Technol. A3(4), 1821, 1985.
29. Linders, J., Niedrig, H., Ram, N. and Koch, H. Rad. Effects (in press) 1985.
30. Whitton, J.L., Hofer, W.O., Littmark, U., Braun, M. and Emmoth, B. Appl. Phys. Lett. 36 (7), 531, 1980.
31. Berg, R.S. and Kominiak, G.J. J. Vac. Sci. Technol. 13, 403, 1976.
32. Stephens, R.B. and Cody, G.D. Thin Solid Films 45, 19, 1977.
33. Thomas, S. and Patterson, E.B. J. Phys. D.3, 1469, 1970.
34. Mattox, D.M. and Sharp, D.J. J. Nucl. Mater. 80, 115, 1979.
35. Ziegler, J.F., Cuomo, J.J. and Roth, J. Appl. Phys. Lett. 30, 268, 1979.
36. Hofer, W.O. Rad. Effects. 19, 263, 1973 and 21, 141, 1974.

37. Ziegler, J.F. Helium – Stopping Powers and Ranges (Pergamon, New York) 1977.
38. Littmark, U. and Hofer, W.O. J. Mater. Sci. $\underline{13}$, 2577, 1978.
39. Yurasova, V.E. in Proc. VII Int. Summer School SPIG. Ed. B. Navinsek (Jozef Stefan Institute – Dubrovnik 1976).
40. Emmoth, B., Chen Hao-Ming., Whitton, J.L., Braun, M., Littmark, U and Hofer, W.O. Proc. of Symposium on Sputtering, Perchtoldsdorf/Vienna, Eds. Varga, P., Betz, G. and Viehbock, F.P. Inst. für Allgemeine Physik, Technical University of Vienna, 1980.

INVESTIGATION OF MICROTOPOGRAPHY INDUCED DURING SPUTTER
DEPTH PROFILING OF Ni/Cr MULTILAYERED STRUCTURES

L. TANOVIC, N. TANOVIC[x] and A. ZALAR[xx]
Electrical Engineering Faculty, University of Sarajevo,
Yugoslavia
[x]Physics Department, University of Sarajevo, Yugoslavia
xx
   Institute for Electronics and Vacuum Techniques, Ljubljana,
Yugoslavia

1. INTRODUCTION

The ion bombardment sputter erosion method is very
useful in surface analysis for obtaining a chemical composition
of the thin films and in-depth elemental composition near surfaces. The electron and mass spectrometry in conjunction with
ion sputter erosion allow one to probe the in-depth elemental
composition near surfaces as well as the interfaces within the
solids. In such a way the so called sputter depth profiles are
obtained (1,2).

The quality of this method of depth profiling is
determined by the depth resolution which is usually defined as
the depth over which the intensity from a step-function profile drops from 84% to 16% of its plateau values (1,2,3).

The advantage of this method for the in-depth profiling
of thin films compared with other analytical methods is in the
good depth resolution lying in the range from 0.3 to 3nm. Many
contributions which influence the depth resolution are usually
sorted in three groups (4):
1. the contributions of the experimental instrumentation,
2. the parameters of the target and
3. the sputter ion beam parameters.
The product of all of them is the surface microtopography
(roughness) which is induced by the sputter ion beam and which
is believed to have the most intensive influence on the depth
resolution. The decrease of the depth resolution due to the
variations of different experimental parameters is expressed
by the change of the "interface width" at the interfaces of
different materials. There are the experimental results which
show the broadening of the interface widths with the increase
of the ion bombardment energy and the increase of the total ion
dose (3, 5). This broadening is usually ascribed to the ion
beam microtopography induced during the removal of the surface
layers while the composition near the surface is analyzed and
determined by the Auger electron spectroscopy (AES) or X-ray
induced photoelectron spectroscopy (XPS). On the other side
there are results that show the intensification of the surface
roughness by the increase of the ion energy and by the increase
of the dose rate and the total ion dose (6,7). However, these
experimental results regarding the influence of the ion beam
parameters on the ion induced topography, are obtained for much

higher values of the ion energies, dose rates and total ion doses compared with the values of the experimental parameters used in the sputter depth profiling.

Therefore our goal in this work was to obtain at least certain qualitative information about the topography induced during a sputter depth profiling and its connection with the distortion of the depth profiles influenced by the increase in the ion energy, dose rate and the total ion dose.

2. EXPERIMENTAL

In our investigation of the ion bombardment induced microtopography during the depth profiling we have used the multilayered Ni/Cr sandwich film deposited on the (100) oriented Si single crystal. The sample was prepared by the sputter deposition at the "Jožef Stefan" Institute - Ljubljana in collaboration with the Surface Science Division of National Bureau of Standards - Washington. This material is known presently as the Standard Reference Material (SRM) for sputter-depth profile calibration (8). The approximate thickness of the individual films on the Ni/Cr multilayered structure is 70 nm except for the outer protective Cr film which is about 35 nm thick. The thickness of the film was measured during the deposition by the quartz oscillator microbalance. The Ni/Cr thin film was made on the substrate of the size 1.00 x 2.54 x 0.04 cm.

In the first set of our experiments, on the first of such samples, 6 shallow craters have been eroded by the ion beam and the second sample had 15 craters eroded by the sputter ion beam. These samples have not been analyzed by Auger electron spectroscopy. The ion beam eroded craters differ among themselves in the erosion depth. For instance at one of the craters the first Cr layer was removed by ion sputtering, that means the first Ni layer was revealed; at the second one the fifth Cr layer was revealed etc. In some of the craters the Si substrate was more or less eroded. This erosion has been performed by the Ar ion beam of 3 keV and the dose rate of 10 μA/cm^2. The total ion dose was different at differently eroded craters. In these experiments the ion beam was rastering the sample during bombardment to provide the most homogeneous sputtering of the sample.

We have observed our samples and differently eroded spots on them by scanning electron microscopy (STEREOSCANN-150). By this microscope we were able to observe only the micro-topography on heavily eroded craters where the bottom of the crater was the Si substrate. We were not able to see any topo-graphy on shallow craters where the observed surface was Cr or Ni, on the bottom or on the sides of the craters. Searching for a better resolution we have then used the transmission electron microscope, applying on our craters the single C replica with and without shadowing. In such a way the microto-pography, with the shapes of the order of the size of a few dozens of nm, has been observed. This possibility of observing some topography encouraged us to prolong the investigation by which we wanted to compare the observed microtopography with the depth profile shapes at the interfaces.

Therefore we have undertaken the second set of experiments in which the Ni/Cr sample with 9 layers of Cr and Ni was divided in three samples. The thickness of in-situ sputtered Si, Cr and Ni layers was 50, 66 and 60 nm, respectively. The sputter erosion of   two   such samples was done by the 1 keV and 5 keV Ar ion beam energies, while the composition of the sample near the surface was analyzed and determined by the Auger electron spectroscopy (SAM-545A, Physical Electronic Industries). We preserved the same dose rate of 8 μA/cm^2 on both samples. We have also obtained a profile of one crater on the third sample with 1 keV Ar ion beam and the higher dose rate of 16 μA/cm^2. During the profiling the incident angle of the ion beam with respect to the surface normal was 42.6? The primary electron beam had the energy of 3 keV, the current of 0.5 μA and the diameter of 40 μm.

The ion induced microtopography on these samples has been examined in the same way as in the first set of the experiments: using electron transmission microscopy and the replica technique to observe the topography.

3. RESULTS

Our observations made on the basis of the electron microscope micrographs and the Auger depth profiles comparisons can be summarized in the following way:

- Hardly noticeable difference in the surface topography is observed for the topographies obtained by the 1 keV and 5 keV ion energies with the same dose rate and on the same level of erosion (Figures 1b and 2b). However, the difference in the Auger profiles at those two energies is very pronounced: the influence widths on the profile are very broadened for the ion energy of 5 keV compared with the interface widths obtained by the 1 keV ion energy (Figures 1a and 2a).

- There is almost no difference in Auger profiles obtained when the surface is removed by 1 μA/cm^2, 1 keV Ar ions and 2 μA/cm^2, 1 keV Ar ions (Figures 1a and 3a). The difference in microtopography induced in these two cases is however observable (Figures 1b and 3b).

- The difference in topography is much more pronounced on the sample bombarded by the ions of the same energy (5 keV) but on two, very different levels of erosion: for instance, on the third Cr layer and the last ninth Cr layer (Figures 2b and 2c). This observation implies the conclusion that the total ion dose is more important than the energy of the incident ions for the ion induced microtopography. The Auger spectroscopy profiles are also significantly deformed by the increase of the total ion dose (Fig. 2a).

These observations on the ion induced microtopography, mostly qualitative, are compared with the interface width change in the Auger spectroscopy profiles. The change of the topography and the interface width change by varying the experimental parameters (the ion energy, dose rate and the total ion dose) is presented in Table I in a very simple way. A bigger or smaller increase of a certain quantity in Table I is presented by a bigger or smaller slope of the arrow.

FIGURE 1 : a) The AES sputter depth profile of multilayered Ni/Cr thin film structures obtained by Ar$^+$ ion beam of the 5 keV energy and of the 8 μA/cm^2 dose rate.
b) and c) The replica transmission electron micrographs of the microtopography on two different levels of erosion on one of such samples.

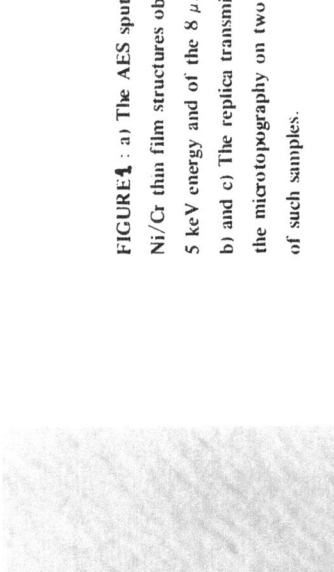

Figure 1b

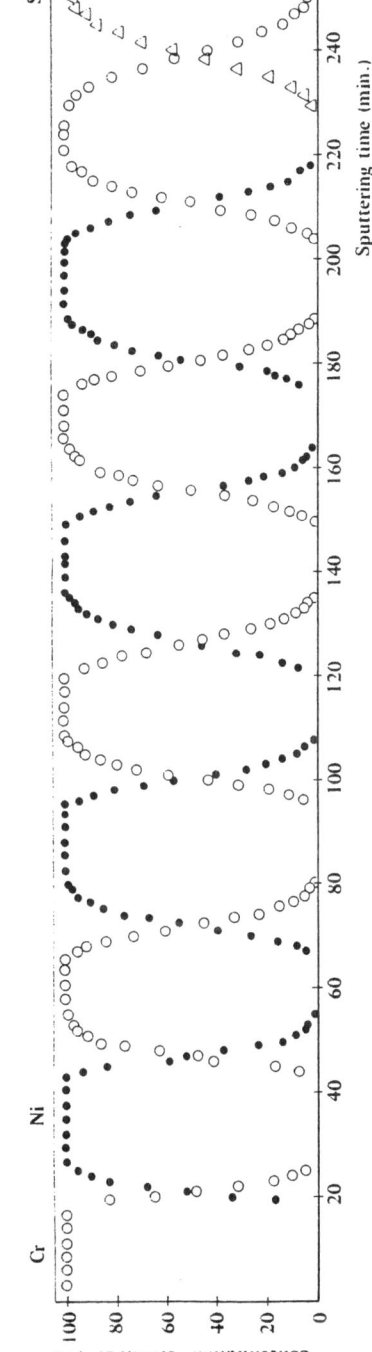

Figure 1a

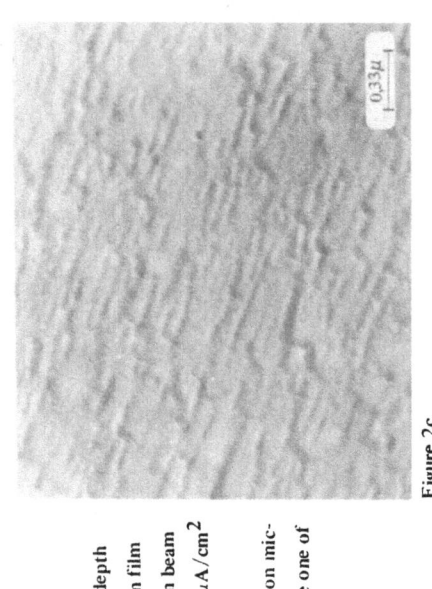

FIGURE **2** : a) The AES sputter depth profile of multilayered Ni/Cr thin film structures obtained by the Ar$^+$ ion beam of the 1 keV energy and of the 8 μA/cm^2 dose rate.

b) The replica transmission electron micrograph of the topography on the one of such samples.

Figure 2c

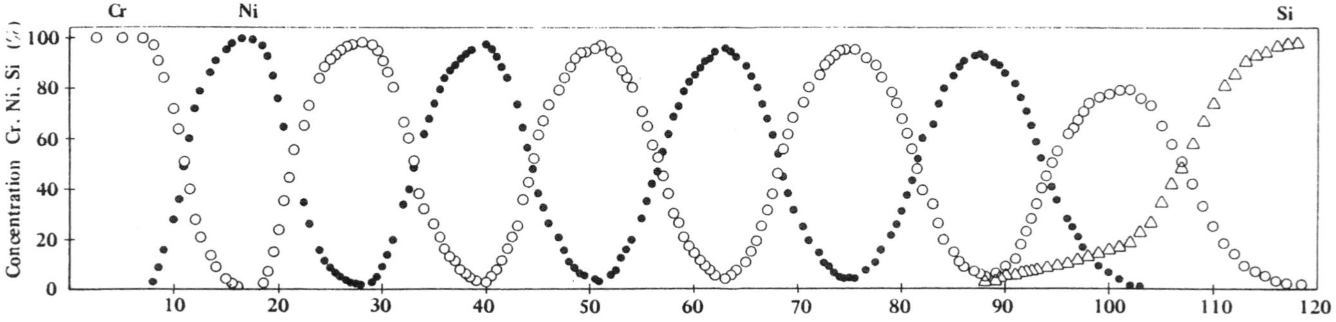

Figure 2b

Figure 2a

FIGURE 3. a) The AES sputter depth profile of multilayered Ni/Cr thin film structures obtained by Ar$^+$ ion beam of the 1 keV energy and of the 16 μA/cm^2 dose rate.

b) The replica transmission electron micrograph of the topography on the one of such samples.

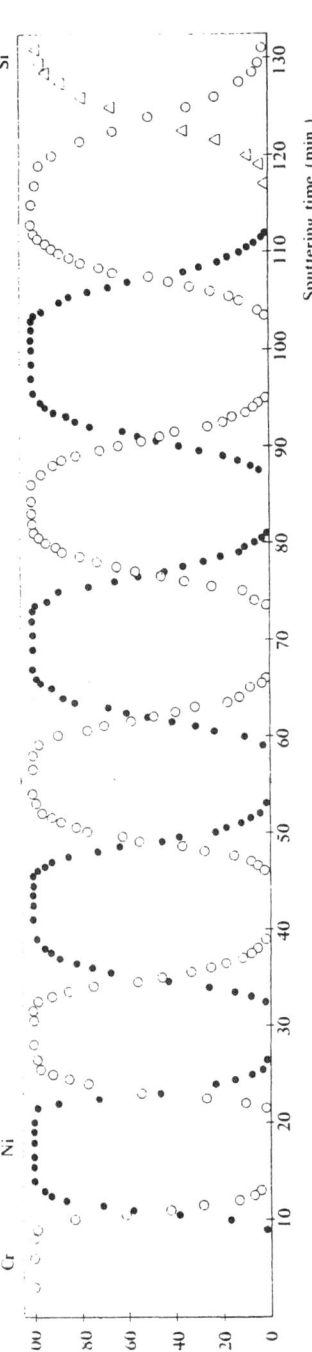

Figure 3a

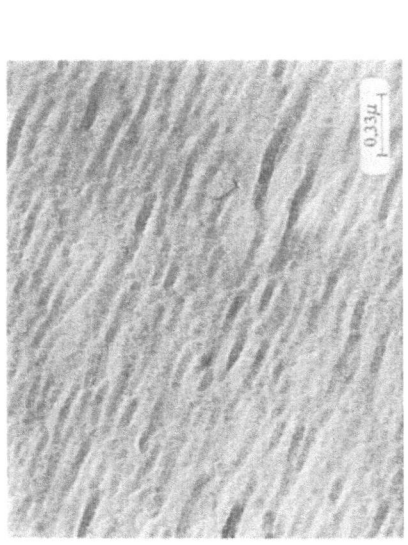

Figure 3b

TABLE I

	Topography	Interface width
Ion energy	↗	↗	↗
Dose rate	↗	↗	↗
Total ion dose	↗	↗	↗

   The difficulty in observing the ion induced topography
of this kind leaves our conclusions about the topography in
this very qualitative level, where we can only say that the
topography is more or less pronounced.
   Further systematic investigations of the ion bombardment
induced microtopography are needed in the range of the experimental parameters interesting for the depth profiling , i.e.
in the ion energy 1-5 keV, and the dose rate range 1-5 $\mu A/cm^2$.
We intend to extend our study in this direction.

ACKNOWLEDGEMENT
This work was supported by the Republic Research and Scientific
Association Sarajevo and the U.S. National Bureau of Standards
under cooperative agreement NBS(G) -282 (YU) through funds made
available to the U.S. - Yugoslav Joint Board on Scientific and
Technological Cooperation.

REFERENCES
1. S. Hofmann, Practical Surface Analysis by Auger and X-ray
   Photoelectron Spectroscopy, Ed. by D. Briggs and M.P. Seah,
   141, J. Wiley & Sons (1983).
2. M.P. Seah, Vacuum 34, 463 (1984).
3. A. Zalar, Thin Solid Films, 124, 223 (1985).
4. J.L. Whitton, SPIG 1980, p.335, Ed. M. Matić.
5. F. Davaria, M.L. Roush, J. Fine, T.D. Andreadis and O.F.
   Goktepe, J. Vac. Sci. Technol., 1 (2), 467 (1983).
6. L. Tanović and B. Perović, Nucl. Instr. Meth., 132,
   393 (1976).
7. L. Tanović, SPIG 1980, p.485. Ed. M. Matić
8. J. Fine and B. Navinsek, SPIG 1982, p.223. Ed. G. Pichler.

EFFECTS OF SURFACE IMPURITIES AND DIFFUSION ON ION BOMBARDMENT-INDUCED
TOPOGRAPHY FORMATION

STEPHEN M. ROSSNAGEL

IBM THOMAS J. WATSON RESEARCH CENTER
YORKTOWN HEIGHTS, NY   10598 USA

A "sputter cone" is a general name given to a class of surface
topographies formed on surfaces undergoing energetic ion bombardment.
Sputter cones are related to ridges, ledges, pits and craters, as well as
rather complicated winding or almost liquid-like surface structures.
Sputter cones were first documented in the early forties by Günterschultze
and Tollmien (1), although the structure was only inferred rather than
observed. In the past 15 years or so, sputter cones have received more
attention, beginning with the work of Mazey, Nelson and Thackery, Nobes,
Carter and Colligon, Stewart and Thompson and also Wehner and Hajicek
(2-13). Since that time, dozens of qualitative and semi-quantitative
observations have been made on these cone-like structures. Three recent
review papers have described the development of the field in great detail
(14-16).

In general, the formation of these surface structures occurs over a wide
range of materials, ion species, and bombardment conditions. For example,
Wehner has described cone formation on 16 materials at varying temperatures
(16). The bombardment conditions can range from energies just above the
sputtering threshold to many tens of keVs with ions from hydrogen to xenon.
Three basic mechanisms of topography formation have evolved to describe
erosional topography. These are: (1) the presence of asperities, defects
or microinclusions, in the near surface region, (2) the presence of
impurities on the surface, and (3) particular microcrystalline
orientations.   There are at least two types of nonerosional topography:
blister formation (and exfoliation) and whisker or related "growth"
formations.   The scope of the present work will deal primarily with the
formation of erosional sputter cones formed due to the presence of surface
impurities. The work by necessity includes aspects of each of the other
mechanisms excluding blister formation.

A typical micrograph of an array of sputter cones is shown in Fig. 1.
These structures are induced by the intentional presence of surface
impurities, as will be described below.   These structures are similar to
many of the features described in the papers listed in the bibliography.
While many of these studies intentionally or implicitly used impurities as
an initiation mechanism for cone formation, numerous other studies have
produced conical or pyramidal features under conditions of very high sample
purity and great efforts to minimize operational contamination. While the
initiation mechanisms of both cone types may involve surface asperities or
defects, the impurity-generated structures have been found to be
differentiated easily from the non-impurity cones in a number of cases
(17).

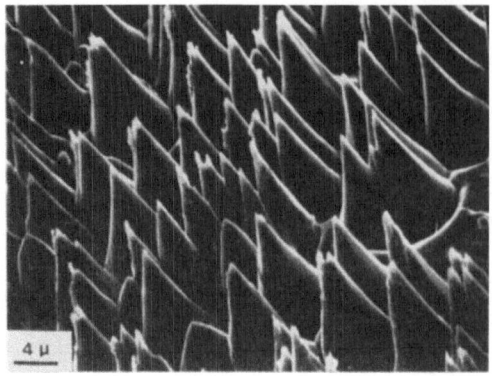

Figure 1. W induced sputter cones on Cu, Dose 2x10^{19} ions/cm,^{2} 1000 eV, 300C

The asperity or non-impurity based cones have a number of general features. They, by definition, form in the absence of measurable impurity concentrations on the surface. Wehner has attributed these cones in all cases to some level of contamination in the pure substrate material (16). The sputter cones often have some form of regular or crystalline features, and are designated "pyramids". And they seem to form only on particular materials, usually of high sputter yield such as Cu, Au, Al and Pb. The density of these structures is typically low as compared to the impurity-based cases. Finally, the most important characteristic is that they appear to form only once, from the original surface. These structures are unstable, and erode away with doses exceeding 10^{19} ions/cm^2. Although there has been some effort here to designate two types of non-impurity cones, in reality they are similar and are simply the result of some local area on the surface having a slightly lower sputter-yield than the surrounding surface for whatever reason. The cone shape is typically attributed to the maximum angle in the sputter yield vrs angle-of-incidence curve. Craters or pits are often the end result of a cone or pyramid which has been sputtered away. The picture of the surface undergoing physical sputtering is a dynamic one, where cones erode, and pits and ridges form and travel across the surface depending on the local variation in the crystal orientation and the angular dependance of the sputter yield.

Impurity based cones have a number of features: (1) the presence of an impurity at levels exceeding 0.1% is necessary for formation and stability; (2) it appears that surface diffusion of the impurities is a prime mechanism in the surface density and physical structure of the cones; (3) quite a number of different materials show cone formation, and there are indications that all vacuum compatible materials may show this structure (18); (4) the sputter cone populations are roughly stable over time as long as the impurity source remains present. From (2) above, the substrate temperature will have a strong effect on cone spacing and structure (18,19,20). This feature is often overlooked or ignored in many sputtering studies. Finally, as the formation of these structures is diffusion dependent, the cones will critically be affected by ion-bombardment-induced diffusion. This topic is, again, often overlooked in numerous sputtering studies or, else introduced in a non-quantitative manner.

The asperity or intrinsic structures and the impurity-induced structures share common relevant phenomena. These include the angle-dependent sputter yield effect causing the conical shape, the reflection of ions down the sides of the cone to form troughs, and some measure of redeposition of material sputtered from the surface impinging on the structure which protrudes above the sputtered surface. In general, both of these structures are classified as erosional topography, i.e., they are structures which remain after the original surface has been sputtered away. There has been extended discussion on this last point, particularly with regards to growth mechanisms and whisker formation. However, the observation remains that these structures in almost all cases are located below the plane of the original surface. One particular exception which will be presented below is the formation of carbon whiskers on graphite surfaces undergoing ion bombardment. These structures do indeed protrude above the original surface, often ten times the height of the material sputtered away, and must be definitely classified as non-erosional or growth features.

The common features between intrinsic and impurity cones have led to a great deal of confusion in the description of sputter cone formation. The rest of this paper deals with the topic of impurity based structures and means for quantitatively characterizing the mechanisms involved. In particular, these mechanisms include thermal surface diffusion and impact-enhanced surface diffusion.

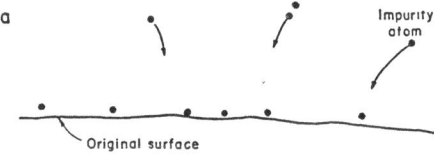

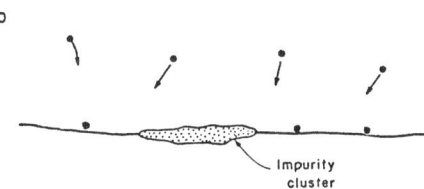

Figure 2. Schematic of impurity-induced cone formation.

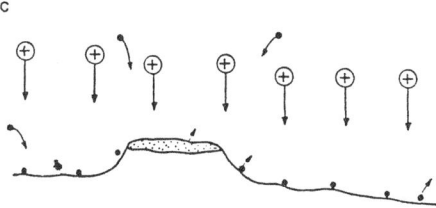

A simple model of impurity-initiated cone formation is illustrated in Fig. 2. In this case, the impurity atoms are initially located on the surface in Fig. 2a. These adatoms have some level of mobility, though, and will diffuse across the surface to eventually nucleate and form islands or clusters of adatoms. The temperature of the substrate will obviously be critical in determining the magnitude of this diffusion. This immediately suggests that experiments done without careful, independent control of the substrate temperature will give no useful information regarding the mechanisms of cone formation. The adatoms on the surface could be present from the bulk material or could be deposited directly onto the surface by evaporation or sputtering from another source. If there is sputtering of the surface occurring during this process, material will be removed from the cluster as well as from the uncovered surface. With a sufficiently high sample temperature to promote adatom mobility, or else a sufficient level of impurities, the supply of impurity adatoms to the nucleating island may equal or exceed the removal rate from the island by sputtering. In this case, the island is dynamically stable and will remain as the surrounding surface is sputtered away. (Fig. 2c). The erosion of the island, as well as the edge of the sputtered area, will tend to form a roughly conical shape corresponding roughly with the maximum in the sputter yield versus angle curve.

Experimentally, the formation of sputter cones is straightforward. Two general procedures are used. The first is simply an ion flux directed at a substrate. The ion flux can be produced in the form of an ion beam, produced either in an accelerator or a broad beam source. The accelerator beams tend to be cleaner, mass analyzed, and highly focussed, but are also generally of low intensity (<0.1 mA/cm^2). Thus, the rate of topography formation is slow. The broad beam ion beams, such as that from a Kaufman ion source, are less clean and not mass analyzed, but can be of very high intensity (>5 mA/cm^2) and can sputter surfaces rapidly. An alternate means of ion bombardment is in a discharge chamber, usually in the form of a triode, where a discharge is initiated by means of an anode and cathode, and an ion current is drawn to the sample by biasing it negative of the anode. These discharges are less clean due to the bombardment of the entire chamber. For intrinsic, or non-impurity structure formation, samples of high purity are sputtered directly. For the impurity case, impurities can be added either to the bulk sample in the form of alloying or inclusions, or deposited onto the surface by evaporation or sputtering. One method of deposition is shown in Fig. 3. In this case, an impurity source is placed in the edge of the ion beam upstream of the sample such that atoms sputtered off the impurity source land on the sample. Calculations of the relative flux in this case show that the impurity flux can be a maximum of a few percent of the incident beam (18). One nice feature of this arrangement is that the relative fluxes are independent of the magnitude of the ion flux and the sample temperature, and depend only on the geometry. Similar arrangements occur often by accident when beam apertures or other "tooling" is exposed to the ion beam upstream of the sample. The use of any aperture in the path of the ion beam may lead to sufficient contamination of the surface to lead to sputter cone formation by the impurity mechanism.

Detailed time-development experiments have been performed on these structures (17). An example of the development of these impurity-based sputter cones is shown in Fig. 4 a-c.

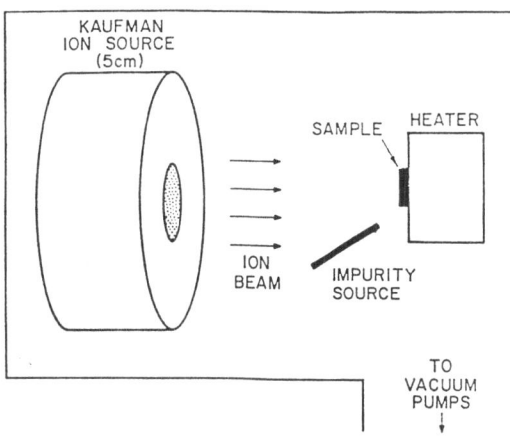

Figure 3. Drawing of experimental apparatus used for impurity induced sputter cone formation. The impurity (seed) source is moveable.

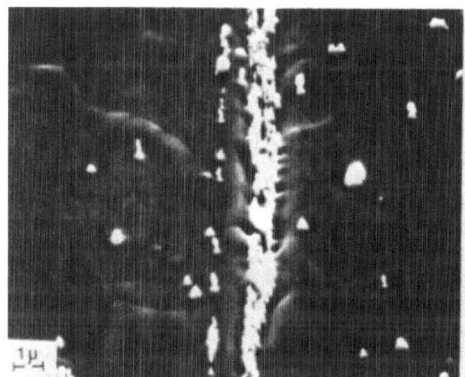
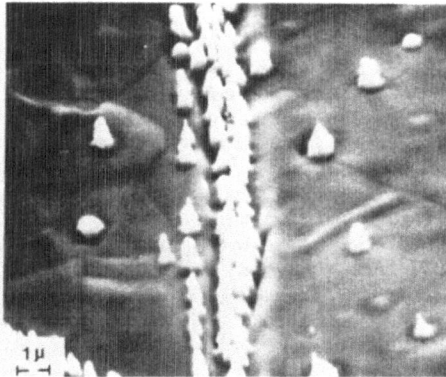

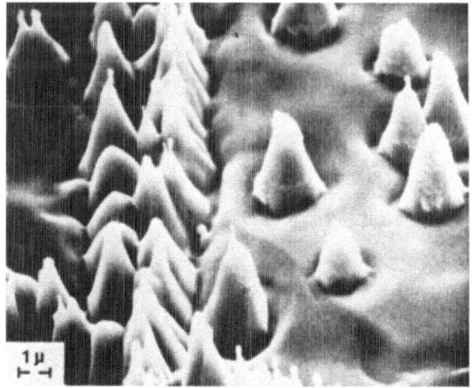

Figure 4. Time development sequence for impurity-induced sputter cone formation. Parameters: Mo impurity on Cu, 300C, 1000 eV. a) 2×10^{18} ions/cm^2, b) 5.6×10^{18} ions/cm^2, c) 2×10^{19} ions/cm^2.

While a detailed description of this time development is not appropriate here, there is evidence in the work of an overlying coating over the cone structure (17). This overlying coating or film apparently protects the surface below it. As the sputtering time is increased, the coating covers a larger and larger area, corresponding to a larger cone size. As the same areas are monitored at larger and larger ion doses, the cone structures enlarge, but their spatial density remains roughly constant. If the impurity source is removed and the ion bombardment continued, the cones actually continue to increase for some time. Eventually, the coating fails or breaks and the cone begins to erode. The failure mechanism is signalled by a break in the coating over the cone, exposing the substrate surface below it. The break increases in size until the coating is entirely removed. The underlying substrate is rapidly eroded due to the high angle of incidence of the substrate material, and the result is soon a broad, flat crater. Chemical analysis of this overlaying coating by means of Scanning Auger Microscopy (SAM) indicates that the coating is not an island of pure impurity material, as suggested by the simple model presented here, but is a dilute alloy, consisting of 2-9% impurity levels in a background of substrate material (21). This feature slightly complicates the clustering model, but in general the basic formalism remains.

Upon closer examination of the impurity/alloy coating overlying the sputter cone, a number of curious structures are observed (21). These features include what appear to be droplet structures on the top of cones, liquid-like drips on the sides of cones, ripples in the coatings, and various agglomerate features which are suggestive of the connection of closely spaced cone structures. These features have been designated "quasi-liquid" because of their liquid-like appearance even though their temperature remains well below even one half of the melting point of the substrate-impurity eutectic. Examples of some of these structures are shown in Fig. 5 a,b and may be seen on closer examination of some of the other figures both in this work and in the work of numerous others. The examples shown here are for Copper substrates whose temperature was held by an adjacent heater to 300C and bombarded with ions at 1000 eV and 1 mA/cm^2 (1 watt/cm^2). The samples were thermally backed to a copper heat sink, which formed part of the heater, and the surface temperature could not have exceeded the measured temperature by more than a few degrees due to the ion bombardment. We are left, then, with the apparent large scale (microns), movement of material on the surface at temperatures well below a melting point.

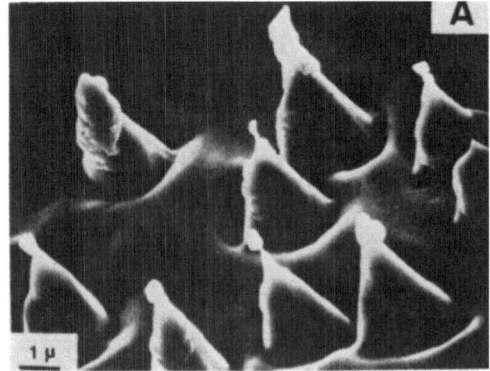

Figure 5a. Carbon induced cones on Cu, 300 C, 7 x 10^{18} ions/cm^2.

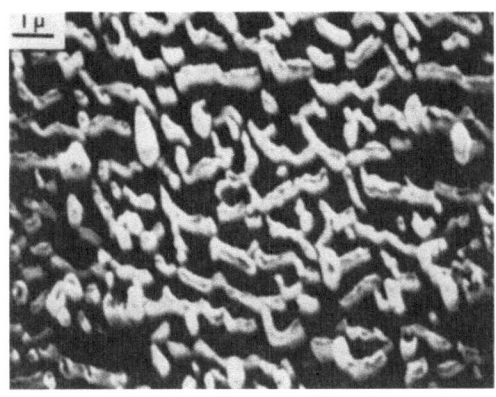

Figure 5b. Tungsten induced structures on Cu, 460 C, 1 x 10^{18} ions/cm.^{2}

One explanation of this quasi-liquid motion of the surface coating is due to the high level of defects induced in the coating by the energetic ion bombardment (21). Each bombarding ion induces defects in the coating. In addition, the after-effects of each ion-induced cascade may lead to increased mobility of these defects within the film. This enhancement of bulk diffusion by increased defect creation and mobility has been demonstrated in a number of cases (22-32). Were the defect level in the coating to become too large due to the ion bombardment, a "partial second-order phase transition" might occur, allowing slippage or apparent flow in the coating (21). The effect of the ion bombardment, then, is to increase the defect level to what it might normally be at temperatures close to the melting point in the absence of ion bombardment. Thus, the film is not melting, but acts as if it is close to doing so.

It should be noted here that one particular type of phenomena described elsewhere in the literature (16) is NOT happening here. Figure 6 illustrates the structure in question. The structure consists of a small conical base extending upwards to a very fine whisker-like neck. Above the neck is a rounded ball many times the size of the neck. Actual measurements have shown the neck to be as fine as 125Å in diameter, with balls of several thousand Angstrom diameter. These structures are physically stable, even though a simple calculation of the strength of the neck would show it incapable of supporting the weight of the ball. Others have described these as melted-down whiskers (16). Were the power incident on the whisker sufficient to melt it, there would be little hope of the melting stopping short of collapsing the entire structure. The power flux on the ball exceeds by sever orders of magnitude the flux on the "original whisker", yet these structures are quite common and cannot be assumed to be chance occurrences or catching a collapsing whisker prior to the total collapse.

Two other observed phenomena will be presented prior to a more in depth look at surface diffusion processes. They are cone bending and crystalline aspects of the cones. Cone bending has been observed in a number of cases and attributed to the high stress on the structure induced by the ion bombardment. Typically, cones have been observed to bend down in a type of spiral, which can be related to a minimization of the total surface area. These bends are random in direction, and not particularly controllable. It is possible, however, to intentionally bend cones in a desired direction (21). This has been done by heating the cone from one side during the last

188

few moments of sputtering. The heating was from a hot filament near the edge of the sample. The result is shown in Fig. 7. The cone tips point regularly away from the heat source, characteristic again of a highly stressed coating. The magnitude of the bending was related to the amount and length of the heating, and was critically dependent on the concurrent presence of the ion bombardment. Without ion bombardment, no bending occurred. If the heat source was on for too long a time, the cones "lay down" on the substrate, and lost most of their structure (Fig. 8). The bent cones, in any case, appeared stable for long periods of time, and were not affected by heating without ion bombardment.

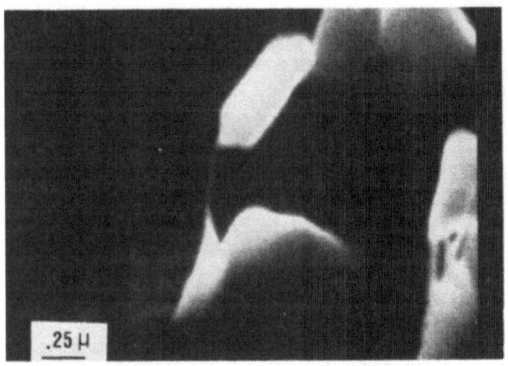

Figure 6. Impurity induced sputter cone formation on Cu, 400C, 1000 eV, 1×10^{19} ions/cm^2.

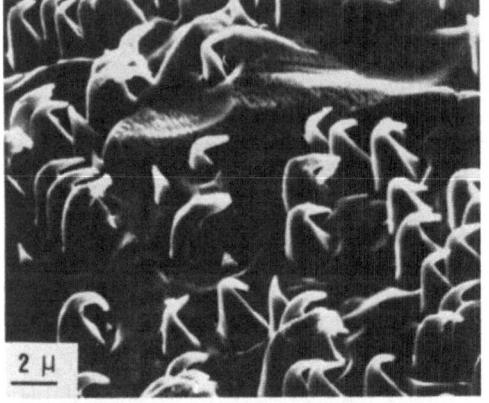

Figure 7. C induced cones on Cu bent by heating from the left at the end of the ion bombardment.

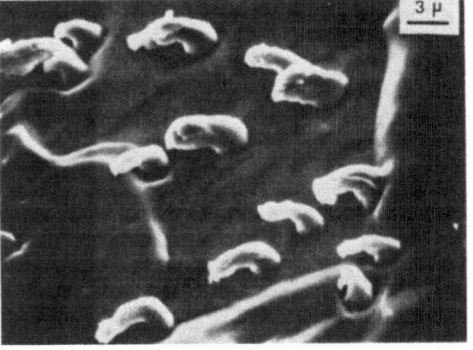

Figure 8. C induced cones on Cu bent by heating from the right for a longer period prior to the end of the bombardment.

The second remaining topic is that of the crystalline aspect of impurity-based cones. Intrinsic, or non-impurity based cones are often designated pyramids because of the presence of facets on the cone sides. It appears reasonable to relate these facets to planes in the substrate material which have lower sputter yields, as the higher yield planes are eroded first. In the impurity case, any structure of the cone is often hidden by the presence of the overlying coating. This coating shows no crystalline features, and often seems as though it is barely solid to begin with (21). At the edge of the coating, near the base of the cone, it is often possible to observe crystalline features on the sputter cones (33). The easiest way to observe this is to view the cones from directly overhead, or in terms of SEM work, at 0 degrees tilt. Figure 9 shows an area of a Cu sample in the presence of Mo impurities. This area shows some evidence of grain boundaries traversing the sample. In each case, there appears to be regular crystalline features in the cone base. Figure 10 sketches out the differing structures on three planes in Cu from single crystal work (33).

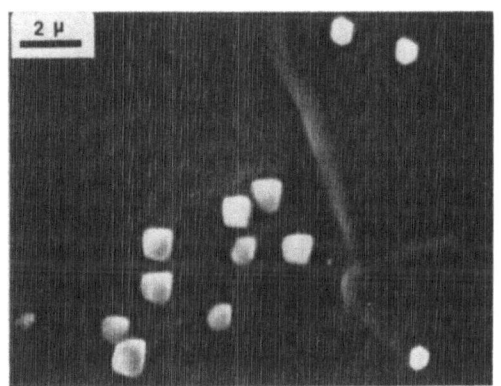

Figure 9. Top view of a multicrystalline Cu substrate showing different cone formations.

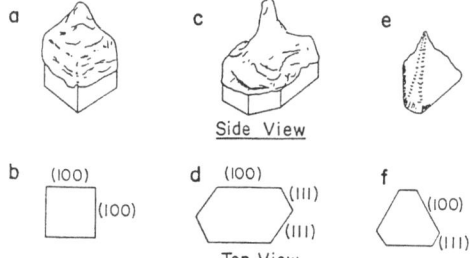

Figure 10. Sketches of cone structure and crystal orientation for cones formed on single crystal Cu.

SURFACE DIFFUSION PHENOMENA

The primary phenomenon by which impurity-induced cones are formed during sputtering is surface diffusion. As described above in Fig. 2, impurities in some way form moderately stable clusters on the substrate surface, and the end result after sputtering is the formation of a cone. It is possible to quantify this process in terms of adatom mobility, surface diffusion activation energy and temperature. An experimental observation, to be quantified below, is that the surface density of cone structures depends inversely on the substrate temperature, i.e., high temperatures produce low densities of cones and vice-versa.

An impurity atom or "adatom" on the surface is located in some adsorption site determined to first order by the local crystal structure. The adatom will oscillate in the potential well described by this site at a frequency given by (18,20).

$$f = (E_d/2ma^2)^{1/2} \qquad (1)$$

where E_d is the height of the potential well, or the activation energy for surface diffusion, m is the adatom mass, and a is the width of the well, or equivalently the distance between adsorption sites. The probability of motion of the adatom is this frequency times $\exp(-E_d/kT)$, where k is Boltzman's constant and T the absolute temperature. The random walk path length of the adatom on the surface will be related to how long it sits on the surface. While the surface is being sputtered the surface lifetime is inversely proportional to the ion bombardment rate. The random walk path length then becomes:

$$r = (E_d/2m)^{1/4} (a/RYs)^{1/2} \exp(-E_d/2kT) \qquad (2)$$

where s is the adatom sputtering cross section, Y the yield (probably 1), and R the incident ion bombardment rate. From nucleation theory, this random walk path length has to be on the order of the spacing between nucleation sites. If the sites were spaced farther away than a few times this distance, most of the adatoms would never reach a site, and the possibilities for stability of the cluster would be small. If the nucleation sites were much closer together than the path length, the clusters would grow rapidly as material is accumulated rapidly. The average spacing of the nucleation sites, then, is on the order of this random walk path length, and should vary linearly with it. The stability of the cluster at the nucleation site also depends on the arrival and removal rate of the impurity atoms. Material removed by sputtering must be replaced by diffusion. There is also likely to be a minimum volume for a stable cluster. This is due to the incident ion energy. If the incident ion brings more energy to the cluster than the sum of its binding energies, it is likely to be blown apart. This translates into a minimum temperature for stable cluster formation (18). That minimum temperature has the form

$$T = E_d / (k \ln (E_d/2m)^{1/2} (Fa/RYr^2 (Y-F))) \qquad (3)$$

where F is an arrival rate ratio of impurity to ion species. Typically, this temperature might be a few hundred degrees C.

The general, underlying relation in this model is that the location of the cone structures corresponds with the location of the dynamically stable impurity surface cluster. The average spacing of the cones, therefore, corresponds with the average spacing of the clusters and is a measure of the surface diffusion. From equation (2) above, the average cone spacing can be related to the temperature if all other parameters are held constant. If the natural logarithm of the average cone spacing is plotted against inverse temperature, a straight line is expected, with slope proportional to the surface diffusion activation energy. Several measurements of this type have been done (18,19,20). A sample plot for the case of Mo-induced cone structures on Al is included (Fig. 11). A number of other experimental systems have been examined with this formalism. The activation energies are listed in the Table for 5 different substrates with up to 11 different impurities. What is observed from this table is that there is little correlation between the activation energy and the impurity material. The activation energy is roughly constant within experimental error for each of the 5 materials studied. This is in contrast with slightly more direct measurements using field emission (22,23). The values found for 4 of the materials are in the 0.2-0.4 eV range, which are quite low compared to the values found in the field emission cases. The apparent dependence on the substrate material and not on the adatom species suggests that other mechanisms may be involved. In particular, the effect of the ion bombardment on the surface diffusion has been ignored. As will be shown below, this effect can be significant and can alter the observed activation energy.

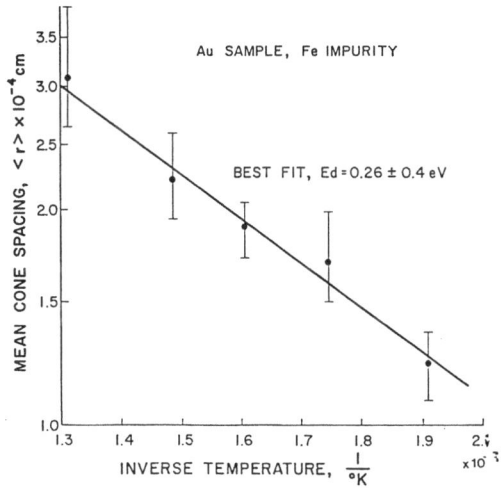

Figure 11. An example of the type of mean cone spacing versus inverse temperature plot used to determine activation energies.

Seed Materials	E_d (eV) Substrates				
	Cu	Ni	Pb	Au	Al
C	0.42	-	-	-	-
Al	0.35	0.30	-	-	-
Ti	0.41	-	0.32	0.28	0.93
Cr	0.43	-	-	-	-
Fe	0.37	0.32	0.30	0.26	1.0
Ni	0.46	-	0.33	-	-
Zn	0.31	-	-	-	-
Mo	0.35	0.36	0.31	0.28	1.20
Ta	0.36	-	-	-	0.97
W	0.37	-	-	-	-
Au	0.39	-	-	-	1.04
Cu	-	0.30	0.32	0.29	-
Statistical Uncertainty	±.06	±.06	±.06	±.06	±.09

TABLE. Measured surface diffusion activation energies.

The above formalism, which has been developed in more detail in earlier work (18-21,33-36), is for the most part independent of the choice of impurity and substrate materials. This is in contrast to the observations of others that cones only form when the impurity material is of higher melting point than the substrate(16). No theoretical formalism is given for those observations. It has been shown by the author and his colleagues that it is possible to form cones on materials with impurities of lower melting point, in direct contrast to the melting point rule. In terms of the above discussion on quasi-liquid formations observed on cone structures, there may be some necessity for this impurity coating to have a significantly different melting point than the substrate, but intuitively one might expect the impurity coating to have a LOWER melting point. In any case, discussion of melting points is at this point purely speculation and probably an artifact of some more fundamental parameters.

The simple surface diffusion described here has for the moment ignored the effect of the ion bombardment on the diffusion. All of the measurements described were for an ion current density of 1 mA/cm^2. The effect of the ion bombardment is to increase the diffusion above the thermal levels, although the diffusion still appears to be characterized in terms of a modified thermal function.

ION IMPACT ENHANCED SURFACE DIFFUSION

Impact enhanced diffusion has been observed in a number of areas. Enhanced adatom mobility has been reported on ion-bombarded field emission tips (24-26) as well as in the bombardment of condensing films (27,28). A larger area of study has been in near-surface bulk diffusion induced by high energy ion bombardment (29-31). Another area of study is ion enhanced CVD, where epitaxial growth is observed at low substrate temperatures in the presence of ion bombardment(32). More recently, there is a growing body of intriguing work in the area of ion-bombardment-induced sub-surface segregation. Two recent reviews have examined this subject in detail (37-38). A companion paper in these same proceedings has also examined this subject in detail (39). A recent review has treated bombardment-induced film modification, although not particularly in terms of surface diffusion (40).

From the surface diffusion formalism briefly described above (eqn. 2), the cone spacing should indeed be dependent on the ion current density. The equation shows an inverse relationship, which is due to a reduction in the adatom lifetime on the surface as a function of increasing bombardment rate. Experimentally, when the ion current was varied at constant temperature, the cone spacing was found to strongly INCREASE with increasing current density. A plot of this is shown in Fig. 12. It should be noted in these experiments that the net power flux to the samples was small. The samples were heat-sunk to a large, temperature controlled heat-sink. Therefore, any heating due to the increasing ion current density could be easily monitored and controlled. The dependence of cone spacing at constant temperature on ion current density appeared to be a general phenomena, independent of the choice of materials. The temperature did also effect the ion bombardment induced diffusion. At higher temperature, the effect was increased, at low temperature (200C), the effect was minimized. At each current density, an activation energy for surface diffusion could be measured by relating the cone spacing to inverse temperature, as seen before. The activation energies found could then be plotted as a function of ion current density, as shown in Fig. 13 for Mo impurities on Cu. That an activation energy could be fitted at all to this data is suggestive that the surface mobility, as described in terms of the cone formation, is still describable in terms of surface diffusion. The activation energy for surface diffusion varies linearly with ion current density over the range studied. The intercept at zero ion current is not zero, however, but 0.17 eV. One possible explanation for this behavior is that the original, unperturbed surface is characterized by the activation energy equal to the intercept, 0.17 eV. The effective activation energy, however, in the presence of ion bombardment is a weighted average of this low initial energy with some additional higher energy adsorption sites induced by the bombardment. As the bombardment is increased, this effective activation energy increases due to the higher percentage of the higher energy adsorption sites. From the data, those sites would have to exceed 0.7 eV.

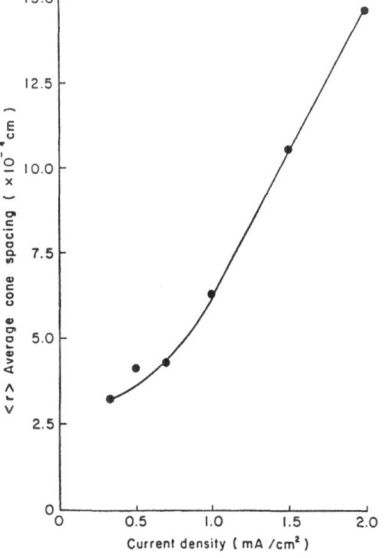

Figure 12. Average cone spacing as a function of ion current density. Sample: Mo induced cones on Cu at 300C, 1000eV ion energy.

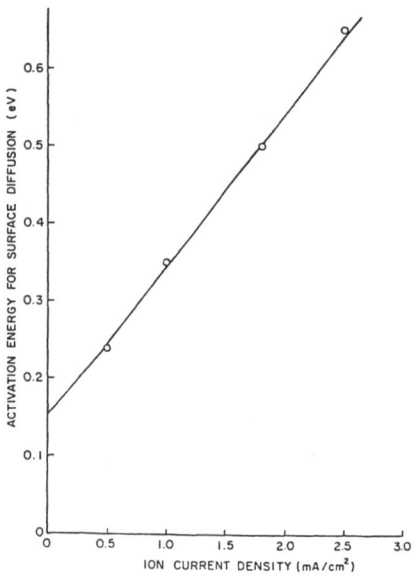

Figure 13. Measured activation energy for surface diffusion as a function of ion current density. Sample: Mo induced cones on Cu at 1000 eV ion energy.

The increase in the effective activation energy for surface diffusion during ion bombardment is in the opposite direction of the data shown in Fig. 12. Increasing the activation energy as a function of ion current density should strongly decrease the observed diffusion as characterized by the cone spacing. Therefore, the pre-exponential or frequency part of the diffusion equation is being modified by the ion bombardment, and this modification must be significant enough to overcome the deleterious effect of the increased effective surface activation energy.

The mean square diffusion length for a random walk is

$$r = a \cdot N_j^{\frac{1}{2}} \tag{4}$$

where N_j is the total number of jumps made by the adatom. The total number of jumps is the sum of the thermally activated jumps plus the ion impact induced jumps. Both processes are summed over the adatom lifetime. If each incident ion is assumed to generate N_i single step jumps, then the total number of jumps is

$$N_j = N_i/nsY + (f/RsY) \exp (-E_d/2kT) \tag{5}$$

where n is the surface adatom density which ranges from 0.1 to 5% of a monolayer. N_i will be some function of the ion current density. From the linearity of the average cone spacing with ion current density (Fig 12) at the higher ion current densities, and using the equation 4 above, the number of induced jumps under these conditions appears to be proportional to the square of the ion current density, suggesting collective effects between nearby ion impacts. An estimate of the magnitude of the impact enhanced diffusion can be obtained by substituting reasonable values into the above set of equations. When this is done, the number of jumps for an individual adatom attributable to the ion bombardment ranges from a few to many thousands (34). The number of jumps found in the similar field emission work was in the 200 range, although the effective current density was lower. That work was also concerned with self-diffusion, which might be expected to proceed more slowly than impurity diffusion.

MECHANISMS FOR ION ENHANCED SURFACE DIFFUSION

There are two basic results in the impact-induced surface diffusion case. The first is the increase in the effective activation energy with increases in the ion flux. The second is the contradictory increase in observed diffusion, under the same conditions. Several mechanisms may be playing a part, but the goal is to find one which explains both of these results. At present, four mechanisms have been examined: defect formation and motion, recrystallization, thermal spike formation, and the effect of surface phonons on adatoms. In examining each of these processes, only the last one does not fail upon examination of the results.

Defects

In the case of volume diffusion, the formation of point defects has been shown to be a dominant effect in ion-impact enhanced diffusion. The impact causes an increase in the vacancy concentration, which causes an increase in diffusion through the vacancy mechanism. Substitutional atoms are also injected into interstitial sites, where they can diffuse rapidly (41). The surface diffusion of impurities does seem to be affected by defect formation. This explains the apparent increase in the activation energy, as on the surface a vacancy-like defect would be a stronger adsorption site. The numbers of these defects would undoubtedly increase with ion current to some power. However, the result remains, that given the increase in activation energy, the diffusion still increases significantly. As there would be only minor modifications in the potential well oscillation frequency at a defect sites versus a non-defect site, the formation and mobility of defects cannot be seen as the driving force in the impact enhanced diffusion.

Recrystallization

Recrystallization of thin films undergoing ion bombardment has been observed both experimentally and in computer simulations (42). The impact of the ion leads to a loss of order in the collision cascade region, which is eventually regained as the cascade volume cools. In the computer simulation work, Harrison and Webb found that little net movement of material occurred. That is, atoms returned to lattice positions quite close, on the average, to their original location prior to the cascade. The movement of impurities at the surface of a magnitude needed to justify the present results is also not possible. Taking a generous size for the surface cascade area, the number of adatoms within it is still only a ten or so atoms, and the spatial region which is in transition allows only a few dozens of jumps. This is well below the magnitude needed to justify the results.

Thermal Spikes

A slightly different aspect of the cascade is the so-called thermal spike formed after the ion impact. The surface of the cascade region can be thought of as a very hot region, although for a very short period of time. The thermal spike effective temperature can be several thousand degrees, for times on the order of 10^{-10} seconds. This increase in temperature at the surface would cause the adatoms to diffuse much faster for that time period. The diffusion is then of the form:

$$N_i = \iint nf \exp (-E_d/k(T + T_r)) \cdot dt \cdot dr \qquad (6)$$

where f is the fraction of the surface influenced by the spike at any given time. This fraction should only depend on the ion current density, and not on the substrate temperature. The temperature increase due to the spike, T_r, depends both on the time dt and the distance dr from the impact point. The temperature effect of this relation (6) is such that the effect should be greatest at low temperatures, where the diffusion is relatively lower. That is, at cold temperatures, the effect of the thermal spike should be significant, at respectively higher temperatures, the effect should diminish. The experimental result is just the opposite. An additional point, similar to the recrystallization case, is that there are only a low number of adatoms in the area affected by the spike, and the observed diffusion requires much higher levels of adatom movement.

Phonons

The discussion so far has centered on the local area of the ion impact, characterized as the cascade region. This area has traditionally been most important in terms of sputtering and recrystallization. However, the energy present in the cascade has to go somewhere. At the fringes of the cascade, where the average kinetic energy drops below the dissociation energy, the energy is partitioned into some large number of phonons and bulk heat conduction becomes the predominant process. This is certainly consistent with the common observation that samples heat up when irradiated. In addition to the bulk phonon process, there is a surface component to the phonon process. Since the impact is at one surface, the result after the cascade will be predominately compressional phonons traveling into the bulk of the sample with a second group of largely transverse phonons traveling across the surface.

The surface phonons are therefore covering a much larger area than the area of the cascade, which in the recent work is desirable because of the large numbers of adatoms encountered. The large area also makes the probability of collective effects, i.e., interactions between adjacent impacts, more probable. The diffusion, as described above, appears proportional to the square of the current density, which implies the interaction of two impacts. Some order of magnitude estimates can be made of the numbers of atoms involved and the energies. The maximum number of phonons can be estimated from the incident ion energy. In this case at 1000 eV, a few tens of thousand phonons are produced, some fraction of them available at the surface. Within a radius of one half micron from the impact point, the observed diffusion can be justified by phonon-adatom iteration probabilities in the 10^{-4}- 10^{-6} range.

This type of interaction is desirable because it preserves the pseudo-thermal aspect to this diffusion experiment. While the activation energy is changing as a function of ion current density, the process still behaves thermally.

WHISKER FORMATION DURING SPUTTERING

Whiskers formed on surfaces undergoing ion bombardment are one of the least reported surface topographies (43, 44). Unlike the sputter cones, which seem to be predominately erosional structures, whiskers are indeed growth features. Wehner has reported a few observations of whiskers formed on surfaces bombarded at low energies and high fluxes and temperatures, and has made the rather broad generalization that all sputter cone features may be based on whisker growth (16). While that topic is still very much in dispute, the formation of graphite whiskers during ion bombardment is definitely established. Graphite whiskers have been grown which are 10-50 times longer than the depth of the material sputtered away. A recent report (44) has described some of the growth dynamics of these whiskers. Unlike sputter cones, which are often associated with impurities, carbon whiskers on graphite form during sputtering without the addition of impurity species. In fact, the addition of impurity material strongly inhibits the growth, and reduced the whisker density by several orders of magnitude. The whiskers formed are 100-10,000$\overset{\circ}{A}$ in diameter, and of length ranging from a few thousand angstroms to 50 microns. The initial growth process appears to be a rather fast process, with the whiskers pointing in random directions with respect to the incident ion beam. With longer sputtering times, the whiskers are observed to all be pointing in the ion beam direction, and have a length and diameter which increase roughly linearly with time. (Figs. 14,15). The unification of direction can be explained in terms of sputtering. Whiskers inclined more than a few degrees from the ion beam direction have a much higher area exposed to the sputtering. The supply mechanism, assumed to be a impact enhanced diffusion, of carbon atoms to the whisker will be insufficient to make up for the losses for whiskers which are too far off the axis of the ion beam.

The density of the whisker structures can be quite high, on the order of 10^8 per square centimeter, and they are efficient absorbers of light. The mechanisms of formation are not well established. No impurities have been observed either with RBS or AES, and crystallographic studies are just beginning to be done (45).

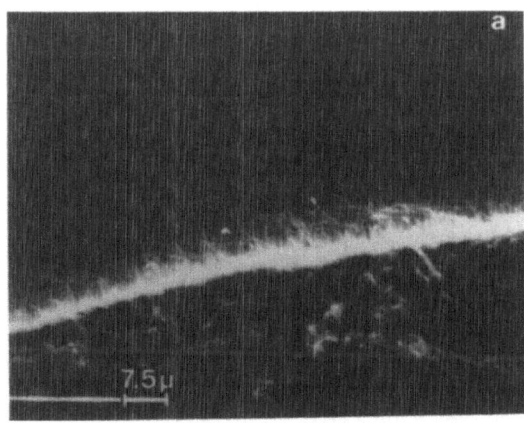

Figure 14. Side view (90 degrees tilt) of whiskers formed on carbon during ion bombardment. Parameters: 475C, 1000 eV Ar, 2 mA/cm^2, 3.7×10^{18} ions/cm.^{2}

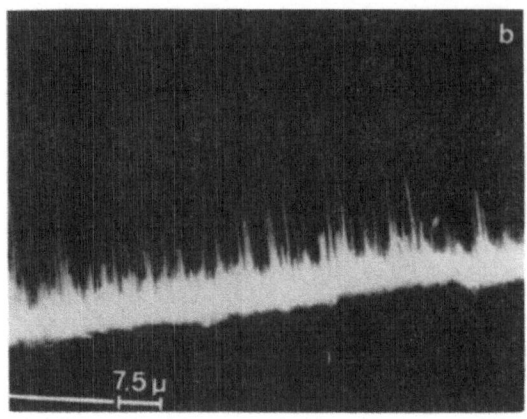

Figure 15. Side view of whiskers formed on carbon at 475 C, 1000 eV Ar, 2 mA/cm , at a dose of 3 x 10 ions/cm

ACKNOWLEDGEMENTS

The author gratefully acknowledgements the collaboration and support of Professor Steve Robinson of Colorado State University. Much of the thermal diffusion works comes directly from him. I am also grateful for the support of Professor Harold Kaufman, formerly of Colorado State University. Much of this work was supported by a number of NASA grants from the NASA-Lewis Research Center.

REFERENCES

1. V.A. Gunterschulze and W. Tollmien, Zeit. Physik 119 (1942) 685.
2. D.J. Mazey, R.S. Nelson and P.A. Thackery, J. Mat. Sci. 3 (1968) 26.
3. M.J. Nobes, J.S. Colligon and G. Carter, J.Mat.Sci. 4 (1969) 730.
4. A.D.G. Stewart and M.W. Thompson, J. Mat.Sci. 4 (1969) 56.
5. G.K. Wehner and D.J. Hajicek, J. Appl. Phys. 42 (1971) 1145.
6. I.H. Wilson and M.W. Kidd, J. Mat. Sci. 6 (1971) 1362.
7. D.J. Barber, F.C. Frank, M. Moss, J.W. Steeds and I.S.T. Tsong, J.Mat. Sci. 8 (1073) 103.
8. M Cantagrel and M. Marchal, J. Mat. Sci. 8 (1973) 171.
9. G. Carter, J.S. Colligon and M.J. Nobes, J. Mat. Sci 8 (1973).
10. R. S. Nelson and D.J. Mazey, Rad. Eff. 18 (1973) 127.
11. P. Sigmund, J. Mat. Sci. 8 (1973) 1545.
12. I.H. Wilson, Rad Eff 18 (1973) 95.
13. W. Hauffe, Phys. Stat. Sol. 4a (1971) 111.
14. G. Carter, B. Navinsek and J.L. Whitton, in Sputtering by Particle Bombardment II, ed. by R. Behrisch (Springer Verlag, Berlin 1984) 231.
15. O. Auciello, in Ion Bombardment Modification of Surfaces ed. by O. Auciello and R. Kelly (Elsevier, NY 1983) 1.
16. G.K. Wehner, J. Vac. Sci. & Technol. A3 (1985) 1821.
17. S.M. Rossnagel and R.S. Robinson, Rad. Eff. Lett. 58 (1981) 11.

18. R.S. Robinson, PhD Thesis, Colorado State University (1979).
19. G.W. Lewis, J.S. Colligon, F. Paton, M.J. Nobes, G. Carter and J.L. Whitton, Rad. Eff. Lett. 43 (1979) 49.
20. H.R. Kaufman and R.S. Robinson, J. Vac. Sci. & Technol, 16 (1979) 175.
21. S.M. Rossnagel and R.S. Robinson, J. Vac. Sci. & Technol, 20 (1982) 506.
22. R. Gomer, Sci. Amer. (Aug 1982) 98.
23. G.L. Kellog, T.T. Tsong and P. Cowan, Surf. Sci. 70 (1978) 485.
24. M. Drechsler, M. Junack and R Meclewski, Surf. Sci. 97 (1980) 111.
25. Zh. I. Dranova and I.M. Mikhailovskii, Sov. Phys. Sol. St. 12 (1970) 104.
26. J.Y. Cavaille' and M. Drechsler, Surf. Sci. 75 (1978) 342.
27. M. Marinov and D. Dobrev, Proc. XVth Czech. Conf. on Electron Microscopy, Prague, Vol B (1977) 569.
28. M. Marinov, Thin Solid Films 46 (1977) 267.
29. A.H. Eltoucky and J.E. Greene, J. Appl. Phys. 51 (1980) 4444.
30. R.L. Minear, D.G. Nelson and J.F. Gibbons, J. Appl. Phys 43 (197 3468.
31. J.E. Greene and S.A. Barnett, J. Vac. Sci. & Technol. 21 (1982) 285.
32. H. Shimizu, M. Ono, N Koyamn and Y. Ishida, J. Appl. Phys. 53 (1982) 3044.
33. J.A. Floro, S.M. Rossnagel and R.S. Robinson, Rad.Eff. Lett. 68 (1982) 57.
34. S.M. Rossnagel, R.S. Robinson and H.R. Kaufman, Surf. Sci. 123 (1982) 89.
35. S.M. Rossnagel and R.S. Robinson, J. Vac. Sci. & Technol. 21 (1982) 790.
36. S.M. Rossnagel and R.S. Robinson, J. Vac. Sci & Technol. 20 (1982) 336.
37. H. Wiedersich in Ion Beam Modification of Surfaces ed. by G. Foti, J.M. Poate and J.S. Jacobson (Plenum, NY 1983), 127.
38. H.H. Anderson in Ion Implantation and Beam Processing ed. by J.S. Williams and J.M. Poate (Academic, NY 1984).
39. Roger Kelly in Proceedings of Erosion and Growth of Solids Stimulated by Atom and Ion Beams ed. by G. Carter and G. Kiriakidis (in press, 1986).
40. J.M.E. Harper, J.J. Cuomo, R. Gambino and H.R. Kaufman, Ion Bombardment Modification of Surfaces ed. by O. Auciello and R. Kelly (Elsevier, NY 1984), 127.
41. S.M. Myers, Nuc. Instrm. Meth. 168 (1980) 265.
42. D.E. Harrison and R.P. Webb, Nuc. Instrm. Meth. (1984).
43. J.A. Floro, S.M. Rossnagel and R.S. Robinson, J. Vac. Sci.& Technol. A1 (1983) 1498.
44. J.J. Cuomo and J.M.E. Harper, IBM Tech. Discl. Bull. 20 (1977) 775.
45. W. Sollberg, S.M. Rossnagel, and J.J. Cuomo (in preparation).

ION TRAPPING AND CLUSTER GROWTH

A. VAN VEEN
Delft University of Technology, Interuniversity Reactor
Institute, Mekelweg 15, Nl-2629 JB Delft, The Netherlands.

During irradiation of a solid surface with energetic ions the fate of an
implanted ion is determined 1. by its location in the lattice, 2. the defect concentration around the implanted atom, and 3. the mobility of the
implanted atom either by thermally activated migration of by interaction
with other mobile defects. At low irradiation fluence the fate of the implanted atom is not influenced by interactions with other collision cascades caused by other ions nor with the mobile defects emerging from these
cascades. The surface acts as the dominant sink for the defects. At increasingly higher fluences the effects of interference between products of
succesive collision cascades become visible. Non-linearity in the growth
of defect populations is then observed, sometimes including clustering of
the implanted atoms. Trapping processes and the early stages of cluster
development are discussed for He, the other noble gas atoms, light impurities H and N, and metallic impurities implanted in metals.

1. INTRODUCTION

The trapping of energetic particles is only one of the many processes
that may occur as a result of ion-bombardment of a solid surface. The
process can not be considered independently of the other processes generated by the ion impact, e.g. damage creation and sputtering. Furthermore
thermally activated processes play a role. Non-linearity in trapped amounts
vs. ion fluence is expected when the accumulated damage exceeds certain
concentration limits.

A good example of the relation that exists between trapping and damage
creation is given by the trapping of helium in tungsten. Low energy helium
injected with an energy below the threshold for damage production will not
be trapped in tungsten except when the crystal has been predamaged before
e.g. by heavy ion bombardment as has been demonstrated by Kornelsen /1/
and Erents et al /2,3/. This result has been explained by a high interstitial mobility of the room temperature injected helium. In this example of
a light ion entering the solid no damage was produced. But in general ion
bombardment conditions are such that only a small fraction of the implanted ions is thermalised in a region without defects. Whether the trapping
gets a permanent character is determined by the interaction of the implanted ion with the surrounding defects. In our example of helium it was
evident that vacancies could act as traps for the implanted helium. In
general it is found that implanted atoms are associated with the bombardment produced vacancies. However in a number of cases (low temperature
bombardment) association with self-interstitials is found to occur.

Methods employed to monitor the implanted atom and its configuration
in the lattice are Hyperfine techniques (PAC and Mössbauer), recently reviewed by Recknagel /4/ and by Pleiter and Hohenhemser /5/, and ion-chan-
neling techniques for atom location (Swanson /6/ and Vogl /7/). Helium

desorption spectrometry has been developed to a technique which via helium decoration of the implanted atom can be used for analysis of implant concentration and configuration /8,9/.

The practical application of ion trapping is evident in the case of ion beam modification of surfaces, where the configuration of the implanted atom is of paramount importance for the desired effects. In other cases, mainly dealing with implantation of (rare) gases e.g. in (fusion)-reactor-materials, sputtering and sputter deposition, accumulation of trapped gas leads to unwanted effects (see contributions in /10/). Since rare gases are highly insoluble in most materials clustering effects are a quite natural phenomenon in rare gas implantation. Clustering and segregation effects are also observed when metallic impurities are involved in the irradiation process.

In sections 2 and 3 different aspects of ion implantation will be described with emphasis on the processes which determine the configuration of the implanted atoms and lead to transport and clustering of the implanted atoms. Sections 4, 5 and 6 deal with the specific properties of different atomic species implanted in metals.

2. ION TRAPPING

2.1. Ion penetration of the surface

The fraction of trapped ions is the product of the fraction of ions that enters the crystal and is thermalised inside, and the trapping probability f of these thermalised ions. An appreciable fraction $(1-\eta)$ is backscattered and plays no role in the trapping. In principle the backscattered fraction could be measured by integrating the angular and energy distributions of backscattered ions and neutrals as obtained in numerous scattering experiments /11,12/, but because of calibration problems this method has never been applied to measure η.

Desorption spectrometry of implanted gases gives the product η f. For high energy heavy ions or for a damaged crystal f will be about unity so that η can be determined. Monte Carlo type computer simulations in the binary collision approach MARLOWE (crystalline solids) /13/ and TRIM (amorphous solids) /14/ have shown to give reliable results. A fully molecular dynamical approach of the ion impact on crystalline surfaces has been applied by Harrison et al /15/. Particularly for low energy impacts this method is more reliable than the binary collision approach because of the simultaneous interaction of more than 2 atoms in this type of collision problems. Threshold energies for entrapment of noble gas ions in W(100) can be derived from measurements by Kornelsen and Sinha /16/. A 0.1% fraction of perpendicular incident ions is trapped at 25 eV for Ne^+, 100 eV for Ar^+, 130 eV for Kr^+ and 200 eV for Xe^+. Thresholds for penetration of Kr^+ into W(100) and W(110) surfaces have been calculated by Smith and Carter following a molecular dynamic method /17/. The calculations arrive at lower threshold values than the experiments which apparently must be ascribed to the fact that after a low energy penetration no stable traps are formed. Carter et al have discussed low energy trapping of noble gas atoms in tungsten and other materials in a review paper /18/. Van Gorkum and Kornelsen /19/ have employed helium desorption on crystals with preexisting defects to measure the energy dependence of the entrance probility for He in tungsten. The threshold energy for the penetration in W(100) was found at 8 eV, which is only little beyond the expected heat of solution (6 eV /20/) of He in tungsten. The difference can be ascribed to energy losses suffered by the helium atom during the dynamical interaction with the surface atoms.

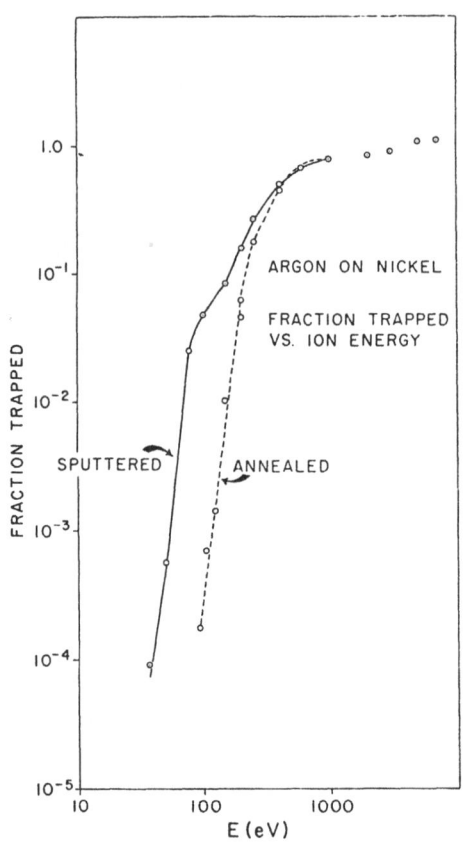

FRACTION TRAPPED

ARGON ON NICKEL

FRACTION TRAPPED
VS. ION ENERGY

SPUTTERED   ANNEALED

E (eV)

The threshold for penetration will strongly depend on the atomic surface configuration and the angle of incidence of the ions. Ions incident on a low index surface with a small elevation angle with respect to that surface will be multiply scattered and not penetrate. In practical situations the energy transferred to the surface atoms will cause surface damage so that newly arriving ions have a higher probability to enter the crystal. The effect of surface damage on the threshold behaviour of low energy argon trapping in Ni(100) is shown in Fig. 1 (ions were injected along the surface normal).

Desorption spectra for noble gas ions injected with energies slightly beyond the penetration threshold show release of the gas at lower temperature than found for ions with keV energy. Van Gorkum and Kornelsen report for W(100) five different surface related argon trapping sites within a distance of .5 nm below the surface /19/.

FIGURE 1. The fraction of incident Argon ions trapped in Ni(100) is plotted as a function of incident argon energy for the sputtered and annealed surface. From Edwards /21/.

2.2. The position of the slowed down implanted ion

An important intermediate step in the trapping process is the configuration of damage and implanted atoms right after most atoms involved have slowed down to thermal energies. By then vacancy rich zones surrounded by self-interstitials have developed. The position of the implanted atoms is still interstitially unless it has lost its last kinetic energy while moving inside a vacancy cluster. In the next phase spontaneous recombination or interaction of vacancies and self-interstitial atoms (SIA) with each other or with the implanted atom may occur. In principle these interactions need not to be activated thermally and depend strongly on the separation between the defect. For Frenkel-pair recombination the recombination radius is of the order of 2-4 lattice units /22/. The same radius can be assumed for SIA-implant recombination if a positive binding exists.

Depleted zones obtained for 10-30 keV ion irradiation of tungsten at temperatures below 15K have been investigated to great detail by Seidman and coworkers with FIM techniques /23/. It was found that with increasing projectile mass the vacancy concentration in this zone can increase to 20%. For projectile/target atom mass ratio larger than unity (not investigated by Seidman et al) English et al find for energies > 30 keV that the vacancies in this zone collapse to vacancy loops /24/. Average numbers of vacancies produced per ion in these studies seem to be consistent with the modified Khinchin and Pease equation /25/.

A large number of vacancies (more than 30%) is surrounded by one or more vacancies at nearest neighbour sites and therefore can be considered to belong to a vacancy cluster /26/. A next step which generally occurs in implantations at temperatures > 50K is the interaction of the self-interstitials (thermally activated migration starts in most metals at T > 40K) with the vacancy clusters, vacancies and also with the implanted atom.

(1) $I + V \rightarrow 0$ recombination
(2) $I + V_n \rightarrow V_{n-1}$ reduction
(3) $I + XV_n \rightarrow XV_{n-1}$ $n \geq 2$
(4) $I + X \rightarrow X$ substitutional $\rightarrow$ interstitial
(5) $I + X \rightarrow IX$ SIA clustering

The upper three reactions are trivial. The fourth reaction is observed for many host-implant pairs by ion beam location studies /6, 7 /. It seems that the implanted atom adopts a so called dumbbell configuration which can migrate as such through the lattice. Dissociation of the pair leads to substitutionality again. The fifth reaction is well known from resistivity recovery studies where interstitial trapping at impurities is seen to be important. At first sight it seems surprising that the implanted atom in the presence of so many vacancies is not captured spontaneously by vacancies or vacancy clusters. Hyperfine interaction measurements on implanted radioactive isotopes (Mössbauer and PAC) show that a major fraction is substitutional (XV) and only traps more vacancies when stage III (vacancy mobility) is passed / 4, 5 /. In Fig. 2 results of Marlowe calculations by Van der Kolk et al /27/ on implanted noble gas atoms in tungsten show that after stage I annealing on the average .3 to .8 vacancies are left within a radius of 2, LU around the implanted atom (LU = lattice unit). The radial distribution was found by allowing SIA-recombination for SIA-vacancy separations smaller than 2 LU.

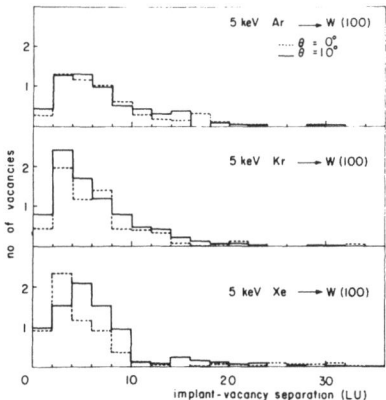

FIGURE 2. Calculated radial vacancy distribution for 5 keV noble gas atoms implanted in a $\overline{W}(100)$ crystal.

A calculation of the vacancies trapped by passing through stage III indicates an average of 2 vacancies for 5 keV and 8 vacancies for 30 keV Kr
ions in tungsten. These results are in agreement with PAC measurements.
Summarising, we might conclude that for implantations beyond stage I the
cascade volume contains vacancies and vacancy clusters, that the equivalent
number of free self-interstitials is available and that the implanted atom
is in substitutional or interstitial configuration. For an ion much lighter
than the matrix atoms (e.g. He and H) the separation between the last vacancy it produced and the position where the ion is stopped is much larger
than for heavier implants. The minimum kinetic energy of the projectile
after a displacement collision with a target atom is given by

$$E_{min} = (m_1 - m_2)^2 / (4m_1 m_2)E_d \qquad (1)$$

Where m_1 and m_2 are the masses of the projectile and target respectively
and E_d the minimum displacement energy. In tungsten (E_d = 40 eV) helium at
the end of its range will still travel with more than 450 eV kinetic energy
without producing further defects. For Ar in tungsten the remaining energy
is > 30 eV. For He in Mo (E_d = 35 eV) E_{min} = 200 eV. Therefore light ions
can travel tens of ångstrøm before they are stopped as an interstitial.
It is clear that only a small fraction of the helium is trapped in its own
vacancy

3. ATOMIC TRANSPORT AND CLUSTERING

3.1. Transport mechanisms

Clustering of implanted atoms requires some means of transport of the
implanted atoms. For light ions H and He in interstitial position mobility
is rather high. For the diffusion of hydrogen in metals see a review by
Volkl and Alefeld /28/. The quoted migration energies vary from 0.2 to 0.5
eV for most metals. Measurements on helium diffusion are rather scarce.
Seidman has measured for He in W E^M = 0.24 eV. Calculations by Wilson et al
/29/ and Baskes et al /30/ yield migration energies of $\cong$ 0.3 eV for bcc
metals and from 0.5 to 0.7 eV for fcc metals. Therefore it is expected
that in experiments at ambient temperatures motion towards a trap is a
quick process.

To some extent also the other light impurities (C, O and N) will move
preferentially as interstitials but with higher activation energies (of the
order of 1-1.5 eV; see Fromm /31/. Some metal impurities in certain hosts
e.g. Au in Pb move also preferentially as interstitials (see e.g. the
review by Warburton and Turnbull /32/ for the behaviour of these so-called
fast diffusors).
The other implants have a strong attraction for monovacancies and will
therefore in most cases adopt a substitutional configuration (denoted by
XV; X is the implanted atom and V the vacancy). Apart from ballistic processes either by a collision with next incoming projectiles or by other
atoms within the collision cascade, the only way for transport is by transformation of the implant in a different configuration:

vacancy capture

$$XV + V \rightarrow XV_2$$

which gives the implant the mobility of a divacancy complex, or by capture
of a self-interstitial

$$XV + I \rightarrow X \quad (XVI)$$

which gives the implant the high mobility of an interstitial. When the implanted atom together with a host atom occupies a vacant lattice site the configuration is called a dumbbell interstital (XVI). Long distance motion of an implanted atom in a dumbbell configuration requires rotation of the dumbbell else the migration is confined to a unit cell of the host lattice. Dederichs et al calculated that the activation energy for rotation might be 4-7 times higher than for 'cage motion' /33/. Motion of a substitutional implant via the vacancy mechanism (XV_2) requires that the implant has a high mobility being inside the divacancy. The number of migration jumps (migration E^M) that can be made once a V or I has been captured is limited because of dissociation of the complex (dissociation energy E^D). The ratio of migration steps and steps leading to dissociation is proportional to $\exp((E^D - E^M)/kT)$. In bcc metals the migration of a V_2 complex involves a change in configuration of the V_2 which nearly leads to dissociation of the complex so that $E^D - E^M$ will be close to zero. For fcc metals however the jumping vacancy remains at first neighbour separation during the migration so that for fcc $E^D - E^M$ might be different from zero. Divacancies not occupied by impurities in fcc are rather stable against dissociation ($E^D - E^M$ ~ 0.5 eV) /34/.

3.2. Clustering

The process of clustering including the above reactions can be described by the following rate diffusion equations:

$$\frac{dn_\alpha}{dt} = D_\alpha \frac{d^2 n_\alpha}{dx^2} \qquad \text{(diffusion)} \qquad (2)$$

$$+ I_\alpha (x) \qquad \text{(implantation/defect-production rate)}$$

$$+ \sum_{\alpha'+\beta'\to\alpha} K^A_{\alpha'\beta'} n_{\alpha'} n_{\beta'} - \sum_\beta K^A_{\alpha\beta} n_\alpha n_\beta \qquad \text{(trapping and association)}$$

$$+ \sum_{\alpha'\to\beta} K^D_{\alpha'} n_{\alpha'} - K^D_\alpha n_\alpha \qquad \text{(dissociation)}$$

where α and β represent defect complexes which contain variable numbers of implanted atom species, vacancies and self-interstitials e.g. $X_i V_j I_k$. Usually most of the defect complexes play a passive role in the clustering process because of their immobility. Self-interstitials generaly are mobile; upon trapping they tend to reduce the number of vacancies in the defect cluster so that $X_i V_j I_k$ is equivalent to $X_i V_{j-k}$ if $k < j$. When $k > j$ $X_i V_{k-j}$ is the remaining defect. The terms containing the implantation rate comprise the number of defects that are formed after low temperature recovery has taken place (spontaneous and correlated recombination of defects).

The reaction constants can be written as

$$K = z \, \nu \exp(S/k) \, \exp(-E/kT) \qquad (3)$$

with E the activation energy of the jump to be made for dissociation or association, ν a frequency factor, S the entropy of the jump and z a geometrical factor which accounts for the size of the defect.

The diffusion coefficient D is written

$$D = 1/6 \, \lambda^2 \nu \, \exp(S^M/k) \, \exp(-E^M/kT) \qquad (4)$$

where λ is the jump distance of the migrating defect.
In most cases the activation energy for migration of the defect is equal
to the activation energy for association of the defect.
  The dissociation energy is

$$E^D = E^M + E^B \qquad (5)$$

(M:migration B:binding)
Values for z in the case of spherical defects are found by solving the
three-dimensional diffusion problem of a spherical sink (radius in LU)
embedded in a larger volume with a certain concentration of mobile defects.
The result is

$$z = \pi R \qquad \text{for migration along lattice sites,} \qquad (6)$$
and $\quad z = 1/3 \ \pi R \qquad$ for migration along interstitial sites.

For more complex geometries of the defect cluster random walk simulations
of the mobile particle in the presence of the defect cluster are used to
find the value of z. Fastenau et al /35/ found for two-dimensional defect
clusters (platelets, interstitial loops) a linear dependence of z on the
radius of the cluster.
It is noteworthy that z depends on the defect concentration /35/. In other
words trapping is non-linear with the defect density. The value of z increases when the defect density increases. The surface plays an important
role in different stages of the defect growth. Firstly as a boundary where
mobile defects disappear, which leads to the boundary condition n(x=0)=0.
In the early phase of defect growth the majority of mobile defects migrates to the surface rather than to associate with other defects. This leads
to a different rate of defect growth for layers close to the surface com-
pated with deeper layers. Furthermore the defect behaviour itself is influenced by the surface. For example a self-interstitial approaching the
surface will at a distance of a few lattice units spontaneously recombine
with the surface instead of continuing the thermally activated migration to
the outer surface. Therefore an effective surface must be defined lying at
some lattice units below the physical surface.
  In the above equations we did not include the effect of a change in surface position either inward by the sputtering action of the projectiles or
outward by swelling due to volume increase of the implanted material below
the surface. Sputtering causes inward motion of the surface so that the
implanted atoms reach (fresh) layers and the most far developed clusters
close to the surface are removed. Therefore stationary state will be present where cluster size and defect density depend highly on the sputtering
rate and implantation depth of the projectiles (see /18,36/.
  The most simple example of clustering is the case when light ions (H)
are injected in a sample containing defects. We assume the injected ions
not to produce damage and to be mobile after thermalisation. Further we
assume that the rate constant for trapping of the mobile particles is not
changed by earlier attachment of mobile particles to the defect. In that
case the concentration of mobile particles during the bombardment is given
by the stationary diffusion equation

$$0 = D_H \frac{d^2 n_H}{dx^2} + I(x) - \sum_i K n_H n_i \qquad (7)$$

with boundary conditions $n_H(0) = 0$ and $n_H(\infty) = 0$ where $n_H(x)$ represents the
concentration of defects containing i trapped particles. The ion flux is
$J = (1/\eta) \ N_0 \int I(x) dx$, where N_0 is the atomic density of the metal.

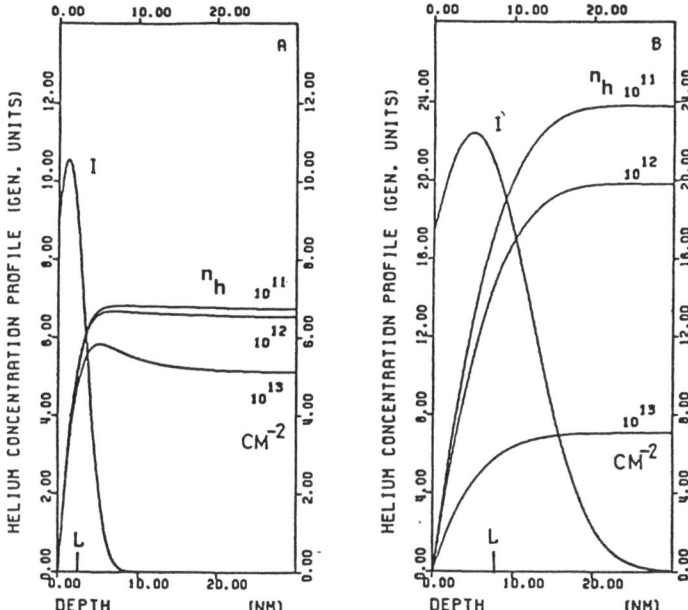

FIGURE 3. The stationary concentration of mobile defects (implantation profile I) corresponding with a) 150 eV He and b) 1000 eV He irradiation of Mo. The areal densities of the defects (vacancies) with depth distribution simular to the profile I in Fig. 6 are indicated.

The total fluence $P = Jt$.

For low defect density (low trapped fraction) the concentration of mobile particles that are stopped at a mean depth L is in a good approximation given by

$$n_H = J/(D_H N_o) \ x \quad \text{for } x<L \text{ and}$$
$$= J/(D_H N_o) \ L \quad \text{for } x>L. \tag{8}$$

In Fig. 3 two examples are given of the calculated stationary concentration of mobile defects (n_H) in the presence of variable amounts of defects. For the concentration of defects n_i containing i trapped particles the following rate equations apply

$$dn_i/dt = -K(n_{i-1} - n_i)n \tag{9}$$

with the solution

$$n_i(x,t) = n(x) \ (1/i!) \ (\mu P)^i \ \exp(-\mu P)$$

where $n(x)$ is the initial defect distribution and $\mu(x) = K/D_o N \ x$ for $x<L$ and $K/D_o N \ \dot{L}$ for $x>L$. The quantity μ can be considered as a cross-section

for the trapping of an injected particle at depth x. Note that the cluster size in this case is distributed according to a Poisson distribution, for which the average size $<i> = \mu P$ and the standard deviation $(\mu P)^{\frac{1}{2}}$.
When damage is produced by the ions, the rate equations are extended with a defect production term. In the case that the helium implantation profile is similar to the defect profile, an analytical expression can be derived for the evolution of the defect population /8/. Van Gorkum and Kornelsen give analytical expressions for the trapping behaviour of mobile defects in a number of casses when the defect concentration is low /37/.

Generally the diffusion and rate equations should be solved numerically. Sometimes simplifications can be made concerning the depth dependence. When the defect concentration is high most mobile particles are trapped in defects rather than released at the surface. The surface can be included by considering it as one of the defects which can trap the mobile particles /38/.

When depth dependence is to be included it can be useful to employ a Monte Carlo type approach of the problem. In that way there can be better dealt with the large variety of defect sizes that will develop during the bombardment and accounted for variation of the defect population as a function of depth. Results for a 2 keV Ar bombardment of Mo(110) are shown in table 1. In the calculation it has been assumed that during the bombardment Ar becomes interstitial by interactions of an SIA with substitutional Ar. Vacancies were assumed to be immobile. At an irradiation dose of $10^{13}Ar/cm^2$ it is seen that a fraction of 7% of the implanted Ar is found in small agglomerates of argon /39/.

Table 1.　Defect populations as a function of dose for 2 keV Ar in Mo(110).

Dose $(10^{11}$ at $cm^{-2})$	Defect peak population $(10^{11}$ at $cm^{-2})$		
	ArV_n $n>1$	ArV	$\sum_{m,n} mAr_mV_n$ $m>1$
4	2.1	0.36	–
20	9.9	2.2(2.5)[a]	0.09
100	32	11	3.1

a) The value in parentheses is derived from experiments /39/.

4. HELIUM AND HYDROGEN IMPLANTATION

As has been pointed out in an earlier section, the trapping of light particles like helium depends strongly on the accumulated amount of trapping centers. Fig. 4 shows the evolution of desorption spectra for increa-1 keV He fluence.

At low fluence the majority of the trapped helium is in single helium occupied vacancies (HeV degasing in H-peak). At higher fluence multiple occupancy is observed (G and F peaks; see table 2) and also the degasing from helium-vacancy clusters and peak bubbles. The peak populations in Fig. 5 show a non-linear behaviour for trapping in vacancies. The curves can be fitted to results of model calculations when it is assumed that 1.6% of the thermalised helium is trapped in its own vacancy. The trapping of the remaining helium goes linearly with the amount of vacancies (thus quadratic with fluence). At a fluence of $10^{14}cm^{-2}$ the total trapped fraction has been

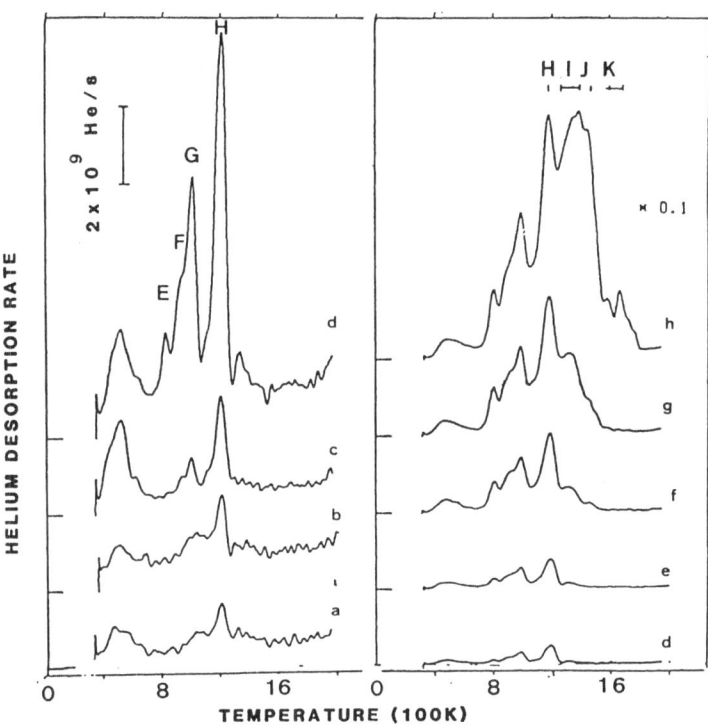

FIGURE 4. Helium desorption spectra obtained for increasing 1 keV He⁺ fluence on Mo(110). Fluences a-h are 2,3,5,10,15,30,50 and 100×10^{12} He cm⁻².

Table 2. Helium release from $He_n V$ precipitates in Mo /40/.

Peak	T_p (K)	n (He)	Mechanism	Experiment v $(10^{13} s^{-1})$	Experiment E^b (eV)	Theory E^b (eV)
E	790	5,6 7,8,9	multiple first order	0.7	2.11 (0.06)	2.45 2.27
F₁	870	4		3	2.4 (0.1)	2.44
F₂	910	3		5	2.6 (0.1)	2.43
G	970	2		30	2.9 (0.1)	2.77
H	1170	1		500	3.8 (0.2)	4.08
M	900-1350	10-16				
M₁	920		MPI-emission	1	2.4	
M₂	1075			1	2.8	
M₃-M₅	1175-1350			1	3.0-3.5	
I₁	1157	14-16		1	3.0	
I₂	1226	17-27		1	3.2	
I₃	1300	28-41		1	3.4	
I₄	1350	42-55		1	3.5	
I₅	1390	56-70		1	3.6	
I₆	1425	70-90		1	3.7	
J	1455-1494	91-150	capture of thermal vacancies			
K	≥1500	≥150	conversion to bubbles; bubble release			

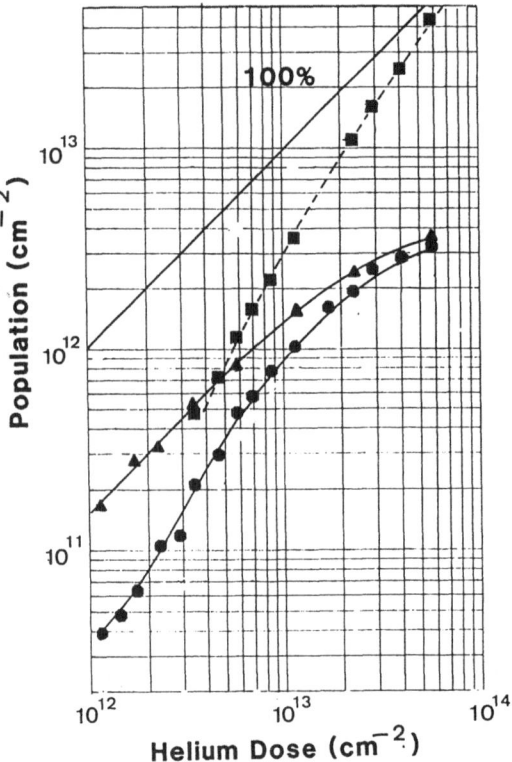

FIGURE 5. The population of various
defects vs. 1 keV helium dose. The
dose indicated is ηP; $\eta = 0.6$; (Δ)
helium filled + empty vacancies, (0)
only helium filled vacancies, and
($\square$) total number of helium atoms
trapped. The solid curves are derived from the model in /8/.

saturated at a level of 100% which
corresponds with the total amount of
thermalised helium.

TDS is a technique which can not
be used at very high accumulated
defect concentrations, because
dissociation products are retrapped
and therefore modify the defect population which was present at the
start of the ramp annealing. In order to observe the further evolution
of defects, which get multiple helium filled during helium implantation, experiments are done in which
the defect concentration is kept low
and the helium is injected at such
low energy that no new damage is
created. The results of an experiment with HeV defects filled with
additional helium up to 2000 He
atoms are given in table 2. It is
observed that up to 8 extra helium

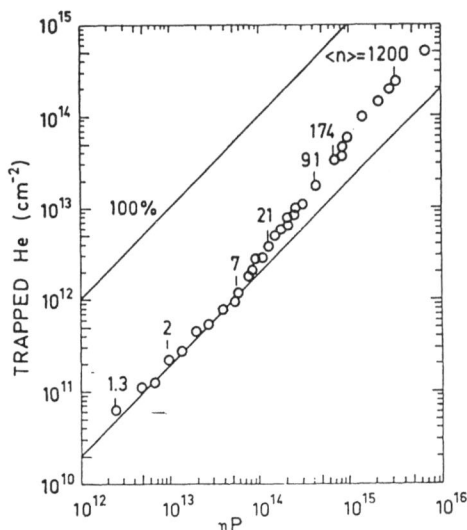

FIGURE 6. The amount of low energy
injected helium that is trapped by
$2 \times 10^{11} cm^{-2}$ HeV defects in Mo(110)
vs. the dose of helium that has en-
tered the crystal. The average number of helium per defect is indicated. The energy of the helium ions
was 150 eV.

atoms in a vacancy the helium is de-
gased at lower temperature than
helium from a single occupied vacancy. But for higher helium amounts
per initial vacancy the degassing
temperature increases steadily.
The increase of degassing temperature (or equivalent: the helium
binding energy) is explained by
atomic displacements around the helium cluster induced by the high
helium pressure inside the cluster.
The displaced atoms (which are considered as bound self-interstitials)
create room for the helium. This process repeats itself when more helium is

attached so that the defect grows in size. This is observed in Fig.6 by an increase in the trapped fraction of helium /40/.

Similar observations have been made for helium trapped at substitutional noble gases in W /41/ and also for substitutional metallic atoms in W /42/. From a very detailled analysis by Kornelsen and Van Gorkum /41/ it can be derived that after attachment of ~ 10 He atoms clustering proceeds independently of the type of defect that initially acted as a trap for helium (Fig. 7).

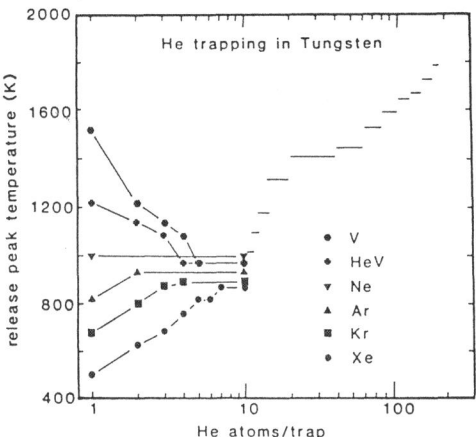

FIGURE 7. Peak temperatures for the six trap nuclei as a function of the number i of helium atoms they contain. Above i=10, the bars indicate approximate-ranges of the multiplicity i involved in peaks at the indicated temperatures which are the same for all the nuclei.

By Kornelsen and Van Gorkum it was assumed that the helium clusters were spherical, but TEM observation of clusters (>500 He) by Evans et al /43/ in Mo revealed that two-dimensional platelets of helium were formed /44/. A model of the cluster based on elasticity theory, where the cluster is considered as a misfit particle in an isotropic medium, gave as a result that for high density helium the formation into a two-dimensional configuration /40/. It follows that the increase in interface energy caused by the transformation is overcompensated by a decrease in strain energy. For relatively small He_nV clusters of a binding energy calculation are plotted in Fig. 8 which shows the experimentally found trend of increasing binding energy only for the 2-dim case. Atomistic calculations on helium clustering in fcc metals /45/ and bcc metals /46/ show qualitatively the same trends. However the results depend on the chosen interaction potentials (pair-potentials).

TEM-work by Evans et al showed that the clusters (platelets) which attain a certain size gain vacancies by emission of prismatic interstitial loops /47/.

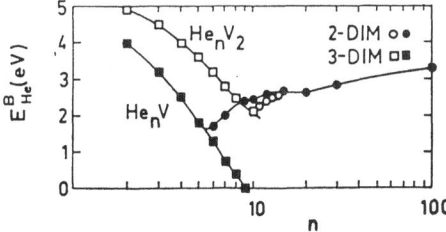

FIGURE 8. Helium binding energies for He_nV and He_nV_2 precipitates calculated with elasticity theory.

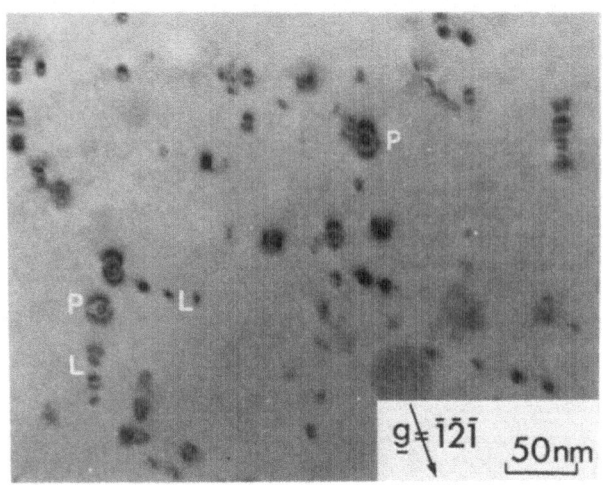

FIGURE 9. Helium platelets (P) and emitted self-interstitial loops (L) in molybdenum. The platelets on {110} planes have been formed by trapping of low energy injected helium (150 eV) at point defects /43/.

For hydrogen it is anticipated that defects shich are large enough so that molecular hydrogen can be accumulated, similar effects, finally leading to high pressure bubbles, will occur. Though molecular hydrogen has about the same equation of state as helium, the hydrogen near the inner surface of the hydrogen filled cavity can dissociate rather easily and be redissolved in the bulk of the metal. Therefore hydrogen bubble development is observed only at high hydrogen flux (compensating the hydrogen release) or at low temperature /48/. Measurements have been done on binding energies of hydrogen to vacancy type defects /49/ but detailed knowledge on multiply filled vacancies/vacancy clusters is still lacking.

5. IMPLANTATION OF NOBLE GAS ATOMS
    Implanted noble gas atoms are generally found to occupy vacancies or vacancy clusters. Mössbauer and PAC measurements for Xe in a variety of metals and also channeling studies locate the majority of the implanted atoms in substitutional position. This is not in disagreement with results of helium desorption studies where noble gas atoms, in particular the heavier ions, are found to be surrounded by additional vacancies. Methods based on hyperfine interactions suffer from 'short-sightness'/4 / so that vacancies at separation distance > 1 lattice unit are not unfluencing the hyperfine parameters. The trapping and detrapping of helium is influenced by the presence of the vacancies. Upon annealing, vacancies are observed to associate with the implanted atoms. Hyperfine measurements are able to identify vacancy clusters up to 4 additional vacancies (see reviews /4 ,5 /).
    In Fig. 10 the desorption of helium attached to implanted argon configurations is shwon. There are distinct desorption peaks due to the release (peak A and B) of helium attached to substitutional atoms and due to the release from implants associated with more than one vacancy (peaks G,G[1] and H,H[1]). Annealing of the implanted atom configuration before decoration with

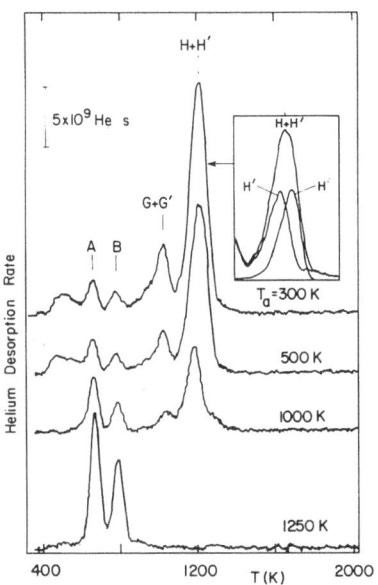

FIGURE 10. Helium desorption spectra from a Mo sample initially irradiated with a dose of 2.2×10^{12} 2 keV Ar ions cm^{-2} annealed to one of the temperatuers shown and then filled with 1.1×10^{13} He cm^{-2} of 150 eV He. The inset shows the 2 single order desorption peaks making up the H+H' peaks.

helium shows the release of the bound vacancies. Fig. 11 shows the different annealing steps. The desorption of the noble gas can also be monitored directly. The result is shown in Fig.12. The low temperature release Ar^S and Ar^I is ascribed to the release of the implanted atoms close to the surface /50/.

The implanted atoms in the bulk are released via the vacancy asssited diffusion process at temperures close to those at which self-diffusion of the metal occurs. It is of interest that the attached vacancies are released before significant diffusion accurs. Apparently the mobility of the XV_n clusters in the temperature region where they are stable is rather low, otherwise the release would have been observed directly after the XV_n had been formed. Similar desorption spectra for fcc metals e.g. Kr in Ni measured by Edwards /21/ give some evidence of diffusive motion of the implants at temperatures well below the temperature of self-diffusion. In Fig. 13 it is observed that apart from the diffusive release peak KD there is another peak 150 K lower which shifts

to higher temperature when the implantation depth increases. This might indicate that KrV_2 is mobile without risk of dissociation. Apparently the migration energy is significantly lower than the dissociation energy of the cluster. The other transport mechanism that can be considered is via the interaction with the irradiation produced self-interstitials. Experiments have been done which show that substitutional atoms can be made mobile by this interaction. The experiment involves two steps: (1) the production of substitutional atoms and (2) the introduction of self-interstitials by low energy noble gas bombardment. The result is shown in Fig. 14. It is observed that a part of the substitutionals is released by the interac-

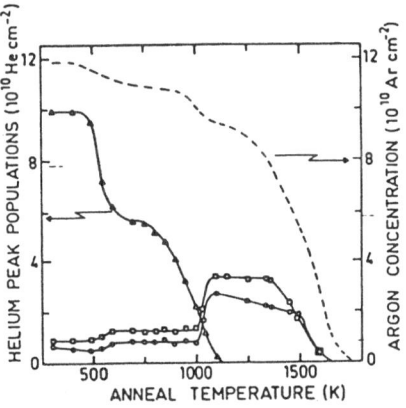

FIGURE 11. The peak population as a function of anneal temperature ($\square$=A peak. O=B peak and Δ=(H+H') peak) for the same irradiation conditions as in Fig.10. Also shown in an integrated Ar desorption spectrum from a similar Ar irradiation.

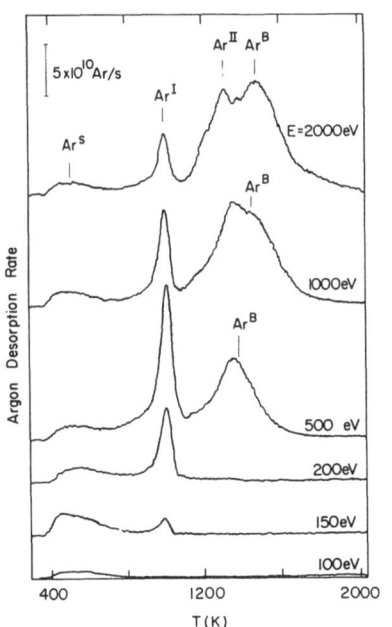

FIGURE 12. Argon desorption spectra obtained after Ar ion irradiation of a (110)Mo crystal to a dose of 1.2×10^{13} Ar cm^{-2} at the energies shown in the figure.

tion. The idea that noble gas atoms in a metal exist as mobile interstitials is not new. Domeij et al /51/ and Centmayer et al /52/ showed that a part of the atoms implanted in channeling directions were able to escape from the damaged zone and to travel long distances. The so-called exponential supertail of the distribution was formed by trapping at background defects. Whereas these authors assumed that interstitials could be formed only by stopping the implanted atoms without displacements of atoms along the channel, the above mentioned experiments show that creation of interstitials is a process that may occur also when channeling conditions are not fulfilled during the implantation. The reaction

$$XV + I \rightarrow X$$

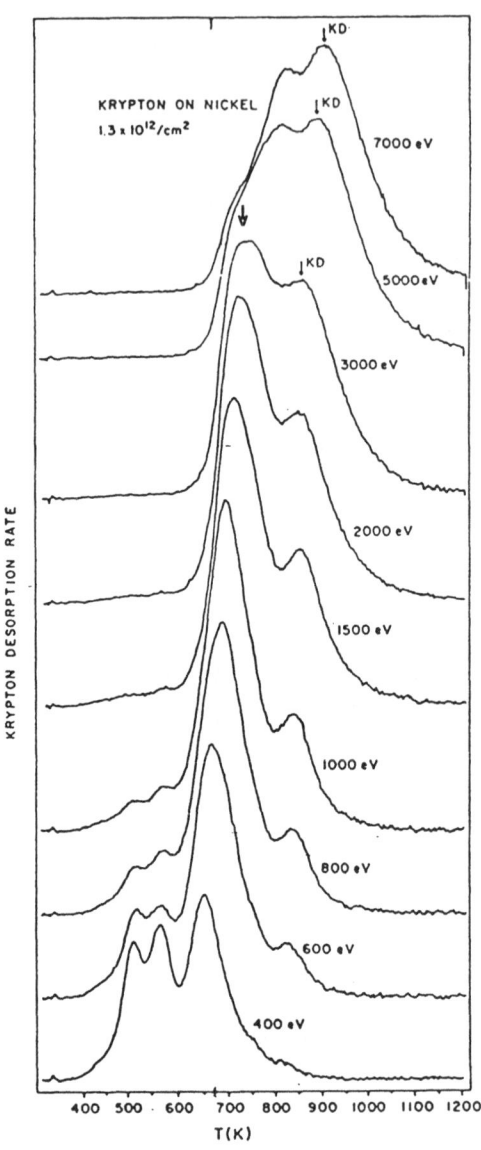

FIGURE 13. The higher-energy krypton desorption at various incident energies showing the motion of the diffusion peaks as the incident energy is increased.

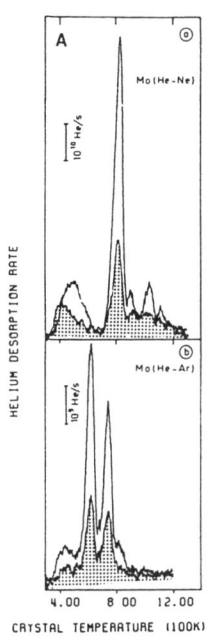 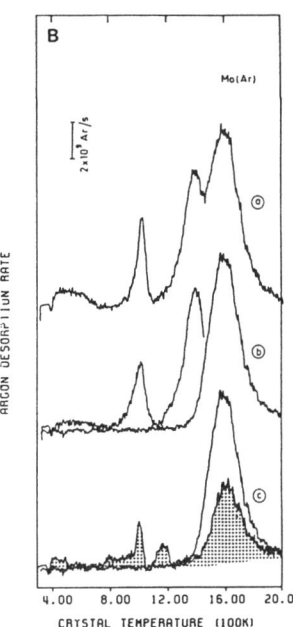

FIGURE 14. A) Helium desorption spectra of helium decorated substitutional Ne (NeV) and Ar (ArV). The shaded spectra are taken after bombardment with 10^{15} cm^{-2} 100 eV Xe$^+$ which causes injection of self-interstitials into the crystal. B) similar as A but now the argon desorption is monitored of ArV (a and b show spectra of the preparation of ArV by 1400 K annealing).

will only occur if

$$E_V^F + E_I^F > E_{XV}^B$$

i.e. the formation energy of a Frenkel pair larger than the binding of the noble gas atom to the vacancy. When this condition is not fulfilled, e.g. for noble gas atoms with atomic volume of the metal atoms of for metals

Table 3. The calculated activation energies for Ar defect reactions

Defect reaction	Energies (eV)	Q$^{b)}$ (eV)
Ar → Ar	(migration)	0.33
V → V	(migration)	1.30
I + ArV → Ar	9.57 + 7.78 > 15.3 + 0.33	-1.97
ArV → ArV + V	9.95 < 7.78 + 3.2 + 1.3	2.35
ArV → Ar + V	7.78 < 15.3 + 3.2 + 0.33	10.70
Ar$_2$V → Ar + ArV	21.54 < 15.3 + 7.78 + 0.33	1.89
Ar$_3$V → Ar + Ar$_2$V	34.59 < 15.3 + 21.54 + 0.33	2.58

with low Frenkel pair energy, the reaction may result in binding of the interstitial to the XV complex

XV + I → XVI

The occurence of binding can be understood in terms of the attraction of two dilatation centers in an isotropic elastic medium /53/. The results of atomistic calculations for Ar in Mo are given in table 3.

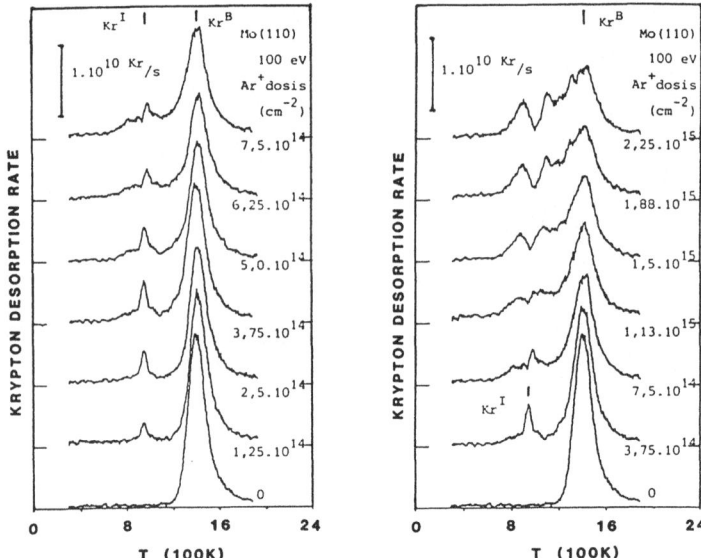

FIGURE 15. KrV defects in Mo(110) irradiated with 100 eV Ar$^+$. Fluences are indicated in the figure.

The results of an experiment in which an increasing dose of self-interstitials is injected in a sample containing KrV defects are shown in Fig.15. The concentration of KrV defects was chosen so that the Kr interstitials formed by the recombination reaction had a reasonable probability to trap at the remaining KrV complexes.
Therefore also clustering reactions

$$KrV + Kr \rightarrow Kr_2V$$
$$Kr_2V + Kr \rightarrow Kr_3V$$
$$Kr_2V + I \rightarrow Kr_2V I$$

are to be expected.
At the start of the experiment a reduction of KrV complexes is observed by a decrease of the peak content of the Kr diffusion peak. A fraction of the formed Kr migrates to the surface where it occupies a sub-surface bound position (appearance of the Surface Peak I). This fracion however is subject to rapid release by the 100 eV Ar bombardment (kinetic energy transfer). Upon further bombardment it is observed that the amount of Kr in the sample is not further decreased and that new low temperature release peaks appear. The explanation is that all of the KrV complexes have been transformed into $Kr_n V I_m$ complexes which are stable against recombination with trapped self-interstitials. In Fig.16 results are given of a computer simulation of the evolution of the defects which support the above interpretation. After a high energy (2 keV) krypton bombardment the krypton desorption spectra

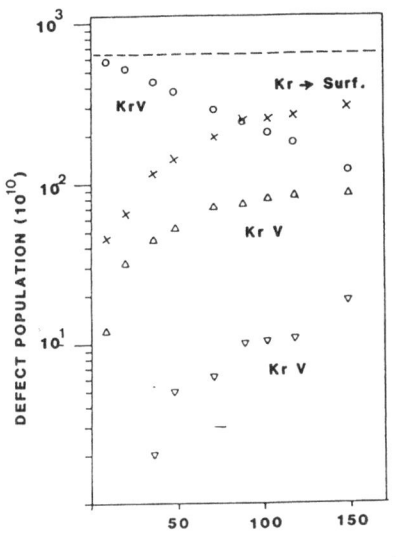

FIGURE 16. Model calculation of the evolution of the defect population during injection with self-interstitials.

med clusters e.g. by the reaction $Kr_2V_3 \rightarrow Kr_2V_2 + V$. New helium desorption peaks are observed due to helium decorated Kr_2V_2 clusters. Because of the required annealing step TDS is not conclusive about the temperature at which the transport of Kr took place. The mechanism could be short range diffusion of KrV_2 complexes or Kr-interstitials formed during the irradiation. At fluences of $\geq 10^{16}$ cm^{-2} in nearly all metals implanted with noble gases small (10-20 nm) bubbles are found /55,56,57/. The bubbles are over pressurized which might be an indication that vacancies play a minor role in the clustering process.

contain again desorption peaks similar as those ascribed to Kr clusters in Fig. 15. The spectra are shown in Fig. 17. Our explanation is that self-interstitials recombining with KrV defects give rise to transport of released Kr to KrV_n complexes. It seems that there are not many vacancies that survive recombination with self-interstitials, else release should have taken place at higher temperatures via formation of small Kr bubbles. The very low temperature release which becomes an important feature at high dose can be attributed to the fact that the sputtering action of the ion beam has moved the surface to depths where the clusters have been formed.

In tungsten clustering of 5 keV Kr$^+$ was observed at an irradiation dose of 10^{13} cm^{-2} /54/. The clusters become detectable with the helium decoration technique after annealing to 1500 K. At that temperature the excess vacancies evaporate from the earlier for-

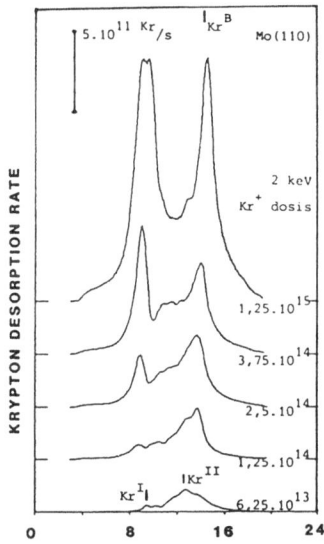

FIGURE 17. Irradiation with a varying fluence of 2 keV Kr$^+$ ions.

6. GASEOUS IMPURITIES

Nitrogen, carbon and oxygen are interstitially mobile in most metals. The activation energy for migration is much higher than for He and H but lower than the self-diffusion energy. In Mo and W nitrogen becomes mobile at about the same temperature as mono-vacancies. It is found that nitrogen binds to vacancies and vacancy-clusters /58,59/. Self-interstitials are able to convert substitutional nitrogen to interstitial nitrogen. Since in most experiments at room-temperature the nitrogen is immobile, the nitrogen is not able tor form clusers. Instead it is expected that self-interstitials are bound to the interstitial impurity so that self-interstitial-clusters are formed. In experiments where the annealing behaviour of damage produced by MeV electrons is studied it is observed that the presence of impurities leads to a rather enhanced fraction of vacancies which have survived Frenkel pair recombination /60/. Experiments with high dose irradiation 3 keV N^+-Mo showed that after dissociation of NV-complexes N_nV_m nitrogen vacancy clusters were formed which had a slightly higher stability against thermal dissociation than the NV-defects /61/.

7. METALLIC IMPURITIES

The metallic implants Ag,Al,Cr,Cu and Mn in tungsten form similar defect complexes as the heavy noble gases in tungsten. Annealing studies by Van der Kolk /62/ however show that vacancy binding energies are much lower than found for the noble gases, e.g. for Ag_2V_2 E^B= 0.5 eV and for KrV_2 E^B= 1.9 eV.

Combined PAC (Perturbed Angular Correlation) and TDS measurements on Ag and I in tungsten were in agreement concerning the observed annealing stages /63/. Self-interstitials are trapped at the substitutional implants during room temperature injection of the self-interstitials. It is concluded that the formation of dumbbell interstitials leads to the segregation of the metallic implants at the surface.

Clustering of the implants is formed for Ag but not for Cu,Al,Cr and Mn. The clustering is observed by the appearance of extra helium de-

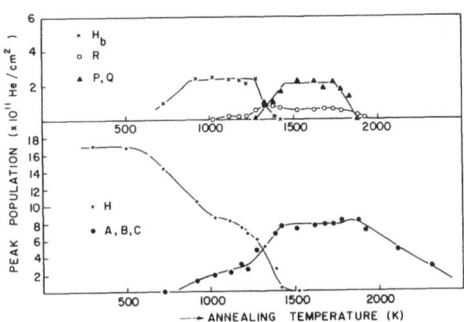

FIGURE 18. Helium peak population as a function of the annealing temperature, for W(100) bombarded with 1.6×10^{13} 10 keV Ag^+ cm^{-2} and injected with 3×10^{12} 250 eV He^+ cm^{-2} after annealing.

Table 4. Peak assignments and dissociation parameters for Ag implanted W (heating rate 40 K/s)

Peak	Temp.(K)	E^d(eV)	ν_o(s^{-1})	Reaction
A	430	1.03	3×10^{12}	$HeAgV \rightarrow He+AgV$
B	650	1.57	3×10^{12}	$He_2AgV \rightarrow 2He+AgV$
R	580	1.40	3×10^{12}	
P	760-1000	1.9-2.5	3×10^{12}	
Q	1000-1260	2.5-3.1	3×10^{12}	$HeAg_nV_n \rightarrow He+Ag_nV_n$ n=2,3
H_b	1390	3.6	1×10^{13}	$HeAg_nV_{n+1} \rightarrow He+Ag_nV_{n+1}$
H	1490	4.5	2×10^{15}	$HeV \rightarrow He+V$

sorption spectra when the implantation dose exceeds 10^{13} Ag^+ cm^{-2}. These peaks are assigned to helium desorption from Ag_nV_m defect complexes. Peak assignments are given in Table 4. Fig. 18 shows the effect of thermal annealing on the implanted Ag atoms. In the lower part of the figure the annea-

ling of AgV$_m$ defects is shown and in the upper part the annealing of the
Ag$_n$V$_m$ clusters. At annealing temperatures beyond 1400 K the substitutional
AgV and the Ag-clusters Ag$_n$V$_n$ have lost the excess vacancies. It is observed
that the clusters dissociate at $\sim$ 1900 K. The substitutional atoms are released by diffusion at a higher temperature.

The cluster found for Ag is consistent with the fact that Ag is an over-
sized atom in W and therefore will form rather stable clusters.

8. CONCLUSION AND FINAL REMARKS

In an early stage of ion irradiation defect complexes are formed which
may act as nuclei for growth of larger defect agglomerates e.g. precipitates, platelets and gas bubbles. Irradiation produced self-interstitials and
vacancies play an important role in the transport of the implanted atoms
which may lead to clustering and segregation effects. Whereas for many
host-implant combinations methods are known to determine configurations
of the implanted atom and the interaction with other defects, for the study
of the early phase of clustering methods are rather limited. Therefore there
is still a big gap between the methods monitoring implanted atoms on a atomic scale and transmission electron microscopy which is able to detect clusters of size ≥ 0.8 nm. High Resolution Electron Microscopy and special positron
annihilation techniques (lifetime measurements) could in future give better
access to the study of these small clusters.

ACKNOWLEDGEMENT
This work was supported by the CODEST XII Programme ('stimulation action')
of the European Community.

REFERENCES
1. Kornelsen, E.V. Radiation Effects 13 (1972) 227.
2. Erents, S.K. and G. Carter, Brit. J. Appl. Phys.(J.PHys D)
   1 (1986) 1323.
3. Erents, S.K., G. Farrell and G. Carter, Proc. 4th Intern.Vac.Congress
   Manchester U.K. Inst. Phys. London (1968) 145.
4. Recknagel, E. and Th. Wichert, Nucl. Instr. Meth. 182/183 (1981) 439.
5. Pleiter, F. and Ch. Hohenemser, Phys.Rev. B25 (1982) 106.
6. Swanson, M.L., L.M. Howe and A.F. Quenneville, J. Nucl. Mat. 67/70
   (1978) 372.
7. Vogl, G., W. Mansel and P.H. Dederichs, Phys. Rev. Lett. 36 (1976)
   1497.
8. Van Veen, A. and L.M. Caspers, Harwell Symp. on inert Gases in Metals
   and Ionic Solids (Sept.1979), AERE Report 9733, 517.
9. Van Gorkum, A.A. and E.V. Kornelsen, Vacuum 31 (1981) 89.
10. Auciello, A. and R. Kelly (eds) Ion Modification of Surfaces, Elsevier
    Publishing Company New York (1984).
11. Boers, A.L., Nucl. Surface Sci. 63 (1977) 475.
12. Buck, T.M., In Methods of Surface Analysis, ed. by A.W. Czanderna
    (Elsevier Scientific Publ. Comp., N.Y. (1975)).
13. Robinson, M.T. and I.M. Torrens, Phys. Rev. B9 (1974) 5008.
14. Biersack, J.P. and L.G. Haggmark, Nucl. Instr. Meth. 174 (1980) 257.
15. Harrison, D.E.,J.P. Johnson, N.S. Levy, J. Appl. Phys. 39 (1968) 3742,

220

and Webb, R.P. and D.E. Harrison, Nucl. Instr. B2 (1984) 660.

16. Kornelsen, E.V. and M.K. Sinha, J. Appl. Phys. 39 (1968) 4546.
17. Smith, A.G. and G. Carter, Radiation Effects 12 (1972) 63.
18. Carter, G.,D.G. Armour, S.E. Donnelly, D.C. Ingram and R.P. Webb, Harwell Symp. on Inert Gases in Metals and Ionic Solids (Sept. 1979), AERE Report 9733, 83.
19. Van Gorkum, A.A. and E.V. Kornelsen, Radiation Effects 52 (1980) 25.
20. Wilson, W.D. and C.L. Bisson, Phys. Rev. B3 (1971) 3984.
21. Edwards. D., J. Appl. Phys. 46 (1975) 1437.
22. Lemnartz, R., F.F. Dworschak and H. Wollenberger, J. Phys. F7 (1977) 2011.
23. Seidman, D.N., M.I. Current, D. Pramamik and C.Y. Wei, J. Nucl. Mat. 108/109 (1982) 67.
24. English, C.A., B.L. Eyre and J.P. Summers, Phil. Mag. 34 (1976) 603.
25. Robinson, M.T. and I.M. Torrens, Phys. Rev. B9 (1974) 5008.
26. Current, M.I., C.Y. Wei and D.N. Seidman, Cornell Materials Science Center Report No 4309 (1982)
27. Van der Kolk, G.J., A. van Veen, L.M. Caspers and J.Th.M. de Hosson, Nucl. Instr. and Meth. B2 (1984) 710 and, Van der Kolk, G.J., Thesis T.H. Delft (1984).
28. Volkl, J. and G. Alefeld, In Diffusion in Solids; recent Developments, Eds. A.S. Nowick and J.J. Burton, Academic Press NV (1975), chapter 3.
29. Wilson, W.D. and R.A. Johnson, In Interatomic Potentials and Simulation of Lattice Defects, Eds. P.C. Gehlen, J.R. Beeler Jr, and R.I. Jaffee (Plenum Press, New York 1972) 375.
30. Baskes, M.I. and C.F. Melius, Phys. Rev. B20 (1979) 3197.
31. Fromm, E. and E. Gebhardt, Gase und Kohlenstoff in Metalen, Springer, Berlin (1976).
32. Warburton, W.K. and D. Turnball, In ref. 28 chapter 4.
33. Dederichs, P.H., C. Lehmann, H.R. Schober, A. Scholtz and R. Zeller, J. Nucl. Mat. 67/70 (1978) 176.

34. Schüle, W. and R. Scholz in Point Defect Interactions in Metals, ed. Takamara et al, University of Tokyo Press, North Holland (1982) 257.

35. Fastenau, R.H.J. and P. Penning, Phys. Stat. Sol. 66 (1981) 613.
36. Carter, G., J.S. Colligon and J.H. Leck, Proc. Phys. Soc. 79 (1962) 299.
37. Van Gorkum, A.A. and E.V. Kornelsen, Radiation Effects, 42 (1979) 93.
38. Baskes, M.I., R.H.J. Fastenau, P. Penning, A. van Veen and L.M. Caspers, J. Nucl. Mat. 105 (1982) 301.
39. Van Veen, A., W.Th.M. Buters, T.R. Armstrong, B. Nielsen, K.T. Westerduin and L.M. Caspers, Nucl. Instr. Meth. 209/210 (1983) 1055.
40. Van Veen, A., J.H. Evans, W.Th.M. Buters and L.M. Caspers, Radiation Effects 78 (1983) 53.
41. Kornelsen, E.v. and A.A. van Gorkum, Radiation Effects, 42 (1979) 113.
42. Van der Kolk, G.J., A. van Veen and L.M. Caspers, Phys. Stat. Sol. (a) 71 (1982) K235.
43. Evans, J.H., A. van Veen and L.M. Caspers, Radiation Effects, 76 (1983) 105.
44. Van Veen, A., L.M. Caspers and J.H. Evans, J. Nucl. Mat. 103/104 (1981) 1181.
45. Wilson, W.D. Radiation Effects 78 (1983) 20.
46. Caspers, L.M., A. van Veen and T.J. Bullough, Radiation Effects 78 (1983) 67.
47. Evans, J.H., A. van Veen and L.M. Caspers, Scripta Met. 15 (1981) 323.
48. Johnson, P.B. and D.J. Mazey, J. Nucl. Mat. 93/94 (1980) 721.

49. Review: Picreaux, S.T., Nucl. Instr. Meth. 182/183 (1981) 413.

50. Van Veen, A., W.Th.M. Buters, G.J. van der Kolk and L.M. Caspers, Nucl. Instr. Meth. 194 (1982) 485.

51. Domey, B., F. Brown, J.A. Davies, G.R. Piercy and E.V. Kornelsen, Phys. Rev. Lett. 12 (1964) 363.

52. Centmayer, F. and R. Sizmann, Radiation Effects 28 (1976) 49.

53. Schroeder, K., P.H. Dederichs and R. Zellen, In Point Defects in Metals III, Springer Tracts in Modern Physics 87, Springer Herlag, Berlin (1980) 171.

54. Van Veen, A., A. Warnaar and L.M. Caspers, Vacuum 30 (1980) 109.

55. Vom Felde, A., J. Fink, Th. Müller-Heinzerling, J. Plüger, B. Scheerer and G. Linker, Phys. Rev. Lett. 53 (1984) 922.

56. Templier, C., C. Jaouen, J.P. Riviere, J. Delafond and J. Grilhé, C.R. Acad. Sci., Paris 299 (1984) 613.

57. Evans, J.H. and D.J. Mazey, J. Phys. F: Met. Phys. 15 (1985) L1.

58. Van Veen, A. and L.M. Caspers, Solid State Comm. 30 (1979) 761.

59. Hautala, M., A. Autilla and J. Hirvonen, J. Nucl.Mat. 105 (1982) 172.

60. Evans, J.H., Acta Met. 18 (1970) 499.

61. Filius, H.A. and A. van Veen, to be published.

62. Van der Kolk, G.J., A. van Veen, J.Th.M. de Hosson and L.M. Caspers, Nucl. Instr. Meth. B6 (1985) 517.

63. Post, K., F. Pleiter, G.J. van der Kolk, A. van Veen, L.M. Caspers and J.Th.M. de Hosson, Hyperf. Int. 15/16 (1983) 421.

BUBBLES, BLISTERS, AND EXFOLIATION

B.M.U. SCHERZER
Max-Planck-Institut für Plasmaphysik, EURATOM-Association
D-8046 Garching/München, FRG

1. INTRODUCTION

Ion bombardment gives rise to modifications of the topography of the bombarded solid surfaces. Inhomogeneous erosion by sputtering is one major reason (see e.g. Carter et al. /1/, G. Carter and J. Whitton, this conference). A second mechanism is due to the implantation of unsoluble gas ions, especially He, but also Ne, Ar, Kr, Xe, H, and N into the subsurface layers of a solid. It is the phenomena observed during gas ion implantation that this paper is concerned with.

1.1 Historical overview

Trapping of gas ions in solids has first been observed as early as 1858 by Plücker /2/ in the study of gas discharges. He found a change in colour of the discharge with burning time and sometimes the discharge extinguished completely, which he explained by a loss of gas to the electrodes. An effect on the surface of the implanted solid was first observed by Stark & Wendt in 1912 /3/. Minerals like rock salt showed microscopic elevations and pits after bombardment with $\sim$ 10 keV hydrogen "canal" rays. Stark & Wendt explained this correctly by accumulation of insoluble gas in small cavities which they called "bubbles" (in german: "Bläschen"). Their work having remained widely unnoticed, about 50 years later Primak et al. /4,5/ published a first detailed investigation of 100 to 140 keV H^+ and He^+ bombardment of corundum and other minerals as well as of Cu and Ni. They observed by interference microscopy overall swelling and in some cases compaction of the irradiated solid and additionally they found evolution of isolated dome-shaped surface elevations which they described as "radiation blistering" in analogy to the well known blistering of metals like Al and stainless steel due to desorption of hydrogen on internal surfaces /6,7/. They developed a first quantitative model of blister formation assuming the internal gas pressure as the driving force. Kaminsky /8/ found by mass spectrometric observation of a Cu surface bombarded by 125 keV D^+ ions that the appearance of surface deformations was accompanied by gas bursts.

The phenomenon of blistering due to hydrogen and helium bombardment of solids received increasing attention since about 1970 because of its implications in the erosion processes occuring at the first wall of thermonuclear fusion devices /9,10/. Erents and McCracken /11/ investigated He-reemission and blister formation in Mo during He bombardment. They observed that the appearance of blisters coincided with the onset of reemission at a critical fluence of $5 \cdot 10^{17}$ He ions cm^{-2} at 20 keV. This showed that the implanted He was essentially immobile and therefore large concentrations of implanted gas could build up in the bulk before reemission occurred. Peak concentrations of 0.5 - 0.6 He-atoms/bulk atoms were found by Behrisch et al. /12/ during He-bombardment of Nb at room temperature in the keV range. At supercritical fluences the whole ion range is saturated up to about this concentration.

The internal structure of materials containing high concentrations of insoluble gas had been investigated earlier by Barnes & coworkers /13,14/ by transmission electron microscopy (TEM). They observed the formation of gas bubbles due to He implantation and subsequent annealing in Be, Al, and Cu. Thomas et al. /15/ observed in the scanning electron microscope (SEM) samples of Pd implanted with 5 MeV He and annealed to 1200°C, of which microtome cross sections were prepared normal to the surface. At the ion range a very porous layer is formed from which cracks extend to the surface. The formation of gas bubbles at room temperature was first observed in He bombarded Mo by Sass & Eyre /16/ and by Mazey et al. /17/. They also found that at low temperatures the bubbles tend to arrange in a sort of lattice oriented parallel to the host lattice. The formation of bubbles implies swelling of the He-implanted bulk volume. This was measured by Blewer & Maurin /18/ on Er and later by St.-Jacques et al. /19/ on Nb. In the subcritical range they obtained swelling rates of about one bulk atomic volume per He-atom. Since this is definitively more than the volume increase due to bubbles visible in the TEM ($\sim$ 10 %) a controversy arose as to whether all He is to be found in these bubbles or whether submicroscopic clusters contain a sizeable fraction of the gas. While measurements of bubble volume by TEM, and of bubble growth due to high energy e^- bombardment support the assumption of He partition between bubbles and submicroscopic clusters /20,21/, He-density measurements by electron energy loss (EELS)-technique give evidence that all He is in observable bubbles /22/. This controversy is still essentially unresolved.

Since volume expansion of an implanted layer is possible only in the direction normal to the surface, the formation of large lateral stress was anticipated upon He implantation. This stress was first measured directly by EerNisse & Picraux /23/ on He implanted foils of Mo, Nb and Al. Blistering was shown to be directly related to stress relief.

The sudden onset of gas release together with blister appearance was first discussed in detail by Evans /24,25/. He assumes that growth of bubbles occurs by punching of interstitial loops due to the large internal pressure until the material between bubbles yields and a small planar cavity is formed by interbubble fracture.

For the further development of blisters two controversial mechanisms have been put forward: In one of them the macroscopic deformation of the surface is caused exclusively by the gas pressure underneath the blister cap. This model, first applied by Primak & Luthra /5/, was adopted later by many authors in order to estimate the critical parameters for blistering.

On the other hand, EerNisse et al. /23/ and Risch et al. /26/ independently developed mechanisms in which the lateral compressive stress is dominating. Their model was later refined by introducing a buckling mechanism by Wolfer /27/. A final decision between both mechanisms is not possible up to now.

The development of blistered surfaces at high fluences was investigated by Martel et al. /28/. They find that He blisters produced by keV implantation disappears by sputtering without consecutive blister generations. In principle this was shown to be true also for energies in the 100 keV range, but here several consecutive blister generations were found before a blister-free equilibrium surface structure is obtained /29/. In some special cases exfoliation of up to 39 layers was observed in a small areal fraction of the implanted spot /30/,31/. But generally blistering can be regarded as a transient phenomenon.

An essential condition for blistering is an implantation profile, peaked at a certain depth beneath the surface. It has been shown by Behrisch et al.

/29/ and by Wilson et al. /32/ that an implantation profile decreasing with increasing depth as obtained by multi-energy or multi-angle implantations may suppress blistering completely.

This must be kept in mind when erosion due to blistering is estimated. Behrisch et al. /29/ have estimated the fluence range in which blistering is expected as a function of energy. Due to the increasing ion range and the decreasing sputtering yield this fluence range increases strongly with energy. A special example of blistering by fusion reactor α-particles was investigated by Bauer et al. /33/. The occurrence of blisters depends in this case on the chosen wall material and its sputtering yield.

Reviews on blistering have been published by Roth /34/, Das & Kaminsky /35/, Erents /36/, St.-Jacques et al. /37/, Scherzer /38/ and Wilson /39/.

1.2 Stages of development of surface modifications

Before entering a detailed discussion of the various stages in the development of gas ion induced surface modifications, it may be helpful to present a short description of the development in terms of increasing fluence of gas ions. This may serve also as an outline of the following discussion.

But, before doing this, I should put some restrictions on the topic, I am going to discuss. In the following I will confine myself to gas-solid systems with essentially zero solubility. Because of the predominance of work done, most of the discussion will be on the behaviour of metals under He-ion bombardment. Heavier rare gas bombardment shows similar effects, but there is much less experimental evidence with respect to surface modifications due to the smaller ranges and the much higher sputtering yield. Hydrogen behaves estonishingly similar to He in some metals even at room temperature (R.T.), in others this similarity shows up when the mobility is suppressed by cooling. Other gases, e.g. O or N, are so little investigated that no clear picture of their behavior can be given.

Another limitation, I imposed myself, is the temperature range. In the following I will only discuss such processes in which the mobility of gas clusters due to bulk or surface diffusion of the host material is negligible.

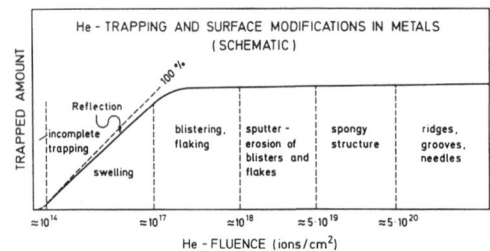

FIGURE 1. Trapped amount of He in a metal as a function of fluence, schematic, with subdivision into ranges of characteristic surface modifications.

In Fig. 1 the trapping of an unsoluble gas like He, implanted monoenergetically into a solid, is described schematically by plotting the trapped amount of gas over the fluence. The initial, very low fluence range ($\lesssim 10^{14}$ cm^{-2}) is characterized by the formation of point defects. Trapping may be incomplete because He-ions coming to rest on interstitial lattice sites are mobile and, as long as the trap concentration is low, may diffuse to the surface and desorb. This range, which is essential for the nucleation of gas bubbles and thus for all the further developments, has been discussed in detail in the preceding talk by van Veen and will not considered here.

We will start with the fluence range $10^{14} - 10^{17}$ cm^{-2}. It is characterized by very effective trapping. Except for the kinetically reflected ions,

all implanted ions are trapped. This is the typical range of bubble formation and growth. At low temperatures the bubbles arrange in a sort of
lattice with very homogeneous size while at high temperatures they are randomly oriented and grow by accumulation of vacancies, bubble coalescence or
some other processes not being discussed here. An essential question concerns the gas density in bubbles and the distribution of gas in small (submicroscopic) clusters in the material between the bubbles. The surface itself stays rather unchanged except for a slight outward shift due to swelling and the formation of extremely large planar stress.

At fluences $\gtrsim 10^{17}$ cm^{-2} a critical situation is reached when at some depth
the trapping capacity of the lattice is exceeded, i.e. when the growth of
bubbles and/or submicroscopic clusters comes to an end because the material
between them becomes thin enough to yield by rupture. This triggers a deformation of the surface causing either a large number of dome-shaped circular blisters as shown in Fig. 2a, or a more or less complete flaking of
the surface. This is accompanied by a simultaneous onset of gas release,
characterized by the saturation of trapping.

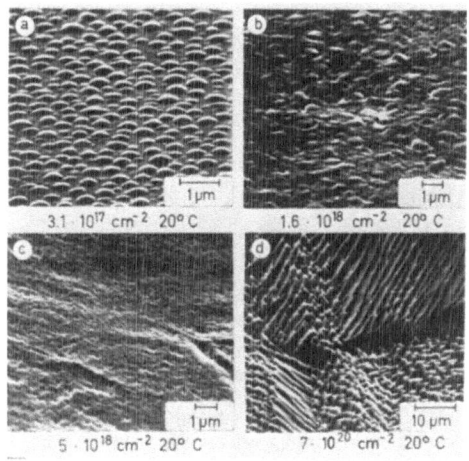

3.1 · 10^{17} cm^{-2} 20° C 1.6 · 10^{18} cm^{-2} 20° C

5 · 10^{18} cm^{-2} 20° C 7 · 10^{20} cm^{-2} 20° C

FIGURE 2. Surface features on polycrystalline Nb after 9 keV He$^+$ bombardment with 4 different fluences at room
temperature and normal incidence (SEM
micrograph by Dr. H. Klingele, München).

These gas produced surface
structures will be gradually
eroded by sputtering as shown
in Fig. 2b and it depends widely on the thickness of the
blister layer and/or on the
number of successive generations
of flakes at which fluence
these surface structures will
be completely removed.

Due to this sputter erosion
depth ranges are exposed at the
surface which have taken up gas
to saturation. On exposure to
the surface these layers show
a spongy structure with a large
density of small holes (Fig.2c).

Finally, at very large fluences inhomogeneous sputter
erosion dominates the surface
topography (Fig. 2d). But still
the observed structures -
ridges, grooves, needles - are
characteristic for a highly
gas loaded surface.

2. BUBBLES AND BUBBLE LATTICES

Implantation of gases in solids in which they are not soluble may nevertheless lead to very high concentrations (on the order of 1 gas-atom/bulk-
atom). In the preceeding lecture van Veen has shown on the example of He how
the gas is initially trapped in point defects and, at higher fluences, trans-
forms into clusters of tens or hundreds of gas atoms with no apparent limit
in number. It has been shown that these small size clusters grow by pushing
of lattice atoms into interstitial positions. At a certain size the production of single interstitials becomes unfavorable energetically compared
to the formation of larger clusters. This may be the limit above which we
can speak of gas bubbles.

Bubbles have been observed in a large number of metals due to He implantation and there is general agreement that at least in the case of He

the bubbles are a necessary precursor of blistering and flaking /38,40/. Bubbles have also been found for other implanted gases with low solubility, e.g. for heavy rare gases Ar and Xe in Al films /40-42/ and in Cu, Ni and Au /43/. Hydrogen bubble formation has been observed in Si /44/ and Cu /45/ and is expected in a number of other metals, e.g. Be, Zn, Ge, Cd, In, Sn, Te, Tl, and Pb.

2.1 Density and pressure of gas in bubbles

In order to understand the modifications of surface topography from the development of gas bubbles due to continuing gas ion implantation, an important quantity is the gas pressure inside the bubbles.

Two situations can be distinguished:
a) the bubble pressure is in equilibrium with the surrounding lattice, i.e. no elastic deformation energy is stored in the lattice due to the presence of the bubble and
b) the bubble is overpressurized and the surrounding lattice is elastically deformed.

In equilibrium the total energy of the system "gas-bubble surface-solid" must be a minimum, i.e. at constant pressure p, temperature T and chemical potentials μ_i of the gas and the solid, the reversible work p dV due to an expansion of the bubble is equal to the work γdA to create a new surface area dA. This leads to the equilibrium condition

$$p = 2\gamma/r \tag{1}$$

where r is the bubble radius and

$$\gamma = f_s - \Sigma \mu_i \Gamma_i . \tag{2}$$

f_s is the specific surface free energy and Γ_i and μ_i the surface atom density and the chemical potential respectively /46/. Only in liquids is γ equal to the surface tension, in solids the stress localized in the neighborhood of an interface is not numerically equal to the surface tension /46/. However, for convenience, the term is often called "surface tension" also for solids. Numerical values of γ for solids have been measured by a number of different methods. There is often confusion about the exact quantity that has been measured, i.e. the surface tension, the specific surface free energy, or the quantity γ as defined above. Very few of the measured data are obtained on clean surfaces and, therefore, the reliability is low. Data are found in the literature (see e.g. Kumikov & Khokonov /47/ and Donnelly /40/) and are typically between 1 and 3 Nm⁻¹ for metals. The relation between pressure and bubble radius according to eq.(1) for $\gamma = 1$ Nm⁻¹ is shown in Fig. 3.

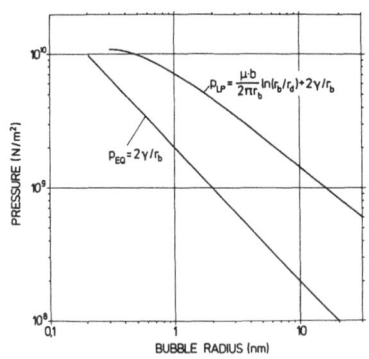

FIGURE 3. Equilibrium gas pressure in bubbles, p_{EQ}, according to eq. (1) ($\gamma = 1$ N m⁻¹), and limiting pressure for loop punching, p_{LP}, eq. (8), as functions of the bubble radius.

In the case of overpressurized bubbles $p > 2 \gamma/r$, and instead of an increase of bubble volume to the point where eq. (1) is again fulfilled, some part of the free energy is stored in lattice strain around the bubble. But the total pressure inside a bubble is limited by the mechanical properties of the bulk

material, especially by the possibilities to punch interstitial loops and
to break the material between adjacent bubbles as will be discussed later.

Recent theoretical work on the energetics of helium bubbles has been reviewed by Trinkaus /48/. He presents detailed evaluations of the bulk free
energy of helium leading to an equation of state for solid and fluid He.
He further evaluates curvature corrections to the bubble-metal interface
free energy and the relaxation energy due to elastic deformation of the
lattice. For bubbles in thermal equilibrium he obtains instead of (1)

$$p \cong \frac{1-\delta_v/r}{1-2\,\delta_{He}/r + 0.6\,v^{1/3}/r} \cdot \frac{2\gamma}{r} \tag{3}$$

For Ni the curvature correction $\delta_v \cong 0.03$ nm and $\delta_{He} \cong 0.1$ nm is estimated,
v being the volume per He atom. The correction term in (3) gives about 30 %
higher equilibrium pressure for r = 1 nm as compared to eq. (1).

Generally it is difficult to determine the pressure inside of bubbles
directly. Therefore more frequently the pressure is obtained from a measurement or estimate of the gas density ρ inside the bubble by means of the
equation of state (EOS) of the gas. Since pressures of $p > 10^9$ N m^{-2} may
easily occur (see Fig. 3) the perfect gas law cannot be applied even for
the rare gases. Other approaches which work also at higher pressures are
the van-der-Waals equation and the hard-sphere approximation. The van-der-
Waals equation can be written in the form

$$p = \rho kT\,(1-b\rho)^{-1} -a\,\rho^2 \tag{4}$$

The term $a\rho^2$ takes account of an attraction between gas molecules and in
the case of He can be neglected for T $\sim$ 100 K. The denominator of the first
term, $1-b\rho$ is a consequence of the assumption that the gas molecules have a
minimum volume b. For $1/\rho = b$, p diverges. In Fig. 4 the van-der-Waals
equation of He is plotted in the reduced variables $Z = p(\rho kT)^{-1}$ which is
called compressibility (Z = 1 for perfect gases) and the packing fraction
$y = \rho\Omega_{He}$ where Ω_{He} is the volume occupied per He-atom at closest packing or
$\Omega_{He} = b$.

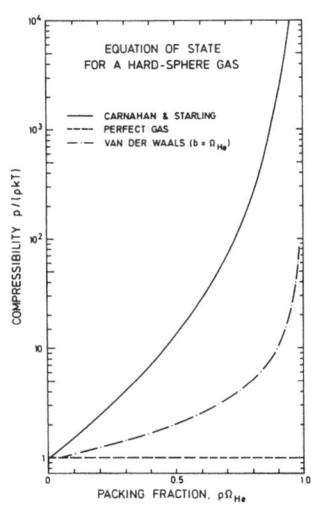

The hard-sphere approximation can be represented by the virial expansion of the
pressure in terms of the density ρ and
is given in the reduced variables by

FIGURE 4. Equation of state for a van-der-
Waals gas, a hard sphere gas, and a perfect
gas in reduced variables. ρ is the number
density, Ω the volume occupied per atom at
closest packing (e.g. for He).

$$Z = 1 + 4y + 10y^2 + 18.36y^3 + 28.2y^4 + 39.5y^5 + 56.5y^6 \ldots \qquad (5)$$

Higher coefficients can either be calculated approximately or closed-form expression be given reproducing molecular dynamics simulation experiments on ensembles of hard spheres. The expression

$$Z = (1 + y + y^2 + y^3)(1-y)^{-3} \qquad (6)$$

by Carnahan & Starling /49/ has been found to give the best approximation. It is also plotted in Fig. 4. It also diverges at $y = 1$.

The assumption of a gas atom acting as a hard sphere at high densities is certainly unrealistic. In Fig. 5 the interatomic potential for He-atoms as a function of the interatomic distance is shown as given by Beck /50/.

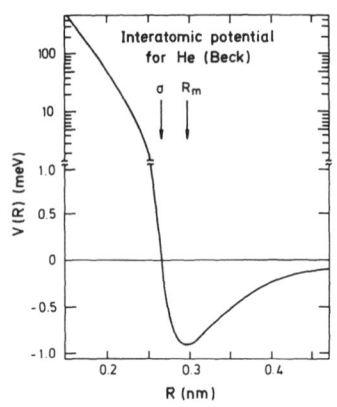

The potential increases with decreasing R at a much softer slope than would be expected for a hard sphere gas.

Several attempts have been made to derive a theoretical EOS on the basis of Beck's potential. Two of them are described in a recent paper by Wolfer et al. /51/. In his approach, Wolfer starts from the hard-sphere equation of Carnahan & Starling /49/, eq. (6), accounting for the finite steepness of the repulsive part of the interatomic potential (Fig. 5) by using a particle diameter which depends on temperature and density of the gas. Trinkaus empirically adjusts the coefficients of the virial expansion so that they agree at low densities with the conventional coefficients (eq. 5) and reproduce the compressibility factor at freezing. He obtains for gaseous or fluid He

FIGURE 5. Interatomic potential for He according to Beck /50/ as a function of interatomic distance R. R_m is the distance where V has its minimum; $V(\sigma) = 0$. Note the change in scale of V(R) for R < 2.5 nm.

$$Z = (1 - \frac{v_e}{v})(1 + \frac{v_e}{v} - 2(\frac{v_e}{v})^2) + (1 - \frac{v_e}{v})^2 \frac{v_e}{v} B(T)/v_e +$$

$$+ (3 - 2\frac{v_e}{v})(\frac{v_e}{v})^2 Z_e - 50(1 - \frac{v_e}{v})(\frac{v_e}{v})^2 \qquad (7)$$

where

$$\frac{v_e}{v} \leq 1 \text{ , where v is the volume per atom and}$$
$$Z_e = 0.1225 \, v_e \, T^{0.555}$$
$$v_e = 56 \, T^{-1/4} \exp[-0.145 \, T^{1/4}] \qquad (\text{in } \text{\AA}^3)$$
$$B(T) = 170 \, T^{-1/3} - 1750/T \qquad (\text{in } \text{\AA}^3)$$

The EOS given by Trinkaus for liquid He is in very good agreement with an empirical EOS by Mills et al. /52/ (MLB) which is based on experimental results up to a pressure of 20 kbar (Fig. 6).

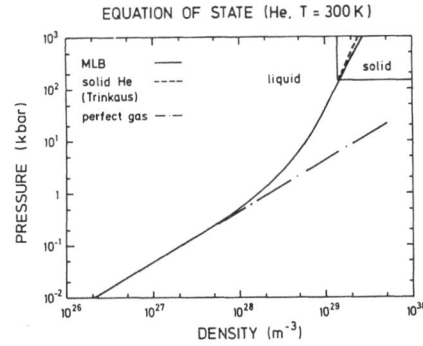

EQUATION OF STATE (He, T = 300 K)

MLB ———
solid He ----
(Trinkaus)
perfect gas —·—

liquid

solid

FIGURE 6. Equation of state for He
at room temperature according to
an extrapolation of experimental
data up to $p \leq 20$ kbar by Mills et
al. /52/ and a theoretical derivation for solid He at $p > 100$ kbar
by Trinkaus /71/.

At pressures above 100 kbar the
He is assumed to be solid. The EOS
for solid He /51/ differs somewhat
from Mills' extrapolation. But, as
discussed by Donnelly /40/, the
existence of solid He in bubbles has not been demonstrated experimentally.
Trinkaus' EOS for gaseous and liquid He (eq. (7)) and the MLB-equation can
for the moment be accepted as a very reasonable fit to theory and experiment.

EOS of other gases for which bubble formation was observed can be found
in the literature, e.g. for H_2 /53,54/, for Ne /55/, for Ar /55, 56/, for
Kr /56/ and for Xe /56, 57/.

2.2 Experimental determination of gas densities in bubbles

There have been numerous attempts to measure gas densities of He in
bubbles by various methods. This work has recently been reviewed by
Donnelly /40/ and only a short survey will be given here. While measurements at high temperatures are mostly consistent with the equilibrium condition, eq. (1), /58/, at low temperatures numerous indications have been
found that bubbles are strongly overpressurized. For the state of saturation we can estimate an upper and a lower limit for the helium density:
Typical saturation concentrations for ^{3}He obtained by nuclear reaction analysis (NRA) are 0.3 and 0.6 He-atoms/lattice atom for f.c.c and b.c.c metals, respectively /12,59/. Assuming a homogeneous distribution of this
gas over the metal lattice gives $\rho_{He} \cong 2 \cdot 10^{28}$ m^{-3} as a lower limit, accounting for swelling by a factor of 1.6 /19/. Assuming, however, all gas to be
trapped in visible bubbles with a volume fraction of 5 - 10 % we obtain a
gas density in bubbles of 3-5 He-atoms per metal vacancy or $\rho_{He} \cong (2.5-5) \cdot$
10^{29} m^{-3} /45/. Most of the many experimental determinations of ρ_{He} fall between these limits, but unfortunately the accuracy of all methods is not
much better than within a factor of two.

It is still an open question, whether the total amount of implanted He is
trapped in gas bubbles as observed in the TEM or whether a major fraction
is distributed in submicroscopic clusters which are lying between the large
bubbles. Since this question is of importance for the surface deformation
and gas release processes it will shortly be discussed here. Johnson &
Mazey /45/ measured bubble radii r and densities C_B for He in Cu by TEM.
Assuming $\rho_{He} \cong 8.5 \cdot 10^{28}$ m^{-3} and an overall $C_{He}/C_{metal} \cong 0.3$ they conclude that most He must be trapped outside the visible bubbles. However,
Jäger et al. /60/ measured $\rho_{He} = 1.2 \cdot 10^{29}$ m^{-3} and $C_{He}/C_{metal} \cong 0.1$ for
He in Al and come to the opposite conclusion that all gas is trapped in
the visible bubbles. Irradiation of helium bubble lattices in Cu and Ni by
1 MeV electrons gives rise to a growth of bubbles without changing their
density /20,21/. This is taken as evidence for additional precipitation of
He from the matrix into bubbles. Van Swygenhoven & Stals conclude from

their measurements of He in Ni that with increasing fluence the He-fraction in the matrix decreases from 0.7 to about 0.5 at the critical fluence. The distribution of such large amounts of He over the matrix requires very high defect concentrations. Johnson & Mazey estimate that assuming He to be trapped in He_6V complexes a vacancy concentration of 0.04 is required to accommodate the helium. This would require a vacancy on every forth site along <110> and every second site along <100> /45/.

Measurements by nuclear magnetic resonance (NMR) /61/ and thermal desorption spectroscopy (TDS) /62/ do not show evidence of simple substitutional sites (HeV) at fluences where bubbles are observed.

2.3 How do bubbles grow?

The growth of a gas bubble in a solid may occur either by pushing lattice atoms into interstitial positions as was already discussed by van Veen. At larger bubble radii it may be energetically favorable that the expelled self interstitial atoms (SIA) arrange in the form of an interstitial loop, a process called loop punching which was first proposed for gas filled bubbles by Greenwood et al. /63/ and which was applied to the blistering process by Evans /23/. Both of these are athermal processes, at least temperature has only a minor influence. At higher temperatures, where vacancies are thermally produced, the bubble may increase by collection of vacancies, thereby reducing the gas pressure to the equilibrium value. We should mention that at high temperatures and a small ratio of injected or produced gas atoms per displacement per atom (dpa) a bias driven growth mechanism takes over. This is typical for void growth and is not of interest in the present context.

For blistering and flaking we are predominantly concerned with the temperature range $T < 0.5\ T_m$ (T_m: melting temperature). The thermal production of vacancies is then negligible and we therefore are primarily interested in the athermal growth processes. We will discuss the loop punching mechanism first and give an estimate where its limits are concerning bubble size as well as temperature.

In bubbles which are small compared to their mutual distance, a stress field builds up locally around each bubble if new gas is supplied at such rates that the equilibrium condition, eq. (1), is exceeded. The excessive stress may be relieved by plastic deformation. There are two possible processes, one of which is the operation of Frank Reed sources producing dislocation loops next to the bubble /64/, the other being the formation of a prismatic dislocation loop (a dislocation loop with its burgers vector normal to the loop plane).

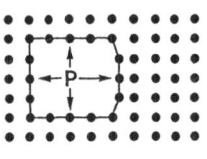

(a) EXCESS BUBBLE PRESSURE DEFORMS SURROUNDING ATOM PLANES.

The latter process is shown schematically in Fig. 7. It can be described as the introduction of an additional layer of atoms between to crystal planes. The diameter of this additional layer being of about same size as the bubble diameter /65/.

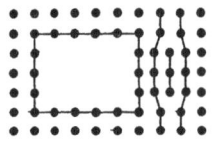

(b) SHUNTING PROCESS ALLOWS EXPANSION OF BUBBLE AND CREATION OF INTERSTITIAL LOOP.

FIGURE 7. Schematic model for punching of interstitial loops from a gas filled cavity /25/.

There is a number of experimental observations of such loops being produced by gas filled bubbles. Wampler et al. /66/ have seen loops emitted from hydrogen filled gas bubbles that formed during quenching of Cu samples loaded with hydrogen gas at high temperature. Recently Evans et al. /67/ found loops punched from He filled platelets in Mo bombarded by low energy He ions. Jäger et al. /68/ observed loop punching from He bubbles formed in tritium loaded vanadium.

The pressure inside a bubble of radius r_0 which is necessary for loop punching to occur can be estimated simply by comparing the elastic energy of the loops to the change in free energy of the expanding bubble. Neglecting the pressure decrease in the expanding bubble and assuming constant chemical potentials of the gas and the gas-solid interface, the change in free energy is $p_0 \, dV$ with $p_0 = p - 2\gamma/r_0$ and $dV = \pi \, r^2 \, b$ (b: Burger's vector). The energy per length of a prismatic dislocation is given by /69/ $\Delta E = \mu b^2 ((1-\nu)4\pi)^{-1} \ln (r_0/r_d)$. μ being the shear modulus and r_d an effective core radius, often taken $r_d \cong b$ /63/, but Trinkaus /48/ found $r_d \cong 0.1 \, b$ by comparison to computer simulation studies /70/. We thus obtain the condition where loop punching may take place:

$$(p - 2\gamma/r_0) \, \pi \, r_0^2 \, b \; \gtrsim \; \frac{\mu b}{1-\nu} \frac{1}{2} \, r_0 \, \ln \left(\frac{r_o}{r_d} \right)$$ (8)

The conditions for loop punching have been discussed in more detail by Trinkaus & Wolfer /48,71/. Two major corrections must be considered. For small bubbles a curvature correction to the free energy of the gas-solid interface must be applied. It increases the equilibrium pressure for a He bubble in Ni with $r_0 = 1$ nm by 30 % compared to the classical value $p = 2\gamma/r_0$. In a second correction the elastic interaction between the bubble and the interstitial loop is considered. As a consequence the pressure difference $\tilde{p} = p_{LP} - 2\gamma/r_0$ for loop punching decreases only for small bubbles with $1/r_0$ while for bubbles with $r_0 \gtrsim 10 \, b$ it stays approximately constant (Fig. 8). In the range $4 < r_0/b < 20$ the threshold pressure for loop punching as given by the approximation

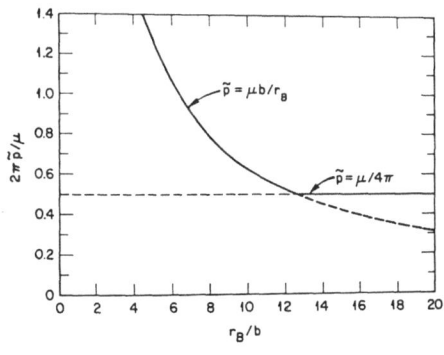

FIGURE 8. Threshold pressure for loop punching, $\tilde{p} = (p - 2\gamma/r_B)$, versus bubble radius r_B, in units of the shear strength $\mu/2\pi$, and the Burgers' vector b, respectively /71/.

$$p_{LP} \; \gtrsim \; (2\gamma + \mu b)/r_0$$ (9)

has been shown to be within 20 % of more detailed calculations. For $r_0 \gtrsim 10 \, B$

$$p_{LP} \; \gtrsim \; \frac{2\gamma}{r_0} + \frac{\mu}{4\pi}$$ (10)

is appropriate. In Fig. 9 the density of He in bubbles in a Ni lattice is given as a function of the bubble radius r_0 for the equilibrium, eq. (1), and for loop punching conditions, eq. (9). The parameters are $\gamma = 2$ N/m, $\mu \cdot b = 20$ N/m and 300 K $< T < 1000$ K and the EOS, eq.(7) is used.

The details of the loop formation process are not understood. At small bubble sizes it seems that interstitials expelled into the lattice arrange

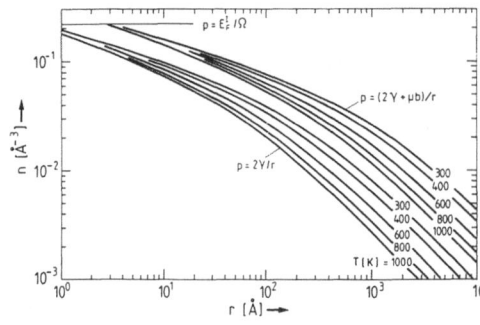

FIGURE 9. Helium density in gas bubbles over the bubble radius for equilibrium bubbles in nickel (p = 2γ/r) and for bubbles at the threshold for loop punching according to eqs. (7) and (9) with temperature T as a parameter. At r $\stackrel{\sim}{\sim}$ 5 Å SIA-emission may be predominant with n ≅ 2·10²⁹ m⁻³ /48/.

in 3-dimensional clusters near the bubble. It was shown /70/ that these clusters form 2-dimensional platelets when the number of self interstitials exceeds 13.

An unsolved problem is the formation of loops in very dense bubble configurations and lattices, where an interaction between dislocation loops from neighbouring bubbles becomes highly probable. Also the assumption of an undisturbed lattice in the derivation of (8) is a very rough approximation in view of the highly damaged lattice which possibly also contains small He-clusters in addition to the visible bubbles.

2.4 Bubble density and size distribution, bubble lattices

The density of gas bubbles as well as their structure and size distributions depend strongly on the temperature of the host material during and after implantation. At low temperatures high concentrations of nearly equally sized bubbles are found after He implantation in metals like Mo, Cu, Ni, Au, stainless steel, and Ti, representing the three metal lattice structures b.c.c., f.c.c., and h.c.p. These bubbles are arranged on superlattice structures which are aligned with the host lattice (Fig. 10) /17,45,72,73/.

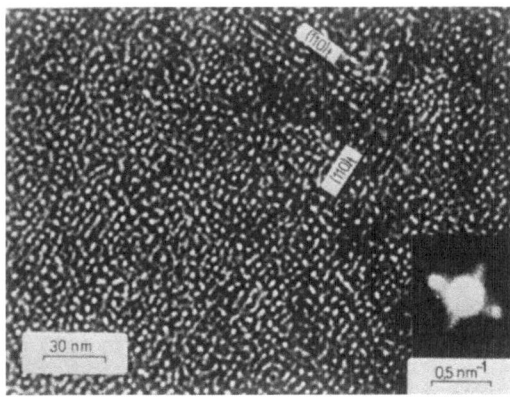

Typical bubble radii of r = 1 nm and superlattice parameters of 5 - 10 nm are observed, with bubble densities, c_B ≅ 10²⁵ m⁻³. The bubble lattice is found to be cubic for b.c.c. and f.c.c. and hexagonal for h.c.p. host lattices. However, while Jäger et al. /72/ exclude an f.c.c. structure of the superlattice in Ni, based on their electron diffraction measurements, Johnson et al. /74/ propose small areas of about 5 lattice constants to be ordered in f.c.c. separated by regions of less obvious ordering. Bubble lattice orientation has also been observed

FIGURE 10. TEM defocus contrast of a He bubble lattice after multienergy He implantation (250 eV $\leq$ E $\leq$ 8 keV) of Ni at 300 K. The 8 keV He-fluence was 10²¹ m⁻². The inset shows extra spots around the (000) transmitted beam in selected area electron diffraction pattern /72/.

in Cu after hydrogen bombardment, however, the bubbles show large size distributions /73/.

Bubble lattice formation has been observed up to temperatures of 973 K in Mo /17/ and 473 K in Ni /72/, respectively. This is above the temperatures where vacancies become mobile. Thermally activated vacancy migration seems to have no influence on the formation of a bubble lattice.

A review of the ordering mechanisms leading to bubble and void lattice formation has recently been made by Krishan /75/. According to this, a complete understanding is not yet available. There are some striking similarities to void lattice formation, which was first observed by Evans /76/ in ion irradiated Mo. However, there are several essential differences: While bubble formation occurs at practically all temperatures, the formation of a lattice is restricted to the lower temperature regime ($\cong 0.3\ T_m$). Voids, on the other hand, occur only at $T \sim 0.6\ T_m$. Further, bubbles grow predominantly due to gas acquisition whereas voids grow due to a small excess flow of vacancies. Thus a void is a practically empty cavity. Also, the ratio of lattice parameters to the cavity radius is $R \cong 4$ for bubbles /17/ while it is $R \cong 11$ for voids /76/. It is widely accepted, that in the early stages of bombardment a random distribution of cavities exists which becomes ordered due to a spatial instability in the concentration of vacancies or interstitials.

A completely different mechanism has recently been proposed by Evans /77, 78/. It has been found, that in b.c.c. metals interstitials are confined to a 2-dimensional motion because they form <110> oriented dumbbell configurations which do not change their orientation between successive diffusion jumps /79/. Evans demonstrated by computer simulation that with this assumption an initial statistical fluctuation in cavity distribution is unstable. As a consequence the cavities are enhanced in some planar sheets while in others they are suppressed. Considering the spatial set of {110}-planes a regular b.c.c. lattice would emerge. Figure 11 shows the result of a simulation viewed in three different crystal directions at different bombardment and ordering stages. The major drawback of Evans' model is that it cannot be applied to f.c.c. or h.c.p. metals where planar motion of interstitials is not observed. A further critical point is how a highly pressurized gas cluster can be extinguished by diffusing interstitials.

2.5 External response of the host lattice

Ion bombardment of solids leads to collision cascades in which Frenkel defects, vacancies and self interstitial atoms (SIA) are produced. These may either annihilate or form clusters in the form of interstitial or vacancy loops. As a consequence, the solid may change its external shape and its total volume. In the case of gas ion implantation the volume change is positive for metals due to the formation of gas filled point defects, clusters and bubbles. These volume increases have already been observed by Primak /5/ by interferometric methods. Large linear expansions of the bombarded surface have been observed for He bombardment of Er /18/ and by St.-Jacques et al. /19/ on Nb by using Tolansky interference microscopy. An example is shown in Fig. 12. The surface displacement is found to be a linear function of the fluence as long as the critical fluence for blistering is not exceeded. From the linear surface displacement an average density of He-atoms/lattice vacancy of $\cong 1$ has been obtained as discussed earlier. The density is lower than that obtained by assuming all He gas to be collected inside the visible gas bubbles and it is also lower than the critical density for loop punching. We therefore conclude that the gas is not evenly distributed over all cavities produced by irradiation.

234

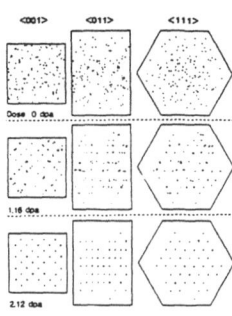

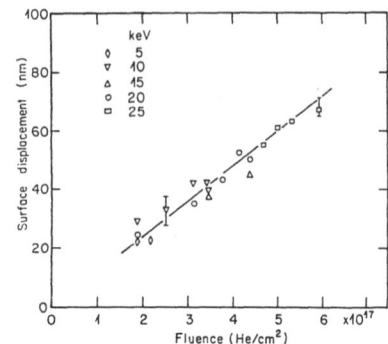

FIGURE 11. Ordering of a statistical bubble distribution due to 2-dimensional interstitial diffusion in a b.c.c. metal. Computer simulation results /78/.

FIGURE 12.   Surface displacement of Nb vs. incident He fluence (5 - 25 keV) /19/.

The swelling in an implanted surface layer can take place only in the direction normal to the surface. Lateral forces are compensated by counter forces from the surrounding unaffected lattice. EerNisse and Picraux /23/ have measured integrated lateral stress S as a function of He fluence in Al, Nb and Mo by a sensitive beam deflection method. A linear relation between S and the fluence is found only up to several 10^{15} He-ions/cm², almost two orders of magnitude below the critical fluence for blistering. Assuming elastic response of the lattice and a volume change $\Delta V/V \sim c(x)$, where $c(x)$ is the He concentration, He densities of 1, 0.42, and 0.26 He-atoms per vacancy are obtained for Al, Mo and Nb respectively. For Mo and Nb these densities are again much smaller than those needed for interstitial emission or loop punching.

The sublinear build-up of stress for $f > 10^{15}$ cm^{-2} is attributed to the fact that new defects are no longer surrounded by perfect lattice /23/. On the other hand, the regime of elastic response to a volume increase is limited to $\Delta V/V \stackrel{<}{\sim} 0.2$ % /28/ by plastic relief processes /17, 28/.

3. LIMITS OF BUBBLE GROWTH

We now have to discuss the question what stops bubble growth and what is the mechanism that limits the concentration build-up of trapped gas. There are basically four different models: 1) The oldest one assumes that bubbles grow by coalescence during bombardment. If the bubble volume during coalescence is conserved, i.e. if the volume V of the newly formed bubble is equal to $V_1 + V_2$, the sum of the original bubble volumes, the coalesced bubble will be more strongly overpressurized than the original ones since the term $2\gamma/r_0$ has decreased while the gas pressure stayed constant. The growth of the bubbles will be stopped, when the diameter has become sufficiently large, so that the surface layer breaks due to the internal pressure. Auciello /80/ has calculated critical fluences by this model which were in good agreement with experimental data on Mo by Mazey et al. /17/. I will not go into more details of this model since experimental evidence does not support the assumption that a blister is formed just from one large bubble. Nevertheless, bubble coalescence does occur, especially at high temperatures as was observed already by Barnes & Mazey /58/ and was investigated in detail by Fenske /81/. But even then the separating layer of a blister or flake is not spherical but rather a rough planar surface.

2) The formation of an interconnected network of inert gas atoms within the host crystal has been investigated theoretically, assuming that the gas atoms are mobile /82/. But since the existence of small bubbles has been confirmed in numerous experiments the percolation model is no longer considered as the blistering mechanism.

3) Another model is built around the hypothesis that mechanical failure is related to stress concentration at structural imperfections /23/. An analytical treatment of such stresses has not been possible. It has been pointed out by Hall /83/ that shear stress in the beam center will be zero and thus blisters that are much smaller than the beam diameter cannot be initialized by shear failure.

4) The model of material failure in the surrounding of overpressurized bubbles was first put forward by Auciello /80/ and quantitatively evaluated by Evans /24/. The basic idea is that in a layer of overpressurized bubbles parallel to the surface there is equilibrium between the forces exerted by the pressure $\tilde{p} = p - 2\gamma/r_0$ in the bubbles and the stress $\bar{\sigma}$ across the material between the adjacent bubbles. The equilibrium condition is

$$\bar{\sigma} = \tilde{p} \pi r_0^2 (d - \pi r_0^2) \tag{11}$$

Assuming fracture for $\bar{\sigma} \gtrsim \sigma_F$ the fracture strength of the material, the critical pressure is

$$p_F = \sigma_F [(\pi r_0^2 C_B^{2/3})^{-1} - 1] + \frac{2\gamma}{r_0} \tag{12}$$

for a cubic bubble lattice of bubble density C_B. Wolfer /28/ has pointed out that it is unrealistic to select the ultimate theoretical strength of materials as the fracture criterion. According to a review by Ashby /84/ he proposes to take the ultimate tensile strength determined by transgranular fracture for metals. This gives a failure strength of about $\bar{\sigma} = 0.009$ μ. For a typical He bubble lattice ($r_0 = 1$ nm, $C_B = 10^{25}$ m^{-3}) and the mechanical parameters $\mu = 10^{11}$ Pa, $\gamma = 2$ Jm^{-2}, p_F will be about 100 kbar. This pressure corresponds to a He density of $\sim 1.2 \cdot 10^{29}$ m^{-3} as obtained from the EOS described above. This value is very close to the densities found in He implanted films by EELS /85/. We can therefore conclude that the interbubble fracture mechanism, assuming a realistic yield strength of the material and a high pressure EOS, is in agreement with experimental evidence. In view of the uncertainty in determining the bubble size a more detailed calculation of critical fluence is not helpful.

4. THE DEFORMATION OF SURFACES
4.1 Conditions for blistering

Assuming interbubble fracture as the triggering mechanism there are two conditions which must be fulfilled for surface deformation to occur:

1) a critical bubble density and size must be achieved to overcome the yield stress of the material between bubbles. This is equivalent to achieving an average critical gas concentration (gas-atoms/bulk atom) which is $0.3 \lesssim c_{crit} \lesssim 0.8$ for He in metals. Since the maximum concentration that can be obtained without diffusion is limited by the reflection of the incident particles and by the surface regression due to sputtering, surface deformation only occurs for

$$(1 - R_N)/Y \geq c_{crit} \tag{13}$$

where R_N is the particle reflection coefficient, and Y the sputtering yield.

2) The critical concentration must be reached first at a depth which is sufficiently large so that the depth range of interbubble fracture does not extend to the surface. Otherwise, no contingent deformation of the covering layer would be possible. Evans /24/ has estimated that interbubble fracture takes place simultaneously in several ($\sim$ 12) neighbouring layers. Therefore the maximum of the range distribution must be at least $\sim$ 6 interbubble distances ($\sim$ 20 nm) from the surface. Indeed Roth et al. /86/ have found a lower limit $t_B \sim$ 25 nm for He-blistering in Nb corresponding to an incident energy of 1 keV.

Both conditions put limits to the appearance of blisters and flakes
a) for heavy ions, because the sputtering yield is high
b) for low energy ions and for broad angular and energy distributions of the incident beam.

Complete suppression of surface deformation can be obtained by implanting the material up to c_{crit} at the surface first and then in successively larger depths /29, 32/.

4.2 Models for surface deformation

Adopting the interbubble fracture as the final and catastrophic stage in the development from single HeV complexes to macroscopic gas bubble ensembles, we must stress that there is no direct experimental evidence for this process. On the other hand there is experimental evidence or theoretical arguments against all other models. Interbubble fracture has therefore been widely accepted.

The interbubble fracture model suggests that the modifications at the surface (blister formation, flaking) occur in two consecutive steps: 1) the rupture of adjacent bubble connections and 2) the rearrangement of the unstable detached surface layer into a new equilibrium position. These processes follow each other immediately, the second being triggered by the first. Again, there has been an interesting controversy on the processes dominating the surface modification. There are two types of forces that may deform the surface layer: gas pressure in the detached layer and lateral compressive stress.

Mechanisms for surface modifications by either forces suffer from difficulties in meeting all experimental evidence and a final evaluation is not feasible today. The gas pressure model assumes that the gas set free by interbubble fracture into the lenticular space between the bulk and the blister cover, plastically deforms the surface layer by the action of gas pressure normal to its surface. In order to explain the blister form observed after 30 keV He bombardment of Mo /17/ Evans /24/ had to assume that gas from a 550 Å thick layer equivalent to 12 bubble lattice layers had to be emitted into the blister cavity. He assumed a plastic thin spherical shell growth where /87/

$$p = 2 \, \sigma_y \cdot t/R \qquad (14)$$

σ_y being the yield stress, t the cover thickness and R the radius of the blister.

A more sophisticated approach was taken by Kamada & Higashida /88/ who calculate by elastic continuum theory the stress field around a penny-shaped microcrack of diameter 2a parallel to a free surface at a depth h. The lenticular cavity formed by the crack is loaded by a gas pressure p. The components of the stress field are found to have square root singularities at the crack tip (Fig. 13). In this region plastic deformation must

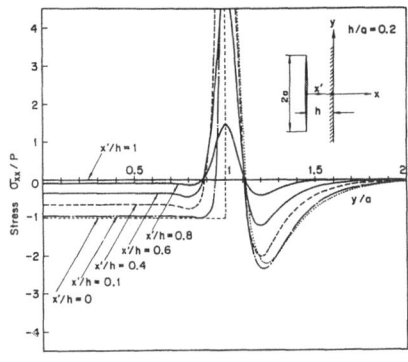

FIGURE 13. Stress component σ_{xx} normal to the surface, normalized by internal gas pressure p, in the layer above a penny-shaped crack (see insert) loaded by a pressive p. Singularity exists at the crack tip /88/.

take place. The size of the plastic zone around the tip is estimated. This plastic zone must spread as the radius of the penny-shaped crack increases at constant depth. The critical condition where the layer over the penny-shaped crack yields and deforms into a dome-shaped blister is assumed when the boundary of the plastic zone touches the surface. By this procedure a relation between the blister radius a and the cover thickness h is obtained

$$f(a/h) = (\sigma_{ys}/p)^2 \tag{15}$$

where σ_{ys} is the yield strength of the material. The parameters a and h are comparatively easily accessible to measurement while σ_{ys} and p can only be estimated with great uncertainty. The dependence of (σ_{ys}/p) on the ion energy and the target material is also unknown.

Further an estimate of the pressure p_G needed to fulfill the Griffith criterion for brittle fracture

$$G \geq 2 (\gamma + \gamma_p) \tag{16}$$

where G is the energy release rate for crack extension, and γ_p is the specific energy for plastic deformation results in unrealistic high values. Therefore, the lenticular cavity considered in the model cannot be formed by a simple crack extension mechanism.

In view of the large swelling observed in gas ion implanted material /18, 19/ and the large integrated lateral stress measured /23/ it has been found necessary to include the action of lateral stress into the models of blister formation. This was first done by Risch et al. /26/ and independently by EerNisse & Picraux /23/. Both used an example from elastic displacement theory where a thin circular plate is clamped at its circumference (Fig. 14) /89/. The elastic displacement w(r) is calculated from the differential equation

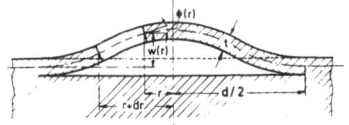

$$M_r + r \frac{dM_r}{dr} - M_t + r (Q_B - S_i \frac{dw}{dr}) = 0 \tag{17}$$

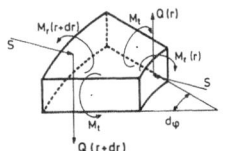

FIGURE 14. Forces and moments acting on a blister filled with gas of a pressure p and compressed by a stress S_i.

The radial and tangential bending moments M_r and M_t, the shear force per length of circumference $Q_B = p\,r/2$, and the compressive lateral stress S_i are depicted in the figure. For $S_i \neq 0$ the elastic displacement $w(r)$ is found to have a singularity for

$$d = 7.66\ E^{1/2}\ (12(1-\nu^2)S_i)^{-1/2}\ t_B^{3/2} \tag{18}$$

where E is Young's modulus, and ν Poisson's constant. At this diameter the elastic model breaks down and similar to Kamada & Higashida /88/ plastic deformation occurs. Since S_i at the critical fluence for blistering has been found independent of the particle energy /23/ a proportionality $d \sim t_B^{3/2}$ is established by eq. (18). As shown in Fig. 15, the experimental uncertainty of measurements of d and t_B in Nb is quite large. Considering the low-energy data by Roth et al. /86/ at small blister diameters a $t_B^{1.5}$-dependence is reasonable while at the higher energies the data give a better fit to $t_B^{1.22}$. In other metals the exponent of r_B was measured by Das et al. /90-92/ and found to vary between 0.85 for V and 1.25 for Be. Similar to the

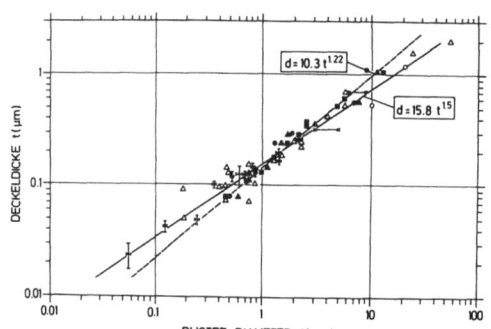

FIGURE 15. Relation between blister diameter and deckeldicke t_B for He bombardment induced blistering of Nb at room temperature /38/.

model of Kamada & Higashida /88/ the growth process of the blister and its final shape cannot be explained by Risch's model. It is, however, interesting to note that in the absence of lateral stress ($S_i = 0$) the elastic displacement increases proportional to p without any discontinuity.

Watson & Wolfer /28/ have developed another elastic deformation model considering the buckling of a simply supported rectangular plate. The plate is only partially attached to the substrate metal by ligaments as may be left over from interbubble fracture. As a result the exponent m in the relation $d \sim t_B^m$ varies from m = 0.75 for very strong to m = 1.5 for weak attachment.

Concluding this section on surface deformation by blistering it should be stated

1) there is up to now no model that can describe blister shape, size, and cover thickness in a quantitative way. Among the reasons are the great uncertainty of yield strength in damaged material, the great difficulties in treating plastic flow and the problems in estimating the gas pressure in the growing blister.

2) Experimental data are not sufficiently exact to discriminate between different models.

3) The controversy on gas pressure or stress induced blister formation cannot be dissolved, but pressure <u>and</u> stress should be included in the evaluation of surface deformation.

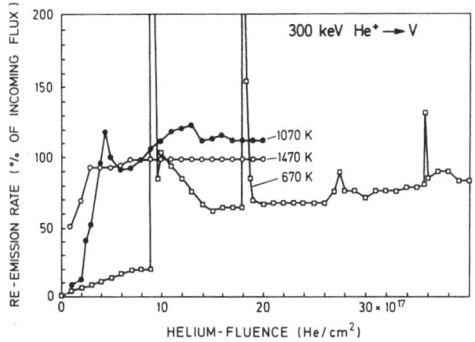

FIGURE 16. Gas reemission as a function of fluence during implantation of 300 keV He in V /95/

5. GAS RELEASE

The emission of implanted gas undergoes a rather sudden change when surface deformation by blistering or flaking occur. As was already mentioned in Sect.1, Kaminsky /8/ first observed that gas bursts during irradiation of Cu with 125 keV D^+ were correlated with the appearance of etch pits at the surface. Erents & McCracken /93/ first measured the dependence of gas reemission on He-fluence during 7-36 keV He-bombardment of Mo. They found 1) a constant low reemission (< 10 %) due to kinetic reflection at subcritical fluences, 2) a sharp break point at the critical fluence f_c, and 3) a rapid increase toward 100 % for $f > f_c$. Bauer and Thomas /94,95/ studied the reemission behavior of Hé in several metals during 300 keV He-implantation. An example is shown in Fig. 16 for V. In the temperature range where strong flaking of the surface was observed (670 K in the example), periodic gas bursts are observed at nearly constant fluence intervals. The number of these bursts could be correlated to the number of exfoliated layers, observed by TEM after the bombardment. The intensity of the bursts decreases with increasing fluence as does the areal fraction of the bombarded spot on which flaking of the corresponding order is found. In the interval between bursts reemission falls below 100 % indicating refilling of the exfoliated part of the bombardment spot with implanted He. Behrisch et al. /12, 96/ measured areal densities and depth distributions of implanted He (1.5 - 15 keV) in Nb. They confirmed that initially the gas is trapped with an efficiency $1 - R_N$ (R_N: kinetic reflection coefficient). At high fluences the areal density saturates at a value that increases almost linearly with energy. The depth profiles of implanted He reach a maximum of ∿ 0.6 He-atoms/Nb-atom independent of energy but have a larger depth distribution at higher energies.

The observed dependence of trapped He on fluence in such cases where negligible flaking occurs has been described successfully by a local saturation model /97/. This model assumes that the implanted gas is collected at or very close to its kinetic range. When saturation concentration is locally reached, additional implanted gas in this range will be reemitted without being trapped in neighbouring unsaturated ranges. The latter are filled up to saturation only by implanted gas ions with the corresponding range. The model has been checked on many examples and has been found applicable not only for He in metals but also for hydrogen in graphite /98/. It also describes correctly the isotope exchange if during implantation the incident beam is shifted from one isotope to another (e.g. H to D).

The physics behind this empirical model are not completely clear. It implies that pathways exist between a saturated range and the free surface.

Such pathways could be interconnected chains of bubbles which were observed to form when the saturation concentration is approached /72/.

Gas reemission patterns as that shown in Fig. 16 can clearly not be explained by the local saturation model. More detailed inspection of the development of implanted He profiles during and after blistering show a decrease of He concentration at the maximum of the range distribution at postcritical fluences. Thus Terreault et al. /99/ find double peaked He-profiles
in Cu irradiated with 25 keV He. Their surfaces as observed by TEM appeared
strongly flaked which makes depth profiling rather uncertain. This difficulty has been discussed by the authors, however, they conclude that the
double peaking is due to the sudden release of gas at the rupture level of
blisters. It is somewhat surprising though, that the minimum between the
two peaks does not fill up again at postcritical fluences but disappears
only when the whole blistered layer has been sputtered. Double peaked depth
profiles have not been found so clearly in other materials /100/. But a
decrease of He concentration at postcritical fluences is often observed.

In a detailed study Ehrenberg et al. /101, 102/ have measured the depth
profiles of He implanted into Ni at 40 keV. They also find a decrease of the
peak concentration observed at the critical fluence when they proceed to a
small postcritical fluence, corresponding to a reemission of about 10 % of
the implanted gas.

In this same fluence range the gas reemission rate shows a peak (Fig.17).
Comparing the amount of gas released in this peak with the total amount of
gas implanted in the areal fraction of the beam spot covered by blisters
it is found that about twice the gas quantity implanted in the very blister
area is emitted (Fig. 18). This is interpreted by assuming a interconnected
system of channels and cracks at the rupture plain of the blisters. Assuming that the decrease in peak concentration corresponds to the gas released
it can be concluded that the interconnected layer extends over the whole
beam spot.

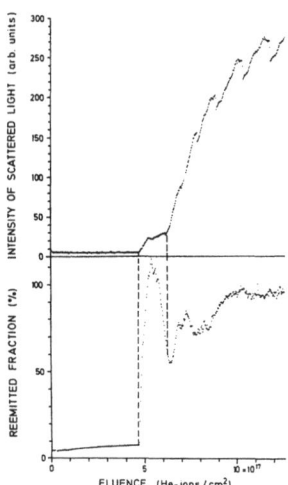

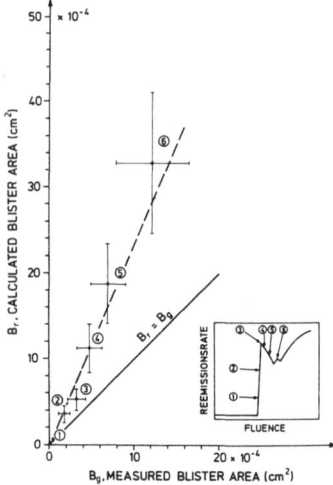

FIGURE 17. Surface deformation
measured by diffuse scattering of
light (upper part) and gas reemission
(lower part) as a function of fluence
in 40 keV He bombardment of Ni.

FIGURE 18. Relation between measured
blister area and the apparent area,
B_r, contributing to gas emission
assuming 100 % emission /102/.

The preceeding discussion shows that the process of gas reemission is complicated by blistering and flaking because, first, rupturing bubbles shed their gas content into some kind of subsurface cavity from which it may escape only when the covering layer fails or is removed. It would therefore be interesting to investigate gas reemission when the saturation is obtained without macroscopic surface deformation. As discussed in the previous section this could be done in a low-energy implant. Although such measurements have not been done to my knowledge for He we know at least from the work on Ni and SS by Jaeger & Roth /72/ what happens to the bubble lattice near the surface. Figure 19 shows in its upper part TEM picture of a microtome normal to a Ni-surface which was implanted by a multienergy distribution of He (0.25 - 8 keV). In the lower part a depth profile of the implanted He taken by nuclear reaction analysis before taking the microtome shows a decreasing distribution from the surface toward the bulk. The conditions for blister suppression are thus fulfilled in this case.

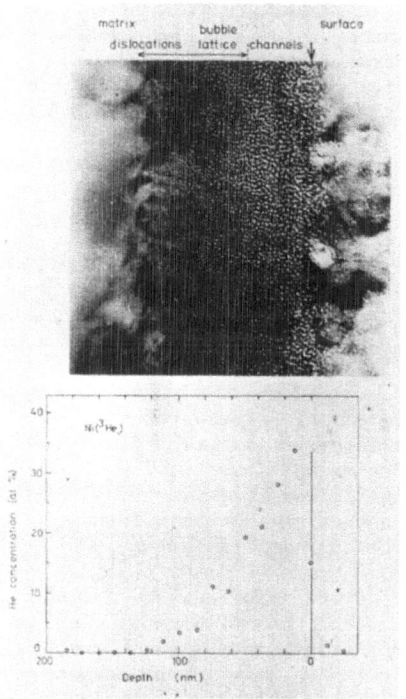

FIGURE 19. TEM micrograph of a transverse section of a He-implanted Ni surface using multienergy implantation (upper part). Depth profile of implanted ^{3}He of the same target (lower part) /72/.

Near the surface, where the saturation concentration ($\cong$ 30 %) has been reached, lengthy structures are seen which seem to form an interconnected network and are interpreted as channels, while at larger depths a typical bubble lattice prevails. It is an interesting point that, although apparently a large fraction of the bubbles has merged into the channel network, the near surface layer still contains the full saturation concentration of He. It is not understood up to now how large gas quantities can be trapped in such an open material structure.

An interesting detail that may bear on these results is found in the thermal desorption spectrum. At postcritical fluences a desorption peak is found in helium-implanted Ni, Cu, Pd, Nb, and Mo /93, 103-105/ which is absent for $f < f_c$. This peak increases with postcritical fluence (Fig. 20), but is also found in unblistered samples that have been implanted to saturation. It was suggested by van Veen /106/ that this peak corresponds to desorption from surface related states which have already been seen by Kornelsen /107/ for very low energy implantation. This could mean that due to the achievement of saturation concentration somewhere in the sample a large increase in surface area occurs. It can be speculated that the gas present in these surface states must be present before the saturation concentration is reached. This would be evidence of some gas at least being

242

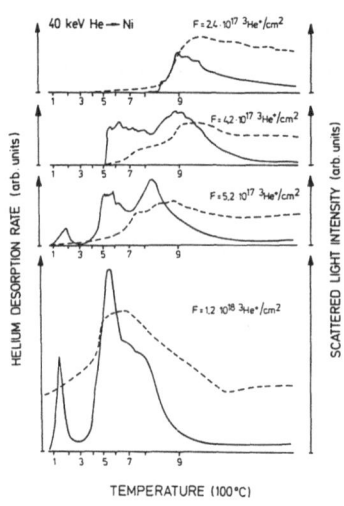

FIGURE 20. Thermal desorption of He from Ni /101/.

trapped outside of gas bubbles.

The low-T peak for implants up to saturation is not specific to He. It is found also for H implanted into metals. In most metals, however, H is mobile at R.T. and the complete trapping corresponding to He occurs only below 160 K /108/. In Be, however, complete H trapping takes place also at R.T. and desorption due to saturation is found at 450 K /109, 110/.

The essential results of the gas reemission and release measurements are:

1) The reemission of implanted He from metals is closely correlated to the appearance of surface deformations (blisters, flakes).

2) A simple local saturation model provides a good general description of the reemission process. Its basic assumptions are trapping of gas ions close to their range, a saturation concentration, and reemission from saturated regions without retrapping in unsaturated regions.

3) Deviations from this model occur for strong blistering and flaking. Buildup of supersaturation in previously undamaged material is observed.

4) The gas releasing area per blister is much larger than the blister area.

5) At saturation concentration a characteristic low energy desorption stage is observed which may be indicative for a large free surface increase.

6. VERY HIGH FLUENCE STRUCTURES

It was first observed by Martel et al. /28/ during He bombardment of Nb (1-5 keV, R.T.) that blisters disappear at a large implantation fluence. When the first blistered layer is sputtered no second layer will be formed because one blister condition is no longer fulfilled. The saturation concentration is not reached first at some distance from the newly formed surface. The sputter erosion proceeds into the bulk material which was saturated previously.

This is apparently not true if the blister covers exfoliate or if flaking predominates. On specially prepared targets, which are well annealed and highly polished, up to 39 repetitions of the flaking process were found in stainless steel and Inconel for 100 and 50 keV He bombardment respectively /30,31/. The fractional area, however, that flakes diminished with each repetition. It has been shown by Behrisch et al. /29/ that even such highly exfoliated layers will finally disappear although at very large fluence. The surface left over after blister disappearance is densely covered with small holes possibly representing the residues of bubble or the surface end of channels. This gives the surface a somewhat spongy appearance.

The contribution of blister exfoliation and flaking to erosion cannot be generally assessed, because the extent of exfoliation depends on many material and ion beam parameters. Bauer et al. /111, 112/ have shown that in a future fusion reactor the possibility of blister formation by 3.5 MeV He-ions from the fusion reaction depends strongly on the sputtering rate by low energy hydrogen particles from the plasma. In general, blister exfoliation is no longer considered as a serious process in fusion reactors mainly because of the energy and angular distributions of the incident particles and because of the complex and rough surface structures at the wall.

In conclusion of this section it should be mentioned that at still higher fluences the surface becomes more and more determined by the sputtering process. The appearance of typical ridges and grooves (Fig. 2d) as well as of needles is observed, although some characteristic details like near surface bubbles which are due to the gas ions still persist.

7. CONCLUSIONS

We have seen that, although most of the processes involved are not completely understood, a quite clear general picture of gas ion induced surface modification exists.

The implanted gas initially trapped at point defects precipitates to larger clusters by trap mutation and finally to bubbles and bubble lattices by interstitial emission and loop punching. The gas densities and pressures insice the clusters and bubbles are sufficient for these processes to occur. The increase of bubble size is stopped when the forces on the interbubble host material become excessive. The process of interbubble fracture initiates gas release and surface deformation. Gas pressure in the fracture zone and lateral stress in the deforming layer plastically deform the material into blister shape. The blister and flake stage is of transient character and disappears on further ion implantation to expose layers saturated with gas to the surface which finally develop typical sputtering structures. There is, however, a number of open questions:

- It is still not clear whether all He is present in the visible bubbles or whether some essential fraction is in submicroscopic traps. In the first case we do not understand how gas is trapped in saturated layers, in the second case we should observe growth of the interbubble clusters.
- The process of loop punching has been experimentally verified on isolated bubbles and clusters. How such a mechanism works in a dense bubble lattice where the superlattice constant is only a few bubble diameters is not understood.
- Details of the surface deformation processes are widely unclear. This starts with the depth at which the rupture of the lattice occurs. The analysis of the shape development is based on elastic continuum theory and the development stops at the onset of plastic deformation. The elastic and plastic properties for highly damaged materials are not well known, and finally we do not understand when blistering ends and when flaking occurs.

REFERENCES

1. Carter G., Navinsek B., Whitton J.L.: Topics in Appl.Phys. 52 (Springer, Berlin 1983) 231
2. Plücker J.: Ann.Phys. Leipzig 105 (1858) 67
3. Stark J., Wendt G.: Ann.Phys. 38 (1912) 921
4. Primak W.: J.Appl.Phys. 34 (1963) 3630
5. Primak W., Luthra J.: J.Appl.Phys. 37 (1966) 2287
6. Kostron H., Z.Metallkunde 43 (1952) 373
7. Nelson G.A., Effinger R.T., Welding Research Suppl., Jan.1955, 12
8. Kaminsky K.: Adv. in Mass Spectr. 3 (1964) 69
9. Behrisch R., Heiland W.: Proc. 6th Symp. on Fusion Technology, Aachen 1970, (Comm.Europ.Communities, Luxembourg 1974, EUR 4593e) 461
10. Behrisch R.: Nucl. Fusion 12 (1972) 695

11. Erents S.K., McCracken G.M.: Radiat. Eff. 18 (1973) 191
12. Behrisch R., Bøttiger J., Eckstein W., Littmark U., Roth J.,
     Scherzer B.M.U.: Appl.Phys.Lett. 27 (1975) 199
13. Barnes R.S., Redding G.B., Cottrell A.H.: Phil.Mag. 3 (1958) 97
14. Barnes R.S.: Phil.Mag. 5 (1960) 635
15. Thomas G.J., Bauer W., Holt J.B.: Radiat. Eff. 8 (1971) 27
16. Sass S.L., Eyre B.L.: Phil.Mag. 27 (1973) 1447
17. Mazey D.J., Eyre B.L., Evans J.H., Erents S.K., McCracken G.M.:
     J.Nucl.Mater. 64 (1977) 145
18. Blewer R.S., Maurin J.K.: J.Nucl.Mater. 44 (1972) 260
19. St.-Jacques R.G., Veilleux G., Martel J.G.: Radiat.Eff. 47 (1980) 233,
     St.-Jacques R.G., Veilleux G., Terreault B.: Nucl.Instr.Meth. 170
     (1980) 461
20. Johnson P.B., Mazey D.J.: Nature 281 (1979) 359
21. Swijgenhoven H.van, Stals L.M.: Radiat.Eff. 78 (1983) 157
22. Manzke R., Jäger W., Trinkaus H., Crecelius G., Zeller R., Fink J.:
     Solid State Comm. 44 (1982) 481
23. EerNisse E.P., Picraux S.T.: J.Appl.Phys. 48 (1977) 9
24. Evans J.H.: J.Nucl.Mater. 68 (1977) 129
25. Evans J.H.: J.Nucl.Mater. 76/77 (1978) 228
26. Risch M.R., Roth J., Scherzer B.M.U.: Proc. Int.Symp. Plasma Wall
     Interaction, Jülich, Oct. 1976, (Pergamon, London 1977) 391
27. Wolfer W.G.: J.Nucl.Mater. 93/94 (1980) 713
28. Martel J.G., St.-Jacques R.G., Terreault B., Veilleux G.:
     J.Nucl.Mater. 53 (1974) 142
29. Behrisch R., Risch M.R., Roth J., Scherzer B.M.U.: Proc. 9th Symp.
     Fusion Technol., Garmisch-Partenkirchen, June 1976 (Pergamon, London
     1976) 531
30. Kaminsky M., Das S.K.: J.Nucl.Mater. 76/77 (1978) 256
31. Whitton J.L., Chen H.M., Littmark U., Emmoth B.: Nucl.Instr.Meth.
     182/183 (1981) 291
32. Wilson K.L., Haggmark L.G., Langley R.A.: Proc. Int. Symp. Plasma
     Wall Interaction, Jülich, Oct. 1976 (Pergamon, London 1977) 401
33. Bauer W., Wilson K.L., Bisson C.L., Haggmark L.G., Goldstone R.J.:
     J.Nucl.Mater. 76/77 (1978) 396
34. Roth J.: Conf.Ser.No. 28 (eds.: G. Carter, J.S. Colligon, W.A. Grant)
     (The Institute of Physics, London 1975) 280
35. Das S.K., Kaminsky M.: American Chem.Soc.-meeting Chicago 1975 (ACS,
     Chicago 1976) 112
36. Erents S.K.: Proc. Conf. Physics of Ionized Gases, Dubrovnik 1976, 312
37. St.-Jacques R.G., Terreault B., Martel J.G., Ecuyer J.L.:
     J.Eng.Mat.Tech. 100 (1978) 411
38. Scherzer B.M.U.: Topics in Appl. Phys. 52 (ed. R. Behrisch),
     Springer, Berlin 1983) 271
39. Wilson K.L.: Nucl.Fusion, Special Issue: Data Compendium for
     Plasma-Surface Interactions (ed. R.A. Langley) (1984) 85
40. Donnelly S.E., Radiat.Eff., in press
41. Templier C., Jaouen C., Rivière J.-P., Delafond J., Grillé J.:
     Comptes Rendus (Acad. des Sciences, Paris), in press
42. Felde A. vom, Fink J., Müller-Heinzerling, Th., Pfluger J., Scheerer B.,
     Linker G., Kaletta D.: Phys.Rev.Letters 53 (1984) 922
43. Evans J.H., Mazey D.J.: J.Phys. F 15 (1985) L 1
44. Johnson P.B., Radiat. Eff. 32 (1977) 159
45. Johnson P.B., Mazey D.J.: J.Nucl.Mater. 91 (1980) 41 and 93/94
     (1980) 721

46. Herring C.: Structure and Properties of Solid Surfaces (eds. R. Gomer and C.S. Smith) (Univ. of Chicago Press 1953) 5
47. Kumikov V.K., Khokonov Kh.B.: J.Appl.Phys. 54 (1983) 1346
48. Trinkaus H.: Radiat.Eff. 78 (1983) 189
49. Carnahan N.F.: Starling K.E.: J.Chem.Phys. 51 (1969) 635
50. Beck D.E.: Molecular Physics 14 (1968) 311
51. Wolfer W.G., Glasgow B.B., Wehner M.F., Trinkaus H.: J.Nucl.Mater. 122/123 (1984) 565
52. Mills R.L., Liebenberg D.H., Bronson J.C., Phys.Rev. B 21 (1980) 5137
53. Michels A., Graaff W. de, Wassenaar T., Levelt J.M., Louwerse P.: Physica 25 (1959) 25
54. Yurichev I.A., High Temperature 17 (1979) 981
55. Finger L.W., Hazen R.M., Zou G., Mao H.K., Bell P.M.: Appl.Phys.Lett. 39 (1981) 892
56. Ronchi C.: J.Nucl.Mater. 96 (1981) 314
57. Syassen K., Holzapfel W.B., Phys.Rev. B 18 (1978) 5826
58. Barnes R.S., Mazey D.J.: Proc. Roy. Soc. A 275 (1963) 47; Donnelly S.E., Lucas A.A., Vigneron J.P., Rife J.C., Radiat.Eff. 78 (1983) 337
59. Risch M.: ph.d.-thesis, University of München, Germany 1978; IPP 9/24, 1978 (Max-Planck-Institut für Plasmaphysik, 8046 Garching, Germany)
60. Jäger W., Manzke R., Trinkaus H., Crecelius G., Zeller R., Fink J., Bay H.L.: J.Nucl.Mater. 111/112 (1982) 674
61. Reed D.J., Harris F.T., Armour D.G., Carter G.: Vacuum 24 (1974) 179
62. Donnelly S.E.: Vacuum 28 (1978) 163
63. Greenwood G.W., Foreman A.J.E., Rimmer D.E.: J.Nucl.Mater. 4 (1959) 305
64. Cottrell A.H.: Dislocations and plastic flow in crystals (Oxford, Clarendon Press, 1953)
65. Parasnis A.S., Mitchell J.W.: Phil.Mag. 4 (1959) 171
66. Wampler W.R., Schober T., Lengeler B.: Phil.Mag. 34 (1976) 129
67. Evans J.H., Veen A. van, Caspers L.M.: Radiat.Eff. 78 (1983) 105
68. Jäger W., Lässer R., Schober T., Thomas G.J.: Radiat.Eff. 78 (1983) 165
69. Kosevich A.M.: Dislocations in Solids 1 (ed.: F.R.N. Nabarro), North Holland, Amsterdam 1979, 33
70. Ingle K.W., Perrin R.C., Schober H.R.: J.Phys. F 11 (1981) 1161
71. Trinkaus H., Wolfer W.G.: J.Nucl.Mater. 122/123 (1984) 552
72. Jäger W., Roth J.: J.Nucl.Mater. 93/94 (1980) 756
73. Johnson P.B., Mazey D.J.: J.Nucl.Mater. 91 (1980) 41
74. Johnson P.B., Mazey D.J.: J.Nucl.Mater. 127 (1985) 30
75. Krishan K.: Radiat. Eff. 66 (1982) 121
76. Evans J.H.: Nature 229 (1971) 403
77. Evans J.H.: J.Nucl.Mater. 119 (1983) 180
78. Evans J.H.: J.Nucl.Mater., in press
79. Jacques H., Robrock K.-H.: J. de Physique, Colloque C5, suppliment au No. 10, 42 (1981) C5-723 and Proc. of the Yamada Conf. on Point Defects and Defect Interactions in Metals, Kyoto 1981 (eds. J. Takamura, M. Doyama, M. Kiritani) (Univ. of Tokyo Press, 1982) 159
80. Auciello O.: Radiat. Eff. 30 (1976) 11
81. Fenske G., Das S.K., Kaminsky M., Miley G.H.: J.Nucl.Mater. 76/77 (1978) 247
82. Baskes M.I., Wilson W.D.: Radiat. Eff. 37 (1978) 93
83. Hall B.O.: Radiat. Eff. 30 (1976) 11
84. Ashly M.F., Ghandi C., Taplin D.M.R.: Acta Metallurgica 27 (1979) 699
85. Jäger W., Manzke R., Trinkaus H., Zeller R., Fink J., Grecelius G.: Radiat. Eff. 78 (1983) 315

86. Roth J., Behrisch R., Scherzer B.M.U.: J.Nucl.Mater. 53 (1974) 147
87. Hill R.: Phil.Mag. 41 (1950) 1133
88. Kamada K., Higashida Y.: J.Appl.Phys. 50 (1979) 4131
89. Timoshenko W., Woinowsky-Krieger S.: Theory of plates and shells (McGraw-Hill, New York 1959)
90. Das S.K., Kaminsky M., Fenske G.: Bull.Am.Phys.Soc. 22 (1977) 382
91. Das S.K., Kaminsky M., Fenske G.: J.Nucl.Mater. 76/77 (1978) 215
92. Das S.K., Kaminsky M., Fenske G.: J.Appl.Phys. 50 (1979) 3304
93. Erents S.K., McCracken G.M.: Radiat. Eff. 18 (1973) 191
94. Bauer W., Thomas G.J.: J.Nucl.Mater. 47 (1973) 241
95. Bauer W., Thomas G.J.: Proc. Conf. Defects and Defect Clusters in b.c.c. metals and their Alloys, NBS Gaithersburg Md. Aug. 1973, Nucl. Metallurgy 18 (ed. R.J. Arsenault) 255
96. Behrisch R., Bøttiger J., Eckstein W., Roth J., Scherzer B.M.U.: J.Nucl.Mater. 56 (1975) 365
97. Doyle B.L., Wampler W.R., Brice D.K., Picraux S.T.: J.Nucl.Mater. 93/94 (1980) 551
98. Roth J., Scherzer B.M.U., Blewer R.S., Brice D.K., Picraux S.T., Wampler W.R.: J.Nucl.Mater. 93/94 (1980) 601
99. Terreault B., Martel J.G., St.-Jacques R.G., Veilleux G., Ecuyer J.L', Brassard C., Cardinal C., Deschêmes L., Labrie J.P., J.Nucl.Mater. 68 (1977) 334
100. Terreault B., Ross G., St.-Jacques R.G., Veilleux G.: J.Appl.Phys. 51 (1980) 1491
101. Ehrenberg J.: Ph.D.-thesis, University of Munich 1981
102. Scherzer B.M.U., Ehrenberg J., Behrisch R.: Radiat. Eff. 78 (1983) 417
103. Thomas G.J., Bauer W.: Radiat. Eff. 17 (1973) 221
104. Pisarev A.A., Telkovskii V.G.: Sov.Atomic Energy 38 (1975) 192
105. Pontau A.E., Layne C.B., Bauer W.: J.Nucl.Mater. 103/104 (1981) 535
106. Veen A. van, Evans, J.H., Caspers L.M., Hosson J.Th.M.de: J.Nucl.Mater. 122/123 (1984) 560
107. Kornelsen E.V.: Radiat. Eff. 13 (1972) 227
108. Möller W.: Nucl.Instr.Meth. 209/210 (1983) 773
109. Möller W., Scherzer B.M.U., Bohdansky J.: IPP-JET-Report no. 26 (Max-Planck-Institut für Plasmaphysik, 8046 Garching) 1985
110. Wampler W.R.: J.Nucl.Mater. 122/123 (1984) 1598
111. Bauer W., Wilson K.L., Bisson C.L., Haggmark L.G., Goldstone R.J.: J.Nucl.Mater. 76/77 (1978) 396
112. Bauer W., Wilson K.L., Bisson C.L., Haggmark L.G., Goldstone R.J.: Nucl. Fusion 19 (1979) 93

EXAMPLES OF ION BOMBARDMENT EFFECTS ON FILM GROWTH AND EROSION PROCESSES - PLASMA AND BEAM EXPERIMENTS

ERIC KAY,
IBM Almaden Research Center, San Jose, California 95120-6099

Many plasma processing technologies owe their unique utility to plasma surface-interactions which lead either to the removal or deposition of material. Plasma etching and film deposition are but two obvious examples.

The purpose of this report is to review some of the effects of energetic particle bombardment both on film growth and on erosion processes as encountered in commonly used plasma processing technologies. Comparisons between plasma experiments and U.H.V beam experiments will be made. Historically, a large variety of thin films have been grown in various glow discharge configurations by sputtering material from a target with energetic inert gas ions and subsequently condensing the sputtered particles on a temperature controlled substrate held in the plasma. High pressure (10^{-1}-10^{-2} Torr) discharges were commonly used to obtain sufficiently intense plasmas to lead to practical rates of sputtering and film deposition, i.e. $\geq$ one monolayer/sec. An inevitable consequence of such a short mean free path environment is a variety of collisional processes in the gas phase involving electrons, ions and neutrals. Some of these collisional processes have large cross sections and are important in the context of subsequent film growth or surface erosion processes. Among the most significant of these are:

a) Short lived electronically excited species.

$$e+Ar \longrightarrow Ar^* + h\nu$$

The characteristic light emission associated with this type of excitation process in frequently used as a powerful diagnostic tool of the plasma.[1-6]

b) Long lived electronically excited metastables.

$$e + Ar \longrightarrow Ar^*$$

In these short mean free path discharges these metastable species are a very effective source for ionizing species in the gas phase other than the sputtering gas itself,[7,8,9] for example the sputtered neutral particles as follows:

c) Penning ionization by long lived metastables

$$Ar^* + X_{(g)} \longrightarrow X^+_{(g)} + e + Ar$$

d) Electron induced ionization

$$Ar + e \longrightarrow Ar^+ + 2e$$

Sputtering gas ions, e.g. argon (Ar^+) formed by electron induced ionization are the major source of positive energetic ions which subsequently bombard any surface whose potential is negative with respect to the plasma potential-such as, for example, the powered sputtering target as follows:

e) Physical sputtering

$$Ar^+ + X_{(s)} \longrightarrow X_{(g)}$$

Any surface which is bombarded with energetic ions in excess of the threshold energy for sputtering, ~20eV, will in fact be sputtered into the gas phase. However, in most discharge configurations only sputtered neutrals will leave the target surface. Secondary sputtered positive ions will not be able to overcome the cathode fall potential of the sheath in front of the target material. With metal or alloy targets, negative sputtered ions are not usually formed except in unique situations.[10]

f) Resonance charge transfer

$$Ar_1^+ + Ar_2^0 \longrightarrow Ar^* + Ar_2^+$$

In this short mean free path regime, ions which are accelerated toward negatively biased surfaces may not reach the full sheath potential because they undergo charge transfer collisions in transversing these sheaths. This is especially true for resonance charge transfer collisions.

The consequence of this is that in this high pressure regime bombarding ions will not be monoenergetic and surfaces will also be bombarded by energetic neutrals resulting from this process. Consequently, the average energy of particles bombarding the target will be much less than total voltage drop across the sheath.

EFFECT OF ENERGETIC NEUTRAL AND ION BOMBARDMENT ON FILM STRUCTURE

In order to minimize this lack of definition of energetic particle bombardment in short mean free path plasmas, more intense discharges can be used in which efficient surface bombardment processes can be obtained at lower pressures thereby minimizing these gas phase collisional processes. Supported discharges such as triodes or alternately magnetrons allow efficient sputtering at $\leq 10^{-3}$ Torr. In the context of film growth or erosion processes both (e) resonance charge transfer and (c) Penning ionization cross sections are negligible at these pressures.

Operating discharges in this larger mean free path condition, $\leq 10^{-3}$ Torr, can therefore be expected to give reasonably monoenergetic ion bombardment by the incident gas ions, but now no Penning ionization of sputtered neutrals will take place as they traverse the plasma.

Substrates on which the sputtered particles are to be deposited can be biased negatively with respect to the plasma potential, thereby allowing monoenergetic positive ion bombardment of the growing thin film surface during deposition. This low pressure approach overcomes many of the usual criticisms about controlled film growth in plasmas in that one is now dealing with essentially monoenergetic ion bombardment of the growing thin film in a practical configuration in which the flux of sputtered particles arriving at the substrate and the flux of monoenergetic bombarding ions at the substrate can be varied over a wide range by controlling the plasma density. However, in spite of this flexibility there is inevitably an interdependence of the ion bombardment processes going on at the sputtering target and at the growing film surface on the substrate since ions are extracted from one common plasma. Also in order to know the absolute energy of substrate bombarding ions it is essential to know the plasma potential under the condition of the externally applied bias voltage to the substrate. This can readily be measured.[10,11,12]

An alternate and totally decoupled approach is the use of a dual ion beam system operated in a long mean free path environment. Here monoenergetic ions are extracted from one plasma source and accelerated toward the target material resulting in sputtering, while monoenergetic ions extracted from a second plasma source are used to bombard the substrate on which the sputtered species from the target are being condensed (see schematic in Fig. 1). However, both the dual ion beam approach as well as the longer mean free path triode or magnetron approach introduce new complexities whose significance has only been qualitatively acknowledged in the past.

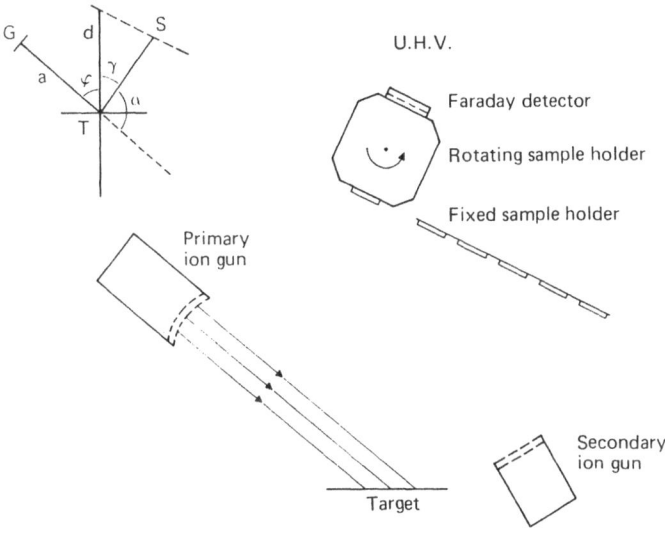

Figure 1. Schematic of UHV Dual Ion Beam System

When an energetic ion beam collides with a surface, only a fraction of the total incident energy is permanently transferred to the target. Some of the incident energy is carried off by the sputtered particles which have non--Maxwellian energy distributions with maxima from ≃3-10 eV. More significantly, in some circumstances, a very significant amount of energy is carried away from the sputtering target in the direction of the substrate by those incident energetic ions which are reflected from the target as Auger neutralized

energetic neutrals. The reflection coefficients and the
absolute energy carried away by these backscattered neutrals
varies dramatically depending on the relative masses involved
in the collision as well as the angle of incidence of the
energetic ions on the target.

The effect of these backscattered energetic neutrals on
film structure can be seen in Figs. 2a and 2b. Here changes
in bulk lattice parameters of a metal film are plotted as a
function of the angle a given substrate subtends with respect
to the target for different energetic ion/target combinations.
In the case of Xe on Cu, all lattice parameters of the
deposited Cu films are very close to bulk values and
independent of the substrate position with respect to the
copper target. This is consistent with the very low reflection
coefficient expected for this mass combination.[13,14] For Ne
on Cu, however, deviations from bulk lattice parameter values
are readily observed as well as an angular dependence. This
lattice distortion effect becomes most pronounced in the Ne
on Pd example where high reflection coefficients of the
incident energetic Ne ions are to be expected. Up to 50% of
the incident energy of those Ne ions which are reflected as
Auger neutralized Ne atoms from the Pd target can be carried
toward the growing Pd film. In this latter example the angular
dependence due to back scattered neutrals at the sputtering
target is again self-evident. Similar conclusions about the
presence or absence of energetic back scattered neutrals can
be drawn from whether or not the deposited film on the substrate is being resputtered as a function of substrate
position. As anticipated in the Xe/Cu system, the deposition
profile of the film followed a perfect cosine distribution.
This is consistent with sputtered Cu particles leaving a
polycrystalline target with a cosine distribution in the
$\leqslant 1000$ eV energy regime and depositing with a sticking
probability of one with no resputtering from the substrate.
In the Pd/Ne system, however, a very significant deviation
of the cosine distribution deposition profile was observed
presumably due to the resputtering of film deposit with back
reflected energetic neutral Ne atoms from the Pd target.

A more systematic approach to energetic bombardment
induced lattice distortions is afforded by controlled monoenergetic ion bombardment of the growing film by ions from a
second ion source as shown in Fig. 1. Here the relative flux
of sputtered metal atoms and energetic ions arriving at the
substrate can be readily controlled. We show the results of
such a study in Fig. 3 where changes in lattice parameter
are reported as a function of energy delivered by the
energetic ions per arriving metal atom at the substrate.

As can be seen in the Ar/Pd case the lattice distortions
go through a maximum beyond which self annealing is thought
to dominate. The lattice distortions introduced by energetic
ion bombardment in Cu are much less pronounced and no obvious
maximum is observed. The reason for these differences for
different metals are presently under investigation. Similar
lattice distortion effects and changes in related physical
properties have also been observed in films produced in triode
configurations.[15] Obviously such anisotropic lattice

distortions can be expected to and do reflect themselves in
many physical properties. Examples of these effects are the
subject of separate publications.

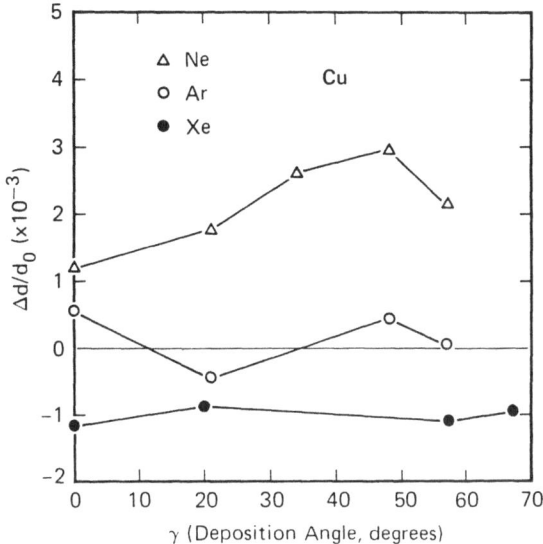

Figure 2a. Lattice parameter changes of films as a function
of substrate position with respect to the target
described by angle (see Fig. 1) for Cu sputtered
by 700 eV Ne⁺, Ar⁺, Xe⁺, Xe⁺ respectively.

EFFECT OF ENERGETIC ION BOMBARDMENT ON SURFACE CHEMISTRY
     The role of energetic ion bombardment of a surface in
a chemically reactive environment is somewhat different than
that associated solely with physical sputtering. In the
chemically reactive environment the key role of ion bombardment can be the conversion of strongly bound (chemisorbed)
layers at or near the surface into new chemical species which
are then very weakly bound, and consequently desorb rather
readily, especially in the presence of subsequent collision
cascade events. This process we choose to refer to as "chemical
sputtering." For some systems of technological interest, the
rates of etching by chemical sputtering are very much higher
than that achieved by pure momentum transfer physical
sputtering or by the spontaneous chemical reaction in the
absence of ion bombardment.
     A typical prototype system which demonstrates the
essential conceptual features of "chemical sputtering" is the
reaction of fluorine atoms with a silicon surface in the
presence of ion bombardment to form volatile silicon fluorides
at room temperature.

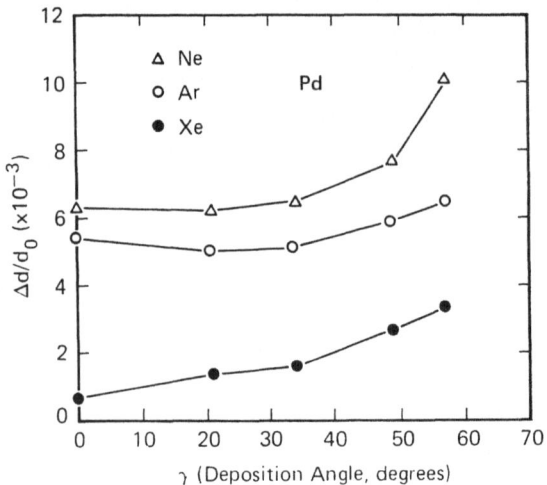

Figure 2b. Lattice parameter changes of film as a function of
substrate position with respect to the target
described by angle γ (see Fig. 1) for Pd sputtered
by 700 eV Ne$^+$, Ar$^+$ and Xe$^+$ respectively

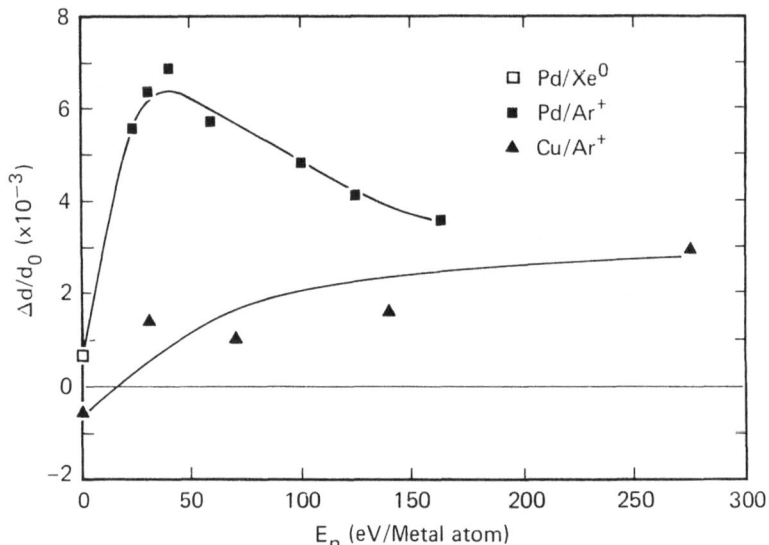

Figure 3. Lattice parameter changes of Cu and Pd films as a
function of normalized energy E_n delivered to
growing film surface by ion bombardment from
secondary ion source.

FLUORINE-SILICON SURFACE CHEMISTRY

The overall surface chemistry of fluorine atoms reacting with a silicon surface as a function of coverage and in the absence of ion bombardment is qualitatively understood. Fluorine atoms chemisorb with high sticking probability and penetrate the Si lattice up to about the equivalent of several monolayer coverage after which the sticking probability of F on such a fluorinated surface drops dramatically. The exact chemical composition of this layer is not yet clear but is most probably best described as an assembly of fluorinated silicon moieties such as SiF, SiF_2, SiF_3 and trapped SiF_4. However, with further fluorine dosage this fluorinated "surface" layer continues to react, but very much more slowly, and converts to volatile silicon fluorides, 85% of which is SiF_4. The remainder is Si_xF_y, where x and y is still controversial. Winters et al.[16] have recently developed in some detail the notion that the mechanism of this type of etching reaction has much in common with the much more thoroughly studied surface oxidation processes where the formation of negative surface ions followed by a Cabrerra-Mott-like field assisted anion-cation replacement process provides a major driving force for mass transport into the lattice and subsequent reaction. Recent ab initio molecular orbital calculations by Bagus et al.[7, 18, 19] are quite consistent with the notion that negative F ions are formed at the Si surface with a strong binding site for fluorine, not only on top of the silicon, but between the first and second silicon lattice layer with no activation barrier for lattice penetration. Early XPS data by Chuang[20] and later by Roop et al.[21] is also consistent with this fluorine surface penetration concept.

ION ENHANCED SURFACE CHEMISTRY

The superposition of energetic ion bombardment dramatically enhances the overall rate of reaction from that described above betweeen fluorine and silicon. This result can be seen clearly from data presented in Fig. 4, where the rate of etching of fluorine at the clean Si surface suddenly increases by an order of magnitude as a 500 eV Ar ion beam is switched on. XeF_2 is used as a convenient way of delivering atomic-like F to the surface. It has been demonstrated that XeF_2 dissociates at the Si surface leaving F atoms free to bond to the surface and, at sufficient coverage, react to form volatile SiF_x with no evidence of Xe being left on the surface. Comparison of data from this laboratory, Tu et al.[22] for XeF_2 on Si with that of Flamm et al.[23] on F atoms on Si shows that near room temperature the reaction probability of XeF_2 is much higher than that of F atoms incident on a fluorinated Si surface which leads to the conclusion that XeF_2 will lead to the formation of volatile SiF_x more efficiently than F atoms. The degree to which ion bombardment can enhance the volatile product molecule formation will in part be related to the reaction probability of a particular etchant specie at various levels of coverage. Why the sticking probability varies from one incident specie to another on any surface is purely speculative at this time, especially in the presence of ion bombardment. As the supply of "F" etchant is cut off (Fig. 4), only Ar+ physical sputtering remains. Clearly the etching rate, when both energetic ions and F atoms are involved at the surface, far exceeds the sum of either individually. Yield measurements by Coburn et al.[24] have demonstrated that one inert gas ion, say Ar$^+$, at a 1,000 eV, incident on a fluorinated Si surface can desorb the equivalent of about 30 Si containing species in the form of SiF_x. This event inevitably exposes a clean Si surface "patch" with high sticking probability for subsequent rapid fluoridation, thereby initiating the sequence of steps as a function of coverage as described above as long as a large enough supply of F containing etchant is available above the surface. We have shown that absolute cross section measurements for this chemical sputtering event give values of $10^{-14}/cm^2$ at fluorine coverages of a monolayer or greater.[24] This is an order of magnitude greater than the cross section associated with pure physical sputtering of SiF_x, at very low coverage (Fig. 5).

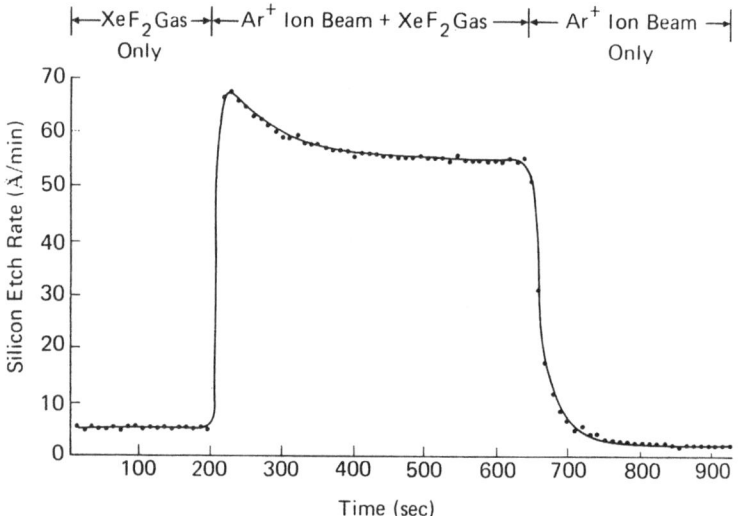

Fig 4. Si etching in XeF_2 with and without ion beam bombardment after J.W.Coburn and H.F. Winters, J. Appl. Phys. 50, 3189 (1979).

Ion bombardment is likely to disorder the receding Si surface so that the initial surface crystallographic orientation is likely to play a much less important role in the etching kinetics in the presence of ion bombardment than in the absence. The degree to which radiation damage, such as the production of vacancies and interstitials and their subsequent diffusion in the presence of ion bombardment influences the observed surface kinetics is not clear at this time. However, initial yields (i.e., SiF_x removed per incident ion) from a fluorinated Si surface which was exposed to a very low ion beam fluence (10^{12} ions/cm^2, i.e., 1 ion/10,000 $\overset{\circ}{A}^2$ of surface) were shown to be equivalent to that from a Si surface which had been subjected to prolonged high current density ion bombardment prior to simultaneous exposure to XeF$_2$ and energetic ions. Furthermore, the reaction rate on Si after prolonged ion bombardment dropped to its original rate as soon as the ion beam was shut off within only 1 or 2 additional monolayers being removed, even though the structural damage produced by the previous ion bombardment must have penetrated for several tens of monolayers. From these observations it would appear that permanent lattice damage resulting from ion bombardment, though undoubtedly present, cannot be the main driving force for the observed enhancement in the reaction rate between F and Si. Whereas physical sputtering is generally believed to arise from the very outermost one or two lattice layers, chemical sputtering is likely to be much less surface constrained and can involve the ion bombardment enhanced formation of volatile product species deeper in the lattice followed by subsequent ion enhanced diffusion to the surface. The degree to which ion bombardment facilitates the penetration of F into the lattice is not clear at this time but qualitatively this enhanced F penetration and compound formation is thought to be as key a factor with ion bombardment as it is with laser bombardment[25,26] but mechanistically for different reasons. Electronic excitation is not thought to play a key role in the ion bombardment case.

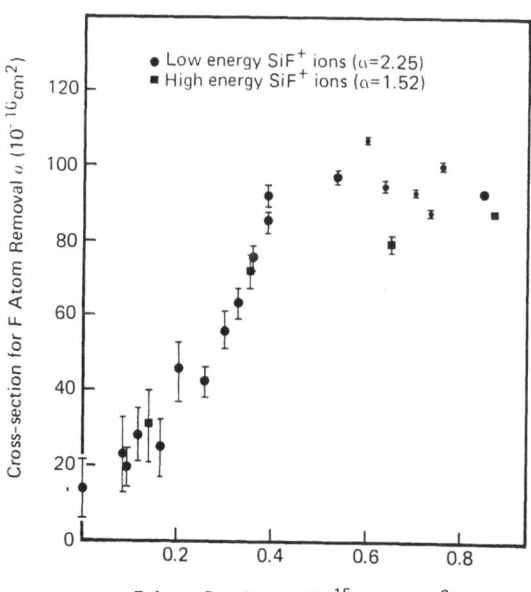

Figure 5. Cross sections for ion bombardment induced F atom removal as a function F coverage. The F removal is measured as emitted SiF$^+$ (SIMS) SiF$^+$ signal was independently calibrated as a f(F) coverage. Ref. 24.

Recent work by Winters[27,28] has clearly shown that
the effect of ion bombardment on the surface reactions between
F and Si cannot be generalized at random. In fact, in numerous
cases ion bombardment actually slows down the etch rate. This
is the case, for example, for the reaction of Cl_2 with Al(100)
and Cu(100).

In both cases it is thought probable that the act of ion
bombardment either prevents the formation or alternately
destroys the stable vaporizing species, which in the case of
Al is known to be Al_2Cl_6 and in the case of Cu it is Cu_3Cl_3,
in the temperature regime of interest. There are other
instances where the reaction probability in the absence of ion
bombardment is so great that ion bombardment does not influence
the reaction rate significantly. XeF_2 reacting with polycrystalline niobium is an example of this situation. Here the
reaction probability is 0.7 in the absence of ion bombardment.
The enhancement due to ion bombardment on the F/Si reaction
is largely due to chemical sputtering as defined earlier. The
induced etching of F on W (111) is also usually dominated by
chemical sputtering mechanisms; however, in the case of W (111)
it can also be shown that lattice damage induced by ion
bombardment can under some circumstances significantly increase
the reaction probability of the XeF_2 with the surface.[28]
Therefore, it can be concluded that, whereas one can often
draw general conclusions about ion induced etching (e.g.,
physical sputtering is not a dominant mechanism) there are
also significant differences from system to system.

ETCHING DIRECTIONALITY

It will become evident that etch feature geometry will be
determined by the balance between neutral etchant versus
energetic ion enhanced surface etchant chemistry together with
selective blocking of the etch feature side-wall chemistry.

The straightforward connections between ion enhanced
etching rates and the concept of directionality can now be
conceptualized using F and Cl interacting with Si as a proto-
type example.

Consider energetic ions only bombarding a Si surface
orthogonally through a mask opening as shown in Fig. 6a. Such
directional ion bombardment alone can be expected to result in
physical sputtering giving rise to a highly directional etch
feature as schematically illustrated in Fig. 6a. If instead
of energetic ions one exposes the Si to neutral etchant species
only, e.g., F atoms, which arrive from all directions from
above the mask opening, then Si would be expected to etch
isotropically as illustrated qualitatively in Fig. 6b. The
rate of this isotropic etching under the mask will be determined
by the neutral etchant flux and the corresponding spontaneous
etch rate at a given substrate temperature and orientation.
This spontaneous F etch rate of Si is considerably higher than
that obtained by pure momentum transfer physical sputtering
alone in the respective F flux or ion flux and energy regimes
commonly encountered in practical systems to be described
later. If now orthogonally directed energetic ions and isotropically directed neutral F atoms arrive simultaneously at

the Si surface through the mask opening, then an etch pattern is obtained as qualitatively represented in Fig. 8c. This etch pattern demonstrates the logical consequence of etching very much faster in the forward direction (in the direction of the incident ions) than the spontaneous isotropic etching. The degree of control over the geometry of the etch feature depends first and foremost over the control we have over the spontaneous isotropic etch chemistry vis-a-vis the ion enhanced directional chemistry. A quantitative way of demonstrating the spontaneous (isotropic undercut) reaction rate due to neutral etchant species vis-a-vis the reaction rate obtained by simultaneous ion bombardment, as a function of etchant flow rate is shown in Fig. 7 for the case of $XeF_2/Ar^+/Si$.

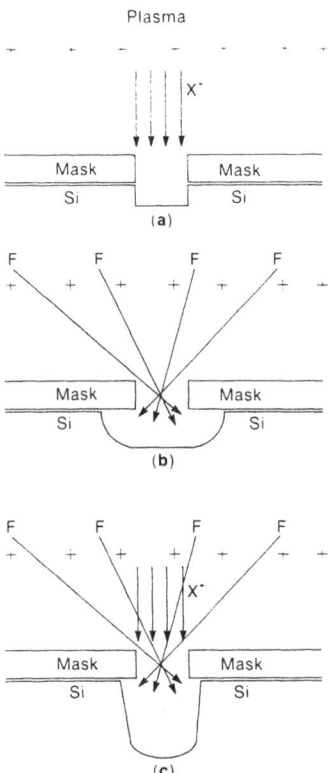

Figure 6. Schematic etch feature geometry for (a) normal incidence ion bombardment only; (b) random incidence F atom etching only; (c) combined ion and atom etching.

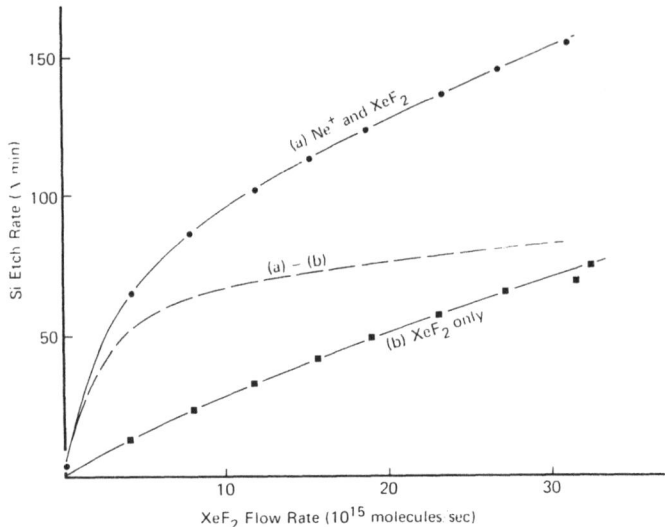

Figure 7. Si etch yield as a function of XeF$_2$ etchant flow
rate with and without 1keV neon laser ion beam
bombardment. Ref. 30.

The overall set of circumstances described above is
conceptually similar to that encountered in the more
complicated plasma environments as commonly used. Ion
bombardment with incident ion direction perpendicular to the
plane of a Si wafer is a condition which exists in reactive
ion etching (R.I.E.) in glow discharge configurations where
the Si wafer is placed on an electrode surface which is at a
negative potential with respect to the plasma potential. In
the most commonly used capacitively coupled R.F. diode
configurations, the powered negatively self-biased electrode
is subject to high energy ion bombardment (usually several
hundreds of electron volts) and the grounded electrode is
subject to low energy bombardment (usually of the order of
several tens of electron volts). The exact energy distribution
at the respective electrodes is determined by many factors of
which surface area of the powered electrode versus total
grounded surface area exposed to the plasma and the plasma
frequency and pressure are key factors. This perpendicular
ion bombardment in a plasma configuration is a key source
conductive to vertical etching. On the other hand, the isotropic etching which determines the degree of sideways undercut etching will be caused by all the neutral plasma etchant
species which inevitably arrive at the Si surface with random
directionality from the plasma.

If the isotropic undercut chemistry could be completely
eliminated, then the etch feature geometry would be determined
primarily by the direction of the incident energetic ions.
Several important situations will be described which tend

toward this latter condition. The most obvious way to reduce
or eliminate the isotropic (undercut) chemistry at the etch
feature sidewalls is either to minimize the arrival of
neutral etchant species or to block the side wall surfaces
with a relatively unreactive barrier layer. This blocking
situation can be approached, for example, using mixed halogen
containing etchants. From earlier work by Pandey et al.[29] ,
it can be deduced that chlorine atoms, whether arriving as
such or formed by chemisorptive dissociation, will cover the
Si surface to saturation after which little further reaction
will take place. From potential energy interaction
calculations by Bagus et al.[18] it becomes clear that Cl atoms,
have to overcome a very large activation barrier of ~10 eV
before they can penetrate the Si lattice to form product[30]
molecules. Cl atoms will therefore tend to block the sidewalls
and thereby slow down further reaction of Si with other
etchant species. On the other hand, we have shown that
wherever energetic ions hit such a chlorinated Si surface, the
reaction will proceed to a volatile end product.[30] Chemical
sputtering of a silicon halide will therefore clearly result
from the bottom surface of the Si etch feature. Whereas the
mechanistic details of this type of ion enhanced reaction are
not yet clear, the phenomenological results of F/Si versus
Cl/Si were clearly demonstrated by beam experiments reported
by Coburn et al., (see Fig. 7 and 8). By comparing Fig. 7 with
Fig.8, it can readily be seen that even with ion bombardment,
the $SiCl_x$ yields in the Cl/Si reaction are not as great as the
spontaneous F/Si reaction. Although the ion enhanced reaction
rate is much slower with Cl than with F, the directionality
can be expected to be greater because of the sidewall blocking
by Cl.

FLUOROCARBON PLASMAS

In commonly used r.f. etching discharges, neither pure
single nor pure mixed halide gases are ordinarily used, but
instead halocarbons are often used, the simplest example of
which is CF_4. In the ground state this neutral molecule is
as inert as Ar and does not chemisorb to a measurable degree.
On the other hand electron impact collisions in a CF_4 plasma
fragment this molecule into several major species: F atoms,
CF_3^+ ions, CF_3 , CF_2 , and CF radicals. All of these have been
identified by the combination of mass spectroscopy and
emission spectroscopy and laser induced fluorescence, see for
example, Refs. 31-34.

In a CF_4 plasma the major ion is CF_3^+ . Whereas CF_2^+ is an
order of magnitude less prevalent, it may be more reactive.
UHV beam experiments by Tu et al.[22] have shown that the
ability of energetic CF_3^+ ions to enhance the F/Si reaction
is essentially equivalent to that previously found for Ar^+
at similar ion energies (Fig. 9). Therefore, to a first order
approximation the enhancement effect on the F/Si reaction is
due to ion bombardment momentum transfer effects between the
incident ion and the fluorintated Si surface and not primarily
due to the chemical nature of the incident ion.

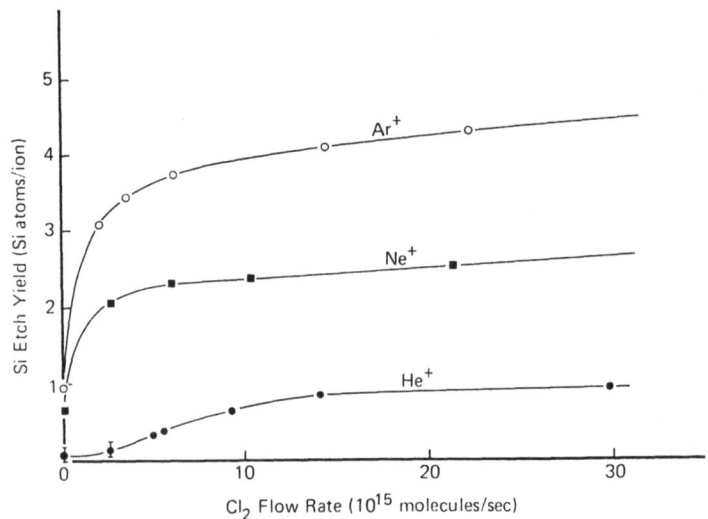

Figure 8. Si etch yield as a function of flow rate of Cl_2 with ion bombardment using helium, neon and argon, 1 keV ion beams. Ref. 30.

From Auger experiments by Coburn et al[35] involving U.H.V., CF_3^+ ion beam bombardment experiments of Si, it becomes clear that C is deposited on the Si surface in this bombardment process. The degree of carbon deposition will depend on the kinetic energy of the incident CF_3^+ since above the physical sputtering threshold two competitive processes, namely C deposition and simultaneous resputtering of previously deposited C will take place until a steady state C coverage will be reached. It has also been demonstrated that CF_3 radicals, which in a CF_4 plasma are expected to be more abundant than the short lived CF_3^+ will very effectively chemisorb and dissociate on a Si surface and thereby deposit C on the surface. The major difference between the CF_3 radicals and ions is that the radicals will not give rise to any physical or chemical sputtering as a competitive process to the C deposition and the radical collection efficiency at an electrostatically biased surface will be less than that for ions. This type of carbonaceous deposition process will serve as a blocking barrier on Si and will tend to slow down further F/Si reaction, especially on surfaces where energetic ion bombardment is expected to be low or absent, such as on the sidewalls of etch features. The more pronounced the carbonaceous deposition on these sidewalls, the more effectively the isotropic F/Si etching will be minimized, thereby leading to more pronounced vertical etch features.

CF_3 radicals and CF_3^+ ions are not the only possible source of a carbonaceous blocking layer. CF_2 and CF radicals are also present in a CF_4 discharge and these species have been shown to be precursors to oligomerization in the plasma gas phase and subsequent fluorocarbon polymer deposition.[32,36,37,38]

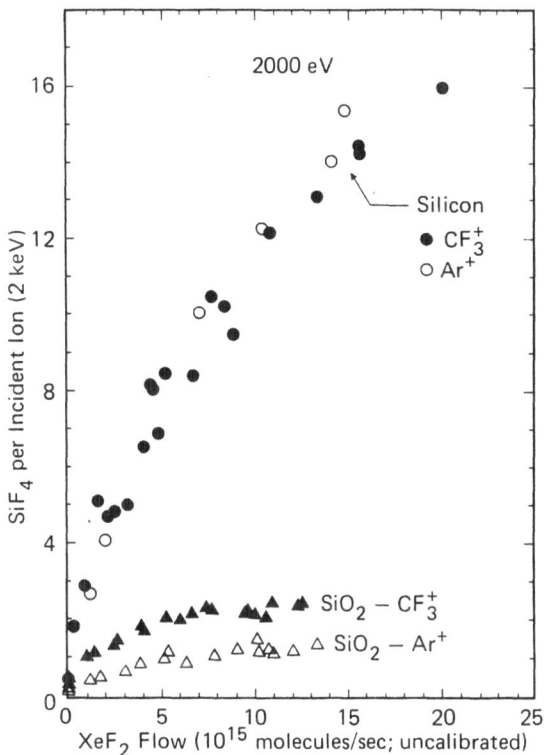

Figure 9. SiF$_4$ etch yield for Si and SiO$_2$ etched by Ar$^+$ and CF$_3^+$ ion beam as a function of XeF$_2$ flux. Ref. 22.

A commonly used plasma etchant gas mixture, where selective etching between SiO$_2$ and Si is required, is CF$_4$ and H$_2$ In this case CF$_2$ species become more and more abundant in the gas phase as the H$_2$ concentration is increased. H atoms react very rapidly with CF$_3$ radicals to convert them to CF$_2$+HF.[39,40] In addition, F etchant atoms are scavenged from the gas phase and pumped out of the system as stable HF molecules. The back reactions, e.g., F with CF$_x$ radicals to ultimately form C$_n$F$_{2n+2}$, is therefore reduced. One can easily observe by optical emission that the F atom concentration goes down as that of CF$_2$ radicals goes up when H$_2$ is added to the CF$_4$ plasma. See, for example, data reported by Millard et al., Fig. 10. CF$_3$ radicals are not monitored by emission spectroscopy. Furthermore, we have shown, directly by optical emission and indirectly by mass spectroscopy, that the fluorocarbon deposition rate is proportional to the concentration of (CF$_2$)$_n$ radicals in the plasma, see Figs. 11 and 12. Therefore, as the polymerization rate increases, so

will its blocking action on the sidewalls of the   etch
features.

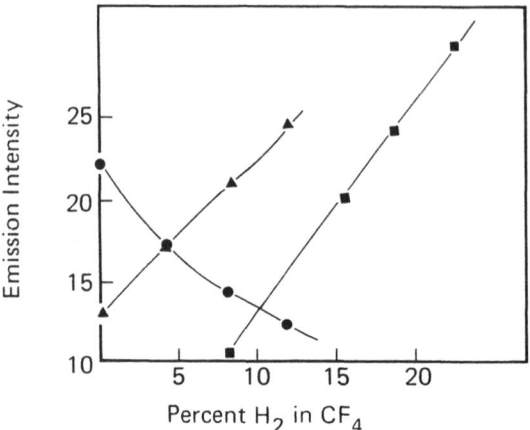

Fig. 10. Emission intensities of F, CF_2, and HF as a function
of hydrogen addition to a CF_4 plasma. Ref. 32.

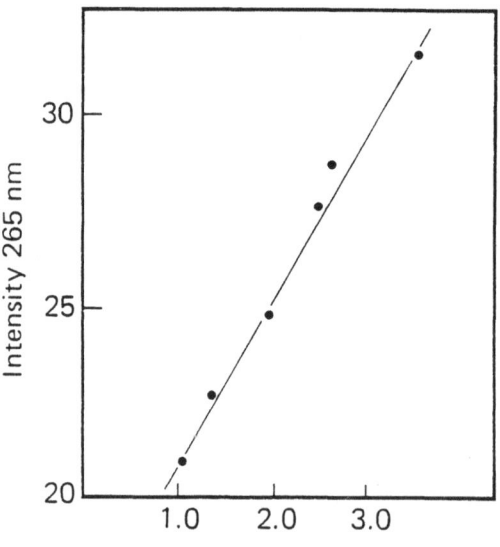

Figure 11. Deposition rate in a parallel plate reactor versus
CF_2 band emission at 265 mm. Ref. 32.

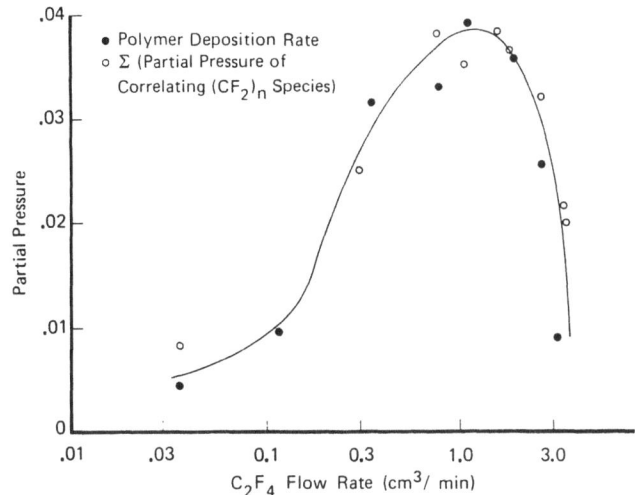

Figure 12. Deposition rate and summation of $(CF_2)_n$ partial pressures in a C_2F_4 plasmas as a function of flow rate of C_2F_4, demonstrating correlation and maximum as a function of residence time. Ref. 36.

The effect of polymer deposition at the bottom of the etch feature, which is subject to ion bombardment, is more complicated and will be discussed in the next section. In order to understand the role of carbonaceous blocking layers not only in the context of minimizing isotropic etching (directionality) but also in the context of selective etching of two different materials such as, for example, SiO_2 versus Si, it is important to consider the consequences of CF_3^+ ion bombardment further. From Fig. 9, we can see that the relative effect of CF_3^+ versus Ar^+ ion bombardment on the etching rate of Si is very similar and not dramatically different on SiO_2. Far more significant is the fact that the absolute ion enhanced rate effect on SiO_2 per incident ion is approximately one order of magnitude lower than on Si. In fact, at room temperature in the absence of energetic ion bombardment SiO_2 will not etch significantly at the fluxes of etchant commonly used.[41,42]

From U.H.V. beam experiments as illustrated in Fig. 13, it becomes immediately evident that "C blocking" of the SiO_2 surface can be expected to be very much less than on the Si surface in the presence of energetic ion bombardment. The overall effect of this selective net deposition of carbonaceous material on Si versus SiO_2 in a typical CF_4+H_2 plasma is self-evident from Fig. 14. Although the ion enhanced etch rate of clean Si is very much greater than that of SiO_2, Fig. 9, the more effective C blocking of Si versus SiO_2 in the CF_4+H_2 situation completely overshadows the etch rates of the pristine materials in Fig. 14.

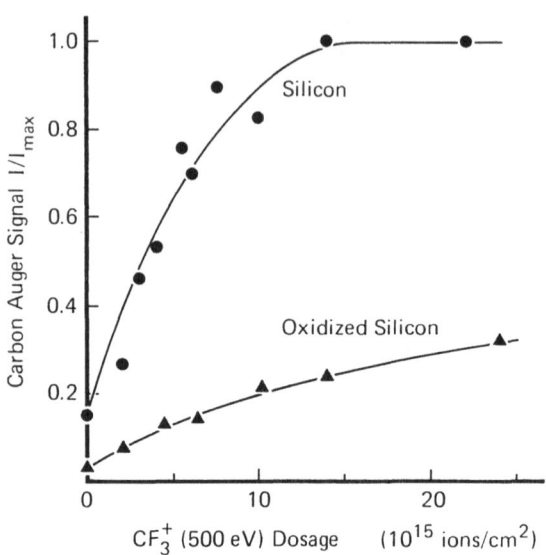

Figure 13. Carbon Auger signal on Si and SiO_2 as a function of CF_3^+ ion beam bombardment dosage. Ref. 35.

Both from the U.H.V. CF_3^+ ion bombardment beam experiments and from mass spectro metric diagnostics of CF_4 plasma effluents, it can readily be shown that in the presence of energetic ion bombardment any source of carbonaceous deposit on SiO_2 is prone to conversion to volatile CO_x and CO_xF_y species; that is to say, ion bombardment helps to free O from the SiO_2 lattice, then tends to oxidize the C to volatile species. The mechanistic details of this process are unclear at this point. Injection of oxygen atoms into the gas phase of a fluorocarbon plasma from any source serves to remove unsaturates, i.e., precursors to polymer film formation, thereby leaving F atoms free to etch with predictable consequences in etch rate and directionality. Furthermore, CF_3 radicals are efficiently[39] converted to F atoms and CF_2O, thereby providing a further supply of F atom etchant species. Obviously, if excessive amounts of oxygen are added, oxidation of the surface to be etched will dominate.

In order to promote the etching of SiO_2 more rapidly than Si, it becomes necessary to deposit as much carbonaceous material on the surface to be etched such that etching of Si will be essentially eliminated by a blocking carbonaceous deposit, whereas etching of SiO_2 will continue at reasonable rates because of removal of the blocking layers as CO_x and CO_xF_y volatiles in the presence of ion bombardment.

Any scheme which deposits carbonaceous material on Si at a greater rate than on SiO_2 in the presence of energetic ion

bombardment, together with etchant species, will accomplish
the same selective etching result in kind, if not in degree,
as in the CF_4/H_2 case. So, for example, any scheme for
increasing the polymerization precursors $(CF_2)_n$ in the gas
phase by adding any other solid or gaseous F atom scavenger,
or by adding C_2F_4 to CF_4 instead of H_2, will decrease the
etch rate of Si much more than that of SiO_2[31] (see Fig. 15).
Recent results[33] have shown that monitoring the CF_2 radical
concentration by laser induced fluorescence in a variety of
etching fluorocarbon plasmas results in a direct measure of
the subsequently found selective etch rates of Si versus SiO_2.
These findings are further confirmation of the correlation of
CF_2 and polymerization and its role in selectivity and
directionality in R.I.E. as outlined above.

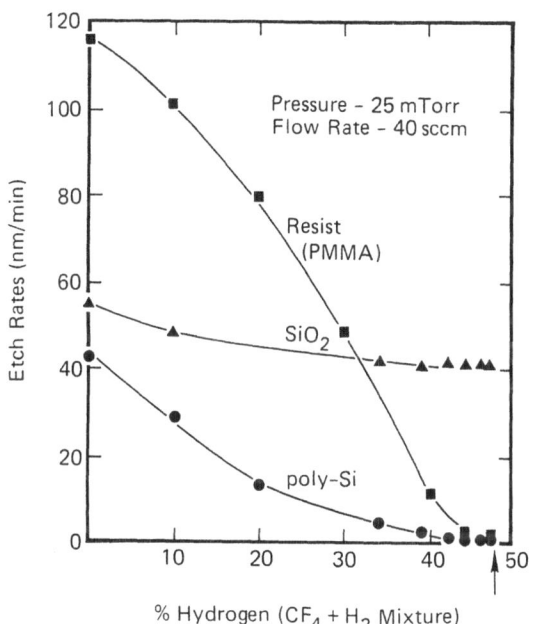

Figure 14. Etch rates of Si, SiO_2 and PMMA resist in a CF_4
plasma as a function of hydrogen injection into
plasma after L. M. Ephrath et al., J. Electrochem.
Soc., 129, 2282 (1982).

The detailed chemical composition of the carbonaceous
deposits on various surfaces in an etching discharge is hard
to predict and therefore the ability to predict its blocking
capability from further etching is also speculative. It should
not be considered surprising that any variable which leads to
a change in the type and concentration of fluorocarbon species
vis-a-vis the main halogen etchant in the plasma will influence
both the degree to which an effective blocking layer is formed

266

as well as its chemical structure. In addition, clear evidence
exists that the degree of ion bombardment of a surface at
which precursors to polymerization are present will greatly
influence the rate of polymer formation as well as the chemical
structure of the resultant carbonaceous deposit. The composition
of the carbonaceous deposit will clearly relate to its
etchability. Data by Kay et al.[43] as represented in Fig. 16
and 17 demonstrates an example of the effect of ion energy on
rate of polymerization, and Fig. 18 shows the effect of ion
bombardment on the chemical structure of the carbonaceous
deposits. In general, the rate of polymerization is enhanced
with ion energy up to a maximum beyond which physical sputtering
together with ion enhanced revolatilization of the polymer
deposit in the presence of F dominates over polymerization. The
structure of the carbonaceous deposit becomes more and more
C-like with increasing bombarding ion energy at the expense
of CF, CF_2, CF_3 component in the film deposit (see Fig. 18).
However, not all the F atoms can be preferentially resputtered,
thereby never reaching pure carbon films. In the case of
hydrocarbon polymerization in the presence of enough energetic
ion bombardment ultimately very hard diamond-like C deposits
are formed.

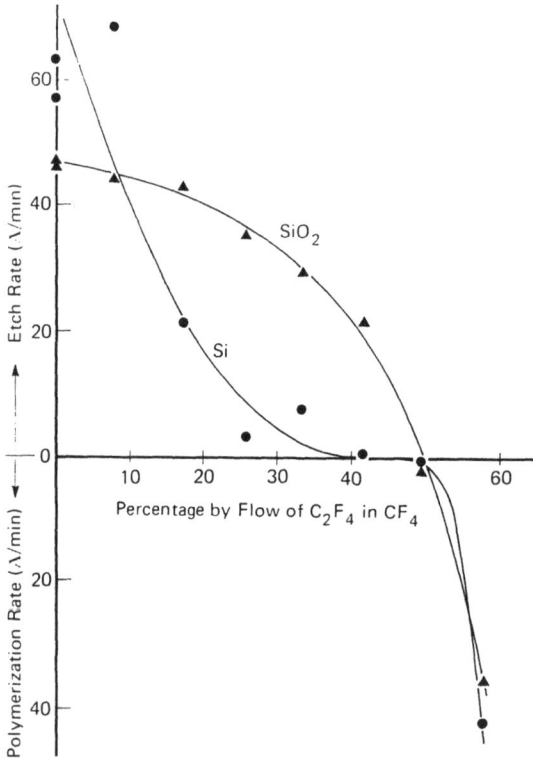

Figure 15. Etch rates of Si and SiO_2 in a CF_4 plasma as a
function of C_2F_4 injection into plasma. Ref. 31.

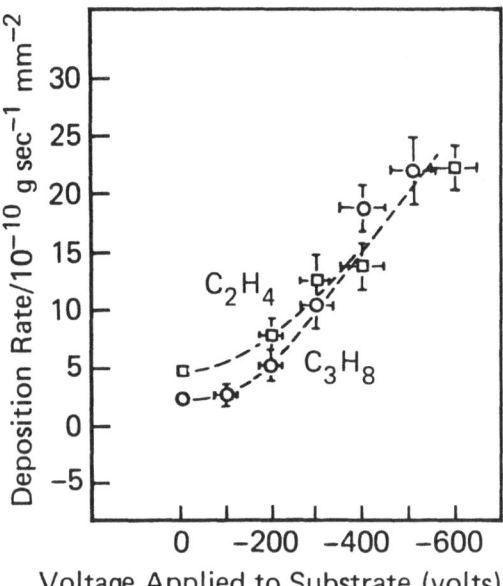

Figure 16. Polymer deposition rate in a C_2H_4 and C_3H_8 plasma as a function of ion bombardment energy during film growth.

BLOCKING LAYERS ON SIDEWALLS VERSUS BOTTOM OF ETCH FEATURES

The mechanisms whereby these ion bombardment effects influence the formation of blocking layers at the bottom of an etch feature relative to the vertical sidewall is only qualitatively understood. At the bottom of the Si etch feature a competition between the following four processes takes place, all involving energetic ion bombardment.

Enhanced etching of Si by F atoms in the presence of energetic ions,

$$Si_{(s)} + F_{(g)} + \text{ion bombardment} \qquad SiF_x(g)\uparrow \qquad (1)$$

Enhanced deposition of polymeric material in the presence of CF, CF_2, etc. in the presence of energetic ions,

$$(CF_2)_n(g) + \text{ion bombardment} \qquad (CF_{2-x})_n(S)\downarrow \qquad (2)$$

where x depends on the energy of the ions.

Enhanced etching of fluorine deficient polymeric deposit by F atoms and CF_2 radicals in the presence of energetic ions,

$$(CF_{2-x})_n(S) + F + \text{ion bombardment} \qquad C_nF_{2n+2}(g)\uparrow \qquad (3)$$

Physical sputtering of material,

$(CF_{2-x})(s)$ and/or $SiF_{(s)}$ +ion bombardment   sputt. species↑ (4)

The energy dependence of these four processes can be expected to be quite different as is their dependence on temperature of the surface. So, for example, the rate of poly-merization by process (2) is far more sensitive to surface temperature than processes (1), (3) and (4). This is due to the observed large drop in sticking probability of the polymer precursor species with very small rise in surface temperature. This result will be verified in the following section.

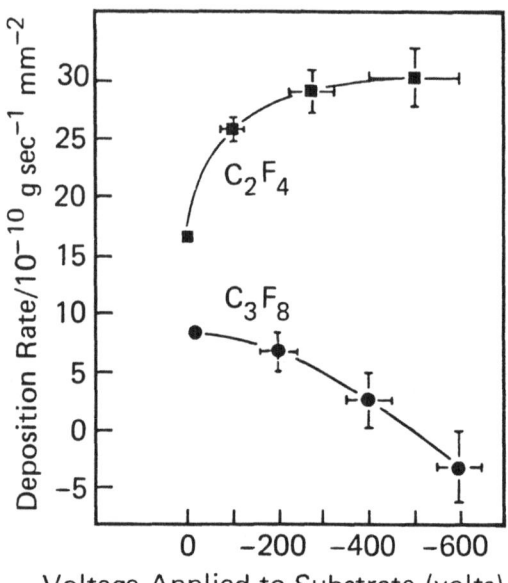

Figure 17. Polymer deposition rate in a C_2F_4 versus C_3F_8 plasma as a function of ion bombardment energy during film growth. In C_3F_8 etching dominates over deposition as ion energy increases.

The sidewalls will be subject to bombardment by that fraction of energetic ions which is backscattered from the bottom of the well. In the context of sidewall etching versus bottom of the well etching, it will be assumed that the energetic particle bombardment of the sidewalls by these backscattered particles will be "low" energy bombardment, whereas bombardment of the bottom of the well by primary ions will be "high" energy bombardment.

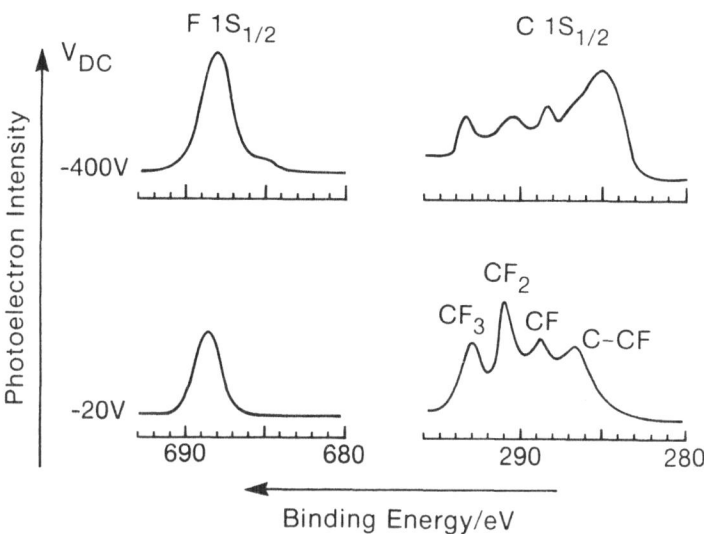

Figure 18. Effect of ion bombardment during film growth on structure of plasma polymerized fluorocarbon films at 20 and 400 eV substrate bias voltage, respectively.

Everything else being equal, low energy bombardment favors deposition mechanism (2) whereas high energy bombardment favors (3) and (4). Therefore, etching dominates at bottom of well whereas carbonaceous deposition slows down sidewall reactions by blocking. See Fig. (19).

METAL CONTAINING POLYMERS BY SIMULTANEOUS ETCHING AND POLY-MERIZATION

These ion energy dependent bombardment processes, 1 through 4, can also be exploited in a totally different context as described below. Kay et al., have recently published a number of papers in which advantage is taken of these competing ion energy dependent etching and deposition processes to produce a new class of materials,e.g., metal containing fluoro-carbons. Injecting C_3F_8 monomer into a capacitively coupled r.f. diode system allows fluoridation to dominate at the electrode subject to high energy ion bombardment whereas poly-merization dominates at the grounded electrode. If the powered electrode is made of Au, no fluoridation can take place and Au will be physically sputtered from the surface due to energetic positive ion impact from the plasma. If the powered electrode is made of material which forms a non-volatile fluoride, then again only physical sputtering can remove these fluorides from the surface and inject them into the plasma. If, on the other hand, volatile fluorides are formed at the powered electrode, then the rate of removal of electrode material as fluorides by chemical sputtering, as defined earlier, may dominate. In all three cases the material ejected from the powered electrode will traverse the plasma and in part be condensed into the polymer forming at the grounded electrode. Species etched from the powered electrode will thus be dispersed in the growing polymer matrix.

Blocking Chemistry

(a) Low Energy Ion Bombardment

$$(CF_2)_x (g) + X^+ \rightarrow (CF_{2-n})_x (s) \downarrow$$

(b) High Energy Ion Bombardment-Chemical Sputtering

$$(CF_{2-n})_x (s) + X^+ + F \rightarrow (C_n F_{2n+2})(g) \uparrow$$

(c) High Energy Ion Bombardment-Physical Sputtering

$$(CF_{2-n})_x (s) + X^+ \rightarrow C_x F_y (g) \uparrow$$

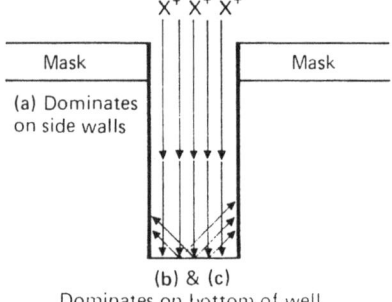

Figure 19. Schematic of blocking chemistry on sidewalls and bottom of etch feature as function of ion energy.

Detailed discussion of this film forming process and film properties is beyond the scope of this article but can be found in several recent publications.[37,38,43,50] Suffice it to say that the electrical characteristics of such metal containing polymers under some circumstances can be dramatically different from conventional dielectric plasma polymerized fluorocarbon polymer film deposits. As a result, all the grounded surfaces being coated with such polymers can lead to electrical changes in the plasma, e.g., the plasma potential, which in turn can impact the four processes, (1) through (4), described earlier, all of which depend on ion bombardment. To this degree these metal containing polymers do relate to plasma processing in general and plasma etching in particular.

Evidence of the lowering of the polymer formation rate as a function of increasing surface temperature has been measured in several ways. The most dramatic example we have encountered can be deduced from data presented in Fig. 20 on metal containing polymers at substrate temperatures differing only by several tens of degree centigrade.

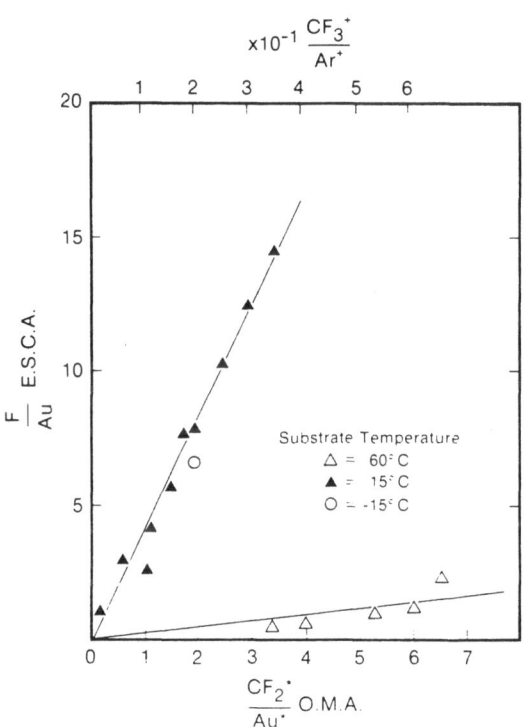

Figure 20. Key gas phase species, CF_2 and Au, arriving at substrate versus F and Au found in polymer films at two different substrate temperatures. Ref. 37.

In Fig. 20, the ratio of the polymer precursor CF_2 and Au atoms arriving at the substrate as measured by optical emission spectroscopy for various plasma conditions and the F and Au content subsequently found by ESCA in the resultant polymer films are plotted at two substrate temperatures. F content by ESCA in the film can only be associated with polymer since fluorides of Au are absent. The straight line behavior further illustrates the correlation between CF_2 polymerization

precursor arriving at the film surface and the rate of polymerization.

Whereas the sticking probability of Au atoms is not
expected to change dramatically between the substrate held at
15° C to that held at 60° C, the degree of polymer formation as
measured by F content is very much less at 60° C than at 15° C.
These results are confirmed further by measuring the deposition
rate on frequency dependent quartz crystal microbalances as a
function of substrate temperature, see Fig. 21. These results
clearly suggest that the sorption of the polymer precursor
on the substrate rather than the subsequent polymer forming
steps is rate limiting as a function of substrate temperature.

The C content is not included in the ESCA data since C
surface contamination in metal rich films confuses the issue
and is not considered important to the theme of this article.

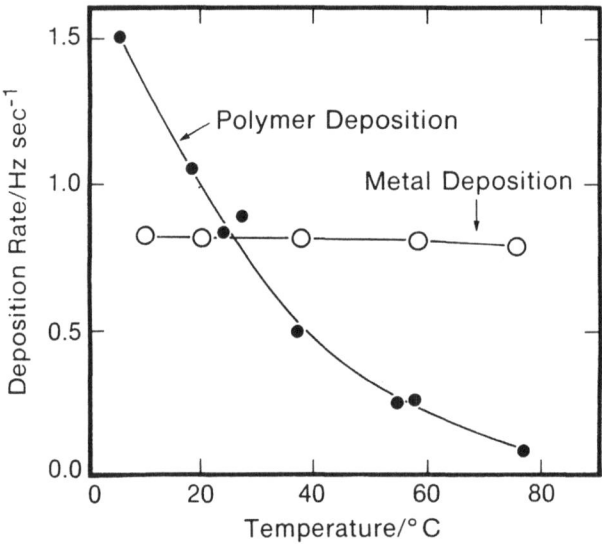

Figure 21. Deposition rate as function of substrate temperature
in pure argon plasma (metal deposition only) and
pure C_2F_4 plasma (polymer deposition only).

ETCHING OF METALS

Whereas a detailed discussion of metal etching is beyond
the scope of this review, a few general comments may be
appropriate. The effects of ion bombardment on reactive ion
etching of several metals was discussed in a previous section.
Several additional types of problems arise with metals of
technological importance such as Cu and Al or their alloys.
Low volatility of certain etch products at room temperature
present a problem together with effective means of removing
reactive etch products from the reaction zone without redeposition. Heating various surfaces, including the surface
to be etched, is not considered a practical solution in many
applications because of the deleterious effects of elevated
temperatures on interfacial diffusion, alloy segregation,

possible degradation of the photoresist material, differential thermal expansion and associated anisotropic film stress leading to interfacial adhesion problems-just to mention a few.

In addition, many metals of interest are susceptible to oxidation and most oxides are non-volatile which interferes with the volatilization of the metal, usually as a metal halide.

Energetic ion bombardment can help to remove the oxide initially but also requires that the partial pressure of oxygen contaminants in the system be kept low so that ion enhancement reoxidation does not interfere significantly with the volatilization of the metal halide. O and H_2O scavengers can be added to the plasma to minimize this problem. In the case of aluminum, chlorine containing etchants such as CCl_4 and oxygen scavenger such as BCl_3 are commonly used. Chlorine is known to react spontaneously and very efficiently with clean aluminum. As discussed earlier, ion bombardment of such clean Al surfaces does not enhance the etch rate quite in contrast to Si. Therefore, isotropic etching will result unless the sidewalls are blocked chemically, for example with polymeric material. Further problems are encountered, resulting from subsequent corrosion of the remaining metallurgy in a device, especially when a chlorine containing etchant is used. Various procedures to minimize this problem are being practiced often involving a post etching treatment with an F containing atmosphere.

REFERENCES

1. R.A. Gottscho and T.A. Miller, Pure & Appl. Chem., 56, 189 (1984).
2. W.r. Harshbarger, in "VLSI Electronics Microstructure Science, Vol. 8, Plasma Processing for VLSI", N.g. Einspruch and D.M. Brown Editors, Academic Press, Inc., Orlando, FL (1984) p.411.
3. P.J. Marcoux and P.D. Foo, Solid State Technol. 24 (4), 115 (1981).
4. J.W. Coburn and M. Chen, J. Appl. Phys. 51, 3134 (1980).
5. R. d'Agostino, F. Cramarossa, S. de Benedictis and G. Ferraro, J. Appl. Phys. 52, 2359 (1981).
6. R. Walkup. K. Saenger and G.S. Selwyn, MRS Symposia Proc. 38, 69 (1985).
7. J.W. Coburn and Eric Kay, Appl. Phys. Lett. 18, 435 (1971), J.W. Coburn and Eric Kay, Appl. Phys. Lett. 19, 350 (1971).
8. J.W. Coburn and Eric Kay, J. Chem. Phys. 64, 907 (1979).
9. J.J. Cuomo et al, IBM J. Res. Develop. 21, 580 (1977).
10. P. Ziemann, K. Kohler, J.W. Coburn and E. K ay, J. Vac. Sci. Technol. B:1 31, (1983).
11. H.F. Winters and J.W. Coburn, Ann. Rev. Mater. Sci. 13, 100 (1983).
12. K. Kohler, J.W. Coburn, D.E. Horne and E. Kay, J. Appl. Phys., 57, 59 (1985).
13. J. Bohiger at al., Rad. Eff.11, 69 (1971).
14. A. van Veen et al., Nucl. Inst. and Meth. 132, 573 (1976).

274

15. P. Ziemann and E. Kay, J. Vac. Sci. Technol. 21, 828 (1982) and J. Vac. Sci. Technol. Al, 512 (1983).
16. H.F. Winters, J.W. Coburn and T.J. Chuang, J. Vac. Sci. Technol.B 1(2), 469 (1983).
17. M. Seel and P.S. Bagus, Phys. Rev. B 23, 5464 (1981).
18. M. Seel and P.S. Bagus, Phys. Rev. B 28, 2023 (1983).
19. P.S. Bagus, Mat. Res. Soc. Symp. 28, 179 (1985).
20. T.J. Chuang, J. Appl. Phys. 51 2614 (1980).
21. B. Roop. S. Joyce, J.C. Schultz, N.D. Shinn and J.I. Steinfeld, Appl. Phys. Lett. 46, 1187 (1985).
22. Y.Y.Tu, T.J. Chuang and H.F. Winters, Phys. Rev. B 23 (2), 823 (1981).
23. D.L.Flamm, V.M. Donnelly and J.A.Mucha, J. Appl. Phys. 52, 3633 (1981).
24. E.A. Knabbe, J.W. Coburn and E. Kay, Surf. Sci. 123 427 (1982).
25. F.A. Houle, J. Chem. Phys. 79, 4237 (1983).
26. D.A. Houle, J. Chem. Phys. 80, 4851 (1984).
27. H.F. Winters and J.W. Coburn, J. Vac. Sci. Technol., in press.
28. H.F. Winters, J. Vac. Sci. Technol. A3, 700 (1985).
29. K.X. Pandey, T. Sakurai and H.D. Hagstrum, Phys. Rev. B 16 3648 (1977).
30. U. Cerlach-Mayer, J.W. Coburn and E. Kay, Surf. Sci. 103, 177 (1981).
31. J.W. Coburn and E. Kay, IBM J. Res. Dev. 23, 33 (1977).
32. M.M. Millard and E. Kay, J. Electrochem. Soc. 129, 160 (1982).
33. P.J. Hargis and M.J. Kushner, Appl. Phys. Lett. 40, 779 (1982).
34. A. Dilks and E. Kay, Macromolecules 14 855 (1981).
35. J.W. Coburn, H.F. Winters and T.J. Chuang, J. Appl. Phys. 48, 3532 (1977).
36. E. Kay, J.W. Coburn and G. Kruppa, Le Vide 183 89 (1976).
37. E. Kay and M. Hecq, ISPC Montreal 2 490 (1983), and J. Appl. Phys. 55, 370 (1984).
38. E. Kay and M. Hecq, J. Vac. Sci. Technol. A2, 401 (1984).
39. K.R. Ryan and I.C. Plumb, J. Phys. Chem. 86, 4678 (1982).
40. I.C. Plumb and K.R. Ryan, ISPC-6 Montreal 2, 326 (1983).
41. D.L. Flamm, C.Y Mogab, and E.R. Sklaver, J. Appl. Phys. 50, 6211 (1977).
42. H.F. Winters and J.W. Coburn, Appl. Phys. Lett. 34, 70 (1979).
43. E. Kay, A. Dilks and U. Hetzler, J. Macromol. Sci. Chem. A 12, 1393 (1978).
44. E. Kay and A. Dilks, J. Vac. Sci. Technol. 16, 428 (1978).
45. E. Kay, A. Dilks and D. Seybold, J. Appl. Phys. 51, 5678 (1981).
46. E. Kay and A. Dilks, Thin Solid Films 78, 309 (1981).
47. M. Hecq, P. Ziemann and E. Kay, J. Vac. Sci. Technol. 1, 364 (1983).
48. J. Perrin, B. Despax and E. Kay, Phys. Rev. B 32, 719 (1985).
49. J. Perrin, B. Despax, V. Hanchett and E. Kay, J. Vac. Sci. Technol., January 1986.
50. J. Perrin, R.L. Siemens and E. Kay, J. Macromol., in press.

ON THE MODIFICATION OF METAL/CERAMIC INTERFACES BY LOW ENERGY ION/ATOM BOMBARDMENT DURING FILM GROWTH

J. M. RIGSBEE[+], P. A. SCOTT[+], R. K. KNIPE[+] AND V. F. HOCK[*]

+ Department of Metallurgy and Mining Engineering, University of Illinois at Urbana-Champaign, 1304 W. Green Street, Urbana, Illinois 61801 USA;
* Engineering and Materials Division, U.S. Army Corps of Engineers CERL, Champaign, Illinois 61820 USA.

1. INTRODUCTION

Elemental Cu and Ti films have been deposited onto ceramic substrates with a plasma-aided physical vapor deposition (ion-plating) process. This paper discusses how the structure and chemistry of the metallic film and the metal/ceramic interface are modified by low energy ion and neutral atom bombardment. Emphasis is placed on determining how low energy ion/neutral atom bombardment affects the strength of the metal/ceramic interface. Analyses of the film, interface and substrate regions have employed scanning Auger microprobe, secondary ion mass spectroscopy, SEM/STEM-energy dispersive x-ray and TEM/STEM imaging and microdiffraction techniques.

One attractive characteristic of ion-plating is that it produces highly adherent coatings (1). Reasons suggested for this effect include: 1) a clean, reactive substrate surface caused by sputtering prior to coating (1-4); 2) heating of the surface due to bombardment with energetic ions and neutrals (5,6); and 3) development of a chemically graded coating/substrate interface due to ion-mixing induced diffusion and recoil implantation (7). Detailed studies of the chemistry and structure of ion plated coating/substrate interfaces are very rare. Two frequently quoted studies of Au on Al (8) and Cu onto steel (9) appear to show a chemically graded interface. These results are somewhat questionable since the microchemical analysis (electron microprobe) technique employed in these studies has relatively poor, by today's standards, spatial resolution. Ongoing research in this investigator's program (10,11) on ion plated metal film/metal substrate couples substantiates the development of a chemically graded interface, even for metals such as Cu and Cr which have essentially no equilibrium solid solubilities near room temperature. Interfacial structure and chemistry studies of ion plated metal/ceramic couples are essentially nonexistent except for a previous paper (12) from this investigator's program. As a general comment, the increasingly complex demands being placed today on the performance of materials promotes the use of combinations of materials (metals, ceramics and polymers) as coatings or composites. Since the interfaces between these dissimilar materials play a critical role in their performance, it is vital that a basic understanding be developed of the capabilities of advanced processing techniques such as ion plating for modifying the structure, chemistry and properties of these interfaces.

2. EXPERIMENTAL PROCEDURE

Elemental Cu and Ti films were deposited onto ceramic (magnesia-alumina-silica) substrates using a dc ion plating process. The metal was electron beam evaporated with a typical pressure of 1 Pa Ar. Prior to film deposition, the substrates were sputter cleaned 10 min with an applied bias of -2000V and current of 40 mA. The effective cathode area was approximately 180 cm². The cathode to which the substrates were affixed was water-cooled and the bulk substrate temperature was maintained at approximately 20C.

The microstructure and microchemistry of the metal film, the interface and the ceramic substrate have been studied by SEM examination of fracture cross-sections, by SIMS and scanning Auger sputter depth profiles in a direction normal to the coating broad face and by TEM/STEM imaging, microdiffraction and EDX analyses on coating/substrate cross-sections. Sample preparation for TEM examination involved development of a procedure for thinning along a direction parallel to the metal/ceramic interface. More details of this technique are given in reference 12.

Adhesion testing, to measure the strength of the metal/ceramic interface, involved epoxy bonding a 2mm diameter stub to the coating surface and measuring the load required to cause failure of either the epoxy, the metal/ceramic interface or the ceramic substrate. The nominal strength of the epoxy and the maximum load of the test device were both approximately 70 MPa. Failures were occasionally found which were combinations of epoxy, coating and substrate failures.

3. RESULTS
3.1. Cu/ceramic

Evaporation of Cu onto the ceramic substrates with zero applied bias resulted in coatings with very poor adhesion. Evaporated coatings on substrates polished with 1/4 micron diamond paste failed at less than 1 MPa. Evaporated coating adhesion was slightly improved (approximately 2 MPa) on rougher substrates, likely due to mechanical keying of the coating in surface crevices. Figure 1 shows how the interfacial failure stress

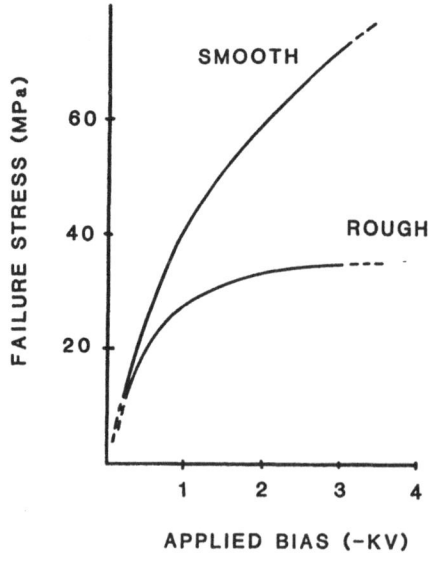

FIGURE 1. Variation in interfacial strength (film adhesion) as a function of applied bias and substrate roughness.

varies with surface roughness and applied substrate bias. For the polished substrate and an applied bias of -3000V, the failure stress exceeded the maximum load capability of the adhesion test device, 73 MPa. For this series of films better than 80% of the tests resulted in no failure. The improvement in interfacial strength for the polished substrate is likely due to the fact that the crevices and porosity of the rougher, as-received substrate serve as stress concentration sites and promote crack initiation and propagation along the interface.

Secondary ion mass spectroscopy (SIMS) and Auger depth analyses of the ion-plated Cu films, previously discussed in reference (12), support the theory that a chemically graded interface has been produced. SIMS analysis shows continuously increasing Al, Si and Mg sputter yields, relative to a constant Cu sputter yield, while sputtering through the coating. At the interface the Al, Si, Mg and Cu sputter yields all increase by about an order of magnitude. This result is interpreted as showing that the elements exist together in an oxygen-rich environment in a chemically mixed Cu/ceramic interfacial region. Auger depth profiles support the existence of this chemically mixed interface since they show a gradual decrease in Cu and increase in O during sputtering through what is believed to be the interfacial region. Care must be taken in interpreting these SIMS and Auger data. The "width" of an interfacial region can be altered by substrate or film roughness, variations in film thickness and mixing during ion sputtering. These SIMS and Auger results will be supported later in this paper by cross-section TEM results.

Figures 2 and 3 are bright-field TEM micrographs for, respectively, evaporated and ion-plated (-5000V, 100mA) Cu films. Figure 2 shows the evaporated films to be polycrystalline with growth twins running approximately normal to the substrate surface. The average grain size was about a micron. It is clear from this micrograph that the Cu grains nucleate abruptly at the ceramic surface (note the sharp line of transition between the Cu grain and the ceramic substrate). Although SIMS and Auger results indicated a chemically graded interface for evaporated films on as-received substrates, cross-section TEM shows this interface to be structurally and chemically quite sharp. This microstructure contrasts sharply with that of the ion-plated film shown in Figure 3. The Cu grains are much smaller (average grain size of about 0.2 microns) and more uniform in size and shape. It was found that the columnar-shaped grains

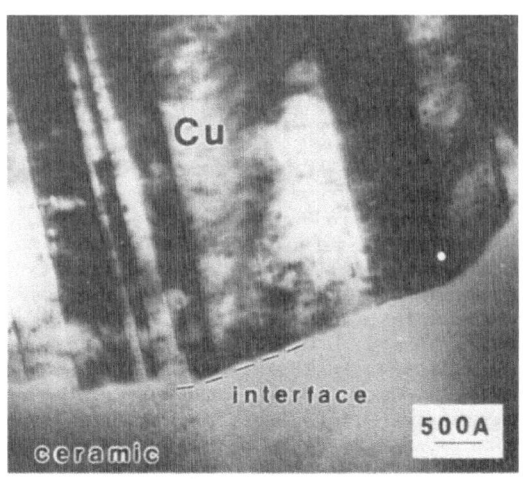

FIGURE 2. Cross-section TEM micrograph showing evaporated Cu film on ceramic.

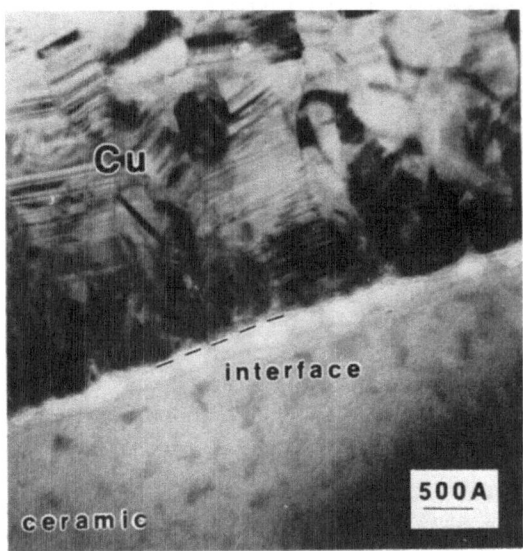

FIGURE 3. Cross-section TEM micrograph showing ion plated (-5 KV bias) Cu film on ceramic.

consisted of an extensive array of annealing twins with the {111} twin plane oriented normal to the growth direction. It is apparent that ion bombardment of the Cu film has modified both the nucleation and growth processes within the growing film. The 100nm thick Cu film region directly in contact with the ceramic substrate had a distinctly different grain structure from the remainder of the Cu layer. This is interpreted as direct structural evidence of the chemically graded interface region detected by SIMS and Auger analysis. EDX microchemical analyses of the near-interface metal and ceramic regions further support the graded interface concept since extensive chemical mixing was evident. It is apparent from these results that cross-section STEM-EDX analysis offers some advantages over surface-sensitive chemical analysis techniques for thin film interfacial structure/chemistry studies.

3.2. Ti/ceramic

The behavior of Ti evaporated and ion-plated films is quite different from that of Cu because of its affinity for oxygen and its natural reactivity with oxide ceramics. The adhesion of films deposited at substrate biases less than -3000V generally exceeded the load capacity of our adhesion testing device (70 MPa). It was found however that when aggressive ion plating parameters were used the film adhesion actually decreased. Using a second anode (+84V and 10.3A) within the ion plating chamber to increase ionization resulted in a cathode current of 250 mA with an applied bias of -2500V. Film adhesion under these conditions was found to vary widely with an average of approximately 30 MPa. Auger analysis of the failed substrates showed that Ti remained on the surface in addition to the ceramic components.

TEM analysis of the interface region revealed that a relatively thick (0.4 microns) and apparently brittle Ti-ceramic reaction layer had formed during the initial stages of film deposition. Figure 4 is a cross-section TEM micrograph showing the Ti layer (very dark due to an ion-milling sputter yield much less than for the ceramic), the ceramic substrate (lower third of the figure) and the Ti-ceramic reaction layer. The layers were identified by electron diffraction and EDX analyses. It is apparent that the strength of the Ti/ceramic interface is actually degraded if ion-plating conditions are used which result in the formation of a thick, brittle reaction layer. The formation of this reaction layer will likely produce large transformation stresses at the metal/ceramic interface which would further degrade film adhesion.

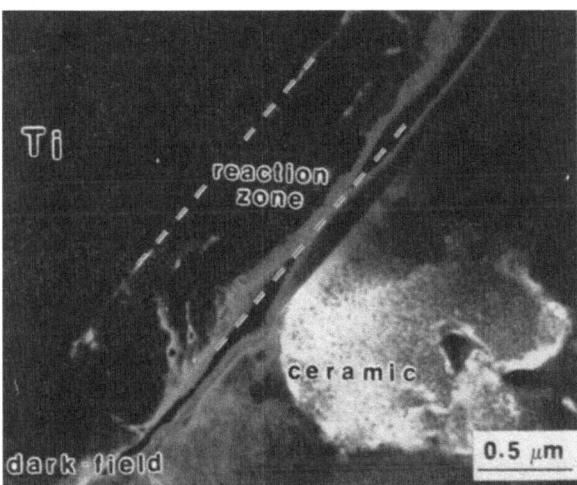

FIGURE 4. Ion-plated Ti on ceramic showing Ti-ceramic reaction zone at the interface.

4. CONCLUSIONS

The strength of the Cu/ceramic interface was found to be strongly influenced by both applied substrate bias voltage and substrate roughness. While evaporated Cu films on substrates polished with standard metallographic techniques had essentially zero adhesion, films deposited with an applied substrate bias showed increasing adhesive strength with increasing bias (in excess of 70 MPa, the limit of the employed epoxy-bonded-stud adhesion test, with -3 KV applied bias). Microchemical analysis indicated that this enhanced adhesion is directly correlated with the development of a chemically graded interface region. The adhesive strength of the ion plated Cu films was also found to be improved with increasing substrate smoothness. Increasing bias also produced a transition in the growth morphology of the ion plated Cu films. With increasing bias from 1 to 5 KV the film microstructure changed from columnar grains with extensive porosity to mixed equiaxed/columnar grains with full density.

The behavior of Ti, because of its inherent reactivity, was found to be quite different from that of Cu. Although the generally superior adhesion of Ti was evident for films either evaporated or deposited with low bias voltage/current, this adhesion actually decreased for films deposited with high bias voltage/current. Interfacial TEM studies show that this effect results from the formation of a compound at the Ti/ceramic interface region. Results indicate that the thickness of this apprently brittle compound plays a critical role in film adhesion.

5. ACKNOWLEDGEMENTS

Partial support of this research by the U.S. Army Research Office under contract DAAG-29-83-K-0151 is gratefully acknowledged. The Center for Microanalysis of Materials in the University of Illinois Materials Research Laboratory, which is supported by DOE under contract DE-ACO2-76ER01198, was used for all materials characterization. C. Loxton, N. Finnegan and J. Baker are gratefully acknowledged for their advice in this work.

REFERENCES

1. D. M. Mattox, J. Vac. Sci. Tech. 10, 47 (1973).
2. H. D. Hagstrum and C. D'Amico, J. Applied Physics 31, 715 (1960).
3. K. L. Chopra, J. Applied Physics 37, 2249 (1969).
4. P. E. Bovey, Vacuum 19, 497 (1969).
5. D. G. Teer, J. Adhesion 8, 289 (1977).
6. D. G. Teer, B. L. Delcea and A. J. Kirkham, J. Adhesion 8, 171 (1976).
7. G. Carter, J. Vac. Sci. and Tech. 7, January (1970).
8. D. M. Mattox, Sandia Corp., Monograph SC-R-65-852 (1965).
9. B. Swarop and I. Adler, J. Vac. Sci. and Tech. 10, 503 (1973).
10. J. M. Rigsbee, D. M. Leet, J. C. Logas, V. F. Hock, B. L. Cain and D. G. Teer, "Ion Plating and the Production of Cu-Cr Alloy Coatings," Surface Engineering, NATO Advanced Studies Institute Proceedings, R. Kossowsky and S. Singhal Eds., Martinus Nijhoff Publishers, Boston, pp. 603-613 (1984).
11. J. C. Logas and J. M. Rigsbee, to be published.
12. J. M. Rigsbee, P. A. Scott, R. K. Knipe, C. P. Ju and V. F. Hock, "Structure, Chemistry and Adhesion of Ion-Plated Metal/Ceramic Interfaces", pp. 206-211 in Proc. of Int. Conf. on Ion and Plasma Assisted Techniques, edited by H. Oechsner, CEP Consultants Ltd., Edinburgh, UK (1985).

PLASMA ETCHING FOR SILICON DEVICE TECHNOLOGY

C.J. HESLOP
British Telecom Research Labs, Martlesham Heath, U.K.

1. INTRODUCTION

The emergence of VLSI (Very Large Scale Integration) in silicon device technology has resulted in the creation of major industries in the USA, Japan and Europe. The influence of VLSI on the design of military hardware, consumer products and telecommunications has stimulated a rapid advance in communication and presentation of information and could even herald fundamental changes in the pattern of employment and social structure. The developments in science and technology on which the industry is based and on which it depends to be economically viable have produced highly refined silicon crystal substrates as the basic raw material, thin film materials and patterning technology (currently reaching the limits set by optical wavelengths), and sophisticated packaging and device testing equipment. The next generation of silicon devices will range from chips capable of storing up to a million bits of information to signal processors with many levels of interconnect and minimum dimensions of about 1 micron.

This paper is concerned with the dry etching of thin films on the silicon substrate which define the circuit elements and interconnect them. This aspect alone involves a great many disciplines, from vacuum engineering, gas discharge physics, through electrical engineering to gas phase and surface chemistry, but nevertheless can only be successful if developed in conjunction with the preceding disciplines of material deposition and lithography, aspects which cannot be covered in this paper.

The task of the etching process is to transfer the patterns determined by the device designer into a succession of material layers grown or deposited on the surface of the silicon crystal. The patterns have to meet very critical dimensional tolerances in order to ensure correct operation of the device and must meet very stringent maximum defect requirements in order to ensure an adequate yield of devices. Wet etching processes have been the mainstay of the industry for many years and have been very successful in meeting all the requirements except where the dimensional tolerances are too stringent to be met by an essentially isotropic etching process. Apart from a few exceptions where a dry plasma process was actually simpler and more cost effective to implement than a wet process, the main stimulus for the development of dry processing is its ability to achieve anisotropic etching under certain circumstances, and hence the ability to replicate patterns accurately where the pitch is comparable to the thickness of the layer being etched. However, as will be seen in more detail later, other requirements of the etching process are more difficult to meet in dry etching and the operation is very capital intensive.

2. COMPARISON OF WET AND DRY PROCESSING

Excellent texts on the basics of wet and dry etching technology are given in references 1-3. The object of this section is to compare wet and dry etching technology in terms of the basic requirements of the etch process without considering the detailed chemistry.

As outlined in the introduction, the task of the etching process is to transfer the required pattern into the deposited layer as shown in fig 1. The information for the pattern is contained in the form of a chrome pattern etched on a glass substrate (the mask) and normally derived from a CAD (computer aided design) system. The pattern is transferred from the glass mask onto the silicon substrate by a photolithographic process shown in figs 1(a) - 1(c). Minimum feature sizes of 1.0 micron can now be achieved by optical equipment over a field size of 14 mm. Although biassing of feature sizes can be used both on the mask and in the printing process, the amount is severely limited when the pitch of the pattern approaches the limit of twice the feature size, and hence the etching process must be capable of replicating as closely as possible the dimensions of the printed pattern in order to achieve the design objectives. Two further requirements of the etching process are minimal attack on the substrate underlying the film to be etched and the introduction of the minimum of defects either in the form of unwanted material remaining, removal of wanted material or surface contamination. Figs 1(d) and (f) illustrate the final two steps in the patterning process, etching of the layer and subsequent removal of the resist mask. Having determined the steps in the process, it is possibly to specify the requirements of the etching process in more detail.

1. Uniformity of etch across a wafer and a batch ($\pm$ 5%).
2. High degree of pattern fidelity (linewidth loss < 0.25 micron).
3. Controlled wall profile (not re-entrant).
4. Selectivity over substrate (> 10:1).
5. Selectivity over mask (> 5:1) and subsequent selective removal of the mask.
6. No contamination of wafer surface.
7. Throughput of > 50 wafers per hour (to be economically acceptable).
8. Automatic handling of wafers.

The figures in brackets are those that might typically be aimed at but the exact requirement will depend on the process. Wet etching processes can meet these requirements well apart from the problem of pattern fidelity, which is illustrated in fig 1(e). It can be seen that a linewidth loss is almost inevitable with an isotropic etch process as even a small amount of overetch can give a loss of up to twice the thickness of the layer, whereas the linewidth loss of a completely anisotropic etch process is determined by the erosion of the resist mask and the cross-section profile of the resist. Although a dry etch process is potentially more suitable for applications where a high degree of pattern fidelity is required, other requirements of the etching process such as selectivity and minimal contamination are often more difficult to meet. If energetic ions are present, then all materials will be eroded at a finite rate, and this is the source of both the advantages of dry etching and the difficulties associated with selectivity and contamination. The art of dry etching is to control anisotropy and selectivity while minimising contamination and maximising throughput.

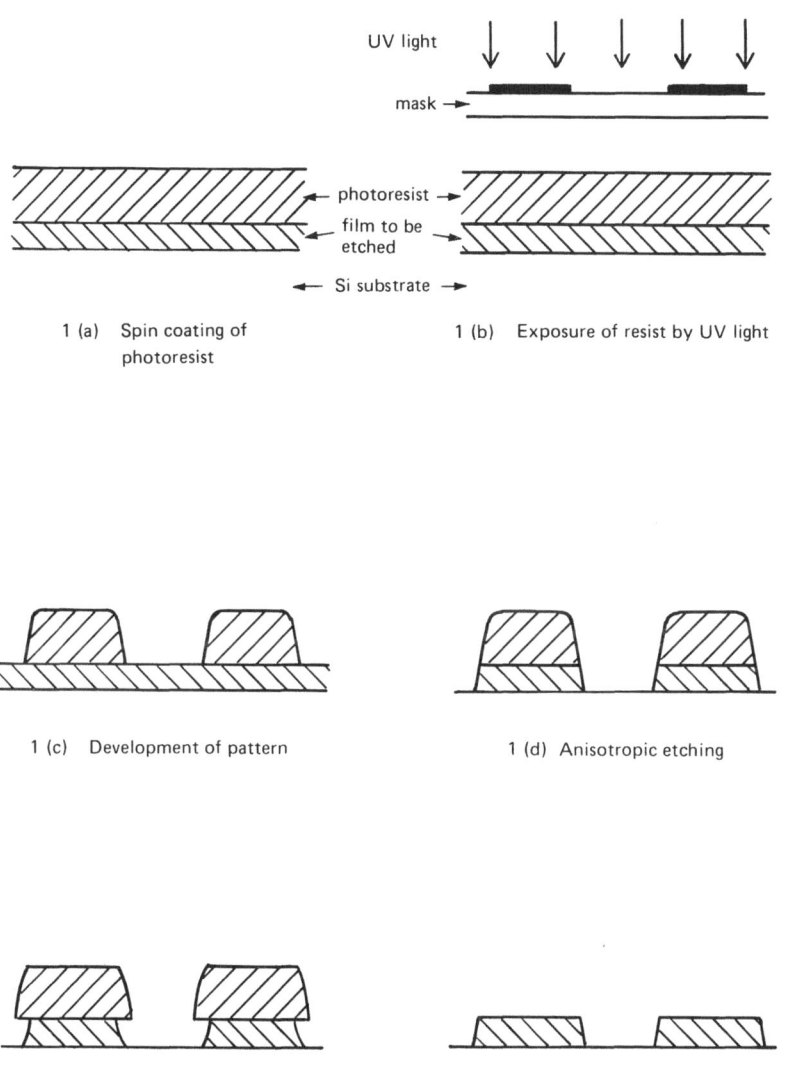

UV light

mask →

photoresist →

film to be etched →

← Si substrate →

1 (a) Spin coating of photoresist

1 (b) Exposure of resist by UV light

1 (c) Development of pattern

1 (d) Anisotropic etching

1 (e) Isotropic etching

1 (f) Resist removal

Fig. 1 Pattern transfer in VLSI

A comparison of wet and dry processing can be made in a symbolic way by looking at the flow of materials, paying particular attention to sources of contamination and methods used to control the etch medium. The etch medium can be defined for these purposes as the immediate environment of the wafer during the etch process. Fig 2 shows such a flow diagram for wet etching, in this case a bulk immersion system but similar principles apply to spray etching, and Fig 3 the flow diagram for dry etching. The parameters which are underlined in the diagram are those which have a major effect on the etch process. In both cases, the silicon wafers are transported into and out of the etch medium. For wet etching, this process is likely to be simply the immersion of the complete cassette of wafers into the etch solution and subsequent removal and re-immersion in a water wash system. For dry etching the transport is likely to be more complicated involving transfer of single wafers by moving belts or arms. The flow of chemical reactants is rather different in the two cases, the dry etching have a continuous flow of gases through the system whereas wet etching involves filling the etch batch and only emptying when the etch medium is judged to be exhausted. However, wet spray etching is more analogous to dry etching. The flow of the etch products is shown in the boxes in the two diagrams.

Having established the materials flow diagram it is possible to look in more detail at the two aspects addressed in this section, the contamination sources and the control of parameters. Turning first to the sources of contamination, these are summarised in table 1. Notice that the cassette and wafer handling would normally be carried out in class 100 or better laminar flow ambient (ie the air is filtered to less than 100 particles per cubic foot of size less than 0.5 micron), but that the presence of the equipment and the handling of the wafers inevitably introduces a disturbance and compromises the quality of the ambient air. This is especially true of human intervention unless adequate clothing and gloves are worn, and moving mechanical parts unless designed correctly with the proper materials. Particulates are troublesome in two ways, firstly they are liable to mask areas of the wafer which are required to be etched and secondly they could get incorporated in the next deposited layer. Note that particles are often extremely difficult to remove once attached to the wafer. During the etch process itself, the two sources of particulates are from the etch chemicals themselves, whether liquid or gas, and from the etching reaction. Although the input chemicals for a wet etching process will have been filtered to given specification during manufacture, care is required in packaging and dispensing to avoid additional contamination. Gaseous chemicals are likely to pick up particulates from the gas lines but these are normally filtered out at point of use. Insoluble etch products will not normally be a problem in wet etching if the correct formulation of etchant is chosen but unwanted reactions with the masking resist must be avoided. However with dry etching, involatile etch products can be a major problem and this highlights one of the major differences between the two types of process. With wet etching it is fairly easy to choose a container with which the etch medium has minimal interaction, typically quartz or polypropylene. This is not so with dry etching where all the chamber surfaces are exposed to an ion flux which sputters the surface, and the film itself may contain either impurities or intentional additions whose volatility might vary considerably from that of the basic film material. This can lead to both residual contamination on the wafer surface after the etch process is finished or local masking of the film surface resulting in spikes of the film remaining on the surface. An example of this effect is

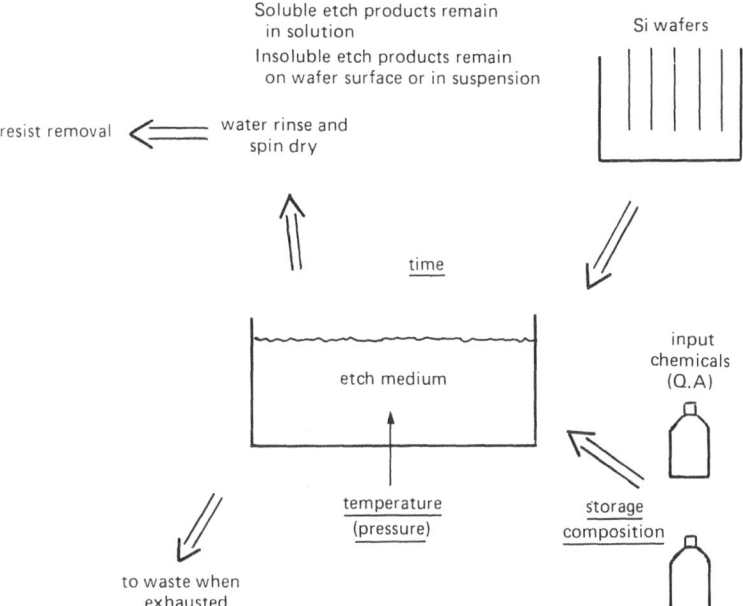

Fig. 2 Principles of wet etching

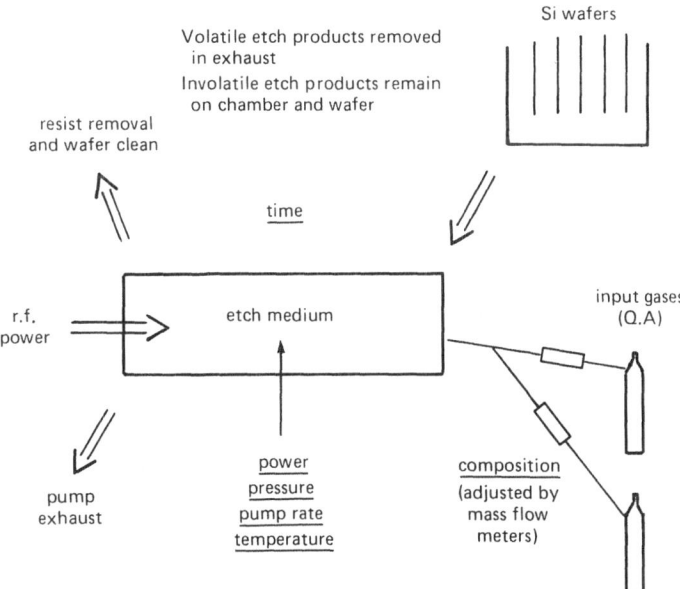

Fig. 3 Principles of dry etching

TABLE 1   SOURCES OF CONTAMINATION

STEP	WET ETCH	DRY ETCH
Wafer input	Cassette handling	Mechanical (belts, moving parts)
Wafer etch	Particulates from chemicals Insoluble etch products	Particulates from gases. Involatile etch products from both wafer and chamber. Particulate contamination from chamber.
Wafer output	Particulates on surface of etchants Cassette handling	Mechanical (belts, moving parts)

TABLE 2   PARAMETER CONTROL

	CONTROLLED PARAMETERS	PARAMETERS IN ETCH MEDIUM CONTROLLING SURFACE INTERACTION	CONTROLLED BY
WET ETCH	1. Etchant temperature 2. Etchant composition 3. Pressure	Wafer/molecular temperature Molecular composition Transport of etch products	1. 2. 1,2,3
DRY ETCH	1. Input gas composition 2. Pressure 3. Rf power 4. Pump rate 5. Wafer temperature (?)	Species absorbed on surface Ion flux and energy Electron flux and energy Photon flux and energy Surface Temperature Mean free path	1,2,3,4 1,2,3 1,2,3 1,2,3 1,2,3,5 2.

shown in fig 4, where a polycrystalline silicon film has been etched
anisotropically in the presence of a non-volatile surface impurity. A
common example of the film material itself causing contamination is that of
Al/Cu alloys where the copper is only marginally volatile in plasmas used
to etch aluminium. Solutions to these problems will be discussed in the
next section. A further source of contamination mentioned in the table is
particulate contamination from the chamber which arises from the fact that
many of the fluorocarbon gases commonly used are not only a source of
active etching species but also a source of polymerising species, the
direction of the reaction being controlled by the conditions in the chamber.
This means that some chamber surfaces can become coated with polymers which
subsequently flake off and contaminate the wafers. Another possibility is
that polymers can form directly on the wafer, and in fact it is often
necessary to do this in order to control either selectivity or anisotropy.

A summary of the parameter control in the two cases is shown in table 2.
The first column shows those parameters which are directly accessible either
electrically or mechanically, and the second column those parameters which
are immediately responsible for the interactions at the surface of the wafer
ie the etch medium. The final column shows the relationship between these
two sets of parameters and it is immediately apparent that for wet etching
there is a simple correspondence between the two whereas for dry etching no
such simple relationship exists. In fact it is doubtful whether even wafer
temperature is truly a controlled parameter unless specially designed heat
sinks are used such as conformable pads. It is worth noting as well that
the materials used to construct the plasma system are likely to play a
large part in determining the concentration of reactive species and ion
energy. However, the ability to predict, or even measure, the parameters
in the second column for a given set of input parameters is only half the
story. To predict an etching reaction we also need to understand the rate
limiting steps and detailed reactions at the surface which arise from the
parameters of the etch medium.

3. INFLUENCE OF EQUIPMENT DESIGN

Having established the parameters of the process important to silicon
device technology, the role of equipment design in determining these
parameters can be discussed. The first part of this section deals with the
influence of design on the etch medium as defined in the second column of
table 2, the second part will deal only briefly with the surface chemistry
as this will have been covered in the talk by Dr Kay, and thirdly the
sources of particulate contamination. Although there are basically three
types of dry etcher, viz the barrel or tunnel configuration, the ion beam
etcher and the planar type, only the latter will be dealt with in any detail
as the beam etcher has yet to be used extensively in silicon processing and
the planar type is more versatile than the barrel. A basic schematic of
the planar etcher is shown in fig 5. The main features are a plasma density
which is orders of magnitude lower than the gas density, an electron energy
much higher than the ion energy within the neutral plasma region, and the
high conductivity of the plasma which allows us to define a plasma
potential. The electrons have a much higher mobility than the ions, and as
a consequence the plasma loses electrons faster than ions and the plasma
potential will rise until the surrounding electric field is sufficient to
balance the loss of electrons and ions. The sheath region of high electric
field is of fundamental importance to the planar etcher, as positive ions

288

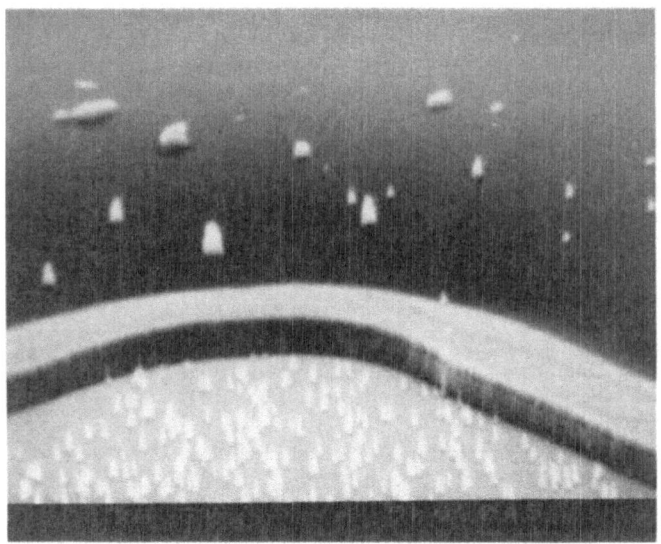

Fig. 4   Spikes on etched surface

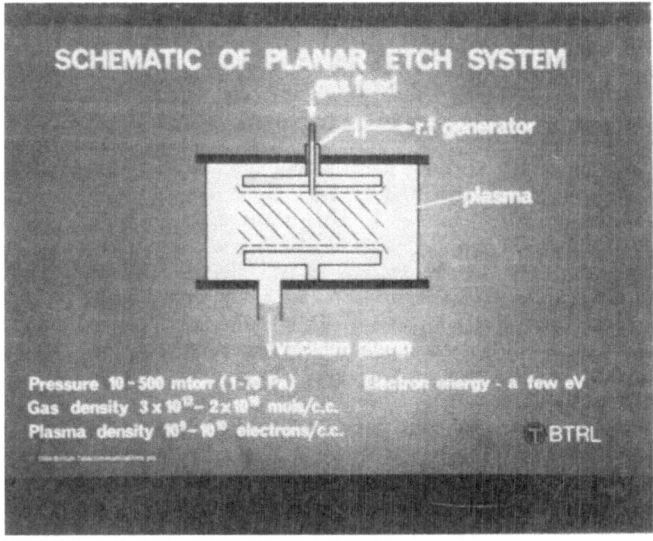

Fig. 5   Schematic of planar etch system

will be accelerated towards the substrate, and assuming that the width of
the sheath region is greater than the feature size on the substrate, the
ion impact will be essentially normal to the surface.

Two basic configurations of the planar etcher are possible, either the
wafer to be etched is placed on the ground electrode (commonly referred to
as plasma etching) or the wafer is placed on the driven electrode (commonly
referred to as reactive ion etching or RIE). Of major importance is the
relative area of grounded and live surfaces as this will be a major factor
in controlling the voltage distribution within the system. This will be
discussed further when considering ion flux and energy.

a. Influence of design on etch medium

The parameters in column 2 of table 2 will be considered here in turn.
The species absorbed on the surface will reflect the composition of neutral
species in the system but weighted according to the relative sticking
coefficients. The chief influence here of course is the input gas
composition and the action of the high energy electrons in decomposing
molecules. The relative amounts of decomposition products will depend on
the electron energy spectrum and the threshold energies for dissociation.
A model for one particular discharge system with fluorocarbon gases has
recently been published by Kushner (4). The electron energy distribution
will not only be of Boltzmann type at the low energy end, but also have
peaks of high electron energy corresponding to the sheath voltages and is
thus influenced by basic geometrical design, pressure, and rf power.
However, it is important not to overlook some subsidiary sources of gas in
the system arising from chamber material outgassing, chamber material
decomposition, reaction products, and the pumping system. Thus, in a
badly designed system, fluorine could arise either from PTFE exposed to the
plasma or backstreaming of fluorinated pump oil. It is not unknown for
silicon dioxide to be etched at an appreciable rate in a nominally pure
oxygen plasma due to these sources. There will also be a background of the
ambient outside the chamber due to the base pressure of the pumping system
and this has to be controlled either by using a vacuum or nitrogen purged
load lock (common with rotary or blower pumps) or a pumping system with a
high performance such as a diffusion pump or cryo-pump. In addition the
etch rate could be limited by the throughput of the pumping system at the
working pressure if this is insufficient to keep the concentration of
reaction products down.

As suggested previously, it is the ion flux onto the surface which gives
rise to anisotropy and this will be considered next. While free radical
reactions at the wafer surface can give rise to useful etch rates,
anisotropy could only occur at long mean free paths and low values of
surface mobility. However, ion induced reactions at the surface will give
rise to anisotropic etching due to the normal electric field at the wafer
surface, and the rate of these reactions will depend on the ion energy and
flux. Thus in a system where volatile products are produced by both neutral
and ion reactions, the anisotropy will depend on the ratio of these two
reactions. As it happens, the use of low pressures (10-30 mtorr is standard
planar systems) both minimises the neutral content and maximises the sheath
voltage. The sheath voltage is also a maximum on the driven electrode due
to the negative biassing of this electrode when capacitively coupled, thus
giving rise to the popularity of low pressure RIE machines for anisotropic

etching. However, this type of machine also has a number of disadvantages. The inherently low etch rates due to limited supply of reactants species means that only large machines capable of accommodating a large number of wafers per run are able to meet the throughput requirements of the technology. The most successful of this class of machine is the hexode etcher originally conceived at Bell Labs, where the required relative electrode areas are met by mounting the wafers on an hexagonal pillar in the centre of the chamber. This arrangement requires very sophisticated wafer handling if the system is automated. It is also necessary to point out that anisotropy can be achieved at very much higher pressures if the correct reactions are used. Single wafer etchers are very attractive from the point of view cost, maintenance and process control but obviously the etch rates must be much higher than conventional RIE machines if satisfactory throughput is to be achieved. Two approaches to this are currently available, either retaining the low pressures and enhancing the production of reactive species by using magnetic field confinement of the discharge, or using higher pressures and achieving the required anisotropy by choosing appropriate chemistry or sidewall passivation techniques. Both these machines usually retain the RIE configuration. Of great importance in all the RIE machines is the ability to minimise the plasma potential with respect to ground surfaces. Whilst it is necessary to maximise the plasma potential with respect to the wafer surface in order to achieve the desired etch characteristics, if the plasma potential with respect to ground rises so that ion impact onto these surfaces gives rise to significant sputtering, then contamination of the wafer surfaces will occur. The two most widely used solutions to this problem are the use of high frequency generators (usually 13.5 MHz) and coating the electrode surfaces with either a material of low sputter coefficient or which produces only volatile products (ie a consumable electrode). The almost universal use of high frequency generators in this application is because the plasma is unable to charge and discharge at a sufficiently high rate to follow the rf, thus the plasma potential stabilises at a sufficiently low value to enable electrons to flow to each electrode at the appropriate part of the rf cycle. The opposite is true of low frequency (< 1 MHz) plasmas where the mean value of plasma potential can be much higher, and is one of the reasons why significant ion energies can be achieved on grounded surfaces in plasma mode machines. However, the penalty to pay is an even higher ion energy on the rf electrode and the consequent contamination problems.

Both electron and photon fluxes will be present at the surface of the wafer but their role in the surface reactions used in silicon technology has not been extensively studied. The electron energy will be low because of the deceleration of the sheath field but the presence of high energy electrons originating from the opposite sheath cannot be discounted. High energy photons will also be present due to the presence of excited species in the plasma, but the energy distribution will obviously depend on the type of gas present.

The surface temperature of the wafer is a difficult quantity to specify because of the interaction of the plasma with any surface mounted sensing devices. Sensing on the back of the wafer will usually compromise the result by decreasing the thermal contact. However, a maximum of about 80 C is probably about right for planar batch machines used at moderate power levels. ' However very much higher temperature (up to 180 C) can be encountered in barrel type etchers where no heat sinking is used. The introduction of single wafer etchers with higher power densities causes more of a problem and a number of solutions have been tried such as comformable heat sinks under the wafer and flushing the back of the wafer with a high conductivity gas such as helium. Apart from control of the reaction rates, an important consideration is flow of the masking resist which will occur if the wafer temperature exceeds the glass transition temperature of the resist.

b. Influence of design on etch chemistry

Two important features of dry etching, selectivity and anisotropy, will be briefly considered here. It is difficult to reach conclusions generally applicable to dry etching as each process has its own unique requirements, but certain common features can be discerned. Selectivity between different materials can be achieved either by enhancing the etch rate of the layer to be etched by increasing the concentration of the appropriate free radicals, or by suppressing the etch rate of the underlying layer by selective deposition of a material which blocks the reaction of the free radicals with the surface. An example of this approach is the use of carbon tetrafluoride (CF_4) for etching silicon and silicon dioxide. Where we require to etch silicon at a high rate relative to silicon dioxide, the fluorine atom concentration is increased by adding a small percentage of oxygen, and selectivities of up to 10 be achieved. However, if hydrogen is added to the CF_4, the fluorine atom concentration is decreased because of the production of HF molecules, and also polymer formation is promoted on oxygen free surfaces such as silicon. Hence in this case, a selectivity of up to 10 in the opposite sense is achieved, ie silicon dioxide etching faster than silicon. A finite selectivity has to be accepted in all practical halogen based systems because of a sputtering contribution at the very least; whether the selectivity is sufficient in any given situation depends on the application.

A similar situation exists for anisotropy, where this can be increased either by increasing the vertical etch rate through the use of ion bombardment or decreasing the lateral etch rate by the use of passivating layers on the sidewall. These two processes are often linked in as much as the passivating layer exists only on the sidewall because the sputtering action of the ion bombardment removes the layer from those areas exposed to the ions. An example of the use of this technique is that of aluminium etching. These etch systems are based on chlorine chemistry because aluminium fluorides are involatile. However, but for the presence of a native oxide, chlorine would attack aluminium even without a plasma, and hence to avoid lateral etching of the aluminium underneath the mask, it is common practice to add a hydrocarbon component to the input gas such as chloroform or methyl chloride in order to promote polymer formation. The ion bombardment will prevent this polymer formation in areas exposed to the ion flux, but on the sidewalls of the aluminium track, the polymer will prevent attack by free chlorine.

The use of polymer forming chemistry as a passivant is not always easy to reconcile with the serviceability of the equipment. It has been claimed that low pressure RIE equipment is cleaner in this respect than high pressure (0.5-3.0 mtorr) equipment but it can only be a matter of degree and the simplicity of cleaning schedules can be an important factor in equipment choice.

An important aspect of selectivity is that required between the layer to be etched and the masking material. This material is usually a photo-sensitive polymer, but the difficulties of ensuring adequate selectivity has resulted in the emergence of a number of multi-level masking techniques where a thin inorganic layer is used as the mask against the plasma etch process, this layer itself being defined by conventional photoresist.

c. Influence of design on particulate contamination

The most vulnerable points in the dry etch process at which particulate contamination can be introduced has already been discussed in section 2. All equipment bought for silicon device production would now have automatic handling of the silicon wafers from cassette to cassette, the wafers being transported either by moving arms or belt and pulley. The general principle of maintaining a clean environment is to minimise the number of moving parts, as any two surfaces rubbing together are likely to be a source of particulates. A properly designed moving arm system is likely to be cleaner than a belt system and the use of versatile robotic arms to pick and place wafers may be common soon. However the requirements of contact between wafer and heat sink (and electrical contact to ensure correct plasma operation) always introduces more complexity into these systems than is desirable.

Another important aspect of equipment is gas flow and the rate at which the load lock gate and, in some cases, the chamber itself moves. Any turbulent flow of vent gas is likely to distribute any particulates which are present directly onto the wafer, and smooth action of any pneumatic systems is essential.

A recent concept which is gaining credibility is the Standard Mechanical Interface (SMIF) box approach in which the box containing the cassette of wafers is only opened within an ultra-clean environment in the machine itself without human intervention. The practical problems of the detailed mechanical design are largely solved but of course it requires all the equipment within a production line to be designed to meet this requirement at the outset and the problems of ensuring that the box remains clean during its life have not yet been fully assessed in a production environment.

As suggested in the previous section, the condition of the chamber surfaces has to be maintained particulate free under conditions where in many cases the deposition of material is inherent in the process. Ideally an oxygen plasma process should be sufficient to maintain the condition of the electrodes, but this is generally not possible if sputtered metal has been incorporated into the polymer. Modern production equipment should be designed to prevent this, for example by the use of the RIE configuration and low plasma potential.

4. SOME PROCESSING PROBLEMS FOR SUB-MICRON TECHNOLOGY

This section will deal with those aspects of patterning technology which become dominant when critcal dimensions of less than 1 micron are used. The influence of alternative printing techniques will be briefly discussed followed by a review of the dry etch parameters necessary to achieve these dimensions.

Although the resolution of optical printing systems can be as low as 0.7 micron, the pratical minimum dimension that can be printed on a silicon wafer is likely to be greater than this and lie between 0.7 and 1.0 micron. In this region of linewidth and below, alternative exposure techniques such as electron beam, ion beam and X-ray lithography become advantageous primarily because of the greater depth of focus combined with high resolution. The beam techniques do not require an intermediate mask and the patterns are written into a sensitised polymer on the wafer surface directly from data tape. Thus the complex task of mask manufacture and handling are eliminated, but because of the serial nature of the writing process, high brightness beams and resists with high sensitivity and contrast are required to make the wafer throughput economic. Flood exposure with X-rays would achieve a higher throughput but the very difficult mask technology has prevented any significant impact so far on device manufacturing. The most advanced of these techniques is electron beam writing because of its additional use in making the chrome on glass masks for subsequent optical printing. However, optical printing techniques have recently been extended by the use of both special resist formulations and deep UV irradiation so that the limit of optical printing can still not be predicted.

One consequence of the search for better resolution and linewidth control has been the lithographic requirement for thinner resist layers and a flat surface on which to print. Variations in resist thickness due to coating on a substrate with topographical features such as metal tracks and contact holes results in variations of linewidth. This is due to spurious reflections from sharp steps and the effect of resist thickness on the required exposure. There are two consequences for this for dry etching technology, one is the improved selectivity to resist that is required, and secondly the introduction of planarisation and multi-level masking techniques. It is interesting to note that although the scaling laws generally applied to devices show how decreases in linewidth lead to big improvements in speed, power consumption and packing density, these gains are only valid if dielectric and conductor thickness are maintained. Thus the difficulties faced by dry etching technology are essentially those of etching smaller structures with thinner resists in materials of the same thickness, ie the aspect ratio is increasing and selectivity has to increase.

The standard solution to the problem of topographical features on the wafer surface is to planarise the surface by the use of spin coating and curing a viscous medium. There are three possible ways in which this technique can be used, illustrated in fig 6. Firstly, the medium itself can be used as a dielectric after suitable curing schedules, and examples of this are the use of polyimides (5) and spin-on glass (6). Secondly, the planarising medium can be applied on top of an existing conformal dielectric, and the planarised surface topography transferred to the dielectric by etching (fig 6(c)). Thirdly, the planarising medium can be used as one component of a two or three layer masking process, (fig 6(d)). The impact on dry etch techniques will be discussed in turn.

The particular problem addressed in fig 6 is how to print a second level of metal interconnect on a surface shown in the first part of 6(a). Here only the first level of metal interconnect and a conformal dielectric coating are shown, the second level of metal will be a conformal coating on the dielectric and thus exhibit all the topographical features of the

294

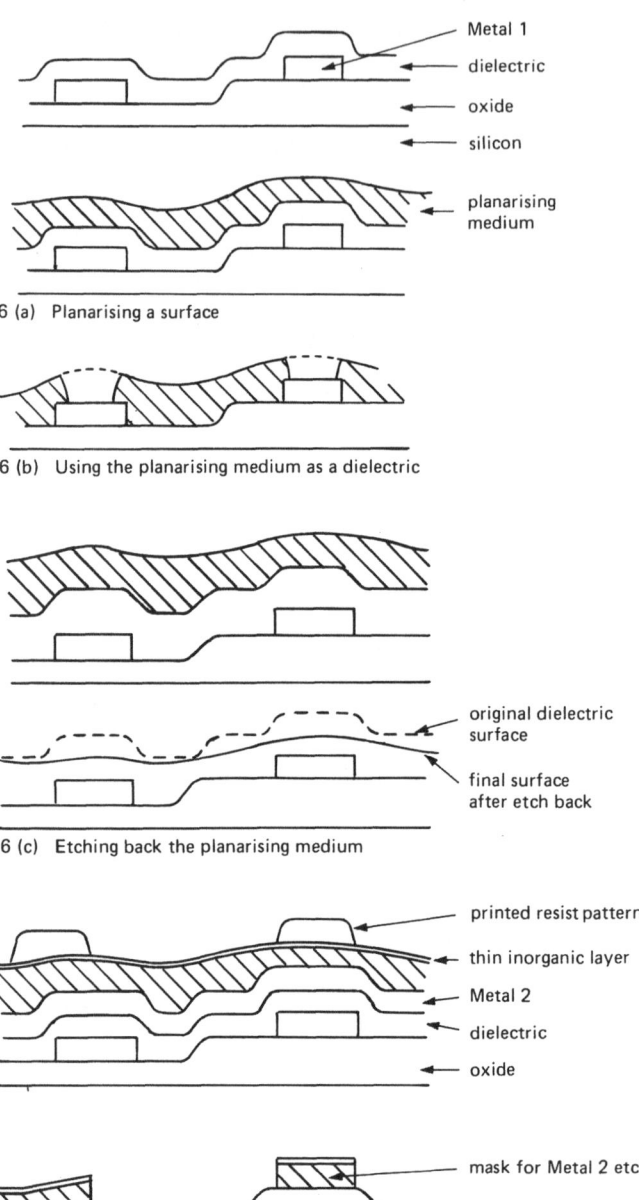

Fig. 6 (a) Planarising a surface

Fig. 6 (b) Using the planarising medium as a dielectric

Fig. 6 (c) Etching back the planarising medium

Fig. 6 (d) Tri—level masking process

Fig. 6   The use of Planarising techniques

dielectric. Where the conformal dielectic is replaced by a planarising dielectric as in fig 6(b), it can be seen that the sharp steps are avoided but significant changes in level are still present, and the uniformity of thickness of subsequent resist coating will depend strongly on the flow properties of the resist. The problems arising from this technique are twofold, firstly the planarisation achieved depends in a very complex way on the pattern itself, in particular the pitch and linewidth. Secondly it can be seen from the diagram that etching of vias in the dielectric is more difficult because of the thickness variation of the dielectric itself, up to 50% variation is frequently encountered. Thus vias with aspect ratios approaching unity have to be etched, while maintaining final via size and a wall profile which can be covered by subsequent metallization, in an effectively very non-uniform layer.

In fig 6(c), a conventional conformal dielectric is used but this is coated with the planarising medium. By using a plasma process in which the selectivity between the medium and dielectric is near unity, the surface contours of the planarising layer are transferred to the dielectric. However, the original deposited thickness of dielectric must be considerably greater than the standard process in order for the final minimum thickness to be acceptable, and the problem of maintaining via size in non-uniform thickness remains.

Fig 6(d) illustrates another alternative where the planarising medium is applied after deposition of the second level of metal. A thin (50-100 nm) inorganic layer is then deposited onto the planarising organic layer and the required interconnect pattern printed on this using conventional lithography. The inorganic layer can then be etched using the photoresist as the mask (wet etching could be used here as the ratio of pitch to layer thickness is large), and the organic planarising layer can then be etched in an oxygen plasma using the inorganic transfer layer as a mask. The final structure prior to etching the metal interconnect is shown in the second diagram of 6(d). Although the etching of the transfer layer is a relatively simple requirement, the etching of the planarising layer is very demanding, requiring high anisotropy and a high overetch capability in order to cope with the variations in thickness. These parameters can usually only be met for an organic layer by using a low pressure RIE machine where the free radical etching component is small and loading effects are small. The latter feature is necessary because considerable lateral etching can occur during the overetch period because of the decrease in radical consumption once the bulk of the film has been removed. Careful design of this type of equipment is required to avoid sputtered contamination on the wafer surface which will result in spikes of material left on the surface.

5. CONCLUSIONS

Dry etch processing has progressed rapidly in the last 15 years to meet the demands of high speed, high density integrated circuit technology. Equipment design has progressed from simple barrel type configuration to batch and single wafer planar type machines with sophisticated wafer handling and diagnostic facilities. While our empirical understanding of the process chemistry has also advanced, the complexity of the etch medium at the wafer surface has limited our understanding of the detailed events leading to volatilization of the wafer surface. This paper has highlighted

those features of the etch process which are most important to the
successful utilisation of the process in the manufacture of integrated
circuits.

ACKNOWLEDGEMENT

Acknowledgement is made to the Director of Research of British Telecom for
permission to publish this paper.

REFERENCES

1. Elliott DJ: Integrated Circuit Fabrication Technology. Published by
   McGraw-Hill 1982.
2. Chapman B: Glow Discharge Processes. Published by John Wiley J Sons.
   1980.
3. VLSI Electronics Microstructure Science, Vol 8, Plasma Processing for
   VLSI. Ed N G Einspruch, Published by Academic Press Inc 1984.
4. Kushner MJ: A Kinetic Study of the Plasma Etching Process; J Appl Phys
   53 (4) April 1982.
5. Suitable materials are manufactured by Du Pont and Hitachi.
6. Suitable materials are manufactured by Allied Chemicals and Merck.

Reactive Ion Etching of GaAs and Related III-V Compounds

Steven Dzioba
Bell-Northern Research
Ottawa, Ontario K1Y 4H7
CANADA

ABSTRACT

Details pertaining to high and low pressure reactive ion etching (RIE) of GaAs, InP and InGaAs are discussed. Attention is given to elaborating the role of ion bombardment and chemical reaction in dry etching of these compound semiconductors. Both high pressure (>0.1 Torr) and low pressure (1-100 mTorr) experiments are outlined using the dark space DC bias voltage to characterize the etch. Surface analysis of etched GaAs and InP using AES, photoluminescence and EDX is used to establish the role of impurities related to changes in surface morphology and chemical changes due to RIE.

1. INTRODUCTION

The growing use of gas plasma discharges to etch III-V compound semiconductors has been stimulated by their high resolution capabilities, process control and potential for automation. In addition, costly and toxic chemicals can be replaced by relatively stable gas sources. Currently, plasma etching of III-V compounds is being developed for high speed digital and analog electronic devices (GaAs metal-semiconductor field-effect-transistors-MESFET's) and optoelectronic devices, in particular diode lasers. For MESFET's, applications include recessed gates, device mesa formation and back side vias [1] while for optoelectronic's the main goal is to fabricate laser facets by dry etching [2] as opposed to mechanical cleaving. In the past, the electronic and optoelectronic components have developed quite independently however there appears to be a realistic potential for integrating the laser with its appropriate electronic driving circuitry. The result is an optoelectronic integrated circuit (OEIC) to be used for high speed optical data transmission [3].

The role of III-V etching in this emerging technology will be an important one provided some basic understanding of the etch process is developed. In contrast to Si plasma etching the current knowledge of GaAs, InP, InGaAs and InGaAsP etching is mostly an emperical one. The following is a non-exhaustive list of some areas of concern in patterning GaAs and its related compounds:

(i) In general, chlorine based gases are used to etch GaAs and therefore post-etch corrosion problems can be encountered particularly in stainless steel vacuum systems containing residual water vapor[4].

(ii) Because $GaCl_3$ and $AsCl_3$ etch products have different volatilities, etched surfaces may become non-stoichiometric [5]. Slight changes in the surface composition of GaAs can effect the Schottky barrier height and ideality factor. This is important for GaAs MESFET device fabrication.

(iii) Depending upon etching conditions, etched structures in GaAs can be anisotropic or crystallographic[6]. In the latter case, preferential etching takes place such that certain crystal planes etch faster than others.

(iv) Thermal effects in GaAs and InP are markedly different from that of Si. Whereas the etching of Si behaves in an Arhennius fashion[7], the same is not true for III-V's. A complex interaction between diffusion and product desorption appears to take place in GaAs[8].

(v) There is a wide variation in GaAs wafer quality both within the crystal, from boule to boule and between vendors. Thus, crystal characterization is required to establish impurity content, dislocation density and subsurface polishing damage all of which can affect plasma etching.

(vi) Further complications arise in III-V alloy hetero-structures such as InGaAs for MISFETs GaAlAs and InGaAsP for laser devices and photodetectors.

In what follows, we will elaborate on these and other issues involved in RIE of III-V compound semiconductors. This work is not intended as a review of the curent research but rather as a sampling of experiments being carried out in GaAs and InP based systems.

2. EXPERIMENTAL SYSTEMS

Etching of GaAs and InP using gas plasmas can be carried out using rf glow discharges, reactive ion etchers (a variation of rf sputtering using reactive gases) and ion beams. Figure 1 shows a general pressure - energy domain for the various types of etching experiments. Typically, plasma etchers (whether parallel plate or barrel configurations) operate at gas pressures in excess of 1 Torr. The material to be etched is in contact with an electrically grounded electrode while the opposite electrode is driven at high frequency (13.56 MHz). The gas discharge is a source of reactive chemical radicals and ions. In this high pressure regime, ions which strike the material do so with energies on the order of a few eV. Further, because of collisions within the discharge the ion momentum is isotropic. Consequently, these high pressure - low energy conditions lead to etching which is dominated by the chemical reaction between plasma species and the crystal surface. As a result, these systems tend to lead to isotropic etch patterns with high selectivity.

Reactive ion etching (RIE) is a variation on the more established rf sputter etching and indeed uses basically the same equipment. Here, gas pressures are between 1 and 100 mTorr and the material to be etched is placed on the rf driven electrode. Although, a counter electrode is often present, this is not strictly necessary since the vacuum chamber walls can be used as the grounded electrode. Provided the area of the rf driven etching electrode is much smaller than the opposite electrode, there will develop a cathode sheath in front of the wafer, referred to as the "dark space". In RIE, an ion will be accelerated out of plasma across this dark space and impact the crystal surface with an energy determined by the DC bias voltage developed across this dark space. Voltages in the range 100-1000 volts are common. Since the mean free path in RIE is on the order of cm, these ions bombard the surface at normal incidence. Hence RIE is

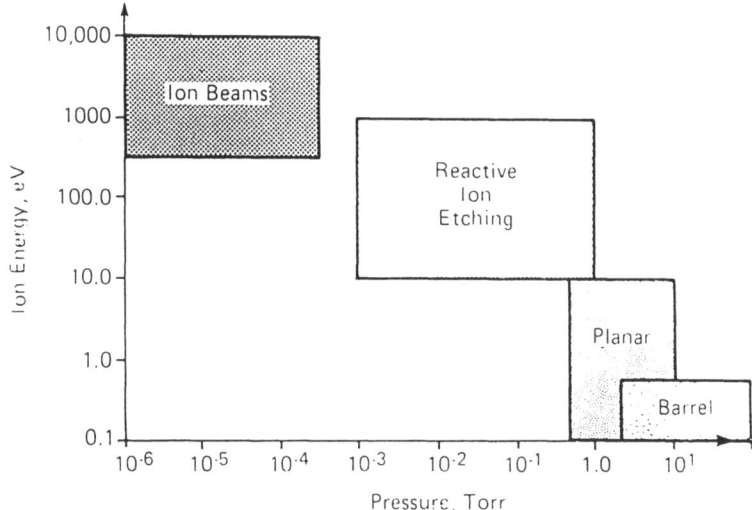

Figure 1. Allocation of various plasma etching techniques based on reactor system pressure (Torr) and energy of bombarding species. Note also there is a correlation with mean free path, i.e., the high pressure regime has low mean free path while the low pressure regime has large mean free path.

characterized by ion induced surface chemistry rather than by purely chemical reactions. This leads to structures with a high degree of anisotropy though with a loss in material etch selectivity.

In high energy-low pressure beam experiments, ions are accelerated out of a source and bombarded the surface with independently controllable ion energy and ion current. Of course ion beam sputtering is well established, though the use of reactive ions is only recently being studied. A variation on reactive ion beam etching (RIBE) is to use inert gases in the ion source and to flood the crystal surface with a highly reactive gas such as Cl_2 or Xe_2F[9]. The interaction leads to "chemical sputtering" and is characterized by very high resolution structures at the nanometer level but with the potential of severe radiation damage effects.

It should be pointed out that these experimental arrangements are only the basis for plasma etchers and many variations are providing important clues to illucidate the role of chemical reactions at the surface and ion bombardment. Systems using downstream arrangements, variable rf frequencies, triode configurations, laser induced plasmas and magnetically enhanced geometries are all being actively explored[10].

The main effort in the results reported in this work is to establish the role and effects of ion bombardment on etching of III-V's with emphasis on RIE at high and low pressures.

3.   III-V Compound Materials

Before proceeding to the etching of GaAs and other III-V's it is worthwhile to mention some details about the single crystals and epitaxial layers grown on these crystals. The importance of residual impurities and wafer quality cannot be underestimated since the formation of involatile impurity compounds leads to contamination of the surface.

Currently, single crystal GaAs is available in 3 inch diameter wafers approximately 500µm thick grown by the liquid encapsulated Czochralski technique (LEC) or in a "D" shaped wafer grown by the horizontal Bridgeman method (HB). Typically semi-insulating undoped LEC wafers have dislocation densities of $10^5 cm^{-2}$ with the major impurities being Cu and Mn at levels below $10^{14} cm^{-3}$. Recently In doped GaAs has been available with dislocation densities as low as $10^2 cm^{-2}$. Electrically conducting bulk GaAs (n-type) is doped with Si to a level of $10^{18} cm^{-3}$. Because of residual sub-surface mechanical polishing damage, up to 10µm of the GaAs surface is removed by etching prior to electronic device processing.

Optoelectronic devices are made by epitaxially growing III-V alloys onto the substrate crystal by liquid phase epitaxy (LPE), vapor phase epitaxy (VPE), metal-organic chemical vapor deposition (MOCVD) and molecular beam epitaxy (MBE). The materials of interest for light emitting devices are GaAlAs, lattice matched to GaAs and InGaAs, InP and InGaAsP lattice matched to InP. The main issues for etching these materials are developing appropriately selective plasma chemistries to pattern one layer over another and to ensure unaltered surface stoichiometry especially for the quaternary alloys.

4.   ETCHING RESULTS & DISCUSSION

4.1 High Pressure RIE - Crystallographic etching

The apparatus used for these experiments is shown in Figure 2 and consists of an rf driven electrode (5 x 5 cm) within a large quartz cylindrical chamber (25cm dia x 35cm long). Pure Ar, Cl_2, $CCl_4 + \chi\%O_2$ mixtures and CCl_2F_2 are introduced into the front of the chamber and exhausted at the rear. Gas flow rates are controlled by a mass flow controller. This system operates at pressures between 0.1 and 1 Torr using a mechanical pump (30 l/min). GaAs crystals are placed on the rf electrode without the use of a thermal bonding agent. In this case the wafer temperature can reach 150 - 200°C on the uncooled stainless steel electrode depending on the applied rf power. A stainless steel counter electrode is fitted to quartz vessel and electrically grounded.

Figure 3 shows the measured relationship between the applied rf power and the induced self bias voltage in front of the wafer. The total pressure for each gas studied is 0.13 Torr. Pure Ar develops the highest bias voltage for a given power level and increases with applied power. For the reactive gases CF_4, CCl_2F_2 and CCl_4, lower voltages are observed. Presumably, the plasma in these cases is more electronegative and, in addition, the electrode surface may have a different secondary electron yield due to chemical changes resulting from radical adsorption on the electrode. Nonetheless, the bias voltage is an accurate indicator of the GaAs etch rate (see section to follow).

Etching of GaAs under these conditions is dominated by a chemical reaction between radicals generated in the discharge and the surface. Figure 4a is a scanning electron micrograph of the GaAs surface etched

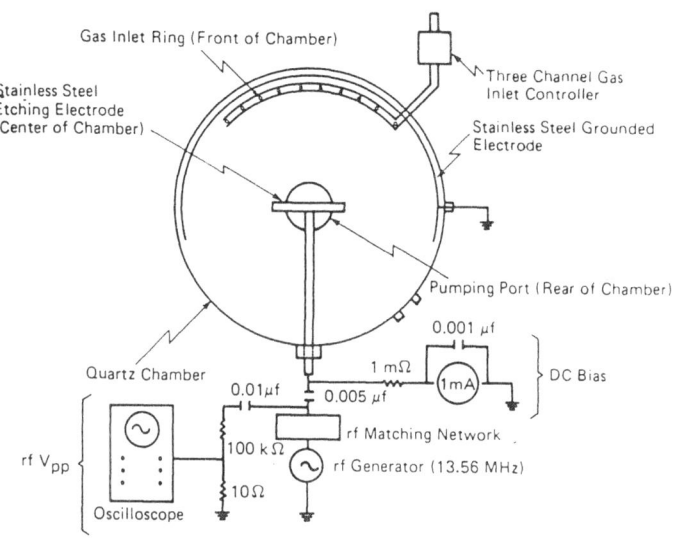

Figure 2. Experimental apparatus used for high pressure (>0.1 Torr) RIE of GaAs.

using CCl_4 + 3%O_2 at a pressure of 0.3 torr and a bias voltage of 100 volts. Etch pits associated with the crystal dislocation network are clearly evident. These pits further indicate the preferential plasma etching of As (111) faces over Ga rich planes. As etching proceeds, there is a lateral movement of the pit walls until they begin to intersect. The result is that the etched surface becomes highly textured as etching proceeds. In the optical micrograph in Figure 4b, a cleaved cross section is shown. Here a thin (1200Å) SiO_2 mask has been used to cover a narrow stripe on the GaAs surface. Because of scattering, the mask is undercut and the GaAs reveals a crystallographic nature due to high pressure etching. We have found that a "cut-off" pressure exists where this crystallographic effect is not observed. That is for etching in CCl_2F_2 below approximately 30 mTorr, the wall profiles are straight i.e. anisotropic. Substantially above 30 mTorr, crystallographic etching is always observed for CCl_4, Cl_2 and CCl_2F_2. The effect appears independent of the DC bias accelerating voltage and therefore is a result of chemical reactivity and product desorption. By orienting the oxide mask in different directions on the GaAs surface, indifferent crystal faces become delineated. To further establish this chemical effect, perpendicular walls were developed by etching at 15 mTorr to a depth of 12μm. The crystal was then dipped in a solution of methanol, phosphoric acid and hydrogen peroxide (3 to 1 to 1 respectively) for 20 secs. This combined dry plasma etch and wet chemical etch gave the same profiles as the 0.3 Torr plasma etch. Clearly, ion bombardment has a minor effect on the overall etching at high pressures except to perhaps enhance etching at dislocations. The total etch rate is therefore a sum of the rates due to dislocations and the rate at areas free of dislocations. Under the above conditions this rate is approximately 6μm/min.

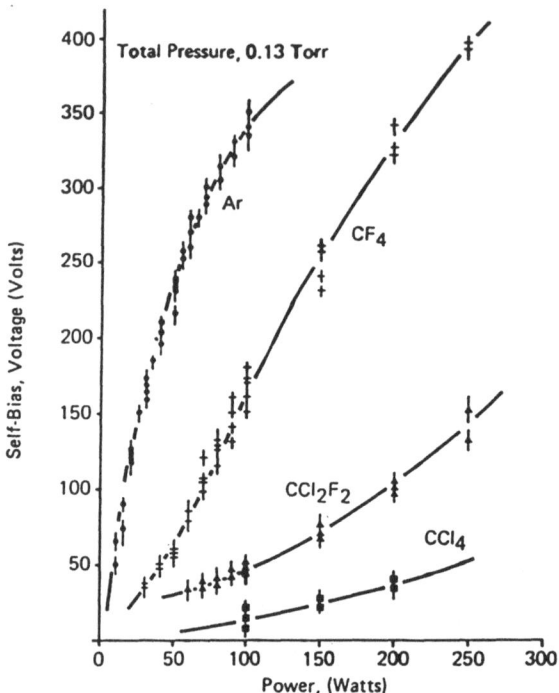

Figure 3. Self bias voltage dependence on applied rf power for CF₄, CCl₂F₂ and CCl₄ reactive gases and the inert gas, Ar.

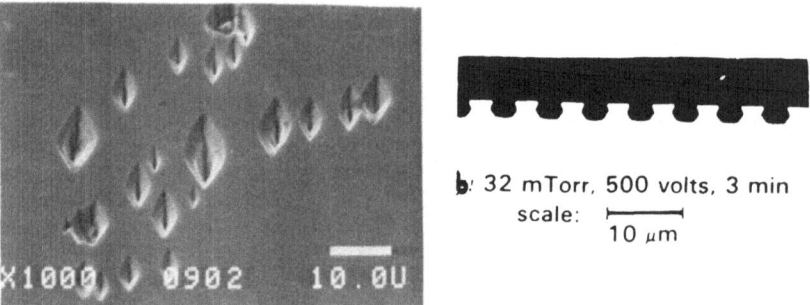

Figure 4. Scanning electron micrograph (a) of GaAs etched in CCl₄ +3%O₂ in high pressure RIE illustrating etch pits associated with dislocation cores. The optical micrograph in (b) is a cleaved cross section of GaAs showing crystallographic etching at 32 mTorr of CCl₂F₂.

In the cases of CCl_4 and CCl_2F_2 etching proceeds through the formation of Ga and As chlorides which are volatile compounds at these temperatures and pressures. However, carbon containing compounds generated in the plasma can deposit on the etched surfaces and thereby impede etching and lead to undesirable surface contaminants. For low ion energies, as in this case, sputtering is ineffective in removing these residues. Through Auger spectroscopy we have established that adding O_2 to the gas supply can lead to effective removal of carbon through the formation of CO and CO_2 which are volatile. The AES spectrum in Figure 5 reveals no detectable carbon however there is evidence of small amounts Fe. The iron, likely in the form of the $FeCl_3$ is due to reactions between Cl and the reactor electrodes. The absence of a Cr signal is due to the fact that the $CrCl_x$ compounds are volatile. The surface $FeCl_3$ is easily removed in deionized H_2O. The AES spectrum also shows a major oxygen peak due to oxidation of the surface. This oxide layer, estimated to be 30A thick, is also easily removed in dilute HCl after etching.

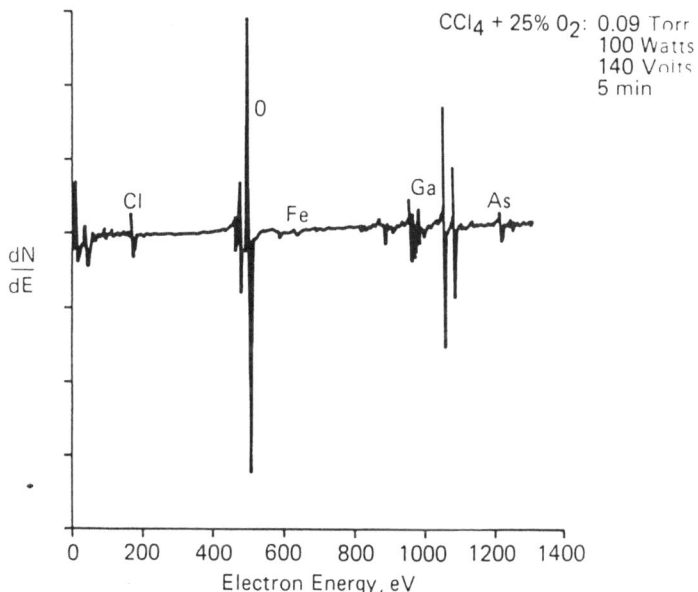

Figure 5. Auger survey of RIE GaAs surface showing effect of carbon removal by oxygen addtions to the discharge.

Thus, in the absence of substantial ion bombardment, high pressure RIE proceeds by chemical formation of volatile compounds. The addition of 25% O_2 to CCl_4, gives a surface free of carbon containing contaminants.

4.2 Low Pressure RIE

For reactive ion etching in the pressure range 1-100 mTorr, we have used the reactor shown in Figure 6. In this case the rf etching electrode (15 cm in dia.) can be water cooled or heated to 400°C. The stainless steel chamber is pumped by a 6" diffusion pump with liquid nitrogen trapping. A counter grounded electrode can be adjusted to limit the size and extent of the plasma. Tyically they separated by 5 cm. The dark space shield also confines the discharge about the electrode. The rf electrode is covered with a quartz plate to minimize deposition of sputtered particles from the cathode. The reactor is also equipped with optical and mass spectrometers. As in high pressure RIE, we use the self bias voltage to characterize the power delivered to the plasma and hence the etching conditions.

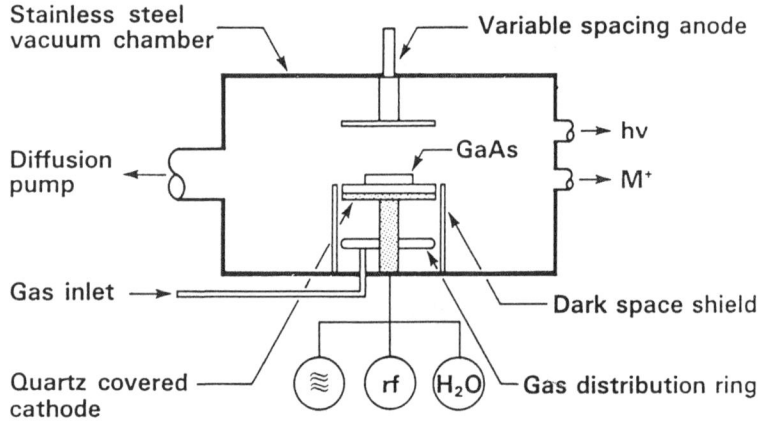

Figure 6. Low pressure (1-100 mTorr) reactive ion etcher.

Reactive ion etching is carried out at pressures such that the mean free path is larger than the dark space width (0.5 to 1 cm depending upon gas type and pressure). Self bias voltages are between 100 and 1000 volts. Figure 7, gives a compendium of results for the etch rates of a variety of materials as a function of the DC bias taken with a cooled electrode. The materials are divided into two groups. One being potential masking materials (W, polyimide, Si_3N_4, Au, SiO_2, TiN) and the other being the III-V's (GaAs, GaAlAs and InP). The gases used are Cl_2, CCl_2F_2 and CF_4 all at a pressure of 15 mTorr. At least for the range of bias voltage from 100-1000 volts the etch rates are proportional to V^n where n = 2.2 except for etching the organic polyimide films. We have also included the calculated Ar sputter yield of GaAs[11]. It is obvious that physical sputtering of GaAs plays a minor role in the absolute GaAs etching rate. However, the data do show a significant enhancement in the etch rates due to ion bombardment. Note also that the slope of the etch rate vs DC bias is greater than that for physical sputtering. Thus, the role of the impinging ion is to induce chemical reactions to form $GaCl_x$ and AsCl or reactions which would otherwise not be kinetically favorable. Such reactions might include product formation or enhanced desorption.

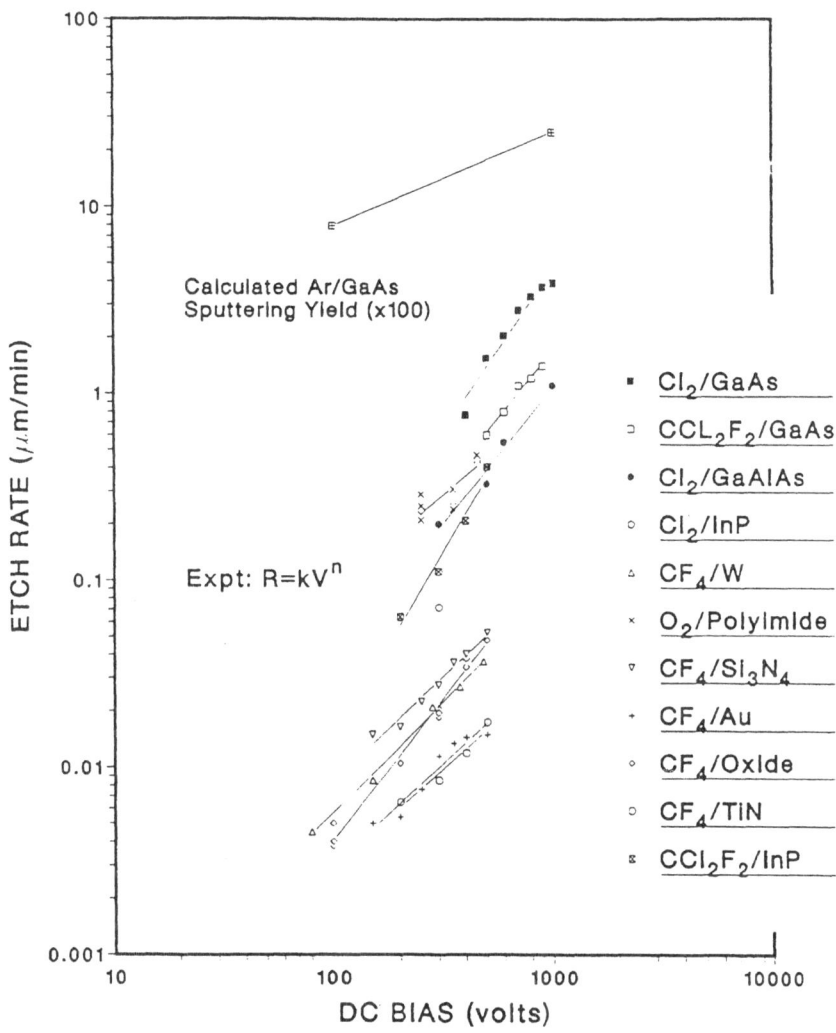

Figure 7. A survey of the etching rates of III-V semiconductors and some selected dielectrics as a fraction of the DC bias voltage.

Because of the inherent directionality of the extracted ions, the etched features are highly anisotropic as shown in the scanning electron micrograph of Figure 8. Nearly ideal vertical walls to a depth of 6μm are observed for this RIE using CCl_2F_2 at a pressure of 12 mTorr and a bias voltage of 300 volts. No evidence of undercutting is observed and the etched surface is planar and specular. The mask material for this etch in GaAs is SiO_2 (1200 Å thick). Because of this directionality features in the micron to submicron range are achievable with conventional photolithographic techniques.

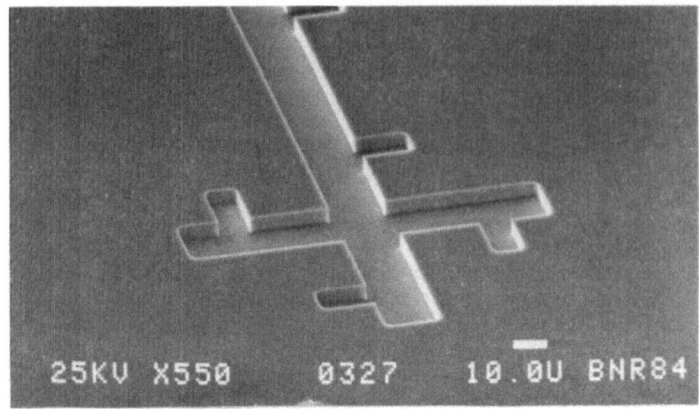

Figure 8. Scanning electron micrographs of GaAs reactively etched
at 12 mTorr CCl_2F_2 with a bias voltage of 300 volts.

Although there are many system parameters which affect the etch rate
of GaAs we will mention only two major effects; gas pressure and substrate
temperature.

The effect of CCl_2F_2 gas pressure on the etch rate of GaAs has been
measured for a constant bias voltage of 300V. In the range 1-25 mTorr the
etch rate is directly proportional to the gas pressure. Thus, the supply
of active radicals and an increase in ion current to the crystal is the
rate limiting step, i.e., first order in pressure.

The effect of substrate temperature on the etch rates of GaAs and InP
are shown in Figure 9. The InP data are due to Gottscho et al[12].
Whereas the Si etch rate behaves in an Arrhenius fashion, clearly the case
for GaAs is not. Using both CCl_2F_2 and Cl_2 the rate increases with
temperature to a maximum value then decreases with further increase in
temperature. The maxima are dependent upon the bias voltage. This energy
dependence suggests a mechanism based on radiation damage. That is, in RIE
the incoming ion disorders the surface to expose "fresh" Ga and As atoms
which are available for reactions. In addition, the deposition of energy
from the ion serves to enhance otherwise inefficient product desorption.
Thus we expect that the amount of disorder is proportional to both the ion
energy and the ion current, i.e. the power delivered to the GaAs surface.
As well, as the crystal temperature increases, we expect some annealing of
this surface disorder. At low energies, the extent of surface damage is
low and therefore a low annealing temperature is required. As the
deposited energy increases, we expect a higher annealing temperature.
Consequently, based on this damage mechanism of RIE, the decrease in etch
rate after the maxima in Figure 9 could be due to annealing of this
disordered surface. It is interesting to note that the same temperature
effect is seen in InP while such phenomenon have not been reported for Si
although damage mechanisms in Si are thought to play a role. The extent of
damage in the III-V compounds can be qualitatively inferred from Schottky
barrier and ideality measurements. In particular, the Schottky barrier
height changes due to RIE do not suggest complete amorphization of the
crystal surface for the energy ranges associated with RIE. In ion beam
studies, at doses in excess of about $10^{14} cm^{-2}$, the surface is rendered
random.

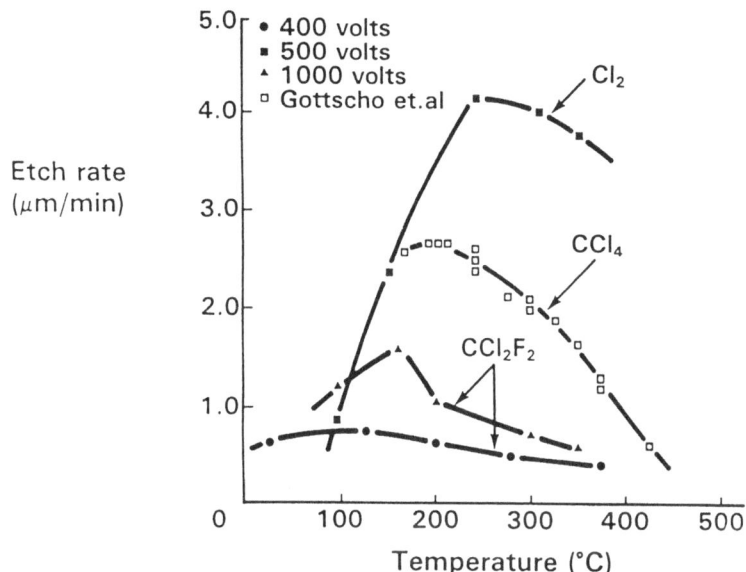

Figure 9. Etch rate dependence on surface temperature. The data for CCl_4, due to ref.[8] are for InP.

Further electrical measurements such as mobility experiments could lead to important information as to the nature of reactively ion-etched GaAs and InP surface disorder.

A major consequence of this anomalous temperature dependence is that cuation must be excercised in etching GaAs wafers for device processing. Usually, wafers are placed on the rf etching electrode without thermal bonding agents. We have found a significant variation (7%) in etch rate across a 2" GaAs wafer when etching on a quartz covered electrode. When the quartz is replaced by a recessed graphite electrode the uniformity increases to 5.3%. Furthermore, when the graphite susceptor is water cooled <u>and</u> the wafer is thermally bonded to the graphite (high vacuum fluid) the etch rate uniformity approaches 2% across 2".

As a direct consequence of these effects, it follows that a control of the dark space voltage is crucial in effectively controlling the rate and extent of surface damage. At constant power input to the plasma, the DC bias voltage can be reduced by decreasing the electrode spacing as demonstrated in Figure 10. Here the DC bias has been measured as a function of the electrode spacing for constant input power levels. For separations greater than 5 cm, the bias voltage is constant. As the spacing is decreased, the plasma becomes physically confined as the DC bias decreases. At a separation of 2 cm and less the plasma is extinguished. This "ignition" point is dependent on the gas pressure. Thus it is possible to maintain a given power level to the discharge, i.e., a constant dissociation yield with a decrease in ion energy. As an alternative, experiments using magnetically confined plasmas can also maintain a high yield of reactive species with an associated low (200 volt) ion energy. It has been also shown that the electrode material (e.g. quartz, metal or oxide) can influence the etch rates by affecting the DC bias. In either case, the control or at least the knowledge of the bias voltage is important.

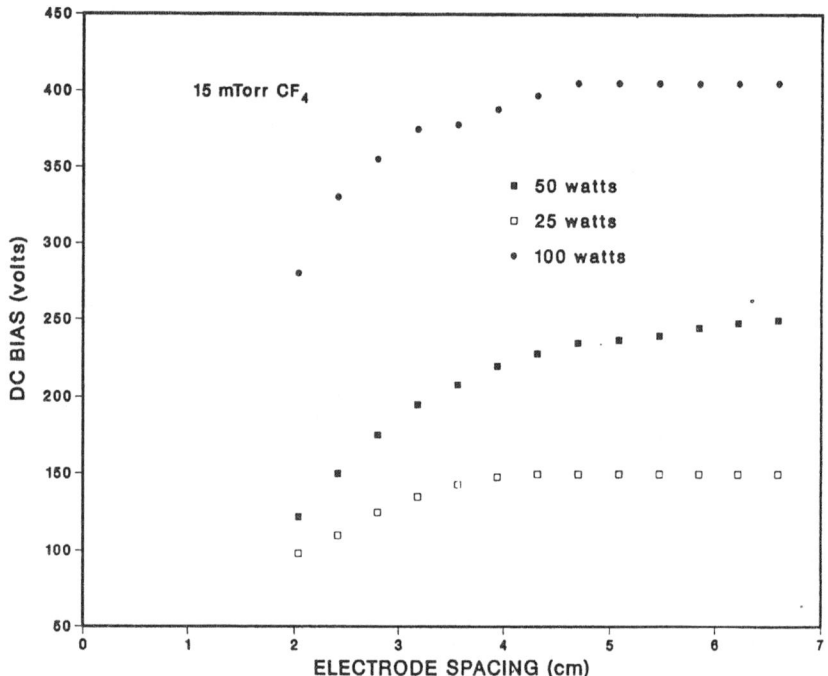

Figure 10. Increase in self bias voltage as a function of anode to cathode spacing for a CF_4 discharge. Both electrodes are stainless steel.

Finally, as an application to GaAs MESFET device processing, Figure 11 shows a series of optical micrographs of an etched 2" GaAs wafer from edge to center. The crystal edges are dominated by striations on the order of 1000 A in height while the central regions of the crystal are free of asperities. Although the exact origin of these striations is still under question, it has been shown that these regions are associated with Cu and/or Mn impurities incorporated during crystal growth. These impurities can act as etching inhibitors thus giving rise to striated surfaces. In addition, subsurface polishing damage extending down to 10µm from the

Figure 11.

Surface morphology of
etched GaAs wafer
showing striations at
the wafer edges (top)
and specular morphology
of the wafer center
(bottom micrograph).

surface has been established for GaAs. This affects the activation of
dopant implants to the extent that material implanted without etching the
surface show activation efficiencies from 0-60% while for material which
has been etch backed to 10µm, 100% efficiency is routinely obtained.

5. SURFACE DAMAGE

The effect of ion bombardment enhanced etching of GaAs on surface
properties can be classed into two categories;

(i) Surface topography due to impurities or other mechanisms
(ii) Electrical changes due to chemical, stoichiometric or crystal
     effects.

The surface topography of RIE GaAs is dominated by impurities which
act as micromasks. Figure 12 shows the formation of pillars on GaAs
surfaces due to carbon impurities within the epitaxial grown material.
These asperities are in the form of cones due to the anistropic etch
behavior of GaAs in chlorine based plasmas. The carbon impurities are
located at the interface between the GaAs substrate and the epitaxially
grown layer.

310

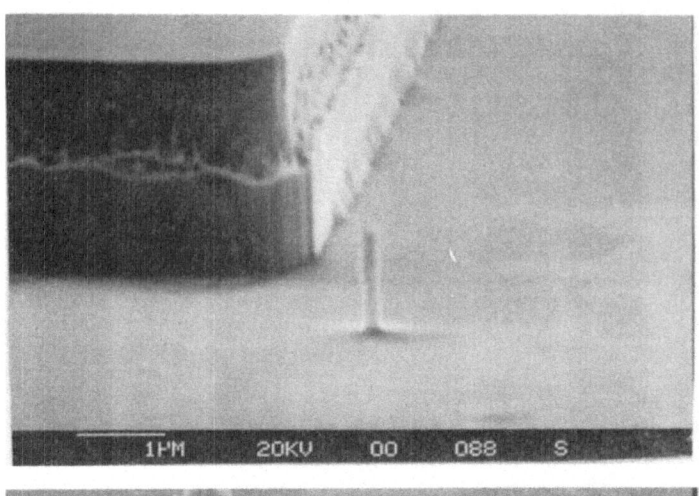

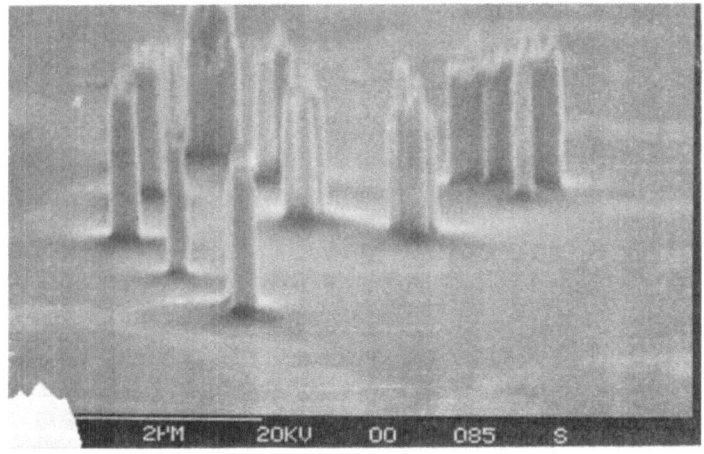

Figure 12. Pillar formation due to micromasks on GaAs.

On the other hand, in InP or InGaAs where physical sputtering mechanisms play a significant role, asperities develop in the form of cone features as shown in Figure 13. The main difference between GaAs and InP in this case is due to the role of chemical reactions versus ion bombardment effects. Whereas GaAs shows typical RIE features, the case for InP or InGaAs is different. Part of this difference can be accounted for in the surface binding energies of the III-V chlorine compounds as shown below;

$$InCl_3 \qquad 1.5eV$$
$$GaCl_3 \qquad 0.72eV$$
$$AsCl_3 \qquad 0.43eV$$
$$Si \qquad 7.8eV$$

Thus we expect the In based compound semiconductors to exhibit more "physical sputtering" behaviour than the GaAs compounds.

Figure 13.   Cone formation on InP due to RIE using Cl_2 + 3% O_2.

TECHNIQUE	ETCHANT	ION ENERGY (eV)	BARRIER (eV)	IDEALITY	REFERENCE
IBE	Ar	250	0.6	1.06	12
		500	0.53	1.22	13
		500	0.64	-	15
		500	0.54	-	16
RIE	CF_4	500	0.82	-	12
	CHF_3	500	0.79	-	12
	CCL_2F_2	-	0.74	-	14
PE	CCL_2F_2 + O_2	—	0.63	1.10	13
IBAE	Ar + Cl_2	500	0.72	no change	17
	Ar + NO_2	500	0.68	1.16	17
RIBE	CF_4 + H_2	600	0.59	1.32	10
rfSE	Ar	-	0.44	1.8	11

Figure 14.   Changes in surface properties in plasma and ion beam etching.

Finally with respect to electrical changes, Figure 14 lists the Schottky barrier hieghts and ideality factors of GaAs as a result of various RIE or ion beam treatments. Most significant are those techniques which employ inert gas ion bombardment namely, ion beam sputtering or rf diode sputter etching. In contrast the reactive techniques, such as RIE show small deviations from the actual barrier height of 0.7eV.

312

In addition to changes in surface properties, i.e. the Schottky barrier height discussed above, significant subsurface damage can occur to depths of several thousands of Angstroms as a result of heavy ion bombardment in low pressure RIE. We have studied the effects of ion bombardment on the electrical properties of n-type GaAs doped with Si either by ion implantation or bulk doped layers grown by MBE. GaAs crystals doped to a level of $2 \times 10^{18} \text{cm}^{-3}$ were subjected to ion bombardment using CF_4 and N_2 in the RIE apparatus. The gas pressure was fixed at 15 mTorr and a constant bias voltage of 300 volts was maintained. Plasma exposures for up to 6 min. were investigated and the change in electrical resistance was measured after each RIE. Figure 15 shows the ratio of initial resistance to final resistance as a function of bombardment time (essentially the ion dose). Both CF_4 and N_2 RIE show significant increase in the sheet resistance after ion etching. However, the degree of damage does not correlate with mass of the bombarding species. Furthermore, since the etch rate of GaAs in CF_4 and N_2 is insignificant, we cannot account for the increase in resistance with removal of surface material. Some insight into damage mechanisms can be gained however from annealing these crystals.

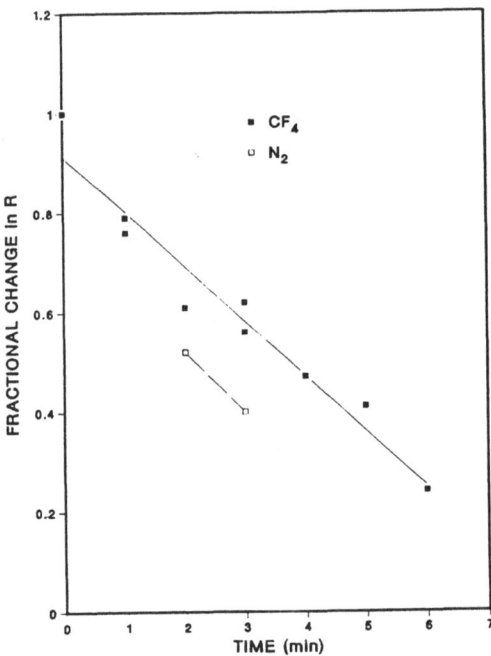

Figure 15. The ratio of initial sheet resistance to resistance after RIE of n-type GaAs using CF_4 and N_2 plasmas. Both cases show a decrease in electrical activation.

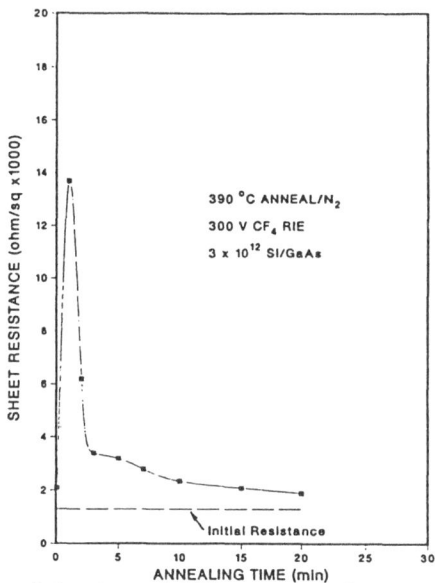

Figure 16. Annealing behavior of RIE n-type GaAs.

In particular, Fig. 16 shows the sheet resistance measured as a function of time at 390°C in flowing N_2 atmosphere. Immediately after RIE and no annealing, the sheet resistance has increased from 1200 ohms/sq to 2100 ohms/sq. However, after a one minute anneal, the resistance has increased even further to 12,000 ohms/sq after which further annealing decreases the sheet resistance. Note however that after annealing for 25 min., the resistance has returned to the value the same as that seen after RIE. That is some damage has remained. The picture that emerges for low voltage RIE is however, somewhat different. In this case, the annealing behavior of GaAs exposed to 50 volt particle bombardment with CF_4 shows an increase in sheet resistance from 1650 ohms/sq to approximately 2600 ohms/sq without annealing. In this case though, a continuous decrease in resistance back to the pre-etch case is observed. As well, no peak is seen for anneals in the range of 1 minute as was the case for the 300 volt RIE. Thus we suggest, based on complete recovery in the 50 volt case, that their is no remnant damage in the crystal.

The following picture could explain these results. First, the high energy heavy ion bombarded surface exhibits some near surface crystal disorder which does not completely anneal at 390°C. Second, the large increases in sheet resistivities of both 50 and 300 volt etches could be explained by the incorporation of hydrogen (protons) into the doped semi-conductor. Specifically, hydrogen would form Si-H bonds thus compensating the electrically active donor. Upon annealing, Si-H bonds are broken liberating the hydrogen and returning the Si to its active donor state.

The source of this hydrogen could be residual H_2 or H_2O in the vacuum system which, upon plasma ignition, generates H^+. This H^+ is accelerated across the dark space along with CF^+_x species and is thus essentially implanted into the GaAs.

6. CONCLUSIONS

We have attempted to outline the role of physical, i.e., ion bombardment effects, and chemical effects in reactive ion etching of GaAs and related III-V compounds. In particular, we have demonstrated that both chemical and physical effects are present with GaAs whereas predominant chemical sputtering effects are present with InP and InGaAs. We have attempted to show that the self bias voltage can be used to accurately evaluate the role of chemical/physical effects in RIE. Further, ion bombardment in RIE leads to changes in the Schottky barrier height of GaAs but also that the incorporation of hydrogen into Si doped GaAs leads to significant changes in the electrical properties of the semiconductor.

7. REFERENCES

1. L.A. D'Asaro, A.D. Butheras, J.V. DiLorenzo, D.E. Iglesias and S.H. Wemple, Inst. Phys. Conf. Ser., 5 (1981) 267.

2. L.A. Coldren, K. Furuya and B.I. Miller, J. Electrochem. Soc., 130 (1983) 1918.

3. H. Matsueda and M. Nakamura, Appl. Optics, 23 (1984) 779.

4. B. Chapman, Semiconductor Int., Nov. 1980, 139.

5. R.E. Williams, Gallium Arsenide Processing Techniques, Artech House Inc., 1984.

6. D.L. Flamm, V.M. Donnelly and D.E. Ibbotson, J. Vac. Sci. Technol. B1 (1983) 23.

7. D.L. Flamm, V.M. Donnelly and J.A. Mucha, J. Appl. Phy., 52 (1981) 3633.

8. R.A. Gottscho, G. Smolinsky and R.H. Burton, J. Appl. Phys., 53 (1982) 5908.

9. P. Sigmund, Phys. Rev., 184 (1969) 383.

10. Y. Yamane, K. Yamasaki and T. Mizutani, Jap. J. Appl. Physics 21 (1982) 2537.

11. K. Yamasaki, K. Asai, K. Shimadi and T. Makimura, J. Electrochem. Soc., 129 (1982) 2760.

12. S.W. Pang, G.A. Lincoln, R.W. McClelland, P.D. DeGraff, M.W. Geis and W.G. Piacentini, J. Vac. Sci. Technol., B1 (1983) 1334.

13. C-L Chen and K.D. Wise, IEEE Trans. Electron Devices, ED-29 (1982) 1522.

14. J. Chaplart, B. Fag and N.T. Linh, J. Vac. Sci. Technol., B1 (1983) 1050.

15. C.S. Sa, D.M. Scott, W-X Chen and S.S. Lau, J. Electrochem. Soc., 132 (1985) 918.

16. G.A. Sarov, Vacuum, 34 (1984) 1027.

17. S.W. Pang, M.W. Geis, N.N. Efremow and G.A. Lincoln, J. Vac. Sci. Technol., B3 (1985) 398.

ION BEAM ENHANCED DEPOSITION AND DYNAMIC ION MIXING FOR SURFACE MODIFICATION

L. CALLIARI[+], F. GIACOMOZZI[+], L. GUZMAN[+], B. MARGESIN[+], and P.M. OSSI[*]

+ Istituto per la Ricerca Scientifica e Tecnologica, I-38050 Povo (Trento), Italy.
* Istituto di Ingegneria Nucleare, Politecnico di Milano, I-20133 Milano, Italy.

1. INTRODUCTION

Ion implantation techniques have proved to be very effective in the production of compounds with predetermined (possibly non-equilibrium) composition, whose characteristics are interesting for the improvement of surface properties (wear-, corrosion-, thermal oxidation resistance) (1). One of the limiting factors for such treatments is the shallow thickness of the modified layer.

A recently developed method to avoid the above drawback is the Ion Beam Enhanced Deposition (IBED) technique : it consists of simultaneous or sequential deposition/implantation steps, which allow for making the depth of the treated region nearly independent on the projected range (R_p) of the implanted ions. Moreover, the ion beam plays a role in increasing the adhesion and uniformity of the deposited film.

By the use of a reactive ion species such as nitrogen the IBED technique may be modified into Reactive Ion Beam Enhanced Deposition (RIBED). Independently controlled atom and ion beams of different kinds, able to form the required compound, impinge sequentially or simultaneously on the sample surface. The ion beam not only contributes the necessary energy and momentum to the process, but it also provides one of the required atomic species. In the first steps of the treatment, mixing of the substrate material with the phase under formation prevails, resulting in good relative adhesion; then the new phase grows, this latter stage allowing to obtain virtually unlimited film thicknesses.

To perform the present kind of experiments, the implantation chamber of an implanter with horizontal beam, was modified by locating under the chamber an electron gun, in a suitable position to sequentially expose the sample to deposition or to implantation, by rotating it around an horizontal axis; if the target is inclined to 45 degrees, it is struck at the same time by the ion beam and the evaporant beam.

In this work three different surface compounds were considered : chromium nitride, which is known to be hard and corrosion resistant, niobium nitride, suitable for superconductivity applications, and the alloy $Cu_x Cr_{1-x}$, which is interesting in view of the possibility to enhance Cu oxidation resistance.

The structure, microstructure and elemental composition of the surface phases were then analysed by Scanning Electron Microscopy (SEM), Auger Electron Spectroscopy (AES) and X-ray diffractometry (XRD); scratch type adhesion testing was performed on CrN films.

2. EXPERIMENTAL TECHNIQUES

2.1. Apparatus

The implants were performed on a 30 keV accelerator equipped with mass analysis, using a chamber suitable also for depositions. The targets were moved back and forth under the beam with a velocity of 0.4 mm s^{-1}, the base pressure being 10^{-5} Pa. The ion beam was focused in a spot, about 3 mm in diameter and then scanned in the vertical plane to form a line about 40 mm long; a measure of the current was possible by collecting in a Faraday cup a fraction of the beam transmitted through a 5 mm wide calibration slit.

The samples were kept above the electron gun evaporator, at a distance of 0.3 m from the source, and were positioned on a rotating and translating holder, driven by stepping motors (2). Film thicknesses and evaporation rates (kept fixed at about 0.1 nm s^{-1}) were monitored by a quartz crystal oscillator. The bulk temperature of the specimens was measured by a thermocouple; it never exceeded 373 K.

2.2 Sample Preparation

All samples were mechanically polished and pre-implanted to sputter clean the surfaces and to induce a defective microstructure, favouring adhesion.

RIBED treated Cr samples were obtained by a sequence of 7 to 20 steps, each one consisting of a 10 nm Cr deposition, followed by a N_2^+ implantation to a dose of $7.5 \ 10^{20}$ $N^+ \ m^{-2}$, determined in order to achieve 1:1 stoichiometry, with a current line density of 2 $\mu A \ mm^{-1}$. Different substrates were used : copper, glass, high speed steel.

RIBED treated Nb samples on copper and glass substrates were obtained using the same technique by 7 sequential steps of 15 nm Nb depositions, each one followed by N_2^+ implantations to a dose of 8.10^{20} $N^+ \ m^{-2}$ as to guarantee 1:1 stoichiometry, with a current line density of 1 $\mu A \ mm^{-1}$;

Cu-Cr mixing by IBED was carried out in two ways :
1) 7 sequential steps of 7 nm Cr deposition on copper substrates, followed by Ar^+ implantations to a dose of 8.10^{19} $Ar^+ m^{-2}$, with a current line density of 2 $\mu A\ mm^{-1}$.
ii) simultaneous implantation/deposition of ions/atoms; the samples were kept at an angle of 45 degrees with respect to both incident beams. The total obtained thickness was about 60 nm, with a total dose of $1.2\ 10^{21}\ Ar^+ m^{-2}$ and a current line density value of 2 $\mu A\ mm^{-1}$.

By partially screening of the ion beam, some only evaporated samples were obtained in every run; we shall refer to these samples as to the reference ones, because they are obtained under identical experimental conditions as the implanted ones.

2.3 Surface Microanalysis and Technological Tests

The samples surface microstructure was examined by SEM; depth composition profiles were determined by AES (electron gun coaxial to a single pass cylindrical mirror analyzer with 0.6% intrinsic relative resolution), combined with 2 keV ion sputter etching. The depth profiles were elaborated with the sensitivity factors method (3). Care should be taken in considering the relative atomic concentration values, particularly as for the C concentration because whenever, like in this case, the Auger lineshape is "carbidic" rather than "graphitic", the use of the sensitivity factor of (3), given there for "graphitic" C, can lead to a large overestimation of the C concentration (4,5).

The structure of the samples was studied by XRD, using Cu K radiation.

The adhesion was measured by means of scratch type testing apparatus (REVETEST Automatic Scratch Tester) (6).

3. RESULTS AND DISCUSSION
3.1 Chromium Nitride

Figure 1a shows the AES sputter profile for as-evaporated Cr on Cu substrate, whereas Figure 1b refers to sequential Cr evaporations and N^+ implantations on a Cu substrate. From Figure 1b, the nearly parallel Cr and N profiles give good evidence of a quasi-stoichiometric compound formation; this finding is further supported by the Cr Auger lineshape (Fig. 1c) which is in reasonable agreement with that reported in the literature (7) for Cr in CrN, as well as with the one we obtained from a reference CrN powder.
Indication of the presence of this compound is given by X-ray diffraction analysis, from which the peaks pertaining to CrN are evident. The evaporated Cr film contains O as well as C due to the rather high reactivity of Cr with the residual gases in

the chamber (H_2O, CO, CO_2) during evaporation. The relevant decrease in the contaminant concentrations after N_2^+ implantation, (see Fig. 1a and 1b) may be attributed to sputter cleaning of the surface.

The comparison of Figs. 1a and 1b also gives evidence of ion beam induced interface broadening, which is an indication of enhanced film-substrate adhesion. This is confirmed by the scratch test results. From Table 1 we note that the critical load value for film rupture (detaching) from the substrate is nearly doubled for the RIBED produced samples with respect to the only evaporated ones. This result appears to be quite independent of the mechanical properties of the substrate (quenched tool steel X150CrMo12KU (AISI D2), HRC 61, and E-copper, HB 37). From the scratch in the SEM micrographs, Cr film appears highly deformed and even cracked for as-deposited

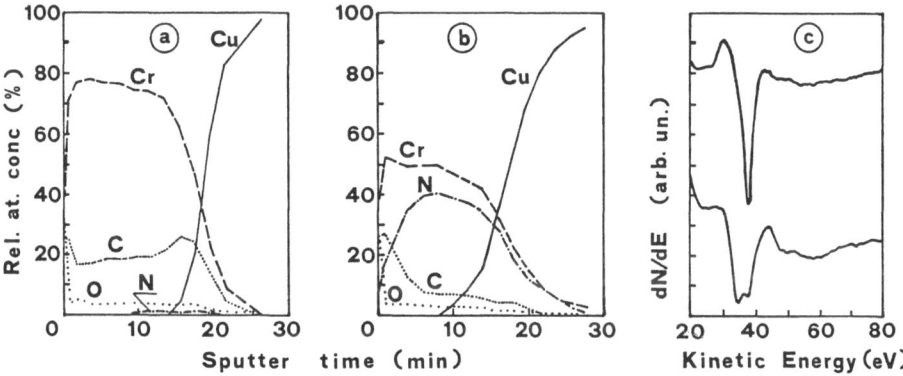

Fig. 1 - Auger concentration depth profiles for a sequentially Cr evaporated and N^+ implanted Cu substrate (b) and for its reference sample (a); (c) Cr (MNN) Auger transition lines for sample a (upper curve) and b (bottom curve).

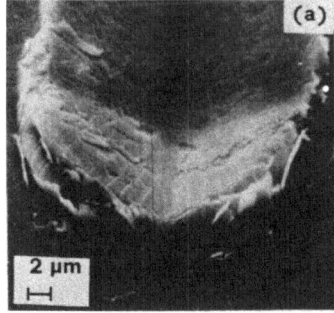
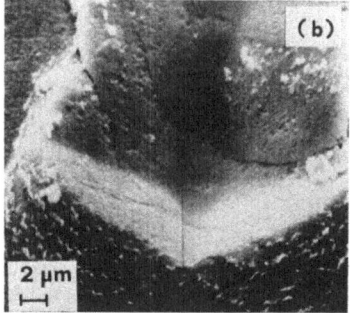

Fig. 2 - SEM micrographs of tracks left by a scratch-type testing with a Vickers indenter (load 10g) for a RIBED-treated CrN-on-Cu sample (b) and for a reference Cr-on-Cu sample (a).

TABLE 1 : Scratch Tests for different specimens.

Specimen	steel/Cr	steel/CrN	Cu/Cr	Cu/CrN
Critical Load L_c (N)	35	71	3	7

----

* performed with Automatic REVETEST equipment; the critical load (L_c) for film detachment is determined by optical microscopy. Diamond stylus radius is 400 μm. The L_c quoted value is an average over 4 measurements.

samples (Fig. 2a), while the CrN film adheres to copper substrate notwithstanding severe plastic deformation (Fig. 2b). Indeed it is already known (8) that the mechanical properties of Cr films obtained by IBED technique are noticeably improved. Due to the low amount of oxygen contaminant at the interface, the improved adhesion is not to be attributed to the presence of an oxide layer.

3.2 Copper-Chromium

Figures 3a and 3b refer to AES depth profiles respectively for a Cr film simply evaporated on a Cu substrate and for an analogous film obtained by IBED (Ar$^+$ was used as the bombarding ion). Figure 3b shows an extended interface, which is attributed to successive mixing steps obtained by a dynamic mixing. Such procedure involves each time freshly deposited material, thus increasing the process efficiency with respect to standard ion-mixing.

Another set of experiments was carried out by simultaneous Cr deposition and Ar$^+$ implantation, which however gave similar

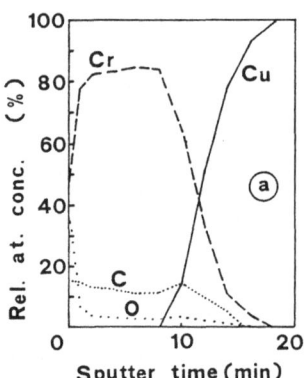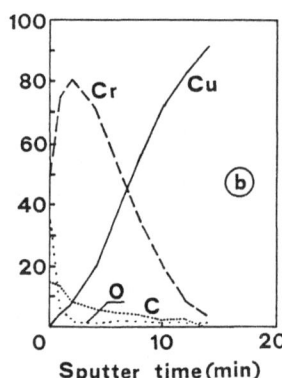

Fig. 3 - Auger concentration depth profiles for a sequentially Cr evaporated and Ar$^+$ implanted Cu substrate (b) and for its reference sample (a).

depth composition results as those of Figure 3. The main difference between the 2 methods lies in the fact that in this latter case the two beams don't impinge normally on the surface. This has to be taken into account by a suitable geometrical factor in the calculation of implanted dose and film thickness. Moreover, the sputtering yield and the sticking coefficient are also changed by a factor which is difficult to quantify. A better control of the process is in any case achieved by using the sequential method.

3.3 Niobium (Carbo-oxy-) Nitride

Fig. 4 shows the Auger concentration profile for a sequential Nb evaporation and N_2^+ implantation on Cu substrates (Fig. 4b) and for the reference sample (Fig. 4a).

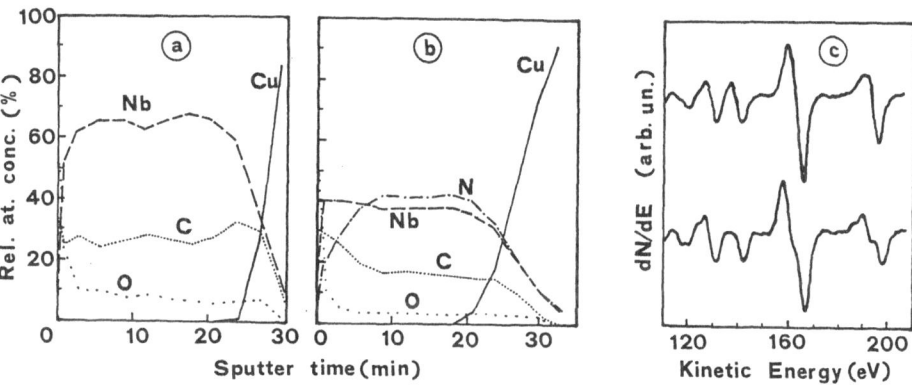

Fig. 4 - Auger concentration depth profiles for a sequentially Nb evaporated and N^+-implanted Cu substrate (b) and for its reference sample (a); (c) Nb (MNN) Auger transition lines for sample a (upper curve) and b (bottom curve).

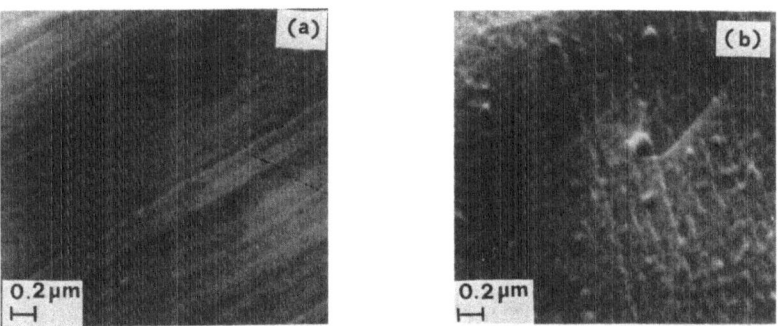

Fig. 5 - SEM micrographs showing morphology of a RIBED-treated Nb-on-Cu sample (b) and its reference sample (a).

The similarity in the depth behaviour of Nb and N in Fig. 4b as well as the good agreement between the MNN Auger lineshape for the RIBED obtained sample (bottom curve in Fig. 4c) and the one reported in the literature (9) for Nb in NbN supports the formation of niobium nitride. Nevertheless the relevant C and O contamination across the whole film thickness, due to high Nb reactivity, may indicate the presence of a multicomponent phase rather than a binary compound.

As for the microstructure of the obtained films, the SEM micrograph for as deposited Nb (Fig. 5a) shows a uniform structureless pattern following the surface morphology of the Cu substrate, allowing to suspect the presence of an amorphous phase. On the contrary, the RIBED treatment induces the granular film structure shown in Fig. 5b.

The XRD analysis confirms the partially amorphous character of the as-deposited Nb film. (Fig. 6a). In the RIBED samples the obtained diffraction peaks (Fig. 6b) are well in agreement with those reported in the literature (10) for a $Nb(N_{0.9}O_{0.1})$ face-centered cubic compound.

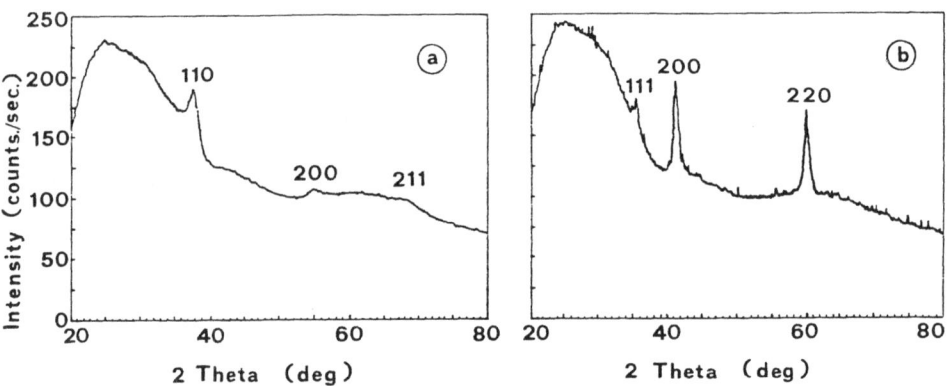

Fig. 6 - X-ray diffractograms of a RIBED-treated Nb-on-glass sample (b) and its only evaporated reference sample (a).

4. CONCLUSIONS

Ion beam enhanced deposition techniques were developed, which proved to be very successful in producing new surface alloys, or compounds, such as CrN, Nb(N,C,O) and Cu_xCr_{1-x}.

Homogeneity of the formed films and their adherence to the substrate are noticeably improved with respect to those of samples obtained by conventional implantation or deposition alone.

Acknowledgements

The authors are grateful to Centre Suisse d'Electronique et de Microtechnique S.A. for scratch testing. It is a pleasure to acknowledge the collaboration of Drs. M. Sarkar, A. Molinari and G. Giunta as well valid help from IRST technical staff.

REFERENCES

1. Harper J.M.E., Cuomo J., Gambino R.J., and Kaufman H.R.: Ion Bombardment Modification of Surfaces, O. Auciello and R. Kelly, eds. Elsevier, Amsterdam (1984) p. 127.
2. Lazzari G., Margesin B., Scotoni I. and Zanini V.: Nucl. Inst. Meth. B6, (1985) p. 210-213.
3. Davis L.E. et al: Handbook of Auger Electron Spectroscopy, Physical Electronics Division, Perkin Elmer, Eden Prairie, MN, (1978).
4. Singer I.L.: Appl. of Surf. Sci. 18, (1984) p. 28-62.
5. Kant R., Sartwell B.D., Singer I.L. and Vardiman R.G., Nucl. Inst. Meth. 87/8, (1985) p. 915-919.
6. Hintermann H.E.: J. Vac. Sci. Technol. B2, (1984) p.816-822
7. Singer I.L. and Murday J.S., J.Vac.Sci.Technol. 17, (1980) p. 327-329.
8. Hoffman D.W. and Gaerttner M.R., J. Vac. Sci. Technol. 17, (1980) p. 425-428.
9. Ramaker D.E., Appl. of Surf. Sci. 21, (1985) p. 243-267.
10. Powder Diffraction File Search Manual, Ed. YCPDS Intnal. Centre for Diff. Data, Swarthmoar, Pennsylvania (1978).

MOLECULAR BEAM EPITAXY AND SELECTIVE MBE DEPOSITION OF GaAs DEVICE STRUCTURES

A. CHRISTOU,
Research Center of Crete and Physics Department, University of Crete, Heraklion, Crete, Greece.

1. INTRODUCTION

Molecular Beam Epitaxy (MBE) is an ultrahigh vacuum technique for preparing thin epitaxial layers of II-VI[1,2] III-V[3-5] and IV-VI[6,7] semiconductors. The growth process consisting of incident molecular beams provides a high degree of control over layer doping, composition, uniformity and thickness. Such control is not achievable with liquid phase epitaxy (LPE) or chemical vapor Deposition (CVD) techniques.

The present status of MBE which includes complex superlattice has developed through the initial research in vacuum evaporation and kinetic condensation of III-V compounds. Stoichiometry can be achieved, even though group V species have vapor pressures which are higher than molten group III metals. The work of Gunther[8], Arthur[9] and Davey[10] resulted in using pure Ga and As loads and evaporating each component from a seperate oven. The III-V compounds typically deposited by MBE are InP[11], GaP,[12] InAs,[13] $Al_xGa_{1-x}As$,[14] $In_{1-x}Ga_xP$[15] and GaAs among others.[16-20]

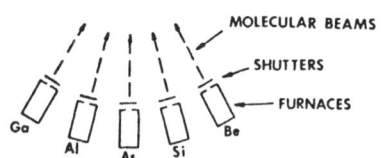

GaAs SUBSTRATE

MOLECULAR BEAMS

SHUTTERS

FURNACES

Ga Al As Si Be

Room Temperature Mobility Enhancement
MBE ≈ 2×
MOCVD ≈ 2×

Low Temperature Mobility
MBE ≥ 200,000 $cm^2/v-sec$
MOCVD ≃ 160,000 $cm^2/v-sec$

High Speed Logic Performance
τ_d ≃ 12.0 psec (77°K) HEMT
τ_d ≈ 60 psec   MESFET

Figure 1. Fundamental features of MBE deposition showing positioning of Knudsen cells.

2. COMPONENTS OF MBE SYSTEMS

The MBE growth process is carried out in an ultrahigh vacuum chamber inorder to minimize collision events with residual gaseous contaminants. Arrays of individual ovens are used as shown in Figure 1. High purity beams of molecules stream from the ovens toward the substrate which is maintained at temperatures of up to 600°C depending on the growth system. The typical growth rate is 2-3 A°sec⁻¹ Very abrupt changes in composition can be accomplished through the manipulation of shutters in front of the ovens.

The incident molecular beam intensity is related to the temperature T (°K), vapor pressure p (torr) and molecular weight M, of each oven charge through equation (1):[21]

$$F = 3.513 \ 10^{22} \ a\pi^{-1} \ h^{-2} \ p \ (MT)^{-\frac{1}{2}} \qquad (1)$$

where a=area of the oven aperture, h is the oven to substrate spacing. The vapor pressure is related to the partial pressure by:

$$\log p = B - AT^{-1} \qquad (2)$$

Therefore $F = F(T)$ only and molecular beam intensities may be regulated by each oven temperature. Direct measurements of F may also be made with ionization gauges. Utilization of a mass spectrum analyzer further provides flux information. The heights of each peak closely duplicates the percentages of element present. In addition, mass spectrum analyzers are usually positioned as monitors of the incorporation and desorption kinetics.

In situ surface diagnostics are used to characterize surfaces grown for a given set of growth conditions. Typically, the stoichiometric growth of III-V semiconductors depends on the group III flux since group V adatoms rapidly desorb from heated substrates. Both Auger spectroscopy and reflection electron diffraction are used to characterize growth mechanisms.

3. SELECTIVE MBE GROWTH

3.1 <u>Microwave Devices</u>. Microwave devices grown by MBE techniques include mixer diodes, IMPATTS and FETS. The selective growth of the active areas by MBE has been accomplished by formation of poly-GaAs regions next to epitaxial MBE regions. Poly GaAs formation can take place by growing over metallic thin film masks.

Mixerdiodes and varactor diodes utilize hyperabrupt profiles which are formed by increasing doping oven temperature while maintaining the growth rate constant. The profile for mixer diodes is n|n⁺|S.I GaAs while for varactor diodes, the doping concentration n is proportional to the depth Z from t raised to the m power.

$$n = AZ^m \qquad (3)$$

Both Schottky barrier and p-n junction varactors have been

built using the above technique. For mixer diodes, a noise figure of 0.6 dB at 94 GHZ is typical.[22] Figures of merit of 42 at 100 GHZ[23] have been reported for MBE Schottky Barrier varactors operating in parametric amplifier circuits.

IMPATTs require highly doped spikes 200-300A° wide, positioned within 2000A° of the Schottky barrier to confine the avalanche zone of low-high-low IMPATTs. Figure 2 shows a typical IMPATT structure. Sn-doped IMPATTs operated at 12GHz have exhibited efficiencies of 18 percent.[24]

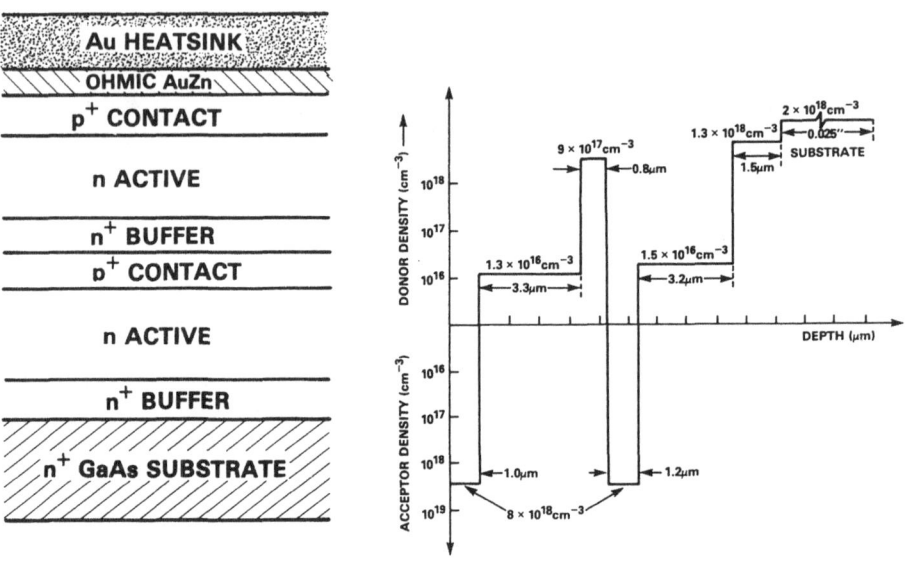

Figure 2. IMPATT profile of devices fabricated by MBE

The above structures can be integrated into MMIC circuits (Monolithic Microwave Integrated Circuits) by utilizing selective MBE growth. The growth sequence is as follows: (1) active first layer, (2) deposition of thin film metal (Ta or W) regions for masking, (3) growth of second MBE layer and (4) etching of Poly-GaAs regions.

III-V compound semiconductors have high electron mobilities and coupling the fabrication of short gate length structures, high frequency field effect transistors have been obtained. Utilizing a thin buffer layer (0.15μm) AlGaAs/GaAs high electron mobility transistor has been obtained[25] Good 77°K Hall mobility ($\mu = 45,000$ cm^2/V-sec) and sheet carrier density ($n_s = 1.0 \times 10^{12}$/cm^2) were obtained in these FETs with a 50 A° undoped AlGaAs spacer layer. Excellent device performance was also achieved with typical room-temperature transconductances of 210 ms/mm and 300 ms/mm for depletion mode and enhancement mode FETs.

The heterojunction or modulation doped FETs (MODFETs) have shown better noise figure performance as compared to conventional MESFETs. A graded aluminum composition has been shown to yield a greater sheet electron concentration[26] A noise figure of less than 1 dB at 18 GHz has been achieved with the MODFET structure. The highest transconductance measured for devices with 1 μm length has been reported to be 300 ms/mm and 350 ms/mm for depletion mode and enhancement mode devices[27]

3.2 Deconfinement of Electrons in MBE MODFETS

Deconfinement of the two dimensional electron gas (at the GaAs/AlGaAs interface) occurs when $\partial^2\varphi/\partial x^2 < 0$ in the potential vs. distance relationship in a FET structure. Electrons will fan out from the 2DEG and will result in a substrate current. Utilizing a simple electron flow model, the electron velocity is given by:

$$\frac{1}{v} = \frac{1}{\mu F} + \frac{1}{vs} \qquad (4)$$

where v_s = saturation velocity and

$$F = \frac{1}{g}\frac{d}{dy}E_F$$

the saturation velocity can be related to the diffusion velocity by:

$$v_s = \frac{1}{4}v \qquad (10^7 \text{ cm/sec}) \qquad (5)$$

In discussing electron flow in the MODFET structure, we define an effective gate length L^* as the distance from $E_x = 0$ to nonconfinement. Then

$$J + \frac{2\varepsilon L^*}{\mu\alpha}U_sJ^{\frac{1}{2}} = \frac{\varepsilon U_s}{\alpha}\left[V_G - V_T - 2kT\ln\frac{J}{J_o}\right] \qquad (6)$$

or

$$J = J^* = \frac{2\varepsilon L^*U_s^2}{\mu\alpha}$$

$$(7)$$

at

$$V_G - V_T = -\frac{4L^*U_S}{\mu}$$

In addition at $J \ll J^*$

$$J = \frac{\varepsilon\mu}{2aL^*}|V_G - V_{T_1}|^2 \qquad (8)$$

at low current densities and at $J \gg J^*$

$$J = \frac{\varepsilon v_s}{\alpha}|V_G - V_{T_2}| \qquad (9)$$

at high current densities. Therefore deconfinement occurs at
the end of the gate and when velocity saturation is present.
The resultant substrate current may be eliminated by a p type
undoped buffer layer.

3.3. Multiple Channel MODFETs

A single heterojunction has a limited saturation velocity
and electron density n_s due to the deconfinement phenomenon
discussed in section 3.2. In order to increase n_s the spacer
layer thickness must be kept to 50 A°or less. However, such
thicknesses are difficult to grow due to degradation from silicon diffusion. The solution presented is a double heterostructure or multiple channel MODFET[28] Such structures have resulted
in n_s of 3 10^{12} cm^{-2} and mobilities of 200.000 cm/v-sec. A
further reduction in source resistance to 1Ω-mm has been
obtained.

In the MBE deposition of multiple channel structures one
must ensure tunneling to the second channel by controlling the
spacer layer thickness between the two channels. The tunneling
probability is given by

$$T = \exp\left[- \frac{2\sqrt{2}}{3h} \sqrt{m^*E} \; d\right] \qquad (10)$$

where $m^* = 0.1\, m_o$ and $E = 0.6$ eV. The thickness d in this case
must be maintained less than 220 A? Further reduction in d will
result in higher transconductance, g_m. For multiple quantum
well structurers a parameter r can be defined as the efficiency
of transferring electrons to the 2DEG layer, where

$$r = \frac{N_s \; (\text{total})}{16N \; d_N} \qquad (11)$$

For $d_N = 330$ A°and $N_s = 4.5 \times 10^{11}$ then $r = 0.11$ and typically r has
been shown[29] to be between 0.1 and 0.2.

4. OPTICAL DEVICES
4.1. Laser Structures

The steps in ΔE (band gap) and Δn (refractive index) for
$Al_xGa_{1-x}As$ provides confinement of the optical flux. The minor-
ity carriers at the $Al_xGa_{1-x}As$/GaAs interfaces are confined to
recombine in a submicron thick GaAs active layer.

The $Al_xGa_{1-x}As$ ternary spans the range of bandgap en-
ergies of $1.42 \leq E_g \leq 2.16$ eV and refractive index of $3.59 \leq n \leq 2.97$
for $0 \leq x \leq 1.0$. In depositing AlGaAs, the reactivity of Al complicates the growth of MBE AlGaAs. However, the employment of
pyrolitic boron nitride crucibles and increasing the growth
temperature (substrate temperature to ≥620°C can solve most
problems associated with AlGaAs deposition. Threshold current densities of
800 Acm^{-2} for double heterostructure lasers[29] have been obtained.
These lasers consisted of $Al_{0.3}Ga_{0.7}As$/GaAs with areas of
375x200 μm² and GaAs thickness of 0.15 μm. The threshold current densities obtained by MBE are comparable in performance to
lasers grown by LPE. The MBE material shows higher uniformity
in thickness and material properties. Figure 3 shows a distributed feedback laser[30] and a taper-coupled laser[31] The distributed feedback from light reflected back into the laser cavity

by the corrugated grating controls the lasing. In this type of
laser the 0.37 μm periodic corrugation was formed by ion
milling.

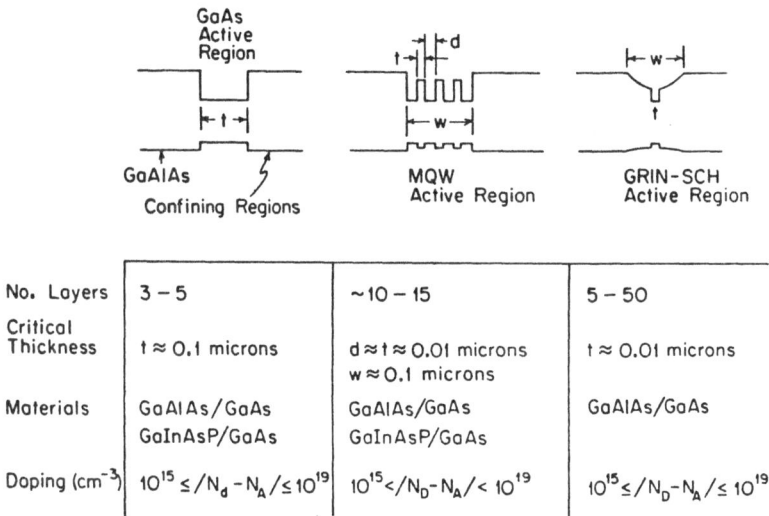

No. Layers	3 – 5	~10 – 15	5 – 50
Critical Thickness	$t \approx 0.1$ microns	$d \approx t \approx 0.01$ microns $w \approx 0.1$ microns	$t \approx 0.01$ microns
Materials	GaAlAs/GaAs GaInAsP/GaAs	GaAlAs/GaAs GaInAsP/GaAs	GaAlAs/GaAs
Doping (cm^{-3})	$10^{15} \leq /N_d - N_A / \leq 10^{19}$	$10^{15} < /N_D - N_A / < 10^{19}$	$10^{15} \leq /N_D - N_A / \leq 10^{19}$

Figure 3. GaAs/GaAlAs lasers showing both the distributed
feedback and taper coupled construction.

Single transverse mode lasers can also be constructed by
MBE utilizing 5.0 μm wide $Al_{0.35} Ga_{0.65}As$ strips embedded in a
high resistivity polycrystalline layer which grew in areas
masked by a thin SiO_2 layer. A long taper coupler shown in
Figure 3 allows for the coupling of light from active to passive regions. The taper coupler can be formed by using a Ta
edge during MBE growth of the 0.3 μm thick GaAs active layer.
Recently, transverse junction $Al_xGa_{1-x}As/GaAs$ lasers were
prepared by MBE on semi-insulating GaAs[32] The lasers showed
threshold currents of 56 mA at 298°K. Transverse junction
lasers can also be integrated with FETs and other signal processing active devices.

The addition of In to GaAs provides the opportunity for
obtaining lasing action at different wavelengths. Material
combinations of $In_{0.53} Ga_{0.47} As/InP$, processed into a double-
heterostructure laser (λ=1.65 μm) operated at $J_{th} = 3.2$kAcm^{-2} at
300°K. Other lasers include $In_{0.16} Ga_{0.84} As/GaAs$ p-n junction
laser (λ = 0.96 μm, $J_{th} = 3.0$ kAcm^{-2} at 77°K[33] and an
$In_{0.53} Ga_{0.47}/GaAs$ DH laser (λ = 0.83 μm).

4.2 Light Emitting Diodes (LEDs) and Solar Cells.

LEDs processed by MBE are a natural extention of Laser
processing. Burrus type LEDs with $Al_{0.3} Ga_{0.7} As/ GaAs$ DH

structures developed output powers of 5 mW at saturation currents of 250-500 mA.[34] A GaSb$_{0.9}$/GaAs LED ($\lambda = 0.995$ µm) using a
compositionally graded GaSb$_{1-y}$As$_y$ buffer has also been successfully processed.

Single shallow homojunction GaAs cells have resulted in
efficiencies exceeding 16 percent.[35] In these cells, recombination losses are reduced because most of the photogenerated
carriers are created below the heavily doped p-n junction layer
located 0.05 µm from the surface. Both p$^+$ n and n$^+$ /n/p$^+$ structures have been reported.

5. SUPERLATTICE DEVICES

5.1 Modulation Doped Superlattices

The quantum mechanical phenomenon observed for Al$_x$Ga$_{1-x}$As/
/GaAs superlattice structures is linked to the energy band
discontinuities in a multiple layer heterojunction sandwich.
Periodicities of 150A° or less are required in the energy band
structure. Superlattice demonstrations of negative resistance,[36]
conductance oscillations from high field domains,[37] direct
optical absorption,[38] shifts in optical band edge[39] and laser
oscillation[39] have been reported.

The basic features of modulation doped superlattices is
that of electron confinement in undoped GaAs layers seperated
from donor impurities located in the Al$_x$Ga$_{1-x}$As layers. Ionized
impurity scattering at low temperatures is reduced and high
carrier mobilities are obtained for layers 50 A° to 400 A° thick.
Negative differential resistance occurs when hot electrons are
emitted by real space transfer from the high mobility GaAs
layers into the low mobility Al$_x$Ga$_{1-x}$As layers.[40] The electron
mean free path in this case must be equivalent or greater than
the superlattice spacing.

The dynamics of electron collection in quantum wells
consists of injection of excess carriers or photogeneration of
excess carriers in the confining layers of a quantum well
heterostructure within a diffusion length of the narrower gap
active region. The diffusion to the well occurs within a time
t$_d$ given by:

$$t_d \sim L^2/D \tag{12}$$

where L is the length of the confining layer and D is the diffusion constant of Al$_x$Ga$_{1-x}$As. Electons can be captured in the
well only by inelastic processes: (a) by optical phonons for
large X and by (b) polar optical scattering for small x. For
high x, intervalley potential scattering is strong and carriers
are easily collected in wells of 100 A° thick.

When an electric field is applied parallel to the layers,
electrons gain energy in the continuum and by momentum transfer
can propagate perpendicular to the layers. The threshold for
electron emission out of the well is close to the threshold of
the Gunn effect. Only a slight heating of the electrons can
lead to large transfers. At electric fields higher than about
200 V/cm real space transfer dominate the IV characteristics.
At higher electric field electrons can move out of the GaAs
layers into the Al$_x$Ga$_{1-x}$As or into neighboring unheated GaAs

layers. Since the mobility in the $Al_xGa_{1-x}As$ is much lower, negative differential resistance can occur due to the real space transfer effect. This effect may generate oscillations similar to Gunn devices.

Field effect transistors using superlattice layers may also be grown by MBE on (100) oriented Cr-doped semi-insulating GaAs substrates at 600°C with V-III flux ratios of 2 to 5. The spacer layers may be the superlattice region. The maximum intrinsic transconductance per unit gate width of FET is then given by:

$$(g_m)_{MAX} = (g\mu n_0 / L) |1 + g\mu n_0 d/\varepsilon v_s)^2|^{-\frac{1}{2}} \qquad (13)$$

where L is the gate length, d is the thickness of the $Al_xGa_{1-x}As$ superlattice layer, μ is the low field mobility and v_s is the saturation velocity. The superlattice thickness d can be expressed as

$$d = d_i + \Delta d + | (2\varepsilon/gN_d) (V_{Bi} - V_{off}) |^{\frac{1}{2}} \qquad (14)$$

where Δd is the effective width of the 2DEG layer ($\Delta d = 50A°$), ε is the dielectric constant of $Al_xGa_{1-x}As$, g is the electronic charge, N_d is the dopant density in the $Al_xGa_{1-x}As$, $V_{Bi} = \Phi_b - \Delta E_c$ is the effective built in Voltage and ΔE_c is the conduction band edge discontinuity at the $Al_xGa_{1-x}As$ / GaAs interface. Table I shows the theoretical and measured transconductance and saturation currents for superlattice MODFETs.

d_i (A°)	g_m (mS/mm)	I_{SAT} (mA)
20	210 (290)	7.5 (27)
40	230 (245)	24 (27.5)
60	235 (240)	23 (26.5)
80	210 (225)	14 (19)
100	145 (170)	8.5 (8.5)

Table I. Measured and theoretical (parenthesis) values of transconductance and saturated current.

The results in Table I are for normally on devices. The large discrepancy for the 20 A° layer is due to the doping layer being too small to be consistent with the other normally-on devices.

5.2 Strained layer Superlattices and Special Techniques

Heteroepitaxial MBE deposition of semiconductors without lattice matching has opened the possibility of depositing small band gap materials on GaAs (such as InGaAs). The thickness of the layers must be small enough to ensure that the lattice mismatch is accommodated by strain without generating misfit dislocations at the interface.[41] The initial work resulted in the GaAs/GaP system and recently both InGaAs/GaAs and Si/SiGe. The incorporation of an insulator BaF_2 on InP has also been accomplished where the mismatch is 5%. The interface has closely spaced deslocations and is therefore not entirely defect free.

Oxide masking, metal and mechanical masking are currently being used for spatially separating MBE structures. MBE GaAs can be layered in polycrystalline GaAs by selectively etching holes in a SiO_2 or Si_3N_4 mask. Growth of self aligned MBE $Al_xGa_{1-x}As$ / GaAs can be obtained in $[\overline{1}\overline{1}0]$ stripe channels. Patterned oxide layers may also be used to define micron-size epitaxial layers and metallizations. Irradiation with an e^--beam which alters the local sticking coeficients may also be used for defining structures in growing MBE layers.

Included as a special technique for the processing of MBE superlattice devices and selectively doped field effect transistors are (1) enhancement mode - depletion mode circuits and (2) preferentially etched gate regions inorder to eliminate trapping in the AlGaAs layers. In accomplishing (1), the starting MBE grown profile is the depletion mode type with the n + layer thickness (AlGaAs) being 500A°thick. In producing the enhancement mode part of the circuit, the d-mode section is masked off and the channel is thinned to 200 - 250 A°inorder to produce e-mode FETs. Mesa isolation or ion implantation is then used to separate the e-mode and d-mode FETs. Figure 4 shows the complimentary pair of N-off and N-on FETs. The structures shown in Figure 4 consisted of 1 μm of undoped GaAs, an undoped $Al_{0.33}$ $Ga_{0.67}$ As spacer layer of thickness d_i, and 500A°of $Al_{0.33}$ $Ga_{0.67}$As doped with Si to a level of $1x10^{18}cm^{-3}$ Measured low field mobilities were between 3300 and 6000 cm^2/v_s at 300°K. Sheet carrier concentrations ranged from $6.7x10^{11}$ to $1.5x10^{12}cm^{-2}$ The FETs were fabricated with a 1x250 μm gate in a a 2 μm channel.

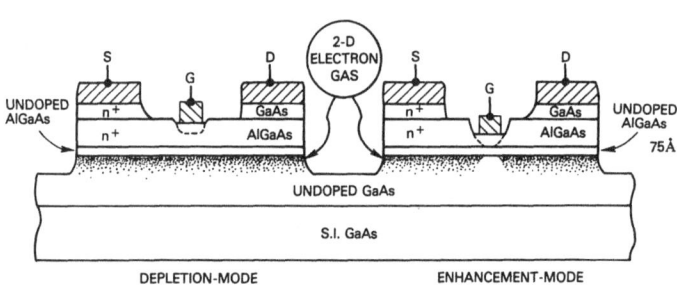

Figure 4. Schematic showing the AlGaAs MODFET with depletion mode and enhancement mode transistors.

The utilization of preferentially etched gate regions has arisen from the observation of I-V collapse and g_m degradation in the dark, and carrier freeze-out at temperature

of 77°K and below. Both of these observations can be related
to trapping in the AlGaAs layer, and in particular the 0.41 eV
trap center. Preferrential etching results in placement of the
gate adjacent to the n⁺ drain in a recessed gate structure.
This results in minimizing the amount of undepleted AlGaAs
between the gate channel and the drain. Table II compares the
transconductance of MODFETs and preferrentially etched self
aligned gate FETs. It is noted that no change in g_m is shown
for the PESAGFET between 300°K and 77°K and likewise between
dark and light. Therefore the effects of AlGaAs trapping may
be minimized by utilizing P.E. gates.

TABLE II. Transconductance per mm of gate showing light and
dark sensitivity in addition to carrier freeze-out
effects.

V/III RATIO = 2.1

	300°K	77°K DARK	77°K LIGHT
MODFET	92 ms/mm 2.5 Ωmm	60 ms/mm 2.0 Ωmm	87 ms/mm 1.0 Ωmm
PE SAGFET	80 ms/mm 4.0 Ωmm	75 ms/mm 3.6 Ωmm	77 ms/mm 3.5 Ωmm

In addition, the P.E. MODFETs have been shown to be
stable to 250°C showing little light sensitivity. Figure 5
shows the I-V characteristics for P.E. MODFETs showing some
limited looping. The dark characteristics were taken at 50 µA
per division while the light characteristics were taken at
100 µA/div and 200 µA/div. At 250°C, some redistribution of
the silicon doping from the n⁺ AlGaAs into the undoped GaAs
layer occurs which results in a 10-20 percent decrease in
transconductance.[42] Additional redistribution of the constituents of the n⁺ contacts has been observed to decrease the noise
figure in low noise MODFETs and to subsequently increase the
total source resistance of the superlattice MODFETs[43]
Electron transfer between heterolayers can offer unique
device structures for the case of impact ionization. The band
edge step (ΔE_C) offer a net enhancement of the electron ionization rate. The band edge discontinuity per hole is much
smaller and therefore the hole-ionization rate is much smaller.
As the electron transfers from AlGaAs to the smaller bandgap
GaAs and therefore sees a reduced ionization threshold. As the
electron transfers from GaAs to AlGaAs a ΔE_C is added to the

334

impact ionization threshold. However, the net effect is an
enhancement of ionization since the ionization threshold
depends exponentially on ionization rate. The control over
ionization rate for electons and holes may be used to improve
impatt diode efficiency and to produce low noise avalanche
photodiodes.[44]

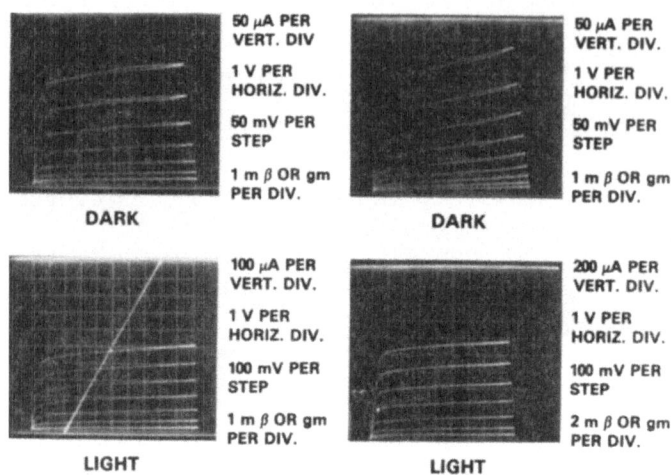

Figure 5. Current-voltage characteristics of preferrentially
          etched MODFETs.

6. MBE GROWTH MECHANISMS

    MBE deposition has been found to proceed via an
adsorption process.[45] The growth of GaAs on a cleaned substrate
of GaAs proceeds in stages that include physisorption,
chemisorption and chemical reaction.[46] Since physisorption
binding is weak, chemisorption and chemical reaction are the
important processes. Nucleation and growth is initiated and
propagates by structural defects.

    The growth of GaAs or GaAlAs is controlled by the Ga and
Al arrival rate since in the temperature range of 500°C-600°C,
As will not stick.[47] The elemental flux interacts with the
surface through a adsorption reaction the adsorption energies
for As_2 and As_4 are 0.38eV and 0.58eV respectively. This
compares with 2.48eV for Ga (in Ga (111)) which is chemisorbed.
However Ga undergoes excessive surface diffusion. The presence
of chemisorption of Ga has been determined from photoemission
studies of Ga on (110) GaAs.[48]

    The incorporation of aluminum with GaAs is accomplished
through an exchange reaction between As and Al. The exchange
reaction has been identified by studying the Al2p and Ga 3d
core states by photoemission spectroscopy.[49] The exchange
reaction results in an interfacial layer of AlAs, releasing Ga
metal into the overlayer.

    Doping effects for III-V compound semiconductors are
complex. Commonly used dopants for n-type are Si and Ge, while
for p-type, Be is predominantly used. Dopants such as Sn have

tendencies to accumulate at the surface[50] Other types of doping
are the interaction of zinc with GaAs and group VI elements
which form neutral donors. Molecular species of O, S, Se, and
Te interact weakly with the surface and results in low doping
levels which are temperature dependent.

Growth and doping by MBE involve nonequilibrium processes.
The doping process is therefore strongly dependent on the flux
of the impurity atoms reaching the growth interface. For
instance, Si, is predominantly a donor with a fairly low degree
of compensation. The evaporation temperature required is
fairly high (1150°C). For acceptor impurities, Be offers the
greatest degree of freedom in that it has an adequate sticking
coefficient and does not suffer from surface diffusion.

7. CONCLUSIONS

MBE science and technology has broadened significantly in
recent years. MBE is supporting basic investigations for semiconductor surfaces, interfaces and superlattices. It is also
the primary technology for two dimensional and one dimensional
electronics. MBE science has resulted in the development of new
device structures in the area microwave and millimeter wave
amplification, optical applications in lasers and LEDs and in
field effect transistors.

Future directions will involve integrating these devices
into structures which include a broad range of bandgap energies.
Also new materials will evolve with optical and electrical
properties derived from periodicity control on a monolayer
level.

REFERENCES

1. Cho, A.Y. and J.R. Arthur, Progress in Solid State Chemistry, edited by J.O. McCaldin and G. Somojai, 10 (1975) 157-
   191.
2. Smith, D.L. and V.Y. Pickhardt, J. Appl. Phys. 46, (1975)
   2366-2371.
3. Chang, L.L. and R. Ludeke in Epitaxial Growth, edited by
   J.W. Matthews (Academic Press, New York, 1975) 37-72.
4. Cho, A.Y, J.Vac. Sci. Technol. 8 (1971) 531-540.
5. Toyce, B.A. and C.T. Foxon, Jpn. J. Appl. Phys. 16, 16-1
   (1977) 17-27.
6. Ota, Y. J. Electrochem. 126, (1979) 1761-1770.
7. Smith, D.L. and V.Y. Pickhardt, J. Electrochem Soc., 125
   (1978) 2042-2050.
8. Gunther, K.G. in the use of Thin Films in Physical Investigations, edited by J.C. Anderson (Academic Press, New York,
   1966) 213-225.
9. Arthur, J.R., J. Phys. Chem. Solids 28 (1967) 2257-2275.
10. Davey, J.E. and T. Pankey, J. Appl. Phys. 39 (1968) 1941-
    -1950.
11. McFee, J.H., Miller, B.I and K.J. Bachman, J. Electrochem.
    Soc. 124 (1977) 259.
12. Farrow, R.C.F., J. Phys. D 7 (1974) L121.
13. Baba, S., H. Horita and Kinbara, A., J. Appl. Phys. 49
    (1978) 3632.

336

14. Drummond, T.J. and Koop, W., Appl. Phys. Lett. 41 (1982) 277.
15. Cho, A.Y and M.B. Panish, J. Appl. Phys. 43 (1972) 5118.
16. Naganuuma, M. and K. Takahashi, Phs. Status solidi, 31 (1975) 187.
17. Sakaki, H and L.L. Chang, Appl. Phys. Lett., 31 (1977) 211.
18. Drummond, T.J., H. Morkoc, K. Lee and M. Shur, IEEE Electron Devices Letters 4, (1983) 171-175.
19. Delagebeaudenf, D., and M. Laviron, Electron Lett. 18 (1982) 103-107.
20. Chin, R., N. Holonyak, G.E. Stillman, J.Y. Tang and K. Hess Electron. Lettrs. 16 (1980) 467-469.
21. Glang, R., in Handbook of Thin Film Technology, edited by L.I. Maissel and R. Glang (McGraw-Hill, New York, 1970) 1-130.
22. Linke, R.A, Schneider, M.V., and A.Y. Cho, IEEE Trans. M.T. T. MTT-26 (1978) 935.
23. Wood, C.E.C., Woodcock, J., and J.J. Harris, GaAs and Related Compounds, 1978 (Inst. Phys. Conf. Ser. 45, London 1979) 28-37.
24. Hierl, T.L., and D.M. Collins, Proc. of 7th Biennial Cornell Conf. (1979).
25. Mimura, T., Hiyamizu , S., Fujii, T., and K. Nanbu, Jpn. J. Appl. Phys. 19 (1980) L225.
26. Mendez, E.E, Price, P.J. and M. Heiblum Appl. Phys. Let 38 (1981) 550.
27. Chen, R.T., Sheng, N.H., Miller, D.L. and C.P. Lee, Conf. Selectively Doped Heterst. Trans., Santa Barbara Ca. Dec. 6, 1984.
28. Inoue, M., Conf. on SDHT, Santa Barbara Ca. Dec. 1984.
29. Tsang, W.T., Appl. Phys. Lett. 36 (1980) 11-18.
30. Casey, H.C., Somekh, S., and M. Ilegems, Appl. Phys. Lett. 27 (1975) 142-146.
31. Reinhart, F.K., and A.Y. Cho, Appl. Phys. Lett. 31, (1977) 457-461.
32. Lee, T.P., and A.Y. Cho, Appl. Phys. Lett. 29, (1976) 164-169.
33. Hiyamizu, S., Fujii, T., Nanbu, K., Maekawa, S., and T. Hisatsugu, Surf. Sci. 86 (1979) 137.
34. Lee, T.P., Holden, W.S., and A.Y. Cho, Appl. Phys. Lett., 32 (1978) 415.
35. Suzuki, T., Konagai, M., and K. Takahashi, Thin Solid Films 60 (1979) 85.
36. Esaki, L., and L.L. Chang, Thin Solid Films, 36 (1976) 285.
37. Esaki, L., and L.L. Chang, Phys. Rev. Lett. 33 (1974) 495.
38. Dingle, R., Gossard, A.C., and W. Wiegman, Phys. Rev. Lett. 34, (1975) 1327.
39. Miller, R.C., and R. Dingle, J. Appl. Phys. 47 (1976) 4509.
40. Shichijo, H., Hess, K., and B.J. Streetman, Solid-State Electron. 23 (1980) 817.
41. Osbourn, G.C., J. Vac. Sci. Technol. 21, (1982) 469.
42. Christou, A., Anderson, W.T., and W. Tseng., I.R.P.S. (1985) April Orlando, Fla.
43. Christou, A., and N. Papanickolaou, Solid State Electronics

   To be published 1985.
44. Capasso, F., Tsang, W.T., and A.L. Hutchinson, Proc. of the
    IEDM, (1981) 284-287.
45. Ranke, W. and K. Jacobi, Surface Science, 63 (1977) 33-39.
46. Joyce, B.A. and C.T. Foxon, Jap. Journal of Applied Physics,
    16 (1976) 17-25.
47. Arthur, J.R., Surface Science, 38 (1973) 394.
48. Bachrach, R.Z. and A. Bianconi, J. Vac. Sci Tech., 15
    (1978) 525-532.
49. Bachrach, R.Z., Bauer, R.S., McMenamin, J.C., and A.
    Bianconi, Proc. 14th Int. Conf. on Phys. of Semiconductors,
    Edinburgh, Scotland (1978).
50. Cho, A.Y., J. Appl. Phys. 46, (1975) 1733.

APPLICATIONS OF ION BEAMS IN MICROLITHOGRAPHY

W. L. BROWN
Bell Laboratories, 600 Mountain Avenue, Murrey Hill,
New Jersey 07974, U.S.A

1. INTRODUCTION

Lithography is a process for formation of two dimensional patterns. In modern electronic technology the fabrication of an integrated circuit may involve a sequence of perhaps a dozen two dimensional patterns precisely registered to one another. Each pattern in the sequence defines the regions in which the next processing step in the IC fabrication procedure is to be applied. Currently, lithography is almost entirely photolithography,(1) carried out as shown schematically in Fig. 1a. Typically a patterned mask is illuminated with ultraviolet light and imaged with a projection optical system onto a thin photosensitive polymer film. The UV exposure of the film alters the polymer in either of the two ways illustrated in Fig. 2. Polymer chains may be cross-linked to render the polymer less soluble in an organic solvent or the chains may undergo scission, rendering them more soluble. These two resist tones are called negative and positive respectively. There is increasing interest in x-ray lithography because of its freedom from the $\sim 0.5\mu$m limitation on spatial resolution that arises from diffraction of the UV light. To avoid the need for focusing x-rays, x-ray lithography is typically performed by proximity printing as shown in Fig. 1b. In each of the cases illustrated in Fig. 1 the lithography is dependent on the mask, which may be an enlarged two-dimensional pattern in the projection printing process or a 1:1 image in the proximity case. The mask itself is typically also made by an optical lithography process, though it is increasingly likely to have been made by exposure

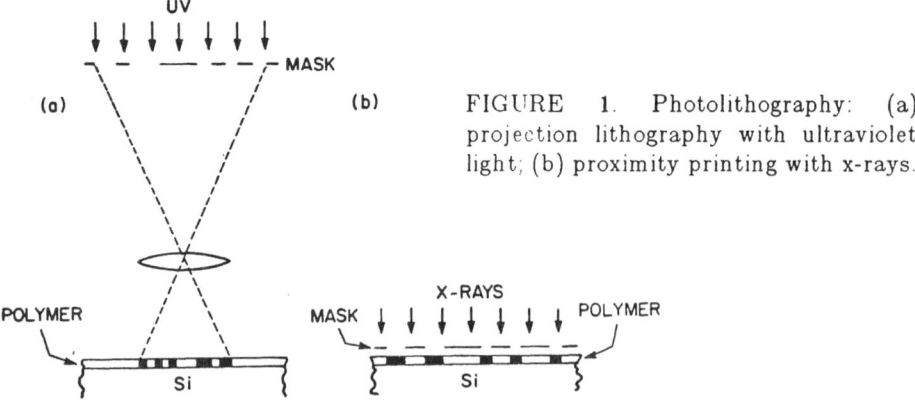

FIGURE 1. Photolithography: (a) projection lithography with ultraviolet light; (b) proximity printing with x-rays.

NEGATIVE RESIST (e.g. PS)

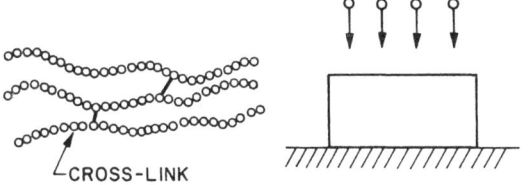

POSITIVE RESIST (e.g. PMMA)

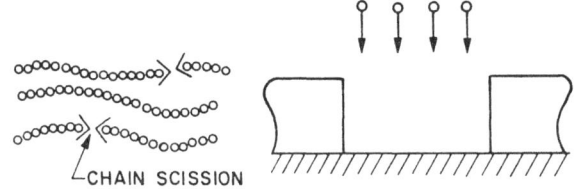

FIGURE 2. The two tones of lithographic resists: (a) negative, in which exposed regions remain after development as in the case of PS (polystyrene); (b) positive in which exposed regions are removed by development as in the case of polymethalmethacralate (PMMA).

of an electron beam sensitive material to a focused and scanned electron beam.(2) In almost all cases the patterns are developed using wet chemistry but multilayer mask materials often involve reactive ion etching to retain resolution in high aspect ratio patterns such as those used in x-ray lithography.

Ion beams are also capable of exposure of polymer (an inorganic) lithographic materials.(3) In fact, there is very high sensitivity in such processes as will be discussed further in section 2. below. Ion beams are not likely to be technologically important in this role, however, at least until the demands of spatial resolution compel a non-optical approach.

It is important to recognize that ion beams play a much wider range of roles in microlithography than as exposure tools for polymer films. Many of these roles are still in very early stages of exploration and development but they can be broadly classified in several major categories.

A. The ion itself is specifically wanted: for example, boron or phosphorus ions for implantation of silicon or Cl ions for reactive ion etching of GaAs.

B. The effect of the ion is wanted:
    B.1 Direct momentum transferring collisions for sputtering
         or defect production.
    B.2 Electronic excitation along the ion tracks.

Categories A and B.1 are almost uniquely the province of ions. The effects in B.2 are different than for electronic excitation by photons, x-rays, or electron beams because the excitations are highly localized, may be very dense and can thus activate processes inaccessible by more lightly ionizing particles. In the sections which follow these properties will be considered in more detail and special implications for microlithography will be discussed.

340

2. ION BEAM EXPOSURE OF ORGANIC RESISTS

Ions have several properties which make them attractive as particles for exposure of organic films. First, their deBroglie wavelengths are extremely short so that diffraction plays no role in the ultimate definition of patterns which they expose. Second, the lateral scattering of ions is low so that the loss of definition that they sustain as they penetrate into a polymer film is small. Fig. 3 illustrates the case for 30 keV H^+ penetrating a 0.1μm film of PMMA overlying a thick silicon substrate.(4) The random trajectories numerically calculated for this figure undergo $<\pm.01\mu$m lateral spreading in the PMMA film. Third, there is very little backscattering from the substrate to broaden the region of exposure in the PMMA. In fact, in the trajectories shown in Fig. 3 there were no large angle backscattered events, in agreement with expectation that $<1\times10^{-5}$ ions will by chance experience sufficiently close encounters to produce backscattering. Fourth, because ions are relatively slow-moving and transfer kinetic energy inefficiently to electrons of the PMMA, there are very few secondary electrons produced in the polymer which have energies larger than $\sim$20eV.(5) Such low energy electrons have short ranges with the result that the exposure of the resist is minimally broadened around the collective tracks of the incident ions. The net effect of points two through four above is that proximity effects (the influence of the exposure in one region on a nearby region) are minimal.

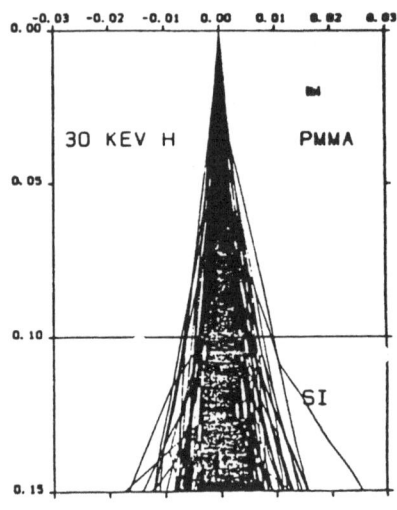

FIGURE 3. Monte Carlo calculation of 30 keV H^+ ions penetrating a 0.1μm thick PMMA film on silicon. The ions are assumed to start as an infinitesimal beam at the entrance face of the PMMA. (after Adesida, et al. (4)).

Finally, ions are very efficient at exposing polymers. Partly this is due to the fact that they lose energy more rapidly as they penetrate a film and that energy deposited is what leads to exposure. However the ion exposure efficiency is typically not simply proportional to the linear energy loss of ions. Fig. 4 clearly illustrates this point for PMMA exposed to H, He and Ar ions.(6) The ion fluence (exposure dose) required for exposure of this positive resist decreases approximately quadratically with increasing linear energy deposition into electronic excitation and ionization (the electronic stopping power). Thus the chemical changes that are stimulated by ionization are dominated by the interaction of two ionization events that take place close together. Extrapolating the non-linearity of Fig. 4 to the electronic stopping power of 20 keV electrons would predict exposure doses for these particles of $\sim 10^{16}$ electrons/ cm^2. Indeed this is approximately what is found.(7) It is interesting to observe that the non-linear behavior applies over such a broad range of electronic stopping powers.

The exposure dose required for the highest stopping power particles of Fig. 4 is only $\sim 10^{11}$ions/cm^2. The statistical consequences of this exposure dose will be discussed in a later section.

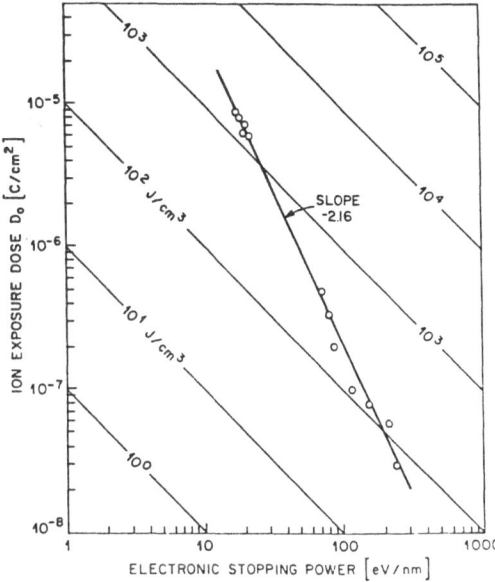

FIGURE 4. The ion dose required to expose PMMA for H, He and Ar ions of different electronic stopping power. Lines of constant volume density of electronic energy deposition are shown in the figure (after Mladenov, et al.(6)).

Fig. 5 compares the conventional expose and develop resist sequence appropriate to PMMA (above) with a more ideal case in which the chemistry of the film when exposed leads to volatile products which self-develop by escaping in the gas phase. Geis and his co-workers (8) have experimented with ion exposure on such a material, nitrocellulose. This polymer has two reaction pathways, shown in

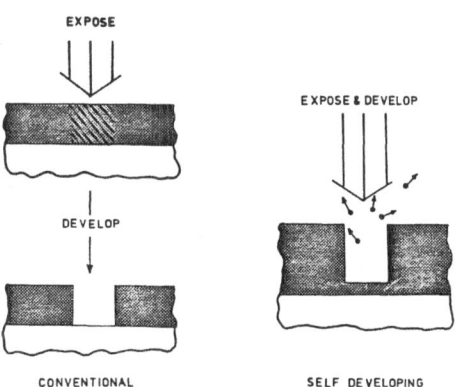

FIGURE 5. Comparison of conventional expose-and-develop resist processing with a self-developing resist process (after Geis (8)).

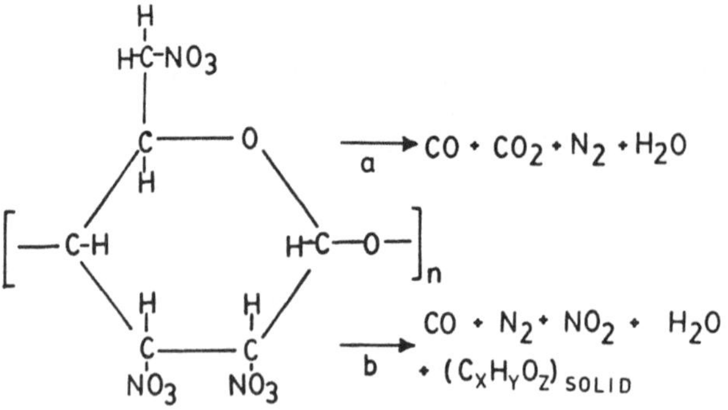

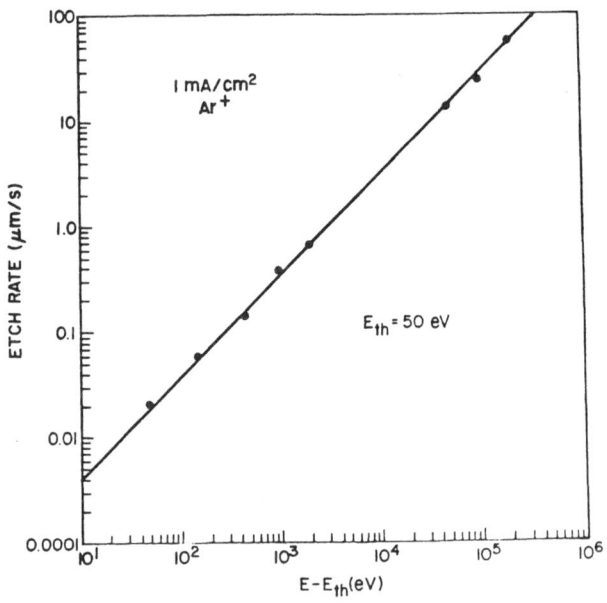

FIGURE 6. The two reaction pathways for nitrocellulose: (a) yielding only volatile products; (b) yielding volatile products but also a non-volatile residue (after Geis (8)).

Fig. 6 one of which gives only volatile products, the other includes non-volatile solids. Clearly, the first reaction path produces a self-developing resist. Ar ion bombardment has been found to result in decomposition of nitrocellulose into totally volatile products.(8) Furthermore, as illustrated in Fig. 7, the exposure development volume per ion is found to be proportional to ion energy,(9) not to the stopping power of the ions (or the square of the stopping power) as might have been expected. This result is not yet understood. Finally, nitrocellulose has been reported to undergo decomposition to form non-volatile residues if the electronic energy density deposited by individual particles is sufficiently low (as it is for electrons, for example). So the clean self-development characteristic does not apply for photons, electrons, or even for high energy protons whose electronic energy loss is low. Once again, the high local density of excitation produced by individual penetrating ions stimulates special chemistry that is advantageous in a lithographic exposure of polymer films.

FIGURE 7. The etching rate for nitrocellulose exposed to 1mamp/cm^2 Ar$^+$ ion beams of different energy. (after Geis (9)).

3. BROAD ION BEAMS AND PARALLEL PROCESSING IN MICROLITHOGRAPHY

Most lithographic processing is carried out using broad and uniform beams exposing 10^8-10^{10} pixels of a two-dimensional pattern simultaneously as is illustrated schematically for ultraviolet and x-ray lithography in Fig. 1. The advantages of parallel processing speed in comparison with serial processing, in which one pixel at a time is processed with a scanning focussed beam, are obvious and enormous. Parallel processing is also the way in which most ion beam processing is carried out. Typically, in either ion implantation or reactive ion etching, a pattern has been defined in a "resist" film on the surface of a substrate. This might be done, for example, by photolithography in a polymer film on a silicon surface. This pattern serves as a mask to limit the regions of the silicon into which dopant ions are implanted or which are selectively attacked by a reactive ion. However, there are several methods still in exploratory states by which a broad ion beam can itself be patterned and no patterned "resist" film is needed on the substrate that is to be processed. These fall into classes of proximity printing or projection printing analogous to the two parts of Fig. 1.

3.1 Proximity Printing

Proximity printing requires a mask that is transparent to the exposing species, in this case, to ions. One possibility is a stencil mask, that is, a mask which has holes completely through the material. Stencils ordinarily do not allow blocking of disconnected regions in a two-dimensional pattern, the center of a ring for example. In addition, stencil masks on a submicron scale are difficult to fabricate. A possible approach to a versatile stencil is that reported by Randall et al.(10) and illustrated in Fig. 8. A background grid of submicron square holes is formed (by lithographic processing) in a 1μm thick silicon nitride film. The desired two-dimensional stencil pattern is then defined by blocking the appropriate arrays of these submicron holes. A perfectly parallel broad beam of ions passing through such a stencil would be patterned with the undesired background grid as well as the desired pattern. However, because of scattering of ions by the ribs of the grid,

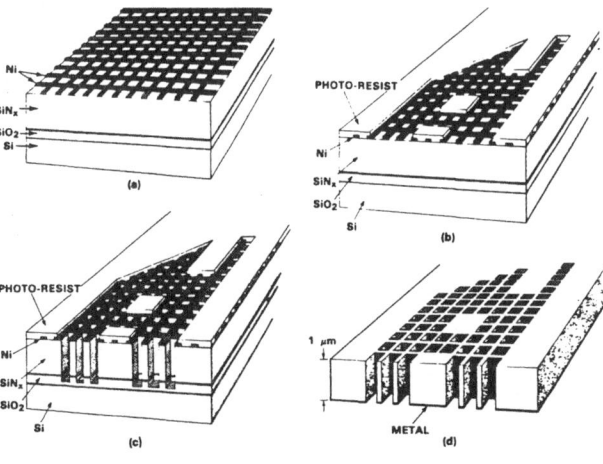

FIGURE 8. A stencil mask technology providing a rigid, uniformly spaced grid on which a pattern is defined by blocked grid squares (after Randall, et al. (10)).

at a short distance beyond the mask the finest features of the beam pattern (the ribs) will be blurred to leave the desired pattern dominant. The spatial uniformity (in angular divergence as well as current density) of the broad ion beam incident on the mask as well as the distance between the mask and the substrate to be exposed are clearly critical parameters in this approach. The background grid structure serves as more than a solution to topographical disconnections; it is also a rigid support that eliminates problems of mask distortion that are dependent on the details of the pattern. This is a difficult mask technology, however, particularly since the spatial scale of the grid pattern must be smaller than the final resolution desired in the patterned beam.

Another approach to proximity printing depends upon channeling of an ion beam through a thin single crystalline membrane.(11) The idea is illustrated in Fig. 9 for patterning a 180 keV proton beam. The open background grid of Fig. 8 is replaced by a 0.7μm thick silicon single crystal with a low index crystal axis normal to the surface. Ions incident parallel to this axis are channeled through the

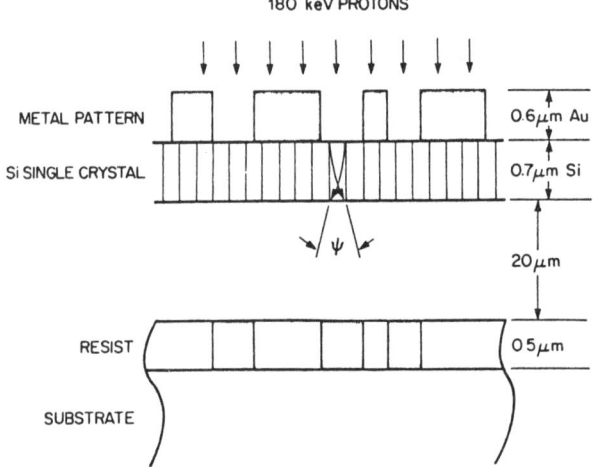

FIGURE 9. Channeled ion lithography illustrated for 180 keV protons. In regions not covered with a 0.6μm Au overlayer the protons, aligned parallel to a low index axis in the single crystal silicon membrane, channel through the membrane and avoid the normal angular broadening due to multiple scattering.

membrane with a reduced energy loss and a greatly reduced lateral scattering. They emerge from the membrane with an angular spread which is related to but less than the critical angle for scattering.(12) The desired two-dimensional patterning of the penetrating beam is produced by a pattern of gold on the upper surface. The gold layer need not totally absorb the beam since it will so broadly scatter ions in those regions that they will not channel in the membrane when they reach it. The highest resolution in the penetrating patterned beam is close behind the mask; the loss of definition grows with the proximity printing distance because of the angular divergence of the outgoing channeled beam.

A submicron pattern produced in PMMA using such a channeling mask is shown in Fig. 10.(11) It is clear that this type of mask technology will never allow broad ion beam exposure of very large areas, whole wafers for example, in current technology. It might however be capable of patterning a beam over the size of individual integrated circuit chips, a large pattern being produced by step and repeat exposure. For ions such as 180 keV protons, the buildup of damage in the mask is very low and a mask has a long life. However, this technique is not attractive for patterning heavy ion beams such as might be desired for dopant implantation. Damage buildup in the mask can be severe, rapidly destroying channeling in the crystal. In addition, the shorter range of heavy ions demands thinner crystalline membranes for penetration.

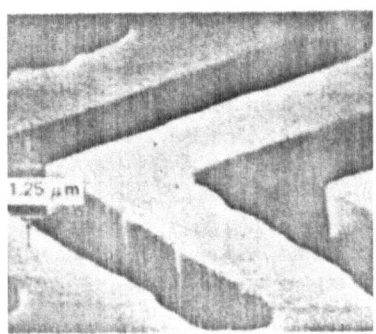 

FIGURE 10. A lithographic pattern produced in PMMA by channeled protons (after Rensch, et al. (11)).

3.2 Projection Printing

A technique for projection printing with ion beams has been investigated with increasing success. A current system is illustrated in Fig. 11, due to Stengle, et al.(13) A broad and uniform ion beam at 5-10 keV floods a stencil mask. The patterned ions from the stencil are accelerated and focused with a 2-lens system to

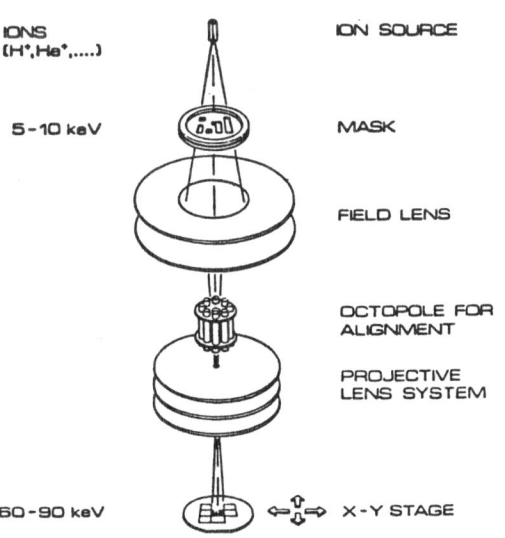

FIGURE 11. A broad beam ion projection printing system with 5-10x linear demagnification. (after Stengl, et al. (13)).

produce a 5-10 times linearly demagnified image of the stencil at an energy of 60-90 keV on the target. The two most critical elements of this system are clearly the mask and the optics. The mask requirements are less severe than those for the 1:1 proximity printing mask discussed in Section IIIa above because of the demagnification. The rib structure can be of larger cell size, for example. The requirements are also less severe because the power loading of the mask by the beam is at a lower power density both because of the demagnification and the lower beam energy at the mask. The ion optical requirements are, however, severe. Retaining high and uniform resolution over a 5-10 mm image is difficult. ·

The principal effort on this system has been in forming H and He ion images for ion damage exposure of masks on wafers. Submicron resolution is clearly evident in Fig. 12.(13) The pattern in this case is in SiO_2, directly bombarded with the patterned beam. Enhanced etching of SiO_2 has been promoted by the ion bombardment as is schematically illustrated in Fig. 13. The enhanced etched rate results in the 5° sidewall taper of the final pattern. Doses required are modest, $\sim 10^{14}/cm^2$, and the potential throughput is thus high. As in the case of channeled beam proximity printing, a step and repeat operation would be required to expose large areas.

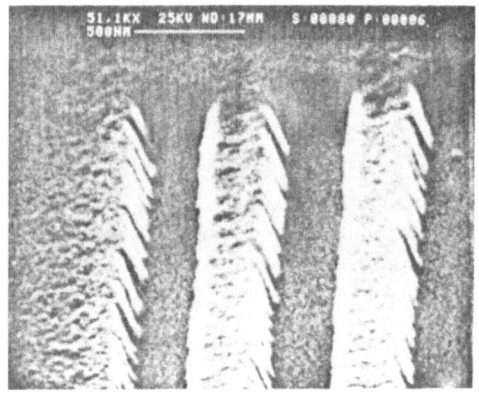

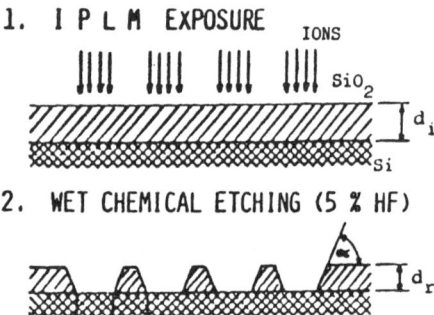

FIGURE 12. A schematic pattern produced by exposure of SiO_2 with the broad beam projection printing system of Figure 11. The damaged SiO_2 etches at an enhanced rate. (after Stengl, et al. (13)).

FIGURE 13. Schematic of Ion Projection Lithography Machine exposure and subsequent wet development leading to the pattern of Figure 12. (after Stengl, et al. (13)).

Of the three methods discussed above for producing broad ion beams patterned with submicron resolution, the projection, demagnification technique is perhaps the most appealing because of its long working distance and its more modest (though still very difficult) requirements for the stencil mask. However, the optical properties of the focusing lenses required to provide $10^8 - 10^9$ pixels (1cm² area with 1.0μm to 0.3μm pixel size) with uniform resolution across the exposure field are extremely demanding. Many potential technological applications depend heavily on its successful demonstration.

4. FOCUSED BEAMS AND SERIAL PROCESSING

For any serious application of focused ion beams in material processing the current density in the focused beam spot is of critical importance. The processing is necessarily serial - one pixel at a time - and hence throughput is inversely proportional to focused current density (the quantitative aspects of this will be discussed in Section 4.2 below). The development of very high brightness ion sources has enormously enhanced the possibilities for technologically significant finely focussed ion beams. These sources will be discussed in Section 4.1 below and the quantitative features (and limitations) of systems using these sources will be discussed in Section 4.2 to follow.

4.1 High Brightness Ion Sources

The new interest in high brightness ion sources is concentrated on two types of source, both of which depend on high electric field to produce ions near a metallic tip through field evaporation or field ionization processes. They are liquid metal ion sources (14,15) and gas field ionization (16,17) sources illustrated schematically in Fig. 14.(3) In a liquid metal ion source (LMIS) a liquid metal wets a solid needle which has an end radius of curvature $\sim 1-5\mu$m. Gallium is one of the simplest LMIS ion sources and for it the needle is tungsten. When a sufficiently large potential is applied to the needle with respect to an annular extraction electrode the electric field forces exceed the surface tension forces of the liquid metal on the needle and the liquid metal is pulled up into a cone. A static solution to the balance of electric field and surface tension forces is a Taylor cone (18) with a full cone angle of 98.6°. The Taylor cone mathematically comes to a point of zero radius at which the electric field would be infinite. However, for fields ~ 1volt/ Å ions are field evaporated from the liquid tip, the Taylor cone is blunted at its end and ions are emitted from a small region of very high field. The details of the steady state of a LMIS involve hydrodynamics, electrostatics, and ion emission and are still unclear. However, the sources emit $\sim$ 2-50 μ amps from a region whose

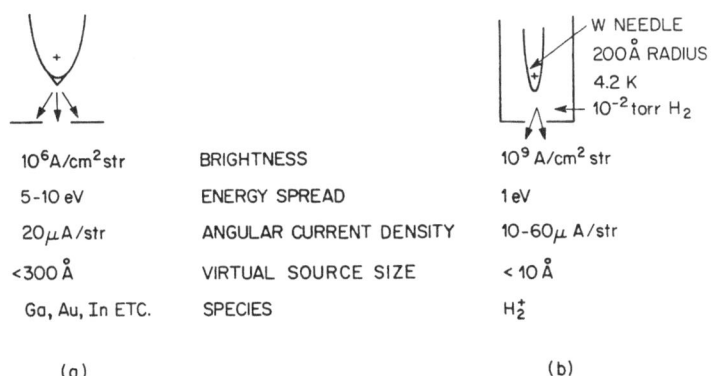

10^6A/cm^{2}str	BRIGHTNESS	10^9 A/cm^2 str
5-10 eV	ENERGY SPREAD	1 eV
20μA/str	ANGULAR CURRENT DENSITY	10-60μ A/str
<300 Å	VIRTUAL SOURCE SIZE	< 10 Å
Ga, Au, In ETC.	SPECIES	H_2^+

(a) (b)

FIGURE 14. Two types of high brightness ion sources: (a) a liquid metal ion source; (b) H_2^+ field ionization source.

virtual source size is $< 1000 \overset{\circ}{A}$. The brightness of such sources is $\sim 10^6$ amps/cm² steradian. They emit with angular current densities $\sim 20\mu$ amps/steradian. More details of these sources will be discussed below.

The H_2 field ionization source has an even higher brightness (16) $\sim 10^9$ amps/cm² steradian at an axial angular current density of 10-60 μ amps/steradian. The source needle has a much smaller tip radius of curvature $\sim 200 \overset{\circ}{A}$ and is operated at a temperature of 4.2 K with H_2 gas at $\sim 10^{-2}$ torr. These conditions provide a high density of H_2 molecules to the highest field region at the tip and molecular ions form by field ionization. The highest brightness occurs where a particularly advantageous structure of metal atoms (maybe only 10 atoms) is formed at the very tip. This may make such sources vulnerable to contamination or to negative ion bombardment and so far very little is known about their operating robustness. These sources have less versatility than LMIS because they operate with H_2. However, their extremely high brightness makes them exciting candidates for lithographic exposure of resists at extremely high resolution or of materials such as oxides or even metals whose chemical reactivity is altered by (presumably) damage produced by hydrogen ions (13).

Liquid metal ion sources have been made with a wide variety of source metals, both elements and alloys. Table I lists some of the major cases. Because LMIS

TABLE 1. Liquid Metal Ions

Ion (Increasing Z)	Liquid Element or Alloy	Reference
Be	Au-Be, Au-Be, Au-Si-Be, Ga-Si-Be, Pd-Si-Be	22, 23, 24, 30, 31
B	Pt-B, Pd-Ni-B, Pt-Ni-B	22, 22, 16
Al	Al	21
Si	Au-Si, Ga-Si-Be, Pd-Si-Be	22, 25, 30, 31
Ni	Pd-Ni-B	22
Cu	Cu	26
Zn	Zn	27
Ga	Ga	14
Ge	B-Pt-Au-Ge	28
As	Pd-Ni-B-As, As-Sn-Pb, As-Pt	22, 28, 29
Pd	Pd-Ni-B	22
In	In	14
Sn	Sn, As-Sn-Pb	26, 28
Sb	Sb-Pb-Au	25, 29
Cs	Cs	26
Pt	Pt-B	22
Au	Au	19
Pb	Pb, Sb-Pb-Au	26, 25
Bi	Bi	20

require a molten metal but also a vapor pressure low enough at the operating temperature to avoid high voltage breakdown, alloys have brought many materials into range that would not be feasible as elements, arsenic and boron for example. In alloy sources all elemental constituents of the alloy are found in the emitted ion beams so that mass selection is essential for utilization of particular ions. Because the total current may be shared between many different species the current at any one species will in general be lower than in elemental (particularly monoisotopic) source materials.

The V-I characteristics of several gallium LMISs are shown in Fig. 15.(19) Sources formed on tungsten needles with tips having different radii of curvature "turn on" at different voltages between tip and extractor electrodes. This behavior is in good agreement with expectations of electric field stress equalling the surface tension of the liquid wetted tip to pull the liquid into a cone and start emission. The shapes of the V-I characteristics reflect different hydrodynamic flow characterists of the tungsten shanks along which liquid flow from a reservoir to the tip. If the source needle is very "smooth" it has a high flow impedance and the source current increases slowly with voltage. If the needle is "rough" (with longitudinal grooves formed by grain boundary etching in the drawn tungsten wire) there is easy liquid flow (low flow impedance) and the source current rises rapidly with voltage.(19) Even in the case of a rough sources, the current rises with voltages much more slowly than for field emission from a solid metal tip, so that flow impedance still has an important influence on the source behavior.

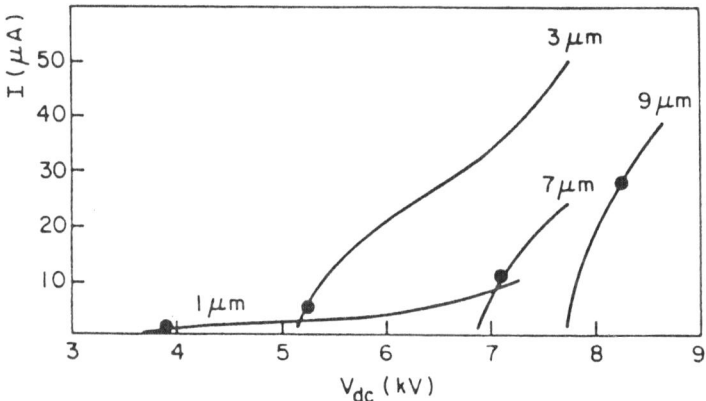

FIGURE 15. Current-voltage characteristics of four different gallium liquid metal ion sources with different tungsten needle tip radii and different roughness to the needle shank. (after Wagner and Hall (19)).

Ions emitted from LMIS have average kinetic energies a little lower than the tip potential (as is true in most field evaporation and field ionization processes). However, there is an energy spread to the ions that is much greater than from a non-liquid source. Full width at half maxima of several different sources is shown in Fig. 16.(20,21) The energy width increases steeply with source current (here plotted as the axial current per unit solid angle) but is ~ 10 electron volts at the lowest currents at which the sources will operate. The origin of this energy spread has been interpreted as due to longitudinal space charge (the Boersch effect)(32) in the very high current density of the ions close to the emitting region of the liquid tip. The measurements of energy spread under a variety of source operating conditions do not agree well (or in some cases at all) with predictions of the simple theory.(33) Nonetheless, the existence of the energy spread is extremely important in the formation of finely focused ion beams. The H_2^+ gaseous source has a much

350

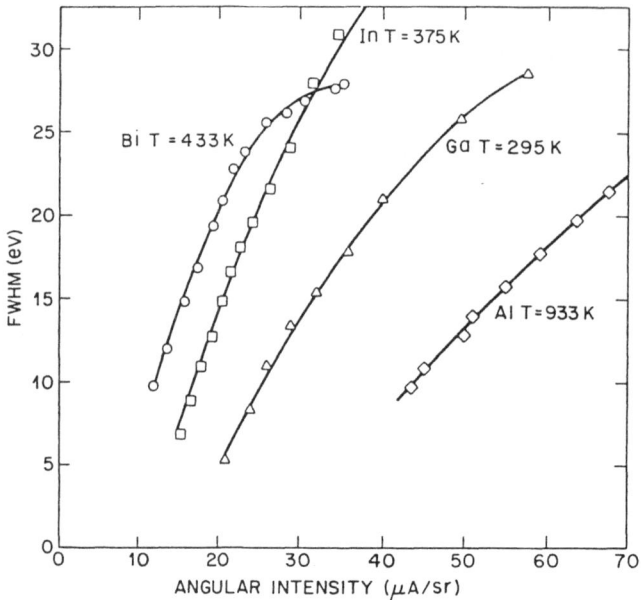

FWHM (eV)

In T = 375 K

Bi T = 433 K

Ga T = 295 K

Al T=933 K

ANGULAR INTENSITY (μA/sr)

FIGURE 16. Energy widths for ion beams from several liquid metal ion sources as a function of the angular current density along the axis of emission. (after Swanson, et al. (20) and Bell, et al. (21)).

narrower energy spread $\sim 2 \, \mathrm{eV}$[16] so that in addition to its intrinsically higher brightness it should be possible to focus its H_2^+ ions to a still higher current density spot. Such results have yet to be reported.

4.2 Finely Focused Ion Beam Systems

Fig. 17 illustrates a simple finely focused ion beam system developed for a liquid gallium ion source.(34) The system measures only 10 cm from the tip of the liquid metal ion source to the target focal plane. Following the ion source extraction electrode there is an aperture that limits the angular divergence of the beam that is admitted to the lens The deflector following the lens is an octapole capable of correcting for astigmatism in the beam. The system is all electrostatic (magnetic optical elements are ineffective for the slow moving heavy ions typical of finely focused ion beam systems). The focal spot size and the current density in the focal spot are dominated in this case by chromatic aberrations of the lens and the energy spread of ions from the liquid metal ion source. This is true for all spot sizes with diameters less than ~ 2 microns. In the chromatic aberration limit the current density i in the focal spot of diameter d is independent of the spot size $d \approx C\alpha(\Delta E/E)$ where C is the chromatic aberration coefficient, ΔE is the energy spread of the ions at average energy E and α is the half angle of the cone of ions accepted by the lens. Then $I = \pi\alpha^2 (dI/d\Omega)$ where I is the total current into the lens and $(dI/d\Omega)$ is the angular current density. This leads to

$$i = \frac{4}{C^2} \frac{(dI/d\Omega)}{(\Delta E/E)^2}$$

It is clear from the final expression that the focal current density is inversely related to the square of the energy spread of ions entering the lens, as well as the square of the chromatic aberration coefficient. Modest improvements in the

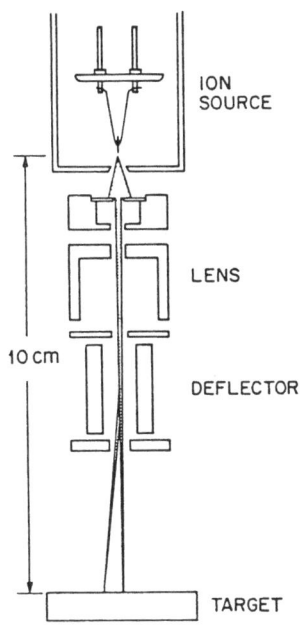

ION
SOURCE

LENS

10 cm

DEFLECTOR

TARGET

FIGURE 17. Schematic of a simple finely focused ion beam system using a single focusing lens and producing approximately unit magnification. (after Wagner, et al. (34)).

effective chromatic aberration coefficient have been made by optimization of lens design,(35) but effort devoted to reducing the source energy spread by optimization of the flow impedance of the source needle has so far yielded no improvement.(33) The net result is focal current density $\sim 1\mathrm{A/cm^2}$ in spot sizes that are as small as 300Å or as large as 1 micron.

It is instructive to calculate the throughput of a finely focused ion beam system with a current density $\sim 1\mathrm{A/cm^2}$. Fig. 18 shows the result. Of course the throughput (for example in $\mathrm{cm^2/hr}$) will be inversely proportional to the fluence (ion exposure/$\mathrm{cm^2}$) desired. It will be proportional to the square of the focal spot diameter when the current density in the spot has a fixed value (as in the chromatic aberration limited case discussed above). The largest throughput line

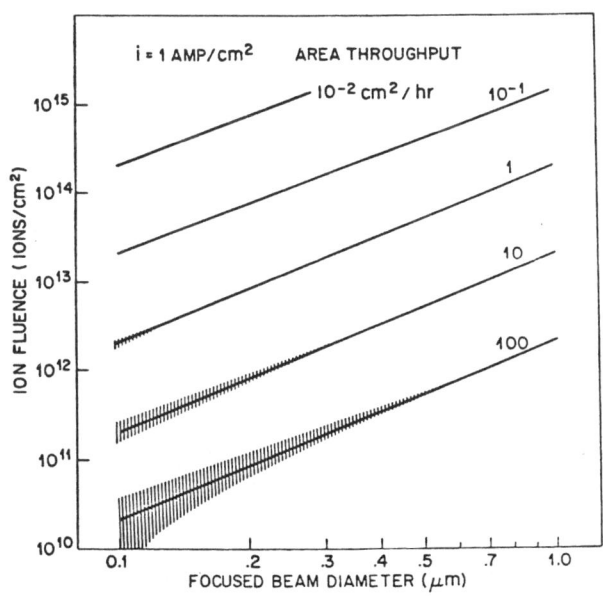

FIGURE 18. Area throughput of a finely focused ion beam system as a function of desired ion fluence and focused beam diameter. A 1amp/$\mathrm{cm^2}$ current density is assumed in the focal spot.

drawn in the figure is for 100cm²/hr, a very modest value with the current technological drive, for example, to use silicon wafers with 100cm² area in a single wafer. There is very little point in developing a focused ion beam technology for use in the .75-1 micron range since material modification methods for this range are already well established. Only at beam sizes ≤0.5 microns is serial ion beam processing likely to be able to make a significant technological impact. At 0.5 micron beam diameter, a throughput of 100cm²/hr can be achieved according to the figure for ion exposure fluences of 3×10^{11}/cm². Fig. 19 shows the approximate ranges of fluence required for a variety of ion beam processes. With the exception of organic resist exposure and the lower end of the implantation doping range, the processes listed require fluences $>10^{13}$ions/cm². For these, throughputs will be at least 30 times lower, 3cm²/hr or less. Highly specialized applications involving very small total processing area are clearly indicated. However, for ion beam exposure of radiation-sensitive materials the required exposure dose can be very low $\sim 10^{10}$ion/cm² and might be even lower if active work specifically on ion beam sensitive resists were undertaken.

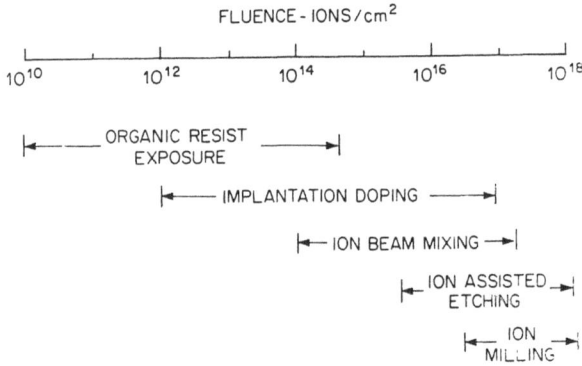

FIGURE 19. Ion beam fluences required for different ion beam processes.

Increasing the sensitivity of materials to ion beam exposure emphasizes another important limitation: statistical exposure fluctuations or shot noise.(3) Since ions are uncorrelated in the ion source or as they reach the focal spot, they arrive randomly in time. Identical exposure times to different pixels of an exposure pattern may actually deliver quite different doses. The fluctuations follow a Poisson distribution. As an illustration, if the focused beam has a diameter of 0.25μm and the average exposure desired is 10^{11}ions/cm², the average number of ions desired in a pixel is only 50. A factor of 2 deviation in the exposure dose may be the maximum acceptable. With 50 ions per pixel as an average, the statistical probability of finding <25 ions in a pixel is 7×10^{-5}, a totally unacceptable error fraction. Higher exposure sensitivity makes required doses lower and the statistical fluctuations still larger. For acceptable failure probabilities the dose per pixel must

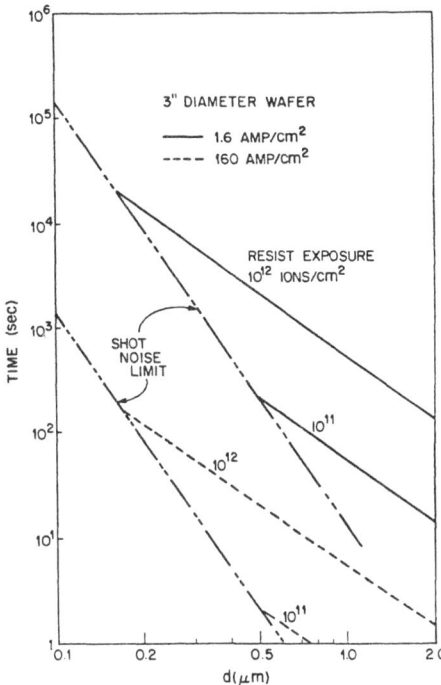

FIGURE 20. The effect of the statistics of shot noise in a focused ion beam current on the exposure time for reliable exposure of 3" wafers. Two required exposure fluences, 10^{11} and 10^{12} ions/cm^2, and two different focused ion current densities, 1.6 and 160 amp/cm^2, are illustrated.

be several hundred to a thousand. The consequence of this shot noise is illustrated in Fig. 20. For the solid lines, corresponding to 1.6A/cm^2 (for numerical simplicity) the time to expose a 3" diameter wafer to an exposure dose of 10^{11} or 10^{12}ions/cm^2 increases with decreasing spot diameter from 2μm as discussed in connection with Fig. 18 above. However, the time increases even more steeply as the two solid curves meet the dot-dashed shot noise line. For spot sizes below this intersection a large enough total number of ions must be delivered to avoid statistical errors no matter what the spot size. This effectively drives the average exposure dose (ion/cm^2) above that required for larger spots. As shown in Fig. 20, for beam diameters $<.2\mu$m there is no benefit in having a resist sensitivity that requires fewer than 10^{12}ions/cm^2. It can not be used with statistical reliability.

The clear conclusion can be drawn from Figs. 18 and 20 that unless the focused current density can be dramatically increased from the $\sim$ 1A/cm^2 value of present focused beam from liquid metal ion sources the time to expose large areas is long and throughput is low. If the H_2^+ gas field ionization source can be developed to provide ~ 100A/cm^2 the benefit for those applications appropriate to H bombardment can be very important. For liquid metal ion sources, efforts to devise achromatic optical systems might have large rewards, since the total current from these sources is $10^4 - 10^5$ times the limited current that can now be utilized because of chromatic aberration. Until or unless large gains in current density are realized, serial lithographic processing by focused ion beams must concentrate on applications that involve *small* treatment areas in *high cost* structures.

5. APPLICATIONS OF FINELY FOCUSED ION BEAMS IN MICROELECTRONICS

5.1 Polymer Exposure for Mask Fabrication

Section 2 above has discussed the exposure characteristics of two very different types of polymers and there are a number of published papers which discuss the phenomenology of many different polymer resists. With the limitation in throughput discussed in Section 4.2 above, it is clear that serial lithographic exposure of resists by focused ion beams can be seriously contemplated for mask fabrication, but not for direct exposure on individual large area device wafers. Primary masks for subsequent broad beam x-ray exposure or ion beam exposure (perhaps in step-and-repeat modes) may benefit from the high resolution and minimal proximity interaction of ion beams in exposure materials.

5.2 Mask Repair

The criteria of applications that involve small treatment areas in high cost structures is well met in the repair of defects in primary lithographic masks. Repair techniques for defects in either x-ray or optical masks have been successfully developed.(36,37) A system constructed for that purpose is illustrated in Fig. 21. Secondary electron images produced by a scanning focused ion beam identify the

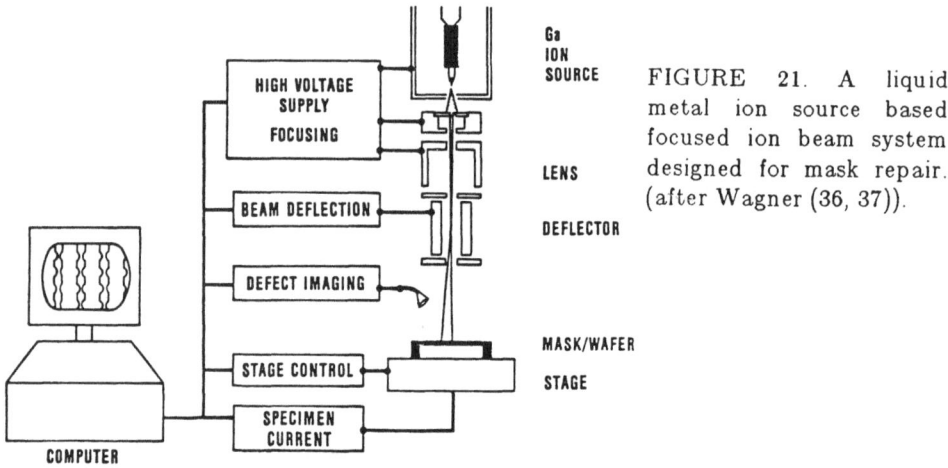

FIGURE 21. A liquid metal ion source based focused ion beam system designed for mask repair. (after Wagner (36, 37)).

defective areas and physical sputtering with the ion beam is used to effect repair. The cross-section of the layer structure in a typical x-ray mask is shown in Fig. 22.

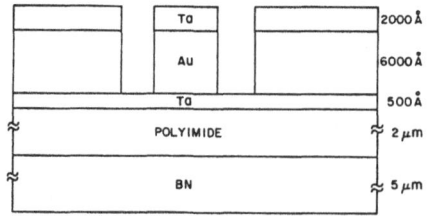

FIGURE 22. A cross section of the layer structure of a typical mask for x-ray lithography.

Defects are primarily due to excess metal in the 8000Å patterned Ta—Au layer. Such a defect is shown in Fig. 23 before and after its removal.(38) The precision of this micromachining is clear from the figure. The time for physical sputtering removal of the heavy mask material is $\sim$ 1sec/μm^2 with a .2μm diameter beam.

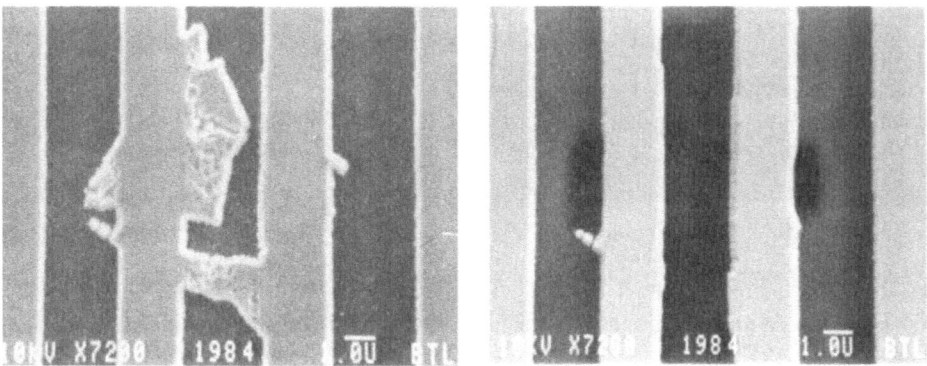

FIGURE 23. A defective x-ray lithographic mask before and after repair with the focused ion beam system illustrated in Figure 21. (after Harriott, Wagner and Fitz (38)).

The cross-section of an optical mask is shown in Fig. 24. Defects in it can be due either to an excess or deficiency of chromium and either can be repaired in masks used for projection lithography. Fig. 25 shows a photograph and a scanning electron micrograph of a mask which has been bombarded in several small

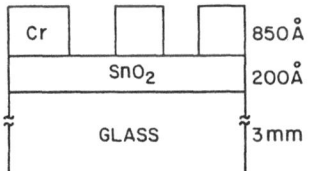

FIGURE 24. A cross section of a mask for optical lithography.

squares.(39) Comparing the photomicrograph and the SEM picture it is clear that originally opaque regions have been made transparent and originally transparent regions made opaque. The former is due to micromachining away excess chromium. The latter, however, is due to micromachining of the glass substrate to form a prism which refracts light illuminating the mask. In optical projection printing, the refracted light is not collected in the optical system and thus the prism region appears dark. To make a prism it is necessary to machine with an ion beam smaller than the total region to be made "opaque" in order to produce a prism topography in the glass.

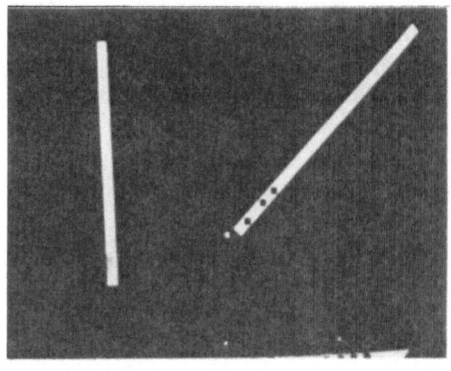 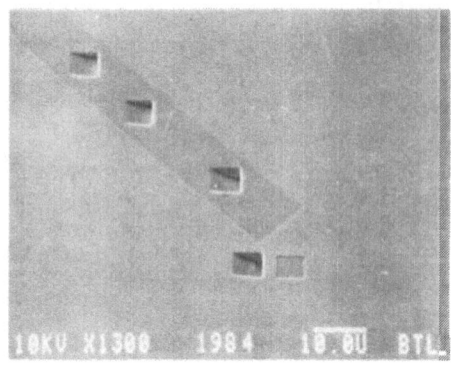

PHOTOMICROGRAPH   SEM

FIGURE 25. A photomicrograph and a scanning electron micrograph of an optical lithographic mask in which originally opaque regions have been made transparent and originally transparent regions have been made opaque using the system of Figure 21. (after Harriott (39)).

5.3 Integrated Circuit Modification

In the development stages of a new microcircuit it may be desirable to make changes in interconnection either because of errors in mask layout or to exercise the flexibility of discretionary connections intentionally provided.(40) Fig. 26a and b shows a scanning ion image and an optical micrograph of a connection before and after it was severed by micromachining with a focused gallium ion beam.(40)

a   b

FIGURE 26. An integrated circuit (a) before and (b) after cutting through a conducting path with the focused ion beam system of Figure 21. (after Harriott (40)).

5.4 Micromachining

The mask repair and integrated circuit modification discussed in sections 5.2 and 5.3 above both involve micromachining with a focused ion beam. Other examples which demonstrate the range of capability of micromachining with finely focused ion beams are shown in Fig. 27, 28 and 29. The precision chevron pattern of Fig. 27 (41) was formed in GaAs with a 100 keV Ga$^+$ beam at a fluence of

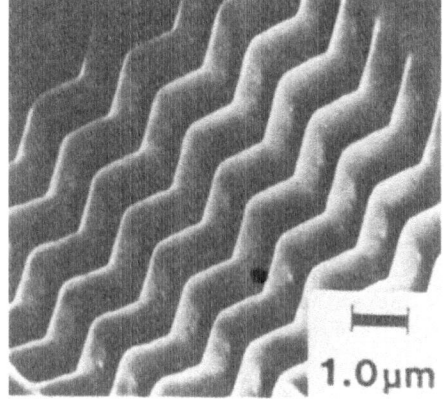

FIGURE 27. A chevron pattern micromachined in GaAs using a 100 keV Ga$^+$ ion beam. (after Kato, et al. (41)).

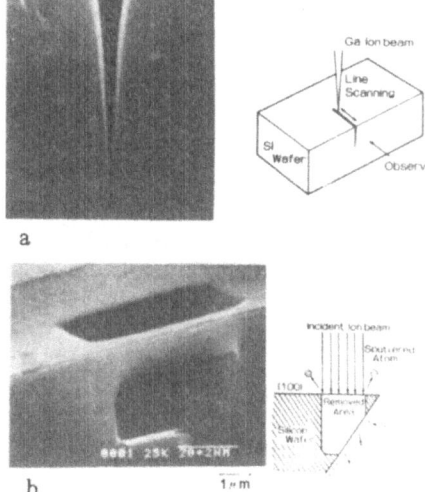

FIGURE 28. Micromachining in silicon with a focused Ga$^+$ ion beam. (after Yamaguchi, et al. (42)).

2.5×10^{17}/cm^2. Fig. 28 (42) shows structures machined in silicon wafers. The cross-section of the sputtered line of Fig. 28a illustrates the tapered sidewalls expected in high aspect ratio grooves because of redeposition of sputtered material. Fig. 28b shows an undercut structure formed by sputtering through to a tapered surface. Fig. 29 (42) shows directly the difference in final topography produced by sputtering depending on the scan mode employed. Redeposition is of dominant importance in the result. This is common experience in sputtering with more conventional ion beams. The net rate of sputtering clearly depends on the details of the structure to be sputtered.

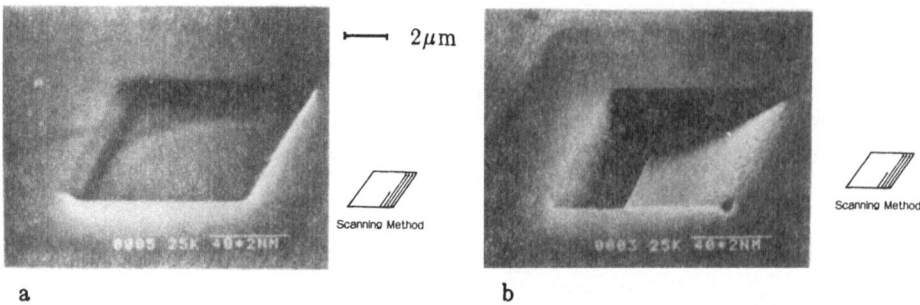

├────┤ 2μm

Scanning Method   Scanning Method

a   b

FIGURE 29. Topography produced by sputtering with a focused Ga$^+$ ion beam in a single raster scan or in multiple scans. Redeposition of sputtered material is of dominant importance. (after Yamaguchi, et al. (42)).

5.5 Ion Implantation

In the examples shown in section 5.1-5.4 above, the chemical identity of the ion being used was of no importance. The ease of operation of a gallium LMIS and the sputtering effectiveness of Ga$^+$ because of its mass made gallium the source of choice. The orders of magnitude dominance of Ga$^+$ compared to all other species from gallium LMIS (Ga$_n^+$ with $n>1$) made it unnecessary to mass select a particular ion to optimize focusing and define penetration depth. However, for ion implantation the particular species is of critical importance. When the dominant species is one of many produced from a liquid alloy LMIS mass selection is essential. Fig. 30 (30,43) is a schematic of a second generation finely focused ion beam system which includes an ExB mass separator between condensing and objective lenses. The focused current densities of Be^{++} are Si^{++} (the dominant Be or Si ions emitted) are an order of magnitude lower (0.1A/cm^2 rather than $\sim$1A/cm^2) than in a simple Ga$^+$ source because of their relatively lower abundance in the source alloy.

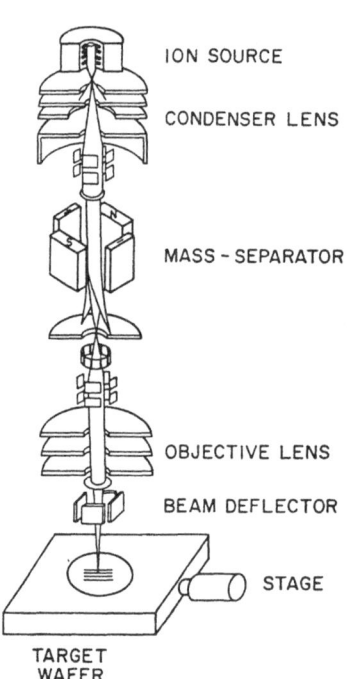

ION SOURCE

CONDENSER LENS

MASS - SEPARATOR

OBJECTIVE LENS

BEAM DEFLECTOR

STAGE

TARGET
WAFER

FIGURE 30. Schematic of a mass selected finely focused ion beam system for use with alloy liquid metal ion sources. (after Miyaguchi, et al. (30) and Anazawa, et al. (43)).

Direct implantation of B⁺ dopant to form transistor structures has been reported by Reuss et al.(44) The cross-section and plan views of their typical devices are shown in Fig. 31. The lateral dimensions in this test structure were 10's of micrometers. Fig. 32 illustrates the emitter-base region and four different lateral base profiles that were implanted with boron using a focused ion beam. The profile control available in this way is unique to focused ion beams and the gain of the transistors fabricated using it showed systematic factors of 2 difference between profiles b and d.

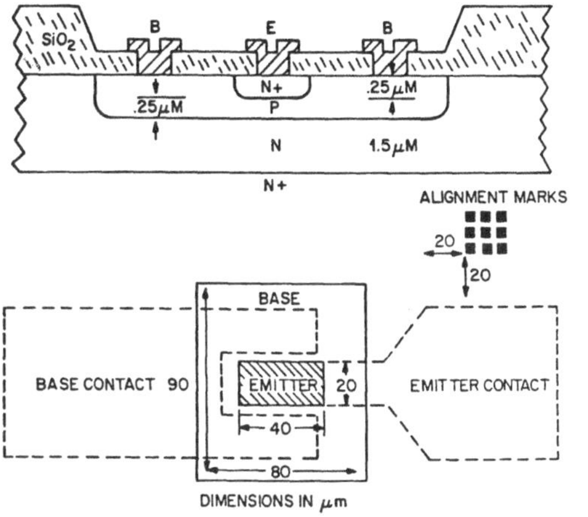

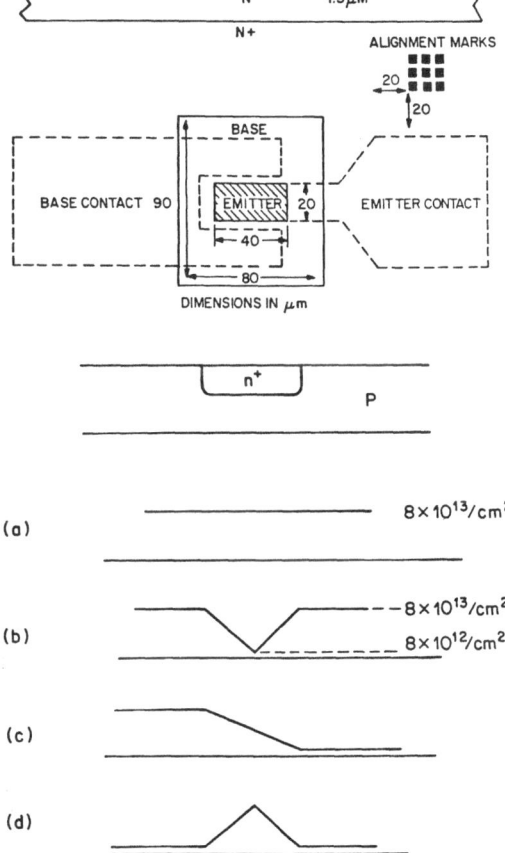

FIGURE 31. Cross section and plan views of transistor structure directly implanted with B⁺ ions. (after Reuss, et al. (44)).

FIGURE 32. Schematic cross section of the emitter-base region in which different lateral doping profiles were produced by B⁺ implantation with a finely focused ion beam

Arimoto et al.(31) have studied submicron isolation by boron implantation into 0.2μm thick layers of n type epitaxial GaAs as illustrated in Fig. 33. Scanning with a 0.1μm B⁺ beam in a line to a variety of final fluences they found resistive

360

isolation of the two sides of the structure for fluences $> 2 \times 10^{14}$ ions/cm². The implanted line did not however retain the 0.1μm width of the focused incident beam and the authors attribute this primarily to scattering of boron ions as they penetrate into the relatively high atomic number GaAs material.

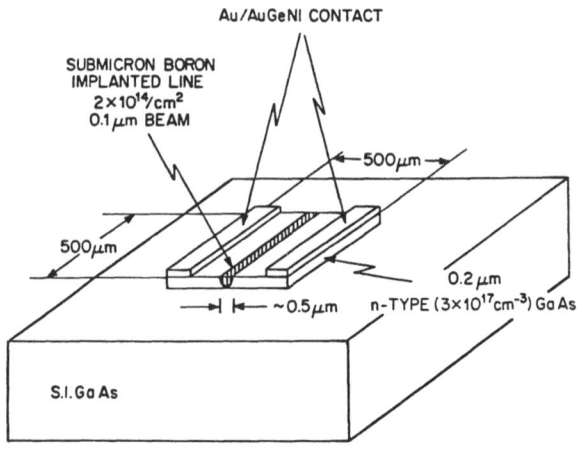

FIGURE 33. A schematic illustration of the geometry used in studies of submicron isolation produced by B⁺ ion implantation in GaAs. (after Arimoto (31)).

5.6 Ion Enhanced Etching

Physical sputtering used in the applications in Section b-d above required relatively large fluences because physical sputtering coefficients are only $\sim 1-10$atoms/ion even when the sputtering geometry is favorable. Experiments to test the feasibility of enhancing of the rate of removal of material have been reported by Ochiai et al.(45) They have used a Ga⁺ focused ion beam on GaAs in the presence of different partial pressures of Cl₂ gas as illustrated in Fig. 34. A small aperture for entrance of the Ga⁺ ion beam provides for a differential pressure between the etching cell and the ion source region. The etching rate is a function of the Cl₂ pressure as shown in Fig. 35.(45) The effective sputtering yield increases from $\sim .2$ cm³/ma min without chlorine to a maximum about 10x that large at ~ 20millitorr Cl₂ pressure. The falloff at higher pressures is believed due to

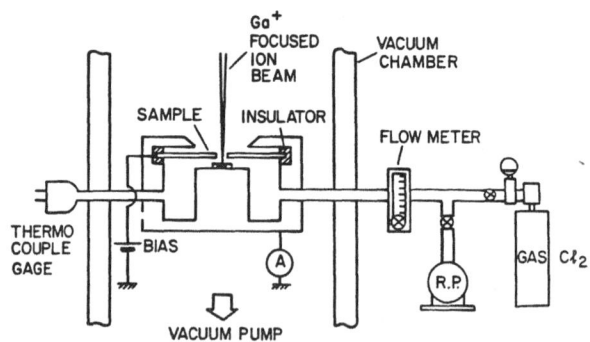

FIGURE 34. Setup for studies of ion enhanced etching of GaAs using a focused Ga⁺ ion beam and Cl₂ gas. (after Ochiai, et al. (45)).

buildup of Cl_2 on the surface which impedes removal of volatile reaction products formed at the bombarded $GaAs-Cl_2$ interface. The ion assisted etching has an important benefit in addition to reaction speed. The formation of volatile reaction products enhances their removal rate from the reaction region and reduces the effect of secondary deposition of "sputtered" material.

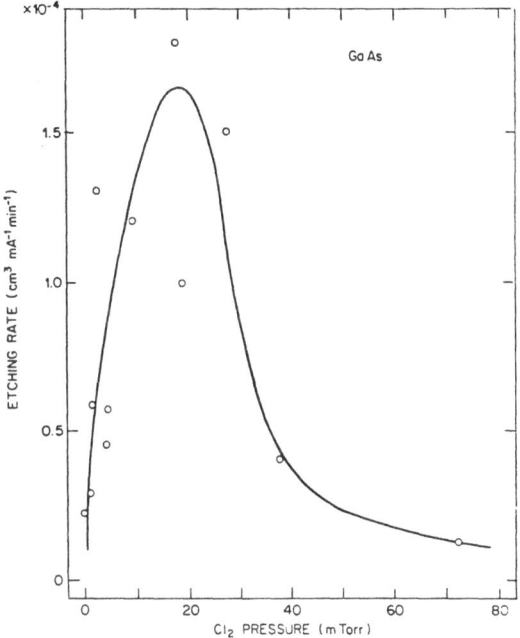

FIGURE 35. The enhanced etching rate observed using the setup of Figure 34. (after Ochiai (45)).

5.7 Scanning Ion Microscopy

Among the array of effects that are associated with focused ion beams scanning over the surfaces of solids, two provide effective diagnostic tools, the emission of secondary electrons and ions. Because focused ion beams can have very small diameters and because slow-moving ions do not excite electrons far from their individual tracks, the spatial resolution of a scanning ion microprobe (detecting ejected electrons) can be impressive. The same incident ions also eject sputtered secondary ions whose identification and detection leads to a high resolution ion scanning microprobe. Levi-Setti (46) has reported results obtained from such an instrument. Fig. 36 shows photographs of typical results. Fig. 36a and b are scans from an integrated circuit, part a taken detecting all secondary ions and part b detecting only $^{27}Al^+$ ions. Fig. 36c shows $^{35}Cl^-$ ions emitted in sputtering of antimony pentachloride intercalated graphite. Using focused ion beams in this type of serial scanning probe mode is a natural exploitation of their special properties.

The principal limitations in using very high resolution for secondary ion microscopy is the small number of secondary ions available from a small area and a thin film. In a $0.1\mu m$ probe diameter there are only 10^5 atoms/monolayer. An ion

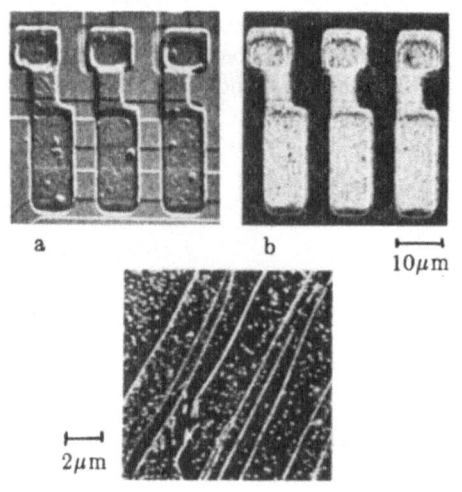

a   b

10μm

2μm

c

FIGURE 36. Secondary ion images produced by sputtering of an integrated circuit with a finely focused Ga+ beam: (a) detecting all secondary ions and (b) detecting ^{27}Al$^+$ only. (c) is the image of graphite intercalated with antimony pentachloride detecting only ^{35}Cl$^-$ ions. (after Levi-Setti (46)).

fraction of 1% and a collection efficiency of 10% lead to an estimate of ∼ 100 collected ions per pixel per monolayer. If major film constituents are being measured this is statistically satisfactory. However, if constituents with abundances $\leq 1\%$ are of interest a monolayer is too thin, and/or the probe area is too small to yield enough secondary ions to be measured quantitatively. These considerations lead one to study grain boundary segregation phenomena for example where impurities concentrate and can be associated with structural defects that can themselves be seen in secondary electron emission. When a finely focused cesium ion beam is more readily available, the negative ion yield of secondaries should be enhanced substantially.

5.8 Formation of Conductors

A final application of finely focused ion beams in microlithography is in the formation of conducting two dimensional patterns by scanning ion bombardment. These regions can be carbon,(47) the residue from ion bombardment dissociation of polymers for example. This is a limiting stage of contamination lithography(48) that has been studied using scanning electron beams. However, the required ion fluence is much lower than the equivalent electron fluence because of the higher energy deposition rate of ions. Required ion fluence is lower still because of nonlinear effects in the heavily ionizing tracks of heavy ions that enhance the decomposition process. Furthermore, the final conductance of carbonized polymers is much higher for ion beam than for e-beam irradiation.(47)

Conducting patterns can also be made from ion bombardment of metal containing organic materials such as palladium diacetate.(49) The ions break metal-organic bonds and free the organic constituents as volatile compounds. The residue is heavily enriched in palladium and conducting. The nonlinear enhancement of decomposition in ion bombardment makes this process much more attractive than for scanning electron beams.

6. SUMMARY

Ion beams already play central roles in the modification of semiconductor, insulator and metallic materials used in current electronic technology. Often these modifications are carried out in two dimensional patterns with very high spatial resolution. In present applications, however, the ion beams themselves rarely define the two dimensional patterns. Instead, broad ion beams, for ion implantation or reactive ion etching for example, flood a material whose surfaces are covered typically by a polymer film that has been patterned by optical lithography. Several techniques for directly patterning broad ion beams with stencil masks are under investigation. These may permit either proximity printing or projecting printing with high resolution.

On the other hand, high intensity focused and scanned ion beams provide for direct patterned writing at resolutions of 0.1 micrometers. Consideration of writing speed at these high resolutions directs the attention of this approach to processing of small total areas, for example as in the repair of defects in masks for x-ray or optical lithography or in alteration of discretionary interconnections in development of integrated circuits.

The broad range of material modification mechanisms that ion beams provide through either the specific chemical nature of a particular ion or through the momentum transfer or electronic excitation associated with ion penetration into materials make them many faceted tools in the continuing evolution of microfabrication techniques. Patterned ion beams or scanning focused ion beams have special roles to play within that evolution.

REFERENCES

1. Optical Microlithography II, Technology for the 1980s, Stover HL, Editor, *Vol. 394* SPIE Bellingham, WA 1983.
2. Electron Beam Technology in Microelectronic Fabrication: Brewer GR(ed), Academic Press, New York 1980.
3. Brown WL, Venkatesan T, Wagner A, Nucl. Instru. & Meth. *191*, 157, 1981.
4. Adesida I, Kratchmer E, Wolf ED, Muray A, Issacson M, J. Vac. Sci. Technol *B 3*, 45, 1985.
5. Hall TM, Wagner A, Thompson LF: J. Vac. Sci. Technol. *16*, 1889, 1979.
6. Mladenov GM, Emmoth B: Apply. Phys. Lett. *38*, 12, 1981.
7. Greeneich JS: Electron Beam Technology in Microelectronic Fabrication: Brewer GR(ed): Academic Press, New York 1980 pg 100.
8. Geis MW, Randall JN, Deutsch TF, DeGraff PD, Krohn KE, Stern A: Appl. Phys. Lett. *43*, 1, 1983.
9. Geis MW, Randall JN, Deutsch TF, Efremow NN, Donnelly JP, Woodhouse JD, J. Vac. Sci. Technol. *B 1*, 4, 1983: Geis MW, Randall JN, Mountain RW, Woodhouse JD, Bromley EI, Astolfi DK, Economou NP: J. Vac. Sci. Technol. *B 3*, 1, 1985.
10. Randall JN, Flanders DC, Economou NP, Donnelly JP, Bromley EI, J. Vac. Sci. Technol. *B 3*, 58, 1985.
11. Rensch DB, Seliger RC, Csonsky G, Olney RD, Stover HL: J. Vac. Sci. Technol. *16*, 1897, 1979.
12. Feldman LC, Bonderup E: Private Communication

13. Stengl G, Loschner H, Maurer W, Wolf P: Energy Beam-Solid Interactions and Transient Thermal Processing 1984, (eds) Biegelsen DK, Rozgonyi BA, and Shank CV; Materials Research Society Symposium Proceedings *35*, 1984 533.

14. Clampitt R, Jefferies DK: Nucl. Instr. and Meth. *149*, 739, 1978.

15. Seliger RL, Ward JW, Wang V, Kubena RL: Appl. Phys. Lett. *34*, 310, 1979.

16. Hanson GR, Seigel BM: J. Vac. Sci. Technol *16*, 310, 1979.

17. Orloff J, Swanson LW: J. Vac. Sci.Technol. *12*, 1209, 1975; J. Appl. Phys. *50*, 6026, 1979.

18. Taylor G: Proc. Roy. Soc. *A280*, 383, 1964.

19. Wagner A, Hall TM: J. Vac. Sci. Technol. *16*, 1871, 1979.

20. Swanson LW, Schwind GA, Bell AE, Brody JE: J. Vac. Sci. Technol. *16*, 1864, 1979.

21. Bell AE, Schwind GA, Swanson LW: Appl. Phys. *53*, 4602, 1982.

22. Wang V, Ward JW, Seliger RL: J. Vac. Sci. Technol. *19*, 1158, 1981.

23. Kubena RL, Anderson CL, Seliger RL, Jullens RA, Stevens EH: J. Vac. Sci. Technol. *19*, 916, 1981.

24. Miyauchi E, Hashimoto H, Utsumi T: J. Appl. Phys. *22*, L225, 1983.

25. Gamo K, Ukegawa T, Inomoto Y, Ochiai Y, Namba S: J. Vac. Sci. Technol. *19*, 1181, 1981.

26. Prewitt et al.: Dublier Scientific Inc., Abingdon, Oxon, England.

27. Okutani T, Fukuda M, Noda T, Tamura H, Watanabe H, Sheperd C: 17th Symposium on Electron, Ion and Photon Beam Technology, Los Angeles, California, May 1983.

28. Gamo K, Ukegawa T, Inomoto Y, Ka KK, Namba S: Jap. J. Appl. Phys. *19*, L595, 1980.

29. Gamo K, Inomoto Y, Ochiai Y, Namba S: Proceedings of the 10th Conference on Electron and Ion Beam Science and Technology, 83-2: The Electrochemical Society, Pennington, NJ 1982.

30. Miyauchi E, Arimoto H, Hasimoto H, Furuya T, Utsumi T: Jap. J. Appl. Phys. *22*, L287, 1983.

31. Arimoto H, Takamori A, Miyauchi E, Hasimoto H: J. Vac. Sci. Technol. *B 3*, 54, 1985.

32. Boersch H: Z. Physik *139*, 115, 1954.

33. Papadopoulos S, Barr DL, Brown WL, Wagner A: Journal de Physique, Colloque C9, supplement au n12, *45*, 217, 1984.

34. Wagner A, Barr DL, Vekatesan T, Crane WS, Lamberti VE, Tai KL, Vadimsky RG: J. Vac. Sci. Technol. *19*, 1363, 1981.

35. Kurihara K: J. Vac. Sci. Technol. *B 3*, 41, 1985.

36. Wagner A: Solid State Technology/May 1983, 97. Solid State Technology/May 1983.

37. Wagner A: Applications of Focused Ion Beams: Nucl. Instru. and Methods *218*, 355, 1983.

38. Harriott LR; private communication.

39. Harriott LR, Wagner A; private communication.

40. Harriott LR, Wagner A, Fritz F, to be published J. Vac. Sci. Technol. Jan/Feb. 1986.

41. Kato T, Morimoto H, Saitoh K, Nakata H: J. Vac. Sci. Technol. *B 3*, 50, 1985.

42. Yamaguchi H, Shimase A, Haraichi S, Miyauchi T: J. Vac. Sci. Technol. *B 3*, 71, 1985.

43. Anazawa N, Aihara R, Ban E, Okunuki M: SPIE, 393 1983.

44. Reuss RH, Morgan D, Greeneich EW, Clark WM, Rensch DB: J. Vac. Sci. Technol. *B 3* 82, 1985.

45. Ochiai Y, Gamo K, Namba S: J. Vac. Sci. Technol. *B 3*, 67, 1985.

46. Levi-Setti R, Crow G, Wong YL, Scan. Elec. Micros. *2*, 535 1985.

47. Venkatesan T, Forrest SR, Kaplan ML, Murray CA, Schmidt PH, Wilkens BJ: J. Appl. Phys. *54*, 3150, 1983: Kaplan ML, Forrest SR, Schmidt PH, Venkatesan T: J. Appl. Phys. *55*, 732, 1984.

48. Broers AN, Molzen WW, Cuomo JJ, Wittels ND: Appl. Phys. Lett. *29*, 596, 1976,

49. Eskildsen SS, Sorensen G: Appl. Phys. Lett. *46*, 11, 1985.

IN SITU ION BEAM PROCESSING FOR JOSEPHSON JUNCTION FABRICATION

J. Salmi
Semiconductor Laboratory, Technical Research Centre of Finland,
SF-02150 Espoo, Finland

J. Knuutila, R. Mutikainen
Low Temperature Laboratory, Helsinki University of Technology,
SF-02150 Espoo, Finland

1. INTRODUCTION
The use of ion beam processing in fabricating Josephson junction tunnel devices introduces several advantages over other plasma discharge techniques both in substrate precleaning and oxidation. The ion flux and ion energy can be controlled independently, the energy spread is narrow, the source plasma is isolated from the substrates and chamber, and the whole procedure is carried out in low pressure environment, thus reducing contamination due to background gas and backscattering. Ion beam oxidation (1) has been used to produce both room temperature diodes (2) and Josephson junctions (3-8). In this work Nb-NbO$_x$-PbInAu Josephson junctions were fabricated with ion beam oxidation preceded by a fixed ion beam cleaning step. The effects of the oxidation time and the ion energy in the oxidation beam on junction characteristics were studied.

2. EXPERIMENTAL PROCEDURE
The junction fabrication was carried out in an e-gun evaporation system with a Kaufman type ion beam source (Ion Tech 2.5 cm source) (Fig. 1). The system was evacuated using turbomolecular -, cryo - and ion pumps to pressures in the 10^{-6} Pa range. Samples were mounted on rotating substrate holders, each of which could be turned between rest, ion beam etch and evaporation positions. The angle of incidence of the ion beam was normal to the substrate, and the source-substrate distance was 65 mm.

The ion beam process consisted of Ar (99.998 %) preclean and Ar/5 % O$_2$ (99.998 %) oxidation. For better uniformity of the beam current density on the substrate, the substrate holders were spun around their symmetry axis displaced 21 mm from the center of the beam. The ion source cathode and neutralizer filaments were made of Ta in order to reduce contamination (8). Both for preclean and oxidation steps the beam current was 2.00 mA and the total pressure 10 mPa yielding an estimated effective current density of 11-17 μAcm^{-2}. Ion process time was controlled by a shutter.

The 10×10 μm^2 junctions were fabricated on oxidized 3-in Si wafers. A Nb layer was first evaporated on substrates held at about 400°C at a pressure ranging between $1.1 \cdot 10^{-5}$ and $1.3 \cdot 10^{-4}$ Pa. The room temperature resistivity of an approximately 300nm thick Nb layer (deposited typically at 2-3 nms^{-1}, $T_c \approx 8.7 \pm 0,3$ K) was 24 $\mu\Omega$cm. The Nb electrodes were patterned using reactive ion etching (RIE) with SF$_6$ using Shipley AZ1470 photoresist. After that, a SiO$_x$ layer was deposited for junction window definition and patterned using chlorobenzene based lift-off. Before evaporating the 300 nm SiO_x layer at a deposition rate of about 1 nms^{-1} the substrate was ion beam etched with 500 eV Ar for 5 minutes. After patterning of the counterelectrode lift-off photoresist, the system was pumped down to about 10^{-5} Pa and the substrates were precleaned for 20 minutes with 500 eV Ar ion beam to remove 11-17 nm from the Nb surface. The preclean was done sequentially for each substrate (5 min etch, 10 min rest), resulting in a temperature increase of $\sim56^\circ$C. The substrates were then left to cool to room temperature for more than 8 hr. Just before oxidation a 2 min etch was done with the same parameters, and the gas was changed to premixed $Ar/5\%O_2$ within some minutes. Oxidation times of 2-20 min and beam energies of 40-500 eV were used. The PbInAu (9) counterelectrode was deposited in the same chamber after pumping down to a pressure less than 10^{-4} Pa. PbInAu was evaporated as a layered structure with PbIn from an alloy of Pb(86wt-%) and In(14 wt-%) with 0.7-0.9 nms^{-1} and Au with 0.1-0.2 nms^{-1} (PbIn 300 nm/Au 7.8 nm/PbIn 300 nm/Au 7.8 nm). A passivation layer of SiO_x(~30 nm) was finally deposited. Another PbInAu layer was processed in a similar manner to form contacts to Nb pads for measurement of the junction parameters.

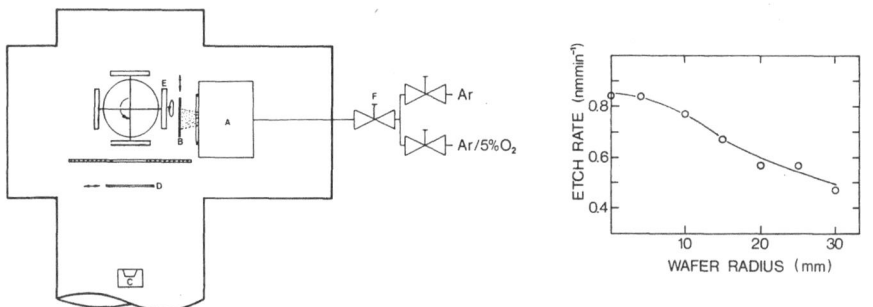

Figure 1. Experimental arrangement of the vacuum system: A) ion source, B) ion beam shutter, C) e-gun evaporator, D) evaporation shutter, E) wafer holder, F) gas inlet.
Figure 2. Spatial etch rate profile of Nb across a wafer with 500 eV Ar ion beam.

3. RESULTS AND DISCUSSION

In order to evaluate the effective current density on the wafer, the current density in the center of the beam vs. the beam total current was measured with a biased electrode. The etch rate for Nb with 1.35 $mAcm^{-2}$ and 500 eV Ar beam was found to be 67 $nmmin^{-1}$. The etch rate of the wafer had cylindrical symmetry as shown in Fig. 2 in the case of a Nb preclean step. If linear scaling between the beam current density and the etch rate is assumed, the effective current density is about 11-17 μAcm^{-2} in the central 50 mm diameter area of the wafers used for measurements.

The properties of the junctions were measured at 4.2 K with the test chips directly immersed in liquid helium. Typical high current density I-V characteristics are shown in Fig. 3. The gap voltage V_g of 2.6 mV is in agreement with other results (10) showing consistency in the PbInAu deposition technique. The maximum dc Josephson current I_1 was inferred from the normal-state resistance R_{nn} (at 5mV) using the relationship $I_1 = (1.6 \text{ mV})/R_{nn}$. Junction quality factors $R_{nn}I_c$ and $V_m = R_jI_1$, where I_c is the measured critical current and R_j the linear tunnel resistance at 2 mV (10), were then evaluated. The values of $R_{nn}I_c$ are in the range from 1.3 to 1.6 mV. The low V_m values of about 5-10 mV suggest that the Ar preclean step does not form any carbide layer on the Nb surface (5). The measured critical currents do not show any systematic spatial dependence across the wafer.

The dependence of the critical current density J_1 on oxidation ion beam energy is shown in Fig. 4. With ion energies between 40 and 100 eV, J_1 decreases with increasing energy. This general behaviour is in agreement with results obtained by using rf oxidation with low ion energies (10). However, samples made with a 500 eV oxidation beam show increased values of J_1. This may be attributed to the increased sputter etching rate, as a competitive process for oxide growth.

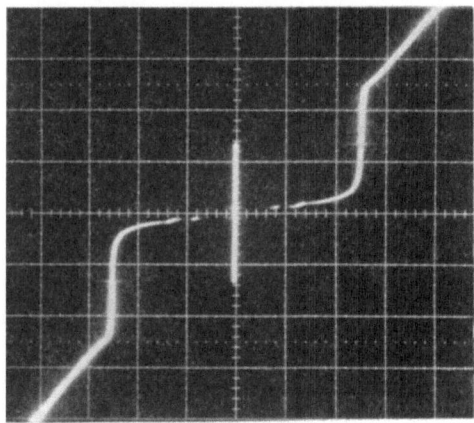
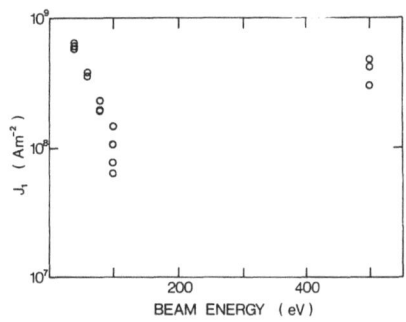

Figure 3.   I-V characteristics of a high current density 10×10 μm^2 Nb-NbO$_x$-PbInAu junction (vertical scale: 2.0 mA/div, horizontal scale 1.0 mV/div).

Figure 4.   Junction critical current density vs. Ar/5% O_2 ion beam energy with fixed preclean procedure, oxidation beam current and oxidation time.

Fig. 5 shows the dependence of J_1 on the oxidation time with two different ion energies. Saturation of J_1 with increasing time is clearly seen for 100 eV points, and thus does not fit the linear dependence on the oxygen dose as shown in Ref. 5. Neither can the saturation behaviour be fitted to simple exponential time dependence as derived from the competition between oxidation and sputter etching (1). The shift of J_1 towards higher values with ion energies lower than 100 eV is still seen with increased oxidation time.

4. CONCLUSION
Ion beam processing offers a controllable and clean method which can be applied to the fabrication of Josephson devices. The future process should include an optimized carbide step after the preclean, possibly with a

mixture of Ar/CH$_4$ (5) to decrease the leakage current. Moreover, the spatial differences of the ion beam current density on the wafers can be further minimized to decrease the J$_i$ spread.

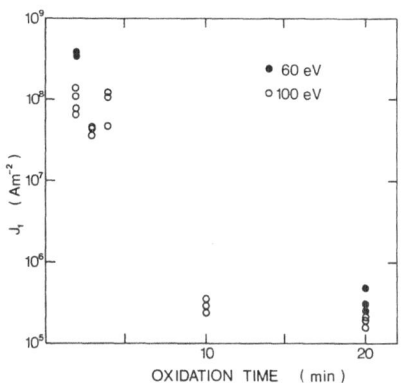

Figure 5. Junction critical current density vs. Ar/5% O$_2$ ion beam oxidation time with fixed preclean procedure and oxidation beam current.

REFERENCES
1.   Kleinsasser, A.W.; Harper, J.M.; Cuomo, J.J., and M. Heiblum. Thin Solid Films 95(1982) 333-342
2.   Harper, J.M.; Heiblum, M.; Speidell, J.L., and J.J. Cuomo. J. Appl. Phys. 52(1981) 4118-4121
3.   Kleinsasser, A.W. and R.A. Buhrman. Appl. Phys. Lett. 37(1980) 841-843
4.   Kleinsasser, A.W. J. Appl. Phys. 57(1985) 2575-2582
5.   Kleinsasser, A.W. IEEE Trans. on Mag. MAG-17(1985) 130-133.
6.   Herwig, R. Electron. Lett. 16(1980) 850-851
7.   Herwig, R. Proceedings of 17th Intl. Conf. on Low Temp. Physics, Karlsruhe, Germany (Eds. U. Eckern, A. Schmid, W. Weber, an H. Wühl, North-Holland, Amsterdam, 1984) 209-210
8.   Pei, S.S. and R.B. Dover. Appl. Phys. Lett 44(1984) 730-705
9.   Lahiri, S.K. and S. Basavaiah. J. Appl. Phys. 49(1978) 2880-2884
10. Broom, R.F.; Raider, S.I.; Oosenbrug, A.; Drake, R.F., and W. Walter IEEE Trans. Electr. Dev. ED-27(1980) 1998-2008

EFFECT OF ION IMPLANTATION ON SUBSEQUENT EROSION AND WEAR BEHAVIOR OF
SOLIDS

C.J. MCHARGUE
Oak Ridge National Laboratory, Oak Ridge, Tennessee   37831

ABSTRACT

The removal of material from a solid surface by mechanical forces is
influenced by material properties (hardness, fracture toughness, yield
strength, surface free energy) as well as system parameters (force, velocity of loading, environment). Ion implantation can modify many of the
material properties either by directly affecting the deformation characteristics or indirectly by affecting the chemical or phase composition at
the surface. The various forms of wear and erosion are analyzed to determine the material and system parameters which control material removal.
The effects of implantation on these critical parameters are noted and
examples of changes in surface topography under various test conditions are
discussed.

1. INTRODUCTION

Ion implantation has received much publicity as a technique for
improving the resistance to wear and erosion and for decreasing the friction of material surfaces. The success of the first studies at Harwell
(reviewed in reference 1) resulted in the study of ion implantation into a
variety of materials but the observed wear and erosion behavior of
implanted specimens was erratic. Eventually it was realized that a
knowledge of the total metallurgical system, that is, the implant species,
the substrate material, and the test environment must be used in designing
the implantation conditions.

Erosion (wear) phenomena are not material properties but are characteristic of an engineering system which involves a complex interplay of
material properties, test procedure, and environment. For ease of
discussion, wear processes can be divided into four main types: (1) adhesive wear, (2) abrasive wear, (3) corrosion or oxidative wear, and (4) surface fatigue or fracture. Each of these may involve several steps which
depend directly upon material properties, and thus may be modified by ion
implantation. In many real systems, two or more wear processes may occur
simultaneously. Since wear is a manifestation of the mechanical and chemical processes that occur at solid interfaces, the specific test procedure
may strongly influence the conclusion drawn from laboratory or accelerated
testing. Many of these features of wear (and erosion) have been recently
reviewed by Yust (2).

This paper will first review the characteristics of the various wear
processes in order to identify the material properties important to each,
and discuss which of these properties might be altered by ion implantation.
A limited number of examples will be drawn from the literature to

illustrate the topographical features of implanted metals and ceramics, and some directions for future research will be mentioned.

2. FRICTION

Historically, the study of friction can be traced to very early periods. Leonardo de Vinci is known to have had an interest in friction. In 1699 Ammontons performed experiments to seek the origin of friction in machines. Coulomb (1785) introduced the concepts of surface roughness, which plays an important part in today's models. Figure 1, after Tabor (3), illustrates Coulomb's view of interlocking surface asperities and their contribution to the friction of sliding bodies. Even today we associate high friction with rough surfaces and low friction with smooth surfaces.

By 1800, the "laws of friction" recognized that the friction force is proportional to the normal load applied to the sliding contact and that the friction force is independent of the area of the sliding surface. Friction is important to the subject of topography of surfaces in contact since it determines the load transfer from one component to another, that is, the stress distribution in the system.

Frictional processes involve the dissipation of energy, a fact not accommodated by the simple roughness model of Coulomb (3). Although Coulomb's simple roughness model was incomplete, the concept of surface irregularities is central to today's understanding of friction (and adhesive wear). Figure 2 illustrates the multiple scales of surface irregularity that are detected at various magnifications. Note the similarities to Fig. 1. At a magnification of 1X, a surface waviness and texture is detected. At 100X, the texture is manifest as a fine scale irregularity called the surface roughness. At a higher magnification, for example 1000X, the roughness contours may also be irregular. The average roughness, R_a, is found by averaging successive ordinate values. Because of this surface roughness, two surfaces nominally in contact actually meet only at surface asperities or high points.

The modern view of friction, summarized by Tabor (3), considers the important factors to be (1) the true contact area, (2) the type and strength of bonds formed at the contact areas, and (3) deformation and rupture at and around the contact areas.

Surface topography and its role in surface interactions has been studied in detail since the mid-1960s and many models for describing surface asperities and their response to the load have been developed. The relationship between the number of asperities and their area and the applied load depends upon the nature of the local deformation at the asperity tip, that is, whether it is elastic or plastic deformation, or brittle fracture.

As two clean pieces of a metal are brought into contact, bonds first form due to van der Waals forces. These increase as the separation distance decreases leading to screened van der Waals bonds. The much stronger metallic bond forms at a separation distance corresponding to about one interatomic distance. The bonding between dissimilar metals has often been discussed in terms of mutual solubility. The suggestion that mutually soluble pairs show strong adhesive bonding while insoluble pairs show weak adhesive bonding is useful as a general guideline but fails upon rigorous inspection.

Buckley (4,5) has determined the force to pull clean metal surfaces apart and correlated these measurements with friction. He discussed the results in terms of the number of d-electrons in the metal and their influence upon the strength of the bond. Metals with strong d-character

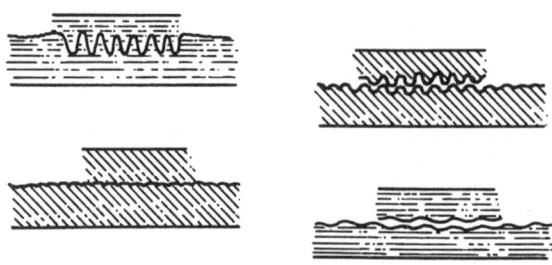

FIGURE 1. Illustrations of Coulomb's view of friction due to surface
asperities (after ref. 3).

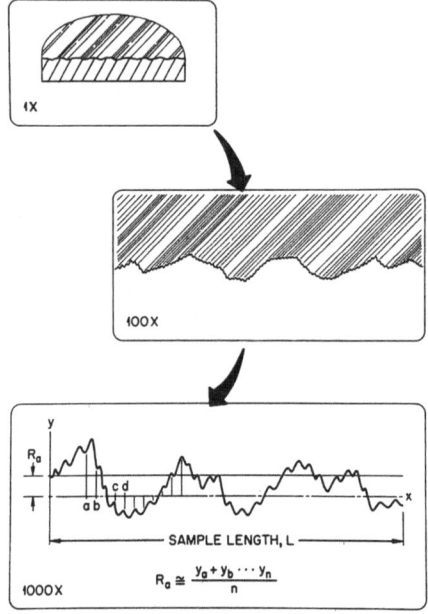

FIGURE 2. Multiple scales of surface irregularity are detected as magnification of the surface is changed. The roughness average, R_a, is approximated by an averaging of successive ordinate values.

are less likely to interact. Figure 3 shows the coefficient of friction as
a function of d-bond character for gold sliding on various metals. The
large value indicates that the shearing force of bonded surfaces is being
measured. Recent advances in the theory of solids suggest that questions
regarding interfacial bonding could now be attacked from a sound fundamental base (6).

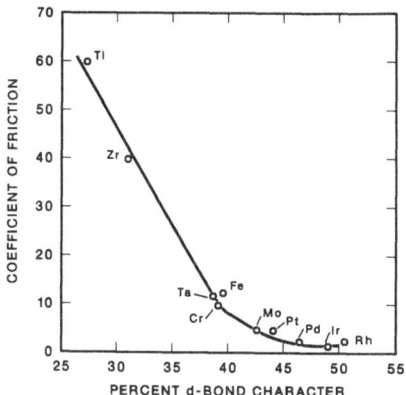

FIGURE 3. Coefficient of friction as a function of amount of d-bonding for gold sliding on various metals. Sliding velocity 0.7 mm/min, load 1 g, 23°C, and 10^{-8} N/m^2 (10^{-10} torr). (After Buckley, ref. 4.)

Since the metallic bond forms only at interatomic distances, the presence of a monolayer of absorbed gas may reduce the interaction between contacting forces to that of residual van der Waals forces, or by a factor of ten or more. The presence of a thin (1 nm) oxide also eliminates the metallic bonding.

The remaining factor to be considered in the discussion of friction is the deformation and rupture at and around the contact areas. The simplistic view considers failure or separation to occur at the weakest point — at the contact interface, randomly in both surfaces for similar pairs, or in the weaker component of dissimilar couples. If the interface junction is brittle, it fractures with little or no plastic deformation, and the work to break the bond should be close to the theoretical bond strength. If the junction is ductile, the work involved in plastically deforming the surrounding region must be considered, and the work-hardening characteristics of the materials become important. The presence of an interfacial layer, of course, adds a further complicating factor.

Following the arguments of Tabor (3), friction arising from interfacial bonds, in the absence of deformation or ploughing, is given as

$$F = A_o \tau = \frac{W}{\sigma_o} \tau \tag{1}$$

where F = friction force, A_o = true area of contact, W = normal force, σ_o = yield strength, and τ = interfacial shear strength. The coefficient of friction μ is then

$$\mu = F/W = \tau/\sigma_o \quad . \tag{2}$$

If there is a ploughing deformation, an additional term is to be added to this expression,

$$F = F_{adhesion} + F_{deformation} \quad .$$

The important feature for the current discussion is the dependence on the material parameters τ_o (shear strength of interface) and σ_o (yield stress).

3. ADHESIVE WEAR

Adhesive wear results from the formation of junctions at contacting asperities on the two surfaces and subsequent rupture of the junctions. It is characterized by the transfer of one material to the surface of the other. Figure 4 contains a photograph of a wear track made by a diamond pin sliding on a TiB_2 disk. The diamond adhered to the TiB_2 and exhibited a high wear rate, whereas only a small amount of TiB_2 was removed. The profilometer trace across the wear track also showed that a wear film had formed (7).

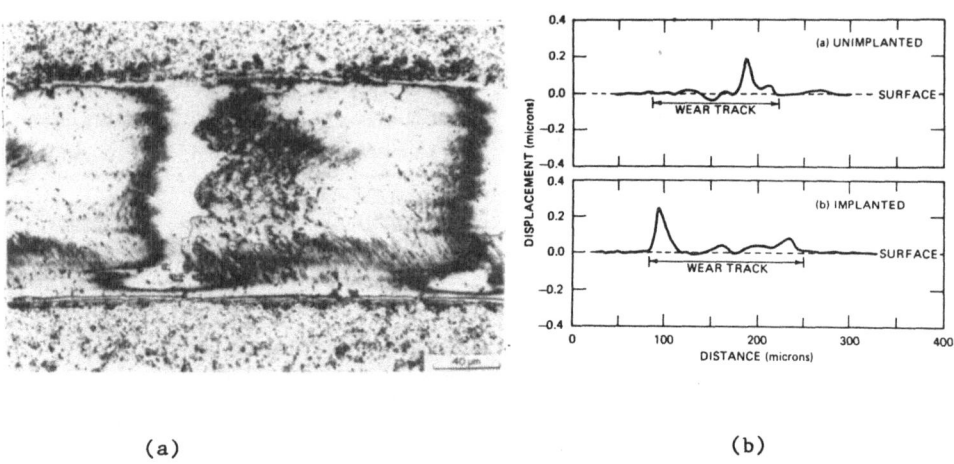

(a)   (b)

FIGURE 4. (a) Optical photograph of an adhesive wear track made by diamond sliding on TiB_2. (b) Profilometer trace of (a). (Reference 7)

Junction formation is attributed to welding due to local temperature increases caused by the sliding of rough surfaces or to local adhesive bonding. Because of the small tip size of asperities and the very high local loads, the instantaneous temperature can be very high. Junction formation due to solid state bonding (adhesion) is the same as discussed above for friction. Thus, adhesive wear should decrease as one progresses from couples of the same metals and those that form solid solutions or intermetallic compounds to incompatible metal couples to nonmetal-metal pairs, although there are many exceptions to this simple view.

Formation of adhesive bonds can also be described in terms of work of adhesion, first defined by Dupré (1869). The work of adhesion per unit area of contact required to separate two solid surfaces is given as

$$\Delta\gamma = \gamma_a + \gamma_b - \gamma_{ab} . \tag{3}$$

In this relation, γ_a and γ_b are the surface free energies for the two free surfaces and γ_{ab} is that of the solid-solid interface.

More attention has been directed to the study of mechanisms involved in breaking the junctions than in forming them. In order to determine the relationship between material properties that can be altered by implantation and material removal mechanisms, several proposed models will be examined. These are (i) shearing of the weaker material (8,9), (ii) minimization of surface energy (10), (iii) critical strain for crack growth (fracture toughness) (11), (iv) surface fatigue (12), and (v) the delamination process (13).

Starting with the premise that sliding causes the rupture of junctions formed due to contact of asperities (8), the volume of material removed, ΔV, should be given by

$$\Delta V = NA_t x \ , \tag{4}$$

where N = number of atoms removed per interaction, A_t = true contact area, and x = sliding distance. The true area of contact is related to the applied load L and yield stress σ_y as $A_z = L/\sigma_y$ (3), such that Eq. (4) becomes

$$\dot{W} = \Delta V/x = NL/\sigma_y \ , \tag{5}$$

where $\dot{W}$ is the wear rate. Archard (8) replaced N in Eq. (5) with K, the probability that a junction between the two surfaces would lead to the formation of a wear particle. Hence,

$$\dot{W} = KL/\sigma_y \tag{6}$$

or, as often assumed, if the hardness $H = 3\sigma_y$,

$$\dot{W} = K'L/H \tag{7}$$

and we have a simple relationship between wear rate and hardness of the deforming (weaker) material.

Rabinowicz (10) proposed that the size of loose particles removed or transferred from one surface to the other is determined by balancing the elastic volume energy against the surface free energy. This approach yields the average loose (removed) particle size as a function of interfacial energy, or

$$\bar{d}_L = \frac{6\gamma_{ab}}{\nu^2 Y \epsilon_m^2} \ , \tag{8}$$

where γ_{ab} = surface free energy of a-b interface, ν = Poisson's ratio, Y = Young's modulus, and ϵ_m = maximum elastic strain. Writing the true area of contact A_t as $n\pi r^2$, where n is the number of contacts and r is the average radius of the contacting asperities, Eqs. (4) and (8) can be rewritten as

$$\dot{W} = N \cdot n\pi \left[\frac{3\gamma_{ab}}{\nu^2 Y \epsilon_m^2} \right]^2 \ , \tag{9}$$

which relates wear volume to interfacial free energy and the elastic properties of the solid. A similar expression can be written for the size of particles that adhere to one surface.

Hornbogen (11) modified Archard's expression in terms of fracture toughness for test conditions under which crack growth determines the removal of material. Archard's formulation gives a simple inverse relationship

376

between wear and hardness. Figure 5 shows that the relationship appears to hold within certain classes of materials but often fails. Hornbogen derived the relationship

$$\dot{W} = K' \frac{\gamma L^{3/2} m^2 Y \sigma_y}{K_I^2 H^{3/2}} \,,$$ (10)

where K' and γ are empirical constants, m the work hardening coefficient, σ_y the yield strength, K_I the fracture toughness, and H the hardness.

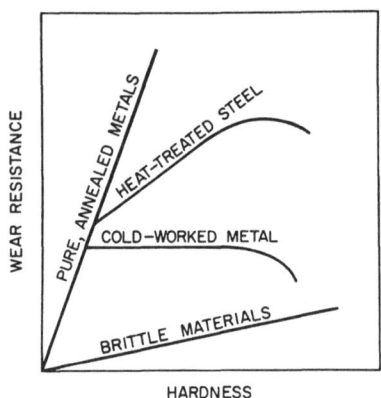

FIGURE 5. Wear resistance as a function of hardness for various classes of material. After Hornbogen (11).

Halling (12) attempted to relate the factor K' to the fatigue properties of the metal, observing that repeated adhesions and stressing should cause failure by fatigue mechanisms. For a particular failure criterion, he proposed that

$$\dot{W} = CL/H_1 \,,$$ (11)

where H_1 is the initial hardness and C is a constant.

Perhaps the most detailed description of the material removal mechanism is given by the delamination model developed by Suh (13). According to this model, the asperities on the softer material are quickly removed by fracture and further wear occurs by asperity (harder material) contact on a plane (softer material). This introduces subsurface deformation which accumulates with repeat loading. Eventually, cracks form and propagate approximately parallel to the surface. When the cracks shear to the surface, long thin wear sheets "delaminate". This process leads to an expression of the form

$$\dot{W} \propto \frac{\mu}{H K_I} \,,$$ (12)

where μ = coefficient of friction.

Certain of the material properties that appear in these relationships can be altered by ion implantation. These are hardness, yield stress, and fracture toughness. Surface energies may be changed by altering the surface composition.

4. ABRASIVE WEAR

Abrasive wear is characterized by the removal of material from one of two surfaces in relative motion caused by the presence of hard protuberances or hard particles either between the surfaces or embedded in one of them. The action is primarily cutting or ploughing and the worn surface contains grooves or scratches, Fig. 6. Following the arguments of Archard (8) and Mulhearn and Samuels (14), the following expression for volume wear rate is obtained:

$$\dot{W} \propto \frac{BF\sigma_a}{H}, \tag{13}$$

where σ_a = applied stress (L/A), B = fraction of contacts that actually remove material, and F = fraction of groove volume removed.

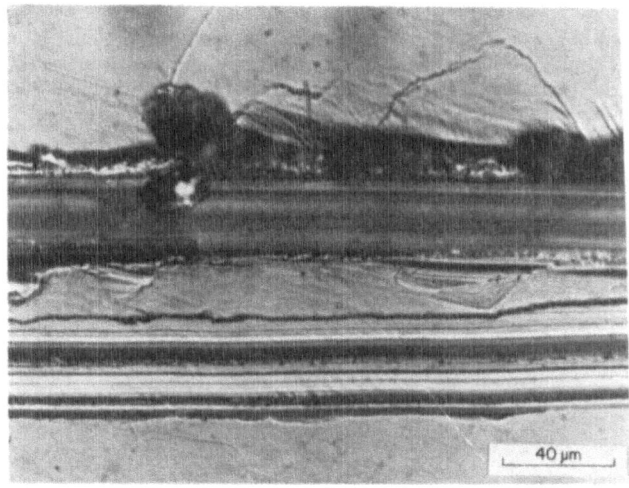

FIGURE 6. Optical photograph of abrasive wear track made by diamond grit sliding on copper. Normal load = 50 g.

Again, there is implied a simple inverse relationship between volume of material removed and hardness; however, the data of Kruschov (15) show variations between classes of material similar to those found by Hornbogen (11) for adhesive wear and shown in Fig. 5. Evans and Wilshaw (16) found that the wear resistance of brittle solids varied as fracture toughness and hardness. There is a large amount of data that shows the importance of the relative hardness of the abrading and abraded materials.

Erosion is a special case of abrasive wear in which the abrading material is in the form of particles carried by a gas or liquid stream. In the case of brittle materials, material removal is the result of cracking

378

due to Hertzian contact stress during impact. Evans, Gulden, and Rosenblatt (17) expressed the wear volume in terms of impact velocity, particle size, particle density, fracture toughness, and hardness. The first three parameters are functions of the particular system and the last two are materials properties.

In ductile materials, wear occurs by a cutting process or by extrusion of lips around impact sites followed by fragmentation of these lips by subsequent impacting particles. A model by Sheldon and Kanhere (18) describes the wear volume in terms of the ratio of vertical to horizontal force on the particle, particle size, velocity and density, and eroding material hardness.

5. SURFACE FRACTURE

Material removal due to direct surface fracture is important for brittle solids and may occur during abrasive or adhesive wear conditions. Figure 7 shows a scratch made in α-SiC. These are many radial cracks which extend into the surrounding material for distances comparable to the groove width. Much of the material is removed in large pieces due to the lateral cracks caused by the tensile component of the stress which is present during unloading. The models of material removal cited above for erosive wear should be applicable, hence, the properties of yield strength, hardness, and fracture toughness should be important. Since the fracture mode of brittle solids is sensitive to the details of the stress state, being particularly susceptible to tensile stresses, the residual surface stresses introduced by ion implantation can affect the wear response.

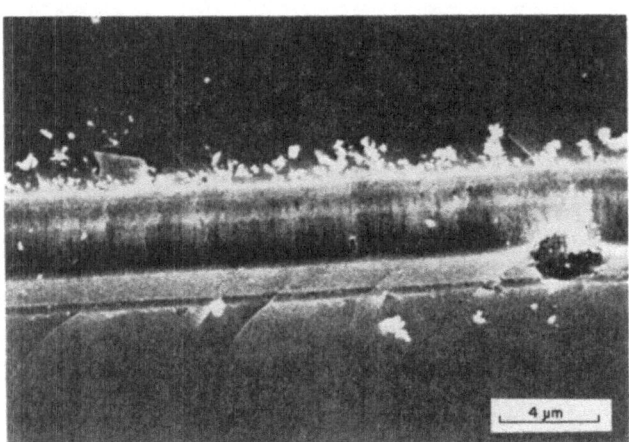

FIGURE 7. Scanning electron micrograph of surface cracking on and around the wear track made in α-SiC.

6. OXIDATIVE-CORROSIVE WEAR

Oxidative (corrosive) wear is characterized by the formation of an oxide or other corrosion product layer and its subsequent removal by mechanical action. The tendency to form adhesive junctions between a metal and an oxide is much less than between metals. Thus, oxidation provides a hard surface layer with little tendency to adhere to an opposing material. On

the other hand, oxides tend to be brittle, and if the softer substrate fails
to support the load applied to the thin oxide layer, the oxide will crack
and perhaps spall. The wear rate will depend upon both the oxidation rate
and the stresses in the oxide-substrate composite.

Dearnaley (19) modified earlier formulations used to describe oxidative wear to give

$$\dot{W} = \frac{k_o tL}{\delta_c \rho H} ,$$

(14)

where k_o = linear oxidation rate, t = time of contact for a single junction
encounter, δ_c = critical oxide thickness for spalling, ρ = oxide density,
and H = oxide hardness. The parameter δ_c depends upon the stress state at
the oxide-metal interface and thus is related to the substrate hardness or
yield strength. These models assume that oxidation occurs due to the temperature rise at asperity contacts.

Equation (14) suggests ways in which implantation may reduce oxidative
wear rates. Surface alloying or compositional changes may reduce the oxidation rate, k_o. The critical oxide thickness, δ_c, may be changed by
reducing growth stresses, increasing fracture toughness of the oxide or
matching the hardness of oxide layer and substrate. The first two might be
changed by compositional effects and the last one by altering the mechanical properties of the substrate.

7. EFFECTS OF IMPLANTATION

The purpose of the above discussion is to identify the properties of
materials which contribute to their wear by mechanical processes. In friction and adhesive wear, the type and strength of solid state bonding, and
the mechanical properties near bonded junctions were identified as important. The mechanical properties of hardness and fracture toughness were
specified as controlling wear during abrasion or erosion. Surface fractures depend upon strength (hardness), plasticity, fracture toughness, and
residual stress states. Oxidative wear is controlled by the oxidation
rate, the mechanical properties of the oxide and substrate hardness. The
types of bonds formed and the oxidation rate may be altered by control of
surface composition. The mechanical properties may be changed by use of
the defect structure produced by the radiation damage, second-phase formation, or interaction of implanted species with defect and microstructural
features.

Before discussing specific examples of the effect of ion implantation
upon the surface topography of subsequently eroded materials, a discussion
of property alteration by implantation will be given.

Roy Chowdhury et al. (20) used a mechanical microprobe technique to
determine the Dupré adhesion energy, γ, for titanium before and after
implantation. They determined γ between a titanium asperity and a titanium
foil. Implantation of about 1.4×10^{17} N/cm^2 (200 keV N$_2^+$) reduced the
value of γ by a factor of three.

Hardness is defined as the resistance of a material to penetration by
a loaded indenter. It has proved to be difficult to relate hardness to
fundamental material properties, although it is generally related to yield
strength in ductile metals. Many of the early attempts to determine the
effect of implantation on hardness employed the Vickers or Knoop technique
in which the depth of the hardness impression exceeded the depth of the
implanted ions. These measurements indicated the direction of change but
failed to give a quantitative measure of the effects. Table 1 summarizes

recent measurements. The data for metallic materials made with the ultra-low load equipment were obtained from load-displacement curves for penetration depths less than the ion range (21).

TABLE 1. Hardness Changes Due to Ion Implantation

Material	Implanted species	Hardness method*	Relative hardness (H_i/H_u)	Reference
Al	N	ULL	4.20	(22)
Be	B	KH5	4.00	(23)
Fe	N	ULL	1.84	(23)
Co	N	ULL	1.40	(22)
Ni	N	ULL	1.04	(22)
Ti-6Al-4V	N	ULL	2.00	(22)
Armco Fe	N	ULL	1.80	(22)
304 Stainless	N	ULL	1.25	(22)
Steel 12T(410)	N	ULL	1.20	(22)
52100	N	ULL	1.00	(22)
18-4-1 Tool Steel	B	VH10	1.80	(24)
Al_2O_3	Cr (300K)	KH15	1.5	(25)
	Y (300K)	VH200	1.4	(26)
	Cr (77K)	KH15	0.6	(27)
		ULL	0.6	(28)
MgO	Cr,Ti	KH10	1.5	(29)
SiC	Cr,N	KH15	0.6	(30)
	N	VH200	0.7	(31)
TiB_2	Ni	KH15	2.0	(32)
ZrO_2	Al	KH20	1.2	(33)

*ULL = ultra-low load; VH10 = Vickers, 10 gf; KH15 = Knoop, 15 gf.

For metallic materials, the relative hardness (implanted ÷ unimplanted) varies from 1.0 to 4.2. The variation results from the different hardening mechanisms activated by the implanted species and implantation damage. In the pure metals implanted with nitrogen, for example, Hutchings et al. (22) argued that the hardening of aluminum was due to nitride formation. The hardening of iron was due to interstitial solid solution strengthening, and the absence of an effect in nickel is due to lack of a hardening mechanism. Similar arguments were made for the metallic alloys. In the case of the ceramics, the softening of Al_2O_3 and SiC is attributed to the change from the crystalline to the amorphous state and a change in deformation mode.

The fracture toughness of implanted materials is very difficult to determine because of the small dimensions of the affected layer. Measurements for brittle materials (ceramics) have been reported for an indentation technique which gives a qualitative indication of the effects (25–30). Increases in the apparent K_I of 15 to 90% have been reported. In some instances, the apparent increase was attributed to the

residual compressive surface stresses that prevented cracks that were initiated in the interior from reaching the surface (26,27).

There have been few direct measurements of the effect of ion implantation upon yield stress because of the difficulties of testing.
Tranchant et al. (34) performed tensile tests on copper single crystals
pulled in the <111> direction. Implantation with aluminum increased the
yield and rupture strengths by 42 and 26%, respectively. Hioki and co-
workers (35) found that implantation of 300 keV nickel increased the rupture strength (in three-point bending) of single crystal Al_2O_3 by 30% and
that 400 keV nitrogen increased it by 50%. Thus, it is clear that the
mechanical properties which are important to wear (erosion) resistance are
altered by implantation.

Phase changes caused by implantation might produce large differences
in response to mechanical forces due to change in deformation mechanisms
or surface reactivity. The literature contains many examples of
crystalline to amorphous transformations, precipitate formation in the surface layer, and a surface compound layer. The cases cited here pertain to
the wear results to be discussed later.

Coimplantation of titanium and carbon or titanium in a steel that contains sufficient carbon (36) or titanium in a vacuum chamber containing
residual hydrocarbons (37) produces an amorphous surface on a wide variety
of steels — ferritic, austenitic, and martensitic. In ceramics, amorphous
surfaces have been prepared by implanting Al_2O_3 at 77 K and α-SiC at
300 K (27). As noted above, the hardness of the amorphous ceramics is 60%
of their crystalline counterpart. Deformation occurs by viscous flow in
the amorphous phases instead of dislocation motion.

Many of the improved mechanical properties of nitrogen-implanted
steels have been attributed to the formation of a fine dispersion of
nitride particles. The compounds Fe_2N, Fe_3N, and Fe_4N have been identified
in ferritic steels (38). High chromium-containing steels (types 304 and
316 austenitic stainless steel) precipitate CrN (39) and martensitic
stainless steels from Cr_2N (40).

Singer et al. (41) reported that a 100-nm thick TiC surface layer
formed during high fluence ($5 \times 10^{17}/cm^2$) implantation of titanium into
iron and M2 steel at substrate temperatures of 600 to 800°C. The carbon
involved was contained in the substrate material for the steel but came
from the specimen chamber atmosphere for the iron targets.

EerNisse developed a technique for determining the stresses lateral
to the implantation surface by measuring the deflection of thin
specimens (42). This technique has been used to determine the surface
stresses introduced into Al_2O_3 (43), steel (44), and thin films of aluminum
and gold (45). Implantation of argon into Al_2O_3 and nitrogen into steel
produced stresses in excess of the yield strength.

8. EFFECTS OF IMPLANTATION ON SUBSEQUENT WEAR

The effects of implantation on the mechanical removal of material
during wear or erosion processes will be illustrated in this section. An
attempt will be made to classify the effects in terms of the various
property changes discussed above. Since several wear mechanisms may be
operative during a single test, results are seldom as clear-cut as the
outline might indicate.

8.1. Change in mechanical properties

Implantation of nitrogen into aluminum reduced the amount of material
removed in a pin-on-disk test (adhesive wear) by a factor of 35 for a
fluence of 1×10^{17} N/cm^2 (46). Figure 8 shows the wear improvement factor

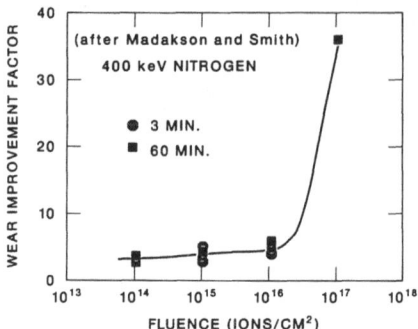

FIGURE 8. The effect of nitrogen implantation on the wear resistance of aluminum. After Madakson and Smith (46).

as a function of ion fluence. Reference to Table 1 shows that a similar treatment increased the hardness of aluminum by a factor of 4.2 due to formation of nitride precipitates. Examination of the wear tracks revealed fewer cracks for the implanted samples. The wear and frictional behavior of aluminum depends on the formation and subsequent failure of the surface oxide film which is hard and brittle. It is proposed that the improvement in wear behavior caused by implantation of nitrogen derives from the increased support provided for the oxide film by the hardened substrate.

Figure 9 shows the increase in wear resistance and hardness as a function of boron fluence for the tool steel 18 W-4 Cr-1 V (24). Both properties increased approximately with fluence in the range of 10^{15} to 10^{18}/cm^2. Implantation did not change the shape of the wear track, only the depth. These changes (170% increase in wear resistance and 40% increase in hardness at 10^{17}/cm^2) were considered to be associated with interstitial solid solution strengthening of the steel by the boron. The implanted ions also occupy defect and dislocation sites to give further strengthening.

As noted above, implantation of Al_2O_3 with chromium at 300 K caused an increase in hardness (50%) and fracture toughness (15%). The SEM of Fig. 10 shows a scratch made by a diamond pin sliding across an Al_2O_3 surface under a normal force of 0.29 N (30 g), corresponding to abrasive or ploughing wear conditions (47). In this case the implanted region is crystalline but both sublattices are characterized by high damage levels. The number of radial cracks and the amount of material removed due to lateral cracks were much higher in the unimplanted region of the specimen. The measured tangential force on the moving pin increased by 20 to 30% as the pin passed from the unimplanted to the implanted region, indicating an increase in the work required to remove a unit of volume of material.

8.2. Change in surface composition

The change in near-surface composition can affect the wear by altering the surface energy, the type or strength of adhesive bonds, or the formation of a surface film.

As discussed above, Roy Chowdhury and co-workers found a reduction in surface energy (work of adhesion) by a factor of three after implanting titanium with nitrogen (20). It has been suggested that this reduction in the work of adhesion may be responsible for the observed reduction in the coefficient of friction for nitrogen-implanted titanium (1).

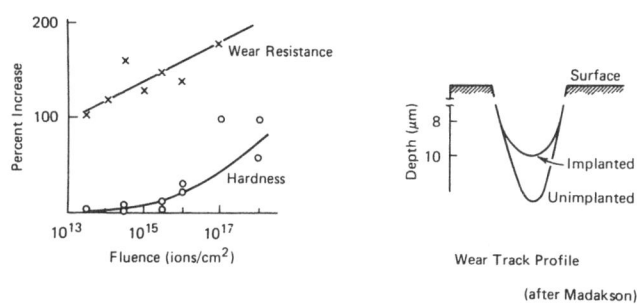

FIGURE 9. The effect of boron implantation on hardness and wear, and wear track profile in type 18-4-1 tool steel. After Madakson (24).

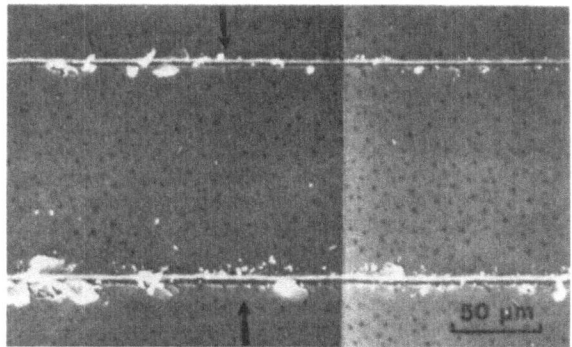

FIGURE 10. Suppression of surface cracks around a wear track by implantation of 2.6×10^{16} Cr/cm^2 into α-Al$_2$O$_3$. Yust and McHargue (47).

The implanted species can have a large effect on the wear process if it promotes the formation of an oxide layer such that the mechanism changes from adhesion or abrasion to oxidative wear. In the early studies by Hartley et al. (48), a variety of ions (Sn In, Ag, Pb, Mo) were implanted to fluences greater than 10^{16}/cm^2. Specimens implanted with tin and molybdenum showed greatly reduced coefficients of friction, Fig. 11, whereas those with lead, indium, and silver had higher values. It was proposed that tin and molybdenum enhanced the formation and retention of the oxide layer.

Hutchings and co-workers (22,49,50) studied the hardness, friction, appearance of wear tracks, and surface composition for nitrogen-implanted Ti-6 Al-4 V. Their results are summarized in Fig. 12. Implantation of 3.5×10^{17} N/cm^2 reduced the wear rate by more than two orders of magnitude. There was also a decrease in friction coefficient from 0.48 to 0.15. Long-term testing resulted in a return to the unimplanted wear rate. Inspection of the wear tracks [Fig. 12(b,c)] showed a marked difference between the unimplanted and implanted areas. The SEMs of tracks on unimplanted regions, or after long-term testing, showed deep grooves and

384

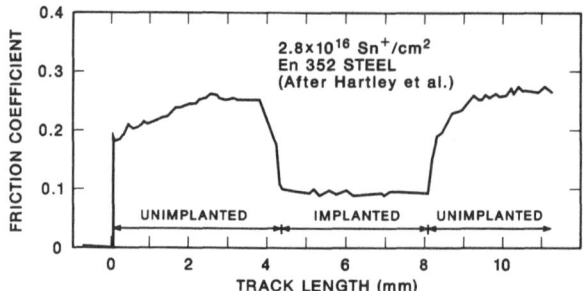

FIGURE 11. The effect of 2.8×10^{16} Sn/cm^2 on the coefficient of friction for E_n 352 steel. After Hartley et al. (48).

were very rough. The photographs of tracks in the implanted areas were smooth and covered by a thin polycrystalline film. The profilometer traces also showed these differences.

Results of an electron probe microanalysis of oxygen and nitrogen in the wear tracks are also shown in Figs. 12(b,c). There were no significant amounts of oxygen or nitrogen in the tracks made on unimplanted areas or after breakthrough of the implanted layer. The traces in Fig. 12(c) were obtained from a track after 1000 turns. From the variation in the oxygen signal and the profilometer trace, it is deduced that an uneven oxide layer had developed in the wear track. In addition, some nitrogen had been lost from the track (i.e., some wear of the nitrogen-containing material had occurred).

Oliver et al. proposed the following description for the wear process. The dominant mechanism of material loss in the unimplanted material is a cutting or machining process. This occurs because of transfer of material from the disk to the pin due to adhesive wear in the initial stage. This layer acts as the cutting edge for further wear. No material transfer was found for implanted disks and the formation of the oxide (or oxynitride) layer suggests wear occurs by an oxidative process. The authors proposed that the effects were largely due to chemical stabilization of the surface film or that chemical effects promote the rapid formation of this film.

8.3. Phase changes

Implantation may alter the phase composition of the near-surface region by formation of metastable solid solutions, formation of second-phase precipitate particles, formation of a compound layer, or promotion of a crystalline-to-amorphous phase change.

The formation of an amorphous surface would be expected to affect the response to the mechanical forces responsible for wear since the physical, chemical, and mechanical properties of amorphous and crystalline phases may be quite different, even for the same overall composition. Figure 13 gives the results of wear tests for 440C stainless steel (a martensitic steel) implanted with 2×10^{15} Ti/cm^2 and 2×10^{15} C/cm^2. For a normal load of 4.85 N (500 g), the coefficient of friction was reduced 40% and the wear depth was reduced 80% by implantation (51). The Hertzian stress in these tests often exceeded the yield stress of the crystalline (unimplanted) steel. These investigators concluded that abrasive wear occurred in the implanted (amorphous) region and adhesive wear occurred in the unimplanted (crystalline) region.

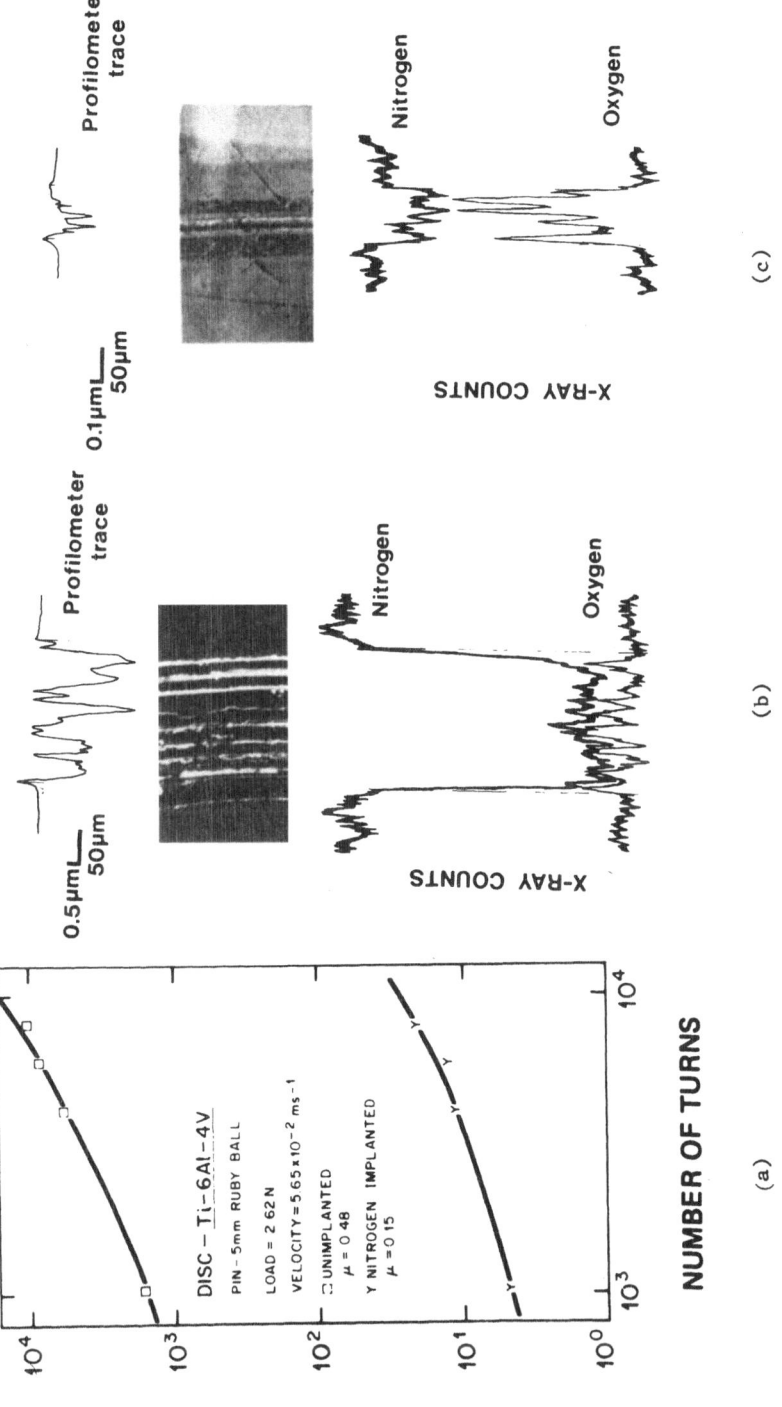

NUMBER OF TURNS

(a)

(b)

(c)

FIGURE 12. (a) The effect of 3.5×10^{17} N/cm^2 on the wear volume of Ti-6 V-4 Al. SEM, profile, and oxygen and nitrogen distribution in wear track of (b) after breakthrough, (c) after 1000 turns. After Hutchings, Oliver, and Pethica (49,50).

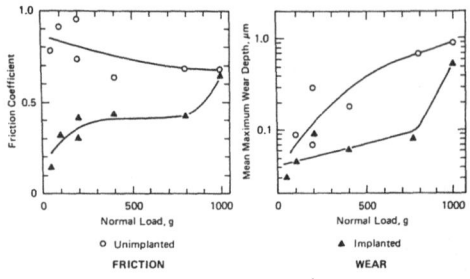

FIGURE 13. Effect of amorphous surface on wear of type 440C stainless steel. (a) Coefficient of friction, (b) wear depth. After Pope et al. (51).

The formation of an amorphous phase in ceramic materials may have two effects on the removal of material during wear. A large volume increase accompanies the crystalline to amorphous transformation in α-SiC. This causes high lateral compression stresses in the near-surface region; the deformation mechanism also changes. Figure 14 shows scratches made by a diamond pin sliding across a single crystal sample of α-SiC. The scratch in the unimplanted region is accompanied by a large number of radial cracks that extend into the surrounding material. Much of the debris consists of angular pieces fractured from the wear track. In Fig. 14(a), there are two regions of large segments removed or loosened due to formation of lateral cracks. In the amorphous region, produced by 2.6×10^{16} Cr/cm^2, the deformation occurred by viscous flow with ductile-appearing chips. A profilometer trace showed that a significant amount of material was present as ridges next to the groove. In this case, it is believed that cracks nucleated in the crystalline region below the wear track were prevented from reaching the surface by the surface compressive stresses, and that the applied stress was relieved by viscous flow (30,47).

Implantation under certain conditions may produce a surface layer of a new compound. Figure 15 contains the wear data of Singer and co-workers (41) and shows the TiC layer on hardened M2 steel to be three to ten times more resistant to abrasive wear than the substrate. In sliding tests where adhesive wear occurred for the substrate material, this TiC layer was free of grooving, cracking, and adhesive debris. The coefficient of friction for the TiC layer in these sliding tests was 40 to 60% lower than that of the hardened steel.

8.4. Surface stresses

Oliver, Hutchings, and Pethica studied the wear of nitrogen-implanted electrodeposited chromium (hard chromium plate) and attributed the improved wear properties to the effects of implantation-induced stresses (50). The wear volume was decreased by a factor of 20 for normal loads of 5.2 N (540 g) and 10.5 N (1080 g), (Fig. 16). There was no significant change in the coefficient of friction, but there was a 40% increase in hardness. Wear tracks in unimplanted material, implanted after 1000 turns at a load of 10.5 N where breakthrough did not occur, and implanted after 515 turns where breakthrough did happen are shown in Fig. 17.

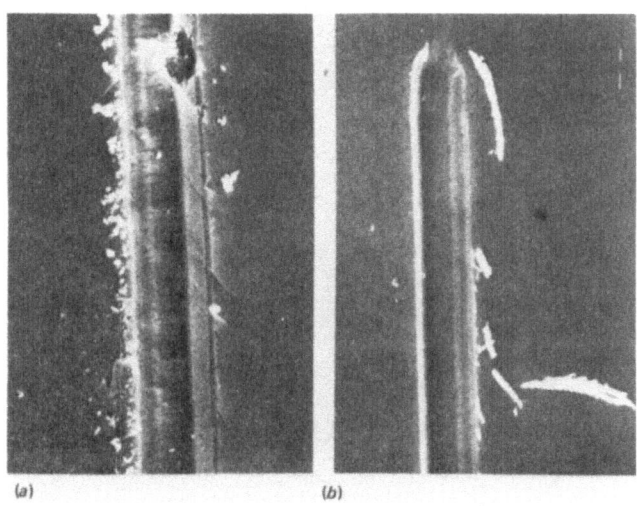

FIGURE 14. Effect of amorphous surface on wear of α-SiC. (a) Crystalline, unimplanted, (b) amorphous, implanted with 2.6×10^{16} Cr/cm^2. McHargue and Williams (30).

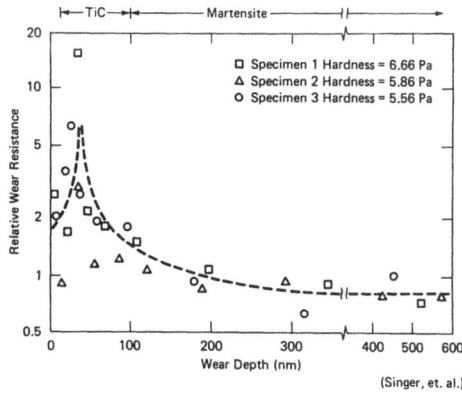

FIGURE 15. Abrasive wear of M2 steel containing a near-surface layer of TiC formed by implanting Ti + C at 600 to 800°C. After Singer et al. (41).

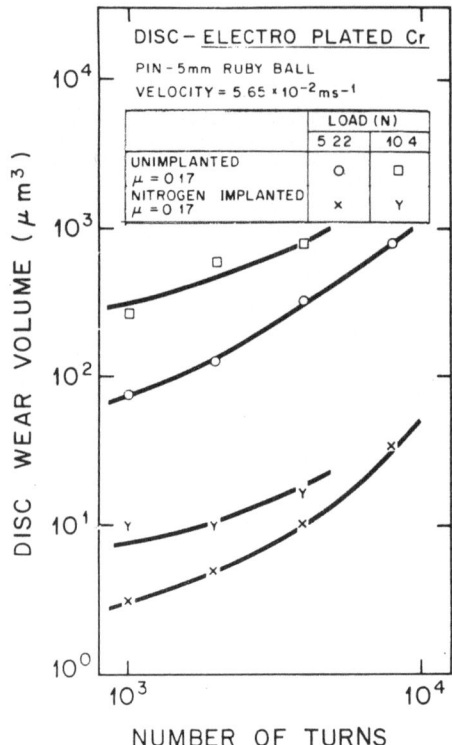

FIGURE 16. Wear rates in electrodeposited chromium before and after implantation with 3.5×10^{17} N/cm^2. After Oliver et al. (50).

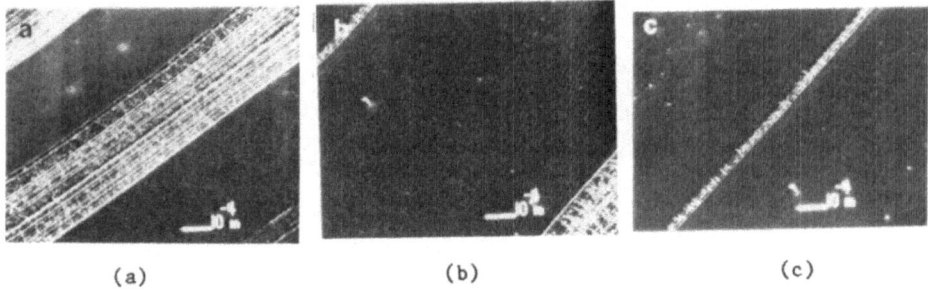

(a)   (b)   (c)

FIGURE 17. Scanning electron micrographs of wear tracks in specimens of Fig. 16. (a) Unimplanted, 1000 turns; (b) implanted, 1000 turns, no breakthrough; (c) implanted, 515 turns, breakthrough. Load 10.4 N.

The wear track in the unimplanted material shows a series of fine features. Higher magnifications revealed these transverse features to be small shear lips which delineate the microcracks which exist in hard chromium plate due to release of the large residual stresses produced by the plating process. In contrast, the wear track on the implanted material (before breakthrough) is much smaller and much smoother, having a maximum depth of around 0.1 μm. The track in implanted chromium which broke through and was terminated after 515 turns was characterized by the appearance of the microcracking at the center of the wear track and fine longitudinal grooving towards the edges.

These results for chromium show that its wear resistance was improved by nitrogen implantation. This improvement was associated with a change in wear mechanism, but in this case it was the role of the microcracks inheren in the plated chromium that was critical. The introduction of a large fraction of nitrogen into a metal lattice causes an expansion of that lattice. For example, the formation of Cr_2N from chromium metal is accompanied by a volume increase of approximately 25%. Such a volume increase will close the microcracks in the implanted region. Thus, the initial wear of implanted chromium occurred by the adhesive wear of two hard materials, and was therefore very slow.

Once the implanted layer was penetrated (or in the unimplanted material), the dominant wear mode changed. As seen in Fig. 17(c), breaking through the implanted layer coincided with the appearance of striations along the wear track. It was suggested that these striations were the product of abrasive wear by fine particles which originate at the shear lips associated with the microcracks. Thus, the wear mode for unimplanted chromium plate may be explained in terms of the shearing of particles from the vicinity of the microcracks combined with abrasive wear by these particles.

Another example in which the residual surface stress state controls the removal of material during wear is given by McHargue et al. for TiB_2 (ref. 7). Figure 18 is a SEM of the wear track made by a diamond pin on a TiB_2 disk. There is a large adhesive component to wear in this system which caused a carbon (diamond) layer to be transferred to the TiB_2. The removal of TiB_2 occurred primarily by the pull-out of the entire grains as a result of grain boundary cracking. It is thought that this cracking was caused by the tensile component of the stress during the unloading stage as the diamond pin passed a given point. The presence of large residual compressive stresses would counter this tensile component. Examination of the wear tracks at high magnification showed profuse cracking in the unimplanted region but very little in the implanted sections.

9. SUMMARY AND CONCLUSIONS

This paper has discussed the major features of wear processes and identified the material properties important to the material removal step. Then these effects of implantation on structure and properties were examined to determine how implantation might affect the wear process. Finally, several examples of wear studies on implanted materials were given to illustrate the range of responses observed.

It is clear that the wear rate for a given mechanism can be altered in a systematic manner. Following the arguments of Rabinowicz (52), Dearnaley (19) suggested improvements in adhesive wear by factors of 2 to 3, abrasive wear by factors of 10^+, and oxidative wear by 10^2 to 10^3 can be achieved by implantation. The most effective treatment is one that changes the wear mode to a less aggressive one. Most studies to date have focused on the mechanical property-wear relationship.

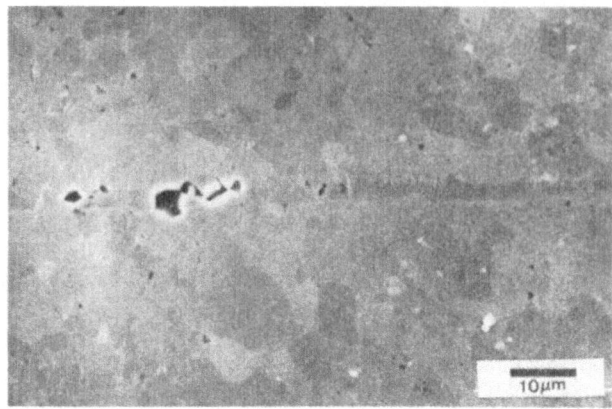

FIGURE 18. Scanning electron micrograph of wear track in TiB$_2$. The unimplanted region is to the left and implanted with 1×10^{17} Ni/cm^2 is to the right. McHargue et al. (7).

A number of cases have been cited in which oxidative wear has been promoted at the expense of adhesive wear. Unfortunately, the manner in which the implanted species causes this change remains unknown. Since this route holds the most promise, there is a great need for careful studies to establish a scientific basis. Likewise, in most instances, the reason for improved wear resistance of amorphous metallic surfaces has not been established.

Few attempts to control wear by alteration of surface composition have been reported. Since chemical species that lower surface energy tend to segregate to surfaces and interfaces, adhesive wear might be addressed by selectively doping, that is, ion implantation.

10. ACKNOWLEDGMENT

This research is sponsored by the Division of Materials Sciences, U.S. Department of Energy, under contract DE-AC05-84OR21400 with the Martin Marietta Energy Systems, Inc.

REFERENCES

1. Hartley NEW: Ion Implantation. Vol. 18, J. K. Hirvonen (ed). New York: Academic Press, 1980, pp. 321–371.

2. Yust CS: Int. Met. Rev. 30 (1985) 141–154.

3. Tabor D: J. Lubr. Technol. 103 (1981) 169–179.

4. Buckley DH: J. Coll. Interface Sci. 58 (1977) 36–53.

5. Buckley DH: Wear 46 (1978) 19–53.

6. Faulkner JS: Prog. Mater. Sci. 27(1–2) (1982) 1–188.

7. McHargue CJ, CS Yust, P Angelini, PS Sklad, and MB Lewis: Proceedings International Conference on the Science of Hard Materials-2, Rhodes, Greece, Sept. 23–28, 1984, to be published.

8. Archard JF: J. Appl. Phys. 24 (1953) 981–988.

9. Bowden FP and D Tabor: The Friction and Lubrication of Solids. Vol. I (1950), vol. II (1964), London: Oxford University Press.

10. Rabinowicz E: Wear 7 (1964) 9–22.

11. Hornbogen E: Wear 33 (1975) 251–259.

12. Halling J: Wear 34 (1975) 239–249.

13. Suh NP: Wear 44 (1977) 1–16.

14. Mulhearn TO and LE Samuels: Wear 33 (1962) 478–498.

15. Kruschov MM: Wear 28 (1974) 69–88.

16. Evans AG and TR Wilshaw: Acta Metall. 24 (1976) 939–956.

17. Evans AG, ME Gulden, and M Rosenblatt: Proc. Roy. Soc. London A361 (1978) 343–350.

18. Sheldon GL and A Kanhere: Wear 21 (1972) 195–208.

19. Dearnaley G: Mater. Sci. Engr. 69 (1985) 139–147.

20. Roy Chowdhury SK, NEW Hartley, HM Pollock, and MA Wilkins: J. Phys. D. 13 (1980) 1761–84.

21. Pethica JB: Ion Implantation into Metals. V Ashworth, WA Grant, and RM Procter (eds). New York: Pergamon Press, 1982, pp. 147–156.

22. Hutchings R, WC Oliver, and JB Pethica: Surface Engineering. R Kossowsky and SC Singhal (eds). Dordrecht: Martinus Nijhoff Publishing Co., 1984, pp. 170–184.

23. Kant R, J Hirvonen, A Knudson, and J Wollam: Thin Solid Films 63 (1979) 27–30.

24. Madakson PB: Mater. Sci. Engr. 69 (1985) 167–172.

25. McHargue CJ, H Naramoto, BR Appleton, CW White, and JM Williams: Mater. Res. Soc. Symp. Proc. 7 (1982) 147–153.

26. Barnett PJ and TF Page: J. Mater. Sci. 19 (1984) 3524–3545.

27. McHargue CJ, GC Farlow, CW White, JM Williams, BR Appleton, and H Naramoto: Mater. Sci. Engr. 69 (1985) 123–127.

28. Oliver WC and CJ McHargue: Mater. Res. Sci. Symp. Proc., to be published.

29. Burnett PJ and TF Page: Mater. Res. Soc. Symp. Proc. 27 (1984) 401–406.

30. McHargue CJ and JM Williams: Mater. Res. Soc. Symp. Proc. 7 (1982) 303–309.

31. Roberts SC and TF Page: Ion Implantation into Metals. V Ashworth, WA Grant, and RM Procter (eds). New York: Pergamon Press, 1982, pp. 135–146.

32. McHargue CJ, MB Lewis, BR Appleton, H Naramoto, CW White, and JM Williams: Science of Hard Materials. R.K. Viswanadham, DJ Rowcliffe, and J Gurland (eds). New York: Plenum Publishing Co., 1983, pp. 451–465.

33. Cochran JK, KO Legg, and HF Solnick-Legg: Mater. Res. Soc. Symp. Proc. 24 (1984) 173–179.

34. Tranchant F, J Vergnol, MF Denanot, and J Delafond: Nucl. Instrum. Methods Phys. Res. 209/210 (1983) 895–898.

35. Hioki T, A Itoh, S Noda, H Doi, J Kawamoto, and O Kamigaito: Nucl. Instrum. Methods Phys. Res. B7/8 (1985) 521–525.

36. Follstaedt DM, FG Yost, LE Pope, ST Picraux, and JA Knapp: Appl. Phys. Lett. 43 (1983) 358–360.

37. Singer IL and RA Jeffries: Appl. Phys. Lett. 43 (1983) 925–927.

38. Marest G, CS Koutarides, Th Barnavon, and J Tousset: Nucl. Instrum. Methods Phys. Res. 209/210 (1983) 259–265.

39. Baron M, AL Chang, J Schreurs, and R Kossowsky: Nucl. Instrum. Methods Phys. Res. 182/183 (1981) 531–538.

40. Whitton JL, GT Ewan, MM Ferguson, T Laursen, IV Mitchell, HH Plattner, ML Swanson, AV Drigo, G Celotti, and WA Grant: Mater. Sci. Engr. 69 (1985) 111–116.

41. Singer IL, RN Bolster, JA Sprague, K Kim, S Ramelingam, RA Jeffries, and GO Ramseyer: J. Appl. Phys. to be published in 1985.

42. EerNisse EP: Appl. Phys. Lett. 18 (1971) 581–583.

43. Kreff GB and EP EerNisse: J. Appl. Phys. 49 (1978) 2725–2730.

44. Hartley NEW: J. Vac. Sci. Technol. 12 (1975) 485–489.

45. Robic JY, J Piaguet, and JP Gailliard: Nucl. Instrum. Methods Phys. Res. 182/183 (1981) 919–922.

46. Madakson PB and AA Smith: Nucl. Instrum. Methods Phys. Res. 209/210 (1983) 983–988.

47. Yust CS and CJ McHargue: Emergent Process Methods for High Technology Ceramics. RF Davis, H Palmour III, and RL Porter (eds). New York: Plenum Publishing Co., 1984, pp. 533–547.

48. Hartley NEW, G Dearnaley, and JF Turner: Ion Implantation in Semiconductors and Other Materials. BL. Crowder (ed). New York: Plenum Publishing Co., 1973, pp. 423–436.

49. Hutchings R and WC Oliver: Wear 92 (1983) 143–153.

50. Oliver WC, R Hutchings, and JB Pethica: Met. Trans. A. 15A (1984) 2221–2229.

51. Pope LE, FG Yost, DM Follstaedt, ST Picraux, and JA Knapp: Mater. Res. Soc. Symp. Proc. 27 (1984) 661–666.

52. Rabinowicz E: Trans. ASME 103 (1981) 188–194.

PROBLEMS, PROSPECTS AND APPLICATIONS OF EROSIONAL/DEPOSITIONAL PHENOMENA

ORLANDO AUCIELLO

North Carolina State Univ., Dept. of Nuclear Engineering, Raleigh, N.C. 27695-7909, U.S.A.

1. INTRODUCTION

The erosion of surfaces due to the interaction of particles (inert or reactive ions/atoms, electrons) and electromagnetic radiation with materials and the deposition of films on different substrates have become subjects of major interest, and therefore the object of extensive studies, as a consequence of the relevance of erosion and deposition processes in different technologies. Recent reviews(1-4) have shown that the processes mentioned above can produce texturing of surfaces due to ion bombardment or vapor deposition. It has been demonstrated in those reviews that textured surfaces already are used in some technologies, have the potential to be useful in others, and are relevant either on detrimental or beneficial bases in different experimental techniques.

Although a relatively short time has passed since the publication of the mentioned reviews(1-4), the activity in this branch of science has already provided us with new insights into different phenomena related to surface alteration by particle bombardment and/or film deposition and the application of the acquired knowledge into new technological developments. Therefore, the main goal of the present review is to update the information relevant to this field of research.

Basic primary processes related to erosional/depositional phenomena in solids are treated in detail in other reviews of these proceedings. Therefore, only secondary effects in particle bombardment-induced modification of surfaces will be briefly reviewed in the first section. These effects (flux enhancement of the primary particle beam by backscattered and/or energetic sputtered species, redeposition of sputtered atoms/molecules, etc.) often occur concurrently with the primary processes (sputtering, material deposition) and may have negative influence on etching/deposition processes and their technological applications. The latter, being the main subject of this review, will be discussed in the following sections.

2. SECONDARY EFFECTS IN ION BOMBARDMENT-INDUCED SURFACE EROSION
2.1. General Information

Secondary effects are often present in ion bombardment-induced topographical changes of surfaces; they may have been present, most probably, since the early observations of morphological modification of solids. However, the influence of these effects was not explicitly recognized until the early seventies, when Bayly(5) and Wilson(6) considered them, qualitatively, as part of the explanation for some of their experimental results; these could not have been explained on the bases of the first-order erosion theory (7), which involve the sputtering process by the primary beam only. Effects discussed by these authors(5,6), and also later by Chapman(8), included: (a) the local flux enhancement of the primary ion beam at the bottom of steep surface features (cones, flaws, steps, etc.), which results from additional fluxes of scattered ions and/or energetic sputtered species from the local inclined surface regions of the mentioned features; (b) redeposition of sputtered species on the surrounding areas of those structures. Later,

Auciello and Kelly(9-13) were able to obtain definite experimental evidence for the existence of <u>secondary</u> (scattered ions and energetic sputtered atoms from cone/pyramid walls), and what they defined as <u>tertiary</u> (energetic sputtered atoms from pyramid-associated pit walls) effects(12) in surface topography evolution. One of their main contributions to this field of research was the demonstration, in a series of systematic experiments(9-13) that sequential bombardments and scanning electron microscopy (SEM) observations of a like area, looking at the same features, was the most appropriate way to obtain reliable information on surface topography evolution. This confirmed a previous single evidence(6) for the power of this method. The reliability of this technique was confirmed also by simultaneous independent studies performed by Carter, et al.(14,15). Auciello and Kelly's work allowed them to demonstrate that not only secondary fluxes originating from primary features (cones, flaws, steps, etc.) were important in the overall topography evolution, but also those (tertiary fluxes) generated at secondary features (pit walls at the bottom of cones/pyramids, trenches at the bottom of steps, etc.) developed during target erosion; pyramids are faceted cones, which generally develop on surfaces of materials with crystalline structure.

The early experimental work and semiquantitative analysis(5-15), on the influence of secondary effects on surface topography evolution, opened the way to more quantitative evaluations, produced recently, on: (a) redeposition of sputtered atoms from one surface region onto another(15-22), and (b) on secondary and tertiary fluxes on cone/pyramid lifetime, pits and trenches formation and evolution, etc.(22-24). The study of primary and secondary effects in bombardment-induced formation and evolution of cones/pyramids evolved from a mere scientific curiosity, since the early report on cone formation in 1942(25), into an extensive research activity due to the introduction of ion bombardment processing in microelectronics, and other solid state device-related technologies(1,2). The production of microrelief is a fundamental part of device fabrication. One of the advantages in studying cones/pyramids evolution was that the time scale for phenomena to take place was considered to be much shorter, so that the principles were more readily discerned. However, new etching techniques, introduced in those technologies mentioned above, have permitted us to increase the erosion rates of materials to such an extent that evolution of any microrelief can now be studied in relatively short times.

Most of the early work, whether experimental(5,6,9-12) or theoretical (17-19) on secondary and tertiary effects on surface topography evolution, described phenomena occurring in surfaces bombarded by normally incident ion beams. Later, results from normal bombardment experiments on cones/pyramids evolution(9-12) were the bases on which predictions were made(26), through phenomenological considerations, about expected asymmetries in the development of cones/pyramids and associated features (pits) on surfaces bombarded at oblique angles of incidence. These predictions(26) were confirmed by subsequent systematic experiments by the author and co-worker(27,28) and another group(29). Features observed include: (a) cones/pyramids axes oriented along the ion beam direction at an angle with respect to the surface normal; (b) a marked asymmetry in the associated pits at the bottom of cones/pyramids; (c) an apparently increasing importance of <u>redeposition</u>, in features development, as a function of bombardment obliqueness; (d) effects due to <u>beam divergence</u>, including interaction with beam defining apertures. Contemporaneously to experiments described above, deterministic two-dimensional models including ion erosion, reflection and redeposition effects, for various angles of ion beam incidence, were developed(21,22) to explain experimental results on topography evolution.

Specific examples where secondary effects play a relevant role in ion bombardment-induced surface topography will be discussed opportunely when considering technological applications of erosional/depositional processes in the following sections.

2.2. Evidence for Secondary Effects From Cones/Pyramids Evolution

Evidence for the existence and the importance of secondary and tertiary effects in ion bombardment-induced surface topography have been frequently obtained by studying cones/pyramids, flaws, steps, etc., evolution. Evidence considered in detail here will refer mainly to the recent systematic work (9-14,23,27-29) on the development of cones/pyramids and associated features, because the results were obtained by performing sequential bombardment and SEM observations following the evolution of carefully identified features; this being a very reliable method for obtaining accurate information.

Fig. 1 shows two representative stages in the evolution of pyramids developed on a polycrystalline Cu surface perpendicularly bombarded with 12 keV Kr^+ ions. Noticeable features are: (a) the shrinking of pyramids (they finally disappear(30)), (b) the development of symmetric pits around the pyramids, and (c) the enlargement of the pyramid apex angles. The origin of cones/pyramids appear to be linked to the necessary formation of initial protuberances (Fig. 2), whatever the mechanism required to create them might be(32). The evolution of protuberances into cones/pyramids and the subsequent shrinking and disappearance of the latter is, in principle, a natural consequence of the erosion process and the dependence of the sputtering yield(S) on the angle of incidence (θ) of the ion beam with respect to the surface normal(32). As is well known, $Y(\theta)$ increases from $\theta=0$ up to $\theta=\hat{\theta}$ (approximately 45-80^0, depending on ion-target material combination(31)), where $Y(\hat{\theta})$ is maximum, and finally decreases rapidly such that it may be assumed to fall to zero at a critical angle $\theta=\theta_c$ (31). θ_c has not been identified experimentally, though from what is known about sputtering(31) and ion reflection(33) as a function of incidence angle, the following ordering can be made(32):

$$\hat{\theta}<\theta_R<\theta_c<\pi/2$$

where θ_R is the angle at which ion reflection becomes total. The inequality $\theta_R<\theta_c$ has been proposed in order to allow for reflected ions still being able to sputter to a limited extent(32). The Y vs. θ dependence is responsible, in first approximation, for the shaping of protuberances, like in Fig. 2, into cones/pyramids(32). However, the pyramid shrinking, with apex angle enlargement, is not predicted by the simple first-order erosion theory(7) where parallel retreat of the pyramid sides is expected to narrow the basal area and contract the pyramid height sympathetically(14). Secondary and tertiary effects, previously defined, appear to be important contributors to the overall pyramid evolution. In particular, redeposition, on the pyramid sides, of low energy sputtered atoms and/or erosion by possible energetic sputtered atoms, both kind of particles originating from the pit inclined walls (Fig. 2), may be important contributors to the observed change in pyramid apex angles, due to changes in the pyramid side slope. Recent calculations by using deterministic models of ion erosion, reflection, and redeposition(22), developed to explain features in microelectronics patterning, show that redeposition can contribute to change the slope of steps (see Section 3.1.1). Similar calculations, still to be applied to the case of pyramids, may put on more firm bases the phenomenological hypothesis proposed above and in previous work(2,32) to account for pyramid shape evolution.

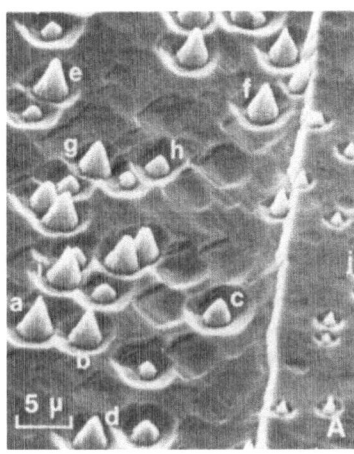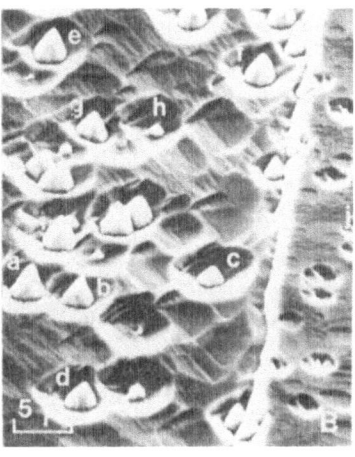

FIGURE 1. SEM micrographs of the same area of polycrystalline Cu at two consecutive stages of a sequential bombardment series with 12keV Kr$^+$ ions (target at room temperature). (A) After 6×10^{18} ions/cm^2. (B) After 8×10^{18} ions/cm^2. The shrinking and disappearance of pyramids as well as the enlargement of the apex angle are readily seen (see text for explanation) (Auciello and Kelly(11)).

Secondary effects were included on semiempirical bases in calculations of velocities of recession of surface features and critical doses for pyramid formation and disappearance(32). Equations relating the necessary dose for formation and final disappearance of pyramids and the sputtering yield were obtained, by establishing conditions for fully pyramidal shape development due to recession of critical points on a protuberance (Figs. 2a and e). The approximate dose necessary to produce a fully pyramidal shape is(32)

$$(It)_1 \approx Nh_1 / [Y(\hat{\theta}) - Y(0)]. \tag{1}$$

By considering the height of the pyramid-associated groove (Figs. 1 and 2d), which is induced by secondary effects, a necessary dose for pyramid disappearance can also be deduced as(32)

$$(It)_2 \approx N(h_1 + j) / [Y(\hat{\theta}) - Y(0)]. \tag{2}$$

Therefore, the total dose $(It)_3 = (It)_1 + (It)_2$, necessary for existence of pyramids, that is, formation and disappearance is(32):

$$(It)_3 \approx N(2h_1 + j) / [Y(\hat{\theta}) - Y(0)]. \tag{3}$$

Values calculated with eq. (3) were compared with experimental results, and reasonable agreement was observed (Table 2 of ref. 32). More recently, experimental studies, by another group(34), on pyramids formation, by using the quasidynamic method described previously, have shown that pyramids formed from asperities of height $\sim 3.5 \mu m$ were fully developed for doses of $\sim 0.2 \times 10^{19}$ ions/cm^2. This value is ~ 3 times smaller than the dose necessary for total existence for similar features (Table 2 of ref. 32) in agreement

398

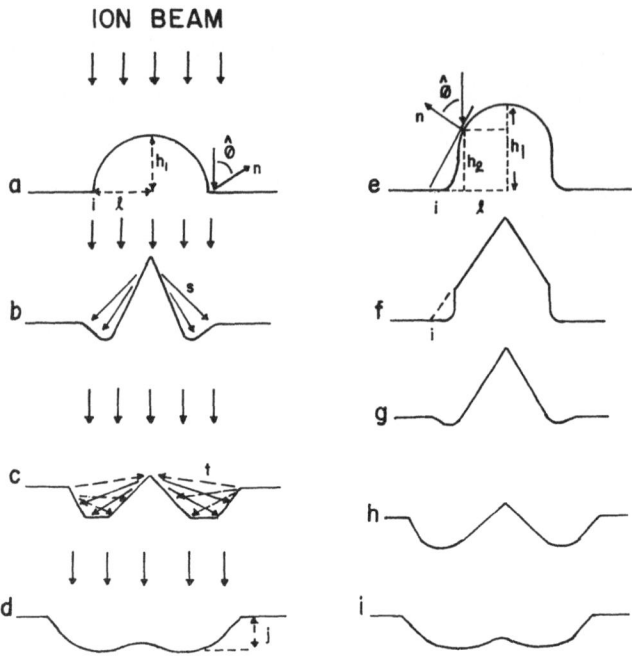

FIGURE 2. Sketches relevant to the bombardment-induced evolution of protuberances into cones/pyramids (depending on material crystallinity). (a-d)
Evolution of a "flat" protuberance into a pyramid and associated pit; secondary (s) and tertiary (t) fluxes may contribute to the experimentally observed
evolution (see text). (e-i) Evolution of a tall protuberance, similar effects as in (b-d) would be operative in this case also, once the shape is
fully pyramidal. The final disappearance of the pyramid leaves an ever-
expanding pit under continuing bombardment. According to first order erosion
theory(7) the local slope for which the ion beam-surface normal angle is $\hat{\theta}$
(angle for maximum sputtering yield) should act as a reference point of
unchanging $\theta = \hat{\theta}$. However, this can be changed due to secondary effects
(Kelly and Auciello(32)).

with differences expected between doses calculated with eqs. (1) and (3).
    Monte Carlo computer simulations performed recently(24), which included
ion reflection, redeposition, and surface diffusion, have reproduced closely
the essential features observed in Figs. 1 and 2. Details can be found in a
recent review(30).
    Still more compelling are the evidence for secondary effects observed
in a development of oblique pyramids. Due to space limitations it is not
possible to show all the evidence in this review. However, they can be
found in work recently published(27,28). Fig. 3 is just to show schematically
the evolution of pyramids developed on polycrystalline copper bombarded at
60° with respect to the surface normal (Fig. 4 of ref. 28). There are various outstanding features that can be attributed to secondary effects. (a)

Flattening of the pyramid top; this has been attributed to an uneven erosion on the pyramid top initiated by secondary fluxes (scattered ions and energetic sputtered atoms) from the beam defining aperture, due to the particular high angle orientation(27,28). Pyramids developed on a surface initially bombarded without a beam defining aperture, to equivalent doses for which flattening appeared in the previous case, did not show this feature. However, the same pyramids developed a flattened top upon further bombardment with the beam defining aperture reinstalled. This effect indicates that care has to be exercised when etching patterns through apertures and/or masks (see following sections), since secondary fluxes originating from them may affect the patterns. (b) The acute angle side of the pyramid on the right of Fig. 3 changes slope as the pyramid recedes (see also Fig. 4 of ref. 28). This can be attributed, tentatively, to a dominant redeposition of sputtered atoms from local areas upstream from the pyramid on the bombarded surface; models developed for microelectronics patterning(22) could in principle be used to study this effect. (c) The pyramid-associated pit develops preferentially on the acute angle side of the pyramid (Fig. 3 and Figs. 3, 4, and 5 of ref. 28). This is immediately understandable in terms of preferentially enhanced erosion, on this side, by secondary fluxes from the pyramid wall. (d) A ridge (tail-like feature) appears on the acute angle side (Fig. 3 and Figs. 3, 4, and 5 of ref. 28). This could in principle be accounted for by first order erosion theory(7). However, modeling(29) performed by applying this theory fail to reproduce accurately both the change of the acute angle side and the exact shape of the tail as experimentally observed(27,28). Fig. 4 is included here to show the close resemblance of some features (pits and tails) associated with oblique pyramids (Fig. 3 of this review and Fig. 7 of ref 28) and patterns for traveling wave devices(35); these features being due mainly to bombardment-induced secondary effects. Further details of oblique pyramids development can be found elsewhere(27, 28,30).

In concluding this section, it can be said that a main theoretical framework, first order erosion theory(7), has been developed to such an extent that many of the main features of ion bombardment-induced surface topography can be predicted with some accuracy. However, secondary effects need to be incorporated within that theory to reproduce and predict more accurately surface topography development. This has been done to a certain extent(18-22); however, more quantitative calculations are still partially hindered because of a lack of experimental information about secondary and tertiary fluxes. Hopefully, future work will provide the necessary information to improve and eventually complete the theoretical framework, in order to understand and predict more accurately ion bombardment-induced surface topography.

3. TECHNOLOGICAL APPLICATIONS OF EROSIONAL/DEPOSITIONAL PROCESSES
3.1 Microelectronics
3.1.1 Comparison of etching processes and problems (influcence of secondary effects and their possible solutions in microcircuit patterning techniques).
(i) "Conventional" inert gas ion-beam etching. The term "conventional" is used here to distinguish ion beam etching techniques in which patterns are produced by substrate irradiation through suitable masks,(1) from the more recently introduced technique in which highly collimated beams are used to produce patterns without the interposition of masks(36,37).

Advantages of this technique are: (a) energy and current density of the ions can be varied independently; (b) since the space around the etched surface is nearly free of electric fields, the substrate can be bombarded at any desired angle. This can help to achieve improved texturing conditions

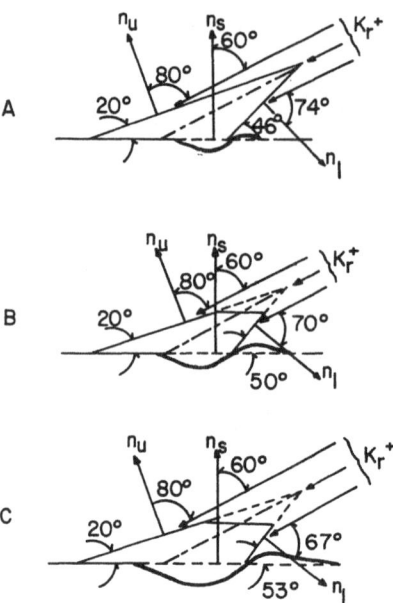

FIGURE 3. Sketches of three stages in the evolution of $60°$ pyramids as observed in Fig. 4 of ref. 28. The angles are mean values obtained from measurments on several pyramids in different side-view micrographs. All are within $\pm 1°$. (A) After 3×10^{17} Kr^+/cm^2. (B) After 1.1×10^{18}. (C) After 1.4×10^{18} (Auciello and Kelly (28)). Compare with Fig. 4.

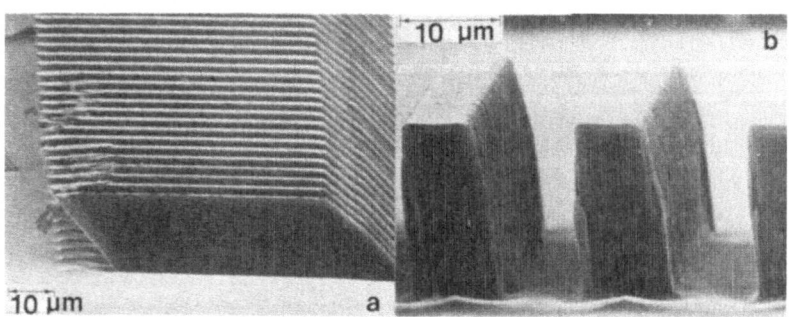

FIGURE 4. SEM of a diamond etched with a 2keV Xe^+ ion beam ($1mA/cm^2$) to a depth of $17\mu m$ at $45°$. (a) Side view, and (b) end-on view (courtesy of M. W. Geis(35)).

while developing the patterns(1), due to the strong dependence of the etch rate of most materials on the angle of incidence (θ) of the ions with respect to the surface normal(31). There are materials, however, like Pt and Ti which present only minor changes in sputtering yield as a function of (θ).(1) This makes them very useful materials for masks, since these should have an erosion as independent as possible of θ; (c) the gas pressure can be kept low enough in the target chamber to provide with a large mean free path for the ions, which can reduce gas pressure-induced secondary effects like back-scattering of sputtered atoms from gas molecules (this can be a non negligible effect in plasma etching, see following subsections); (d) insulators can also be etched, provided that positive surface charging created during bombardment is neutralized by an electron bombardment.

Disadvantages of this method include: (a) secondary effects, of the kind already analyzed in Section 2. Scattered ions and energetic sputtered atoms from steep pattern walls can enhance erosion at the bottom intersection between the walls and local flat regions of the substrate. This generally results in development of trenches along the pattern walls (Fig. 5). Even more undesirable is the possible effect of redeposition of sputtered material on the steep pattern walls of resists or metallic masks used for patterning (1), which may lead to interconnection of isolated regions and shortcircuits (see Fig. 12 of ref. 1). Obviously, the quality of any etching method depends on the accuracy with which mask patterns are transferred onto substrates, this process being dependent on the mask material used. Different photoresists and metallic materials are currently being utilized as reliable masks(38).

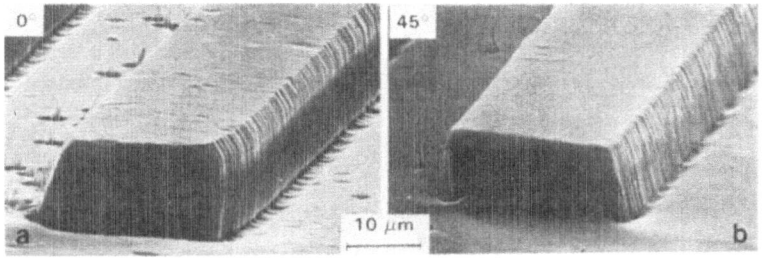

FIGURE 5. SEM of a pattern developed on diamond etched to a depth of 10μm by 2keV Xe⁺ ion beam bombardment (1mA/cm²). Sample bombarded at (a) normal incidence and (b) at 45° with respect to the surface normal. Notice the faceting on top of the pattern and the trenching at the bottom (courtesy of M.W.Geis(35)).

Fig. 6 shows a clear example of redeposition effects and faceting. The latter is attributed to the primary erosion process, for which the sputtering yield vs. angle of ion incidence dependence has a well-defined maximum at some angle in a wide range of 40-80° depending on bombarding ion-material combination(1,31). Due to this angular dependence, facets may be initiated at the upper corners of the side walls. Redeposition effects, on the other hand, have been simulated lately(21,22)(Fig. 7, for example, see also R. Smith's chapter in this proceedings) with a certain degree of accuracy. In general, the build-up of material is more uniform for larger groove widths.

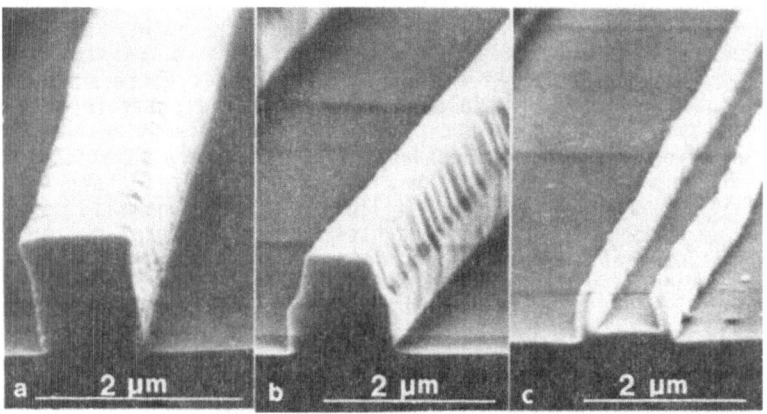

FIGURE 6. SEM micrograph illustrating the phenomena of faceting and redeposition during ion etching. (a) AZ-1350-photoresist mask on a Si substrate prior to ion beam etching. (b) After etching to a depth of 1300Å. (c) The redeposited material left after dissolution of the photoresist in a solvent (courtesy of H. I. Smith (16)).

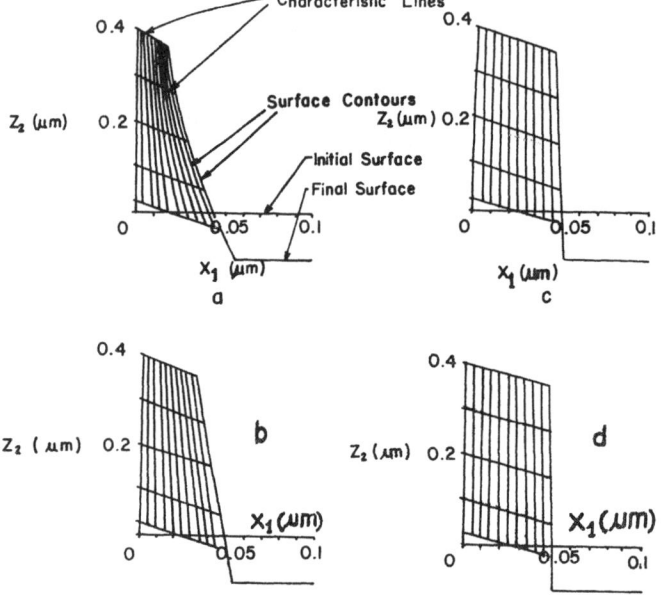

FIGURE 7. Theoretically calculated redeposition profiles for rectangular grooves of different widths and for different depths of erosion using the cosine distribution for the sputtered particles. The depth of erosion corresponding to each surface contour is 10nm and the groove height is 0.4μm. Groove widths are: (a) 0.5μm, (b) 1.0μm, (c) 3.0μm, (d) 6.0μm (courtesy of R. Smith (21)).

The surface contours, although calculated for equal erosion times, are not evenly spaced because as material is being redeposited the groove width is continuously contracting. The average angle of inclination of the groove walls to the horizontal changes with the depth of material eroded. Results show that this angle diminishes, as a function of groove width, from 90° for larger groove widths. Additionally, the simulations have shown that the ratio between the amount of redeposited material at the bottom of the groove and the erosion depth is generally larger for smaller groove widths. The results of this simulation(21,22) do not include a nonnegligible secondary effect, that is, as the redeposited material builds up, the groove walls are no longer perpendicular to the surface. Thus, these walls themselves are subjected to erosion by the bombarding ion beam, and the redeposited material can be subsequently sputtered and secondary redeposition occurs. The cited model(21,22) for deposition accounts for the continuously changing geometry of the profiles analyzed, which is an improvement over previous calculations (18,19). However, the improved calculations still do not include the effect of simultaneous erosion of redeposited material by ion beam etching, or the effects of ion reflection, which may alter the redeposition by affecting the bottom of the groove (source of redeposited material) due to trench formation (1,38,39); the trench being formed at the bottom of the groove walls by combined primary beam-scattered particles and energetic sputtered atoms bombardment.

Different masking techniques have been developed lately, which allow minimizing or eliminating the redeposition problem. Fig. 8 shows, schematically, in a self-explanatory form, some of these methods.

Another problem associated with inert gas ion beam etching is the smaller selectivity that can be achieved, in the erosion process, when compared with chemical etching (following sections). The reason is because in the present case the etching is mainly produced by physical sputtering (momentum transfer), and although different materials sputter at different rates, the difference is not as marked as in the case of chemical sputtering present in plasma etching.

(ii) "Conventional" Reactive Ion-Beam Etching. One of the most relevant aspects of an etching process, for microcircuit fabrication, is high fidelity in pattern transfer from masks. The two most important requirements to achieve this are anisotropy and selectivity in the etching process. Inert gas ion-beam etching can have the first but it lacks good selectivity. One way to achieve both, simultaneously, is to use ion beams where the ions are themselves chemically active with respect to the etched materials. Thus, reactive ion beam etching (RIBE) emerged, lately, as a reliable alternative for microdevice fabrication. Systems used in RIBE and etching mechanisms related to this technique are extensively discussed in recent reviews (3,4,41).

(iii) Plasma and Reactive Ion Etching (RIE). Plasma etching produces a selective erosion of materials by formation of volatile molecules due to chemical reactions between low energy (<1eV) chemically active species, created in the plasma, and surface atoms of the etched material. Plasmas commonly used consist of ionized gases containing highly reactive ions, free electrons, and free radicals. These plasmas are generally characterized by their relatively low temperatures ($50-250^{\circ}C$), high pressures (13-400 Pa), and etch gas flow rates of 50-500 cm^3/min. Two methods are currently being used to excite the plasmas, namely, a d.c. or a r.f. power-induced discharge. The latter is the most generally used. Unlike RIE(41), no bias potential is present in plasma etching, therefore physical sputtering due to energetic ions is in principle negligible, and the etching is produced mainly through chemical reaction processes as described above. Some of the main advantages

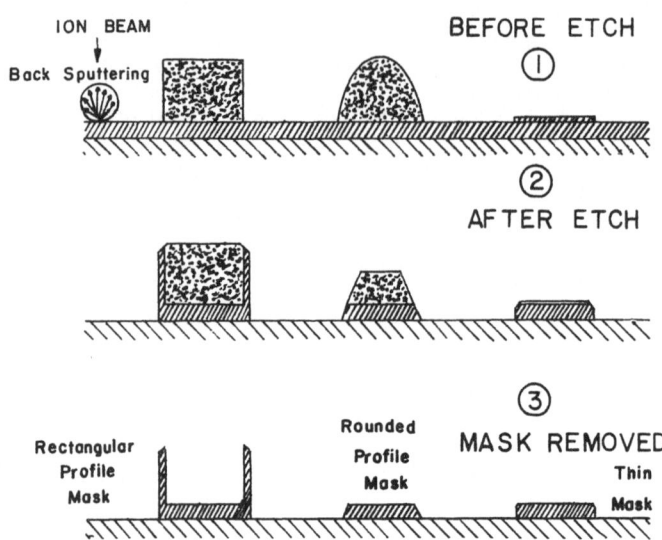

FIGURE 8. Different masking geometries for ion beam etching. Rectangular profile masks often lead to severe redeposition problems after etching. Rounded masks or thin masks overcome this problem (R. E. Lee(40)).

of this technique are: (a) the relatively simple equipment needed; (b) the possibility of etching large samples (with diameters of 30 cm or more); this being extremely important in production applications. Disadvantages include: (a) etching by species striking the surface at normal incidence mainly, due to the particular potential distribution around the target, therefore inhibiting the possibility of etching at oblique angle of incidence, which in some cases help to avoid some secondary effects; (b) the discharge is produced in a relatively high pressure background of gas. This can lead to a non-negligible effect of sputtered particles diffusing back to the target after colliding with plasma species; an effect which can affect the etch rate and also contribute to redeposition of sputtered material, this in turn affecting the pattern dimensions. In fact, some studies(42) have shown that the back diffusion effect may limit the application of RIE whenever resolution and line width of the order of 1μm are required; (c) finally, plasma etching produce some undercutting at the mask-substrate interface. This undesirable effect has been attributed to the high surface mobility of etching species adsorbed on the surface.

Reactive ion etching (RIE) has the advantage, over plasma etching, of being able to produce an enhanced anisotropy in the etching process, while keeping a high chemical etching selectivity. A variation of RIE (where the ions are themselves chemically reactive) is the bombardment of materials by inert gas ions (Ar^+, Kr^+, etc.), while simultaneously exposing them to sub-eV chemically active species. An intensive research activity is presently being carried out in order to understand the mechanisms responsible for etching in both RIE and inert gas ion-enhanced chemical etching. Current experimental results and models suggest that synergistic effects, due to the interaction of energetic ion (inert or reactive)-induced effects and chemically reactive species are responsible for an enhanced erosion observed. Synergism is defined as a phenomenon whereby the combined effect of independent processes is different than the linear superposition of the same effects when occurring separately. Recent reviews on synergistic effects(43,44) indicate that this seems to be a general phenomenon which occurs in erosion of chemically reactive materials like Si, Ge, GaAs, etc., when exposed to fluorinated, chlorinated, etc., environments, in microelectronics processing, and C, TiC, SiC, etc., when exposed to hydrogenated environments, in fusion technology, for example. In the case of semiconductors, synergism is desirable because it produces an enhanced-highly directional erosion, which can be used to obtain patterns with vertical side walls and little or no undercutting of masked features(41). Several mechanisms have been proposed to explain the enhanced erosion due to synergism in semiconductors, and they are described in detail in a recent review(43), therefore no further discussion will be presented here, but a brief schematic explanation in Figs. 9a and b, of the two leading hypotheses proposed to explain ion-enhanced etching. However, whatever the valid mechanism might be, synergistic-induced erosion of semiconductors seems to be a good alternative for the production of high quality patterns in microelectronics.

Additionally, it may be of interest mentioning that recent results on synergistic erosion(43) indicate that even with low energy ions (<300eV) the synergistically-induced enhanced erosion appears to be substantial. This could be advantageous to ion-induced etching processes, in that surface damage could be substantially diminished. In some applications, the surface damage introduced during the etching process has to be subsequently annealed(2), which introduces an undesirable extra step in the fabrication of microcircuits.

(iv) Maskless Ion-Beam Etching. One of the problems associated with "conventional" ion bombardment-induced etching techniques, described previously, is related to the use of masks to produce a desired pattern on an appropriate substrate. A promising new technique is currently being developed based on the ion beam etching method. The outstanding characteristic of this method is that a highly focused beam of submicrometer (1000-5000Å) diameter can be scanned over a surface and engrave a pattern without using any mask (see Fig. 11.5 of ref. 2). Besides having the advantages of the "conventional" ion beam etching methods described previously, the present maskless technique has the extra advantage of high lateral resolution. In addition to its potential use for controlled etching of patterns, this technique can be applied to create active microdevices (transistors, diodes, etc.) by doping submicrometer areas in microcircuit patterns. Since the time of publication of a recent review(2), where this technique was described, a large number of publications related to this subject have appeared in the literature, which indicates the great interest that this potentially powerful technique has aroused in the semiconductor processing field of research. Details of this method can be found in a recent review(2), and therefore no further discussion will be presented here.

3.1.2 New Emerging Etching Techniques for Microcircuit Fabrication
(i) Combined Etching/Deposition Techniques for Formation of Microfeatures on Surfaces.
(a) Planarization of Patterned Surfaces. The existence of steps in the topography of patterned surfaces presents several obstacles to the achievement of higher speed and higher packing density of integrated circuit devices(46). The minimum size of features in lithographic techniques is severely restricted by resolution and line width control limitations associated with the presence of steps; poor coverage of stepped surfaces may lead to unreliable metallization interconnections. Vertical stacking of integrated circuits is almost impossible to achieve on a stepped topography.

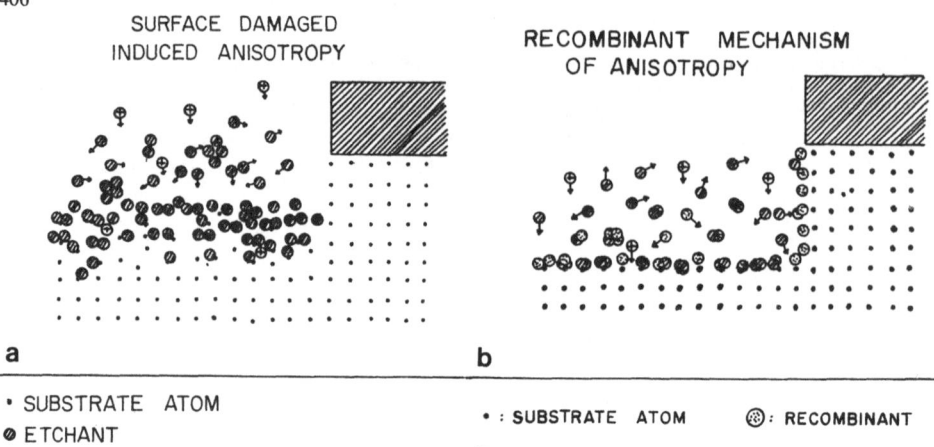

FIGURE 9. Mechanisms for ion + reactive species bombardment-induced enhancement in erosion of semiconductors. (a) Damage (bond breaking and/or atoms displacement) produced by normally incident ions leads to accelerated chemical reaction between etchant species and atoms on horizontal surfaces, which produce anisotropic etching. (b) Recombinant mechanism; recombinant highly bound radicals adsorbed on the vertical walls of the pattern inhibit reaction between etchant species and surface atoms, while ion-induced recombinant desorption or ion-enhanced substrate gasification proceeds on horizontal surfaces (Flamm and Donnelly(45)).

Minimum feature size and packing density are also limited by the lateral oxidation characteristics of local oxide formation on patterned surfaces. These limitations would be removed if the recesses on a patterned surface were filled with an inert material and the surface planarized prior to subsequent processing operations(46).

Planarization can be accomplished by the technique shown schematically in Fig. 10. A patterned surface S is covered by a film of inert material B followed by a film of liquid-like material C so that the top surface C is planar. C and the excess material B could in principle be removed by plasma etching, inert gas ion beam etching, RIE, or RIBE, in such a way that a planar surface will be preserved throughout the etching process, provided that the etch rates of B and C are practically the same. In general, the normal incidence etch rates of B and C are different, which would preclude the applicability of this technique. Fortunately, there are some materials, like SiO_2 and AZ-1350 photoresist, which have the same erosion rate when bombarded at a particular angle (59° with respect to the surface normal for SiO_2 and AZ-1350). It has been demonstrated that these irradiation conditions can produce planarization of SiO_2 covered patterned Si surfaces(46). The condition mentioned above indicates that only ion beam etching methods would be suitable for application in this technique, because of the controllability of the bombarding ions angle of incidence, which in principle cannot be easily obtained for plasma and reactive ion etching. However, it has been shown recently(47) that directional reactive ion etching can also be accomplished at oblique angles. This is achieved by placing a grid, tied to the cathode (target) potential, in front of the tilted substrate to be etched, in such a way that a planar dark space is formed between the sample and the plasma. In practice, the grid forms a portion of the top surface of an

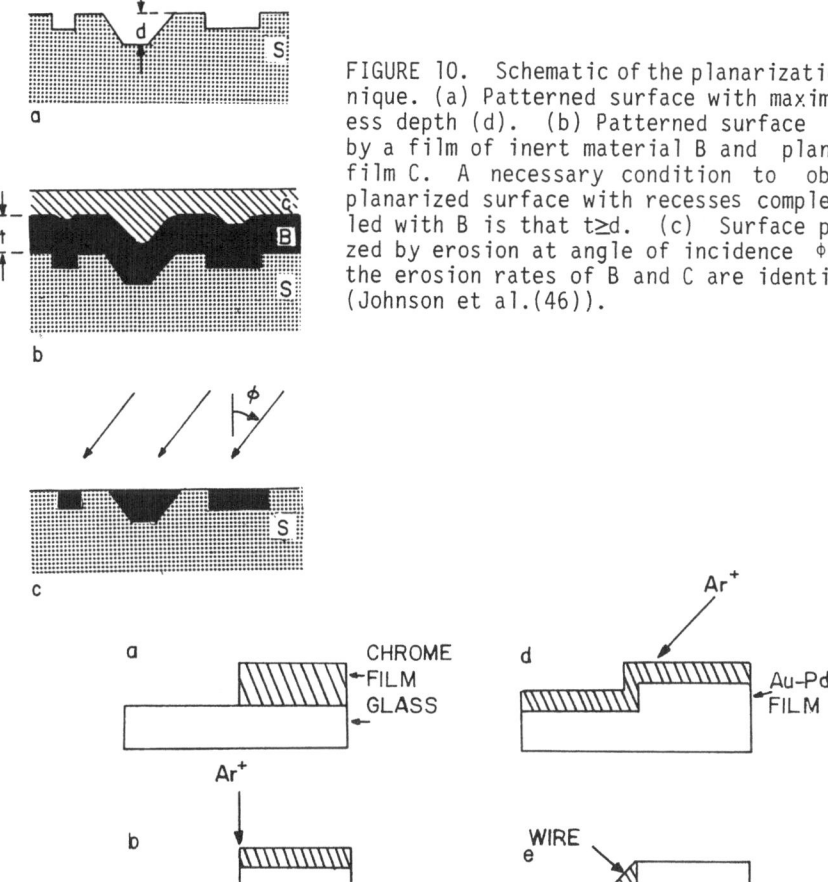

FIGURE 10. Schematic of the planarization technique. (a) Patterned surface with maximum recess depth (d). (b) Patterned surface covered by a film of inert material B and planarizing film C. A necessary condition to obtain a planarized surface with recesses completely filled with B is that t≥d. (c) Surface planarized by erosion at angle of incidence φ where the erosion rates of B and C are identical (Johnson et al.(46)).

FIGURE 11. Fabrication procedure for formation of fine metal wires. The substrate used for this particular example was microscope cover glass, which side view is shown schematically. Procedures a-c are the same for both process A and B. (a) Half the substrate is covered with a thin Cr film. (b) The substrate is ion etched to produce a square step. (c) The Cr film is removed with a chemical etch. Process A: (d) The substrate is coated with the metal film (e.g.Au-Pd) and ion etched at an angle until stage (e)is obtained. Process B: (f) A second metal film evaporation is produced parallel to the substrate to coat only the step edge. A subsequent ion etching normal to the substrate may be required to remove the light coating on the rest of the substrate (Prober, et al.(48)).

equipotential cathode Faraday cage which is placed over the substrate, to eliminate any distorting fields or plasma within the box. Thus, ions from the plasma are accelerated across the dark space, pass through the cathode grid at normal incidence and drift, presumably unimpeded, to the tilted substrate. Care must be exercised to minimize deposition on the substrate of grid sputtered material. Details on this and other aspects of this technique can be found elsewhere. Gratings with smooth walls, without faceting or trenching at the bottom, have been obtained on etched fused silica.

In conclusion, the planarization technique assisted by the well-known etching methods can become a powerful alternative for microcircuit fabrication, particularly for very large scale integrated (VLSI) circuits.

(b) Fabrication of Microscopic Metal Lines on Surfaces. The formation of fine metal wires, 300 Å wide and 0.5mm long, on surfaces has recently been reported(48). The wires were produced by using a combined deposition photolithographic-ion-milling technique. Fig. 11 shows a schematic and brief explanation of the whole procedure. Although glass was used as a substrate to demonstrate the technique, other materials (semiconductors, ceramics, etc.) could in principle be used as substrates. With the processing methods outlined in Fig. 11, the yield of uniform and continuous wires is high. This is largely due to the excellent uniformity of the ion-milling process and its application in a fully self-aligning procedure(48). Both processes (A and B) described in Fig. 11 can be used with a variety of substrates metal film combinations. The procedures described above cannot produce high density or complex patterns. However, they are suitable for production of individual conducting wires for studies of one-dimensional effects in superconducting, magnetic, or normal metals, and in Josephson devices.

(ii) Laser-Induced Chemical Etching and Deposition For Device Fabrication. First studies(49-51) demonstrated the feasibility of producing photon enhanced chemical reactions by using lasers. However, it was only recently that several groups started more systematic research in this field, encouraged by its potential application in microelectronics. It was observed that laser photons with wavelengths in the range from UV to the infrared could be readily focused onto solid surfaces and therefore be suitable for promoting surface reactions with high spatial resolution. Also, the monochromacity, the coherence and the high photon flux of the laser light were thought to be advantageous features. Indeed, laser-induced chemical etching(52,53), doping of semiconductors(54), and chemical-enhanced vapor deposition(54,55) have been demonstrated. Laser-enhanced electrochemical plating and etching for liquid-solid systems have also been reported(56).

Recent work has been directed at trying to identify the mechanisms responsible for photon-induced enhancement in chemical etching. The most favored are briefly described in the following.

(a) Chemical Etching Activated by Molecular Vibrational Excitation. It has been shown that laser irradiation was able to excite reactive molecules into highly vibrationally excited states and thereby enhance the process of dissociative chemisorption and subsequent surface reactions to form volatile products(57,58). This mechanism was demonstrated for the $Si-SF_6$ system, where reaction occurred only when SF_6 molecules were vibrationally excited. It was shown that the surface reaction yield as a function of laser intensity used in the experiment follows the relation $EY=I^{3.5}$, indicating that three or more photons were likely to be involved in promoting SF_6 molecules into high vibrational levels to overcome the activation barrier for reaction. Furthermore, it was observed that deactivation collisions of excited molecules with the gas reduced the lifetime of the excited state, therefore localizing the excited molecules to a small region just above the

Si surface.

(b) Chemical Etching by Photon-Generated Radicals. Photons can also dissociate molecules to produce reactive radicals either by multiple photon excitation of the ground electronic state or single photon photolysis involving excited electronic states. This effect was also observed with the $Si-SF_6$ system. The main difference between this and the vibrational excitation mechanism is that even fluorine atoms generated several millimeters away from the surface can still survive collisions with other gas phase molecules and diffuse to the surface to react over it.

(c) Chemical Etching by Photon Excitation of Solids. Surface reactions can also be induced by the solid excitation alone. This effect has been observed, as an example, in the interactions of XeF_2 gas with Si, SiO_2, Ta, and Te films. The gas molecules do not absorb infrared photons of the laser used for this case, and the only effect of the radiation is to cause lattice excitation and heating so that the F atoms landing on the surface can rearrange to form SiF_4 molecules which subsequently desorb into the gas phase.

Other reactive gas-semiconductor systems have been studied also, e.g., $Cl_2 \rightarrow Si(52-b)$, $HCl \rightarrow Si(52-b)$, $CH_3Br \rightarrow GaAs$ and $InP(52-a)$, and results similar to those described above have been obtained. In summary, laser-induced enhancement of chemical etching can be used to produce rapid/high-resolution etching of semiconductors to create complex patterns for microcircuit application. The technique allows production of maskless definition of surface relief structures or maskless separation of active regions on semiconductors. A spatial resolution of $\approx 1 \mu m$ has been achieved in semiconductor patterning.

Laser-induced deposition of material has also been investigated(54). As an example, gases of metal-alkyl-compounds trimethylaluminum [(TMAℓ)/Aℓ/ $(CH_3)_3$] and dimethyl cadmium [(DMCd)/Cd/$(CH_3)_2$] were irradiated with a CW argon-ion laser with a wavelength of 257.2mm after filtering. The gas molecules were dissociated by flash photolysis in front of a substrate, which resulted in the deposition of the metallic atoms over it. Laser photodeposition can be used to generate directly metal patterns without the intermediate photolithographic steps required in conventional microelectronics fabrication. The highly focused laser beam can produce molecules dissociation in a narrow region, which have resulted in narrow deposition patterns. Deposits narrower than $\sim 2 \mu m$ have been produced by UV laser-induced photodissociation of organometallic compounds. Furthermore, by applying laser photodecomposition to two-component gas mixtures, it may be possible to deposit compounds(54).

Laser-induced etching may help to overcome one of the major disadvantages of ion-beam etching, that is, the residual surface damage introduced during the etching process; this involves the ejection of atoms from the surface as a result of the collision cascade initiated by momentum transfer between the incident ion and substrate atoms(59). The implications of a damaged surface layer depend on the application, some are unaffected by surface disorder while others require complete elimination of damage. A notable example of a device extremely sensitive to bombardment-induced damage is the metal-oxide-semiconductor field effect transistor (MOSFET), one of the cornerstones of modern microelectronics technology. Low energy (<1keV) Ar^+ bombardment of the gate surface of a Si-MOSFET, for example , produces a high density of interface states, low carrier mobility in the conducting channel, and surface leaking current resulting from redeposition of metal atoms(60). Yet all of these detrimental effects are eliminated by removing the 50-100 Å thick damaged layer with a brief chemical etch(60).

Much work is still necessary to better understand the laser-induced etching mechanisms, which in turn may help improve the etching rates achieved with this technique (for example, $\sim 6 \mu m/s$ for Ar-ion-laser-induced etching of

Si exposed to Cl_2 or HCl gases) and increase its potential for semiconductor and eventually other materials processing.

3.2 Surface Optical and Acoustical Technologies

3.2.1 Surface Optical Devices. Ion beam applications in optical technology involves the construction of: (a) integrated optical circuits(61), (b) semiconductor-diode lasers(61), (c) interference gratings for spectroscopic applications(62-65), and (d) textured surfaces for optical storage of information(66-68).

The physics and methods for pattern formation of integrated optical circuits are similar to those used in microelectronics, the main difference being that in the optical case the circuits are produced by etching patterns on multilayer films composed of a high refractive index glass or organic polymer film placed between lower refractive index glass or polymer films. The light is confined in the circuits (waveguides) by the total internal reflection phenomenon, which is present whenever a light wave travelling through one material strikes a flat interface separating it from another material that has lower refractive index(60). Semiconductor-diode laser construction involves similar principles as those used, in general, in integrated optical circuits fabrication(60).

Perhaps interference grating fabrication is the ion-induced etching application to optical devices where secondary effects show a more noticeable influence. A schematic illustration of effects taking place near a vertical mask wall is shown in Fig. 12 and have been observed experimentally (see Fig. 14 of ref. 1). Surface gratings are commonly prepared by producing a grating pattern in a thin layer of photoresist and transferring the relief pattern onto the underlying substrate by ion beam etching, chemical etching, or plasma etching techniques. The versatility of ion beam etching to produce high quality gratings of any period is demonstrated in Fig. 13, which shows the process to doubling a triangular grating of period Λ. Grating periods of 750Å have been produced by this technique(64). Damage produced by ion-beam etching is also a drawback in grating processing and subsequent steps to the etching process have to be applied in order to eliminate the damage. Further details on gratings processing by ion-induced etching can be found elsewhere(62-65).

Recording media for use in high-speed/high density optical information storage systems appears to be a new promising application for ion bombardment-induced textured surfaces. Much of the attention in this field of research has been focused on the production of read-only storage materials with direct read-after-write capabilities(69). One of the primary concerns in this field of research is the development of materials with long-term stability and low power writing requirements(66). Initially, light absorbing films have been used for this kind of application. A laser beam is modulated by an electrical signal, which carries the information to be stored (i.e., either digital or analog information). The modulated beam is focused onto a small (<1μm diameter) area on the recording medium and burns a hole on the light absorbing film, which is deposited on a reflective substrate(70). The information is thus stored as a series of "spots" along a spiral track. The read-out of the stored information is afterwards obtained by passing an unmodulated less intense laser beam (to avoid further recording) over the path previously traced by the recording beam. The light reflected from the recorded "spots" is then transformed, via a photodetector, into an electrical output.

An alternative to absorbing films, which have a tendency of peeling off from the substrate as a drawback, is a surface covered with a dense array of columnar structures. Reactive ion etching can be used to produce these sur-

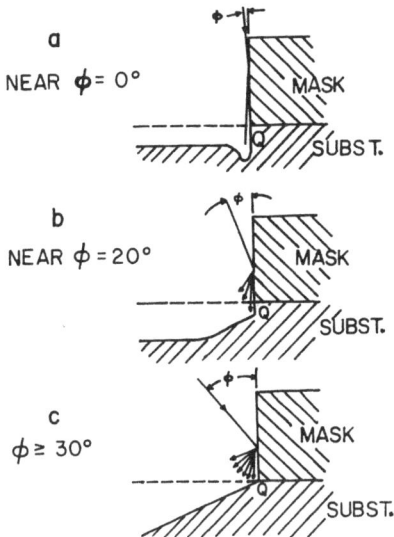

a

NEAR $\phi = 0°$

b

NEAR $\phi = 20°$

c

$\phi \geq 30°$

FIGURE 12. Schematic illustration of the change in erosion behavior of the substrate near the base of a vertical mask as a function of angle of incidence of the ion beam. At ϕ near $0°$, ion reflection from the wall leads to enhanced erosion or trenching. With increasing angle of incidence, sputtering from the wall and redeposition on the substrate lead to reduced erosion near Q. (L. F. Johnson(65)).

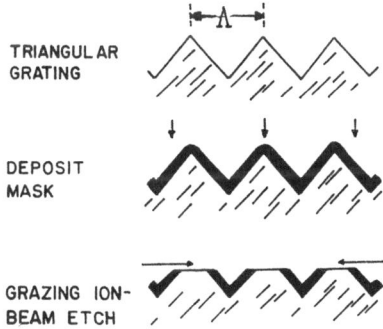

TRIANGULAR GRATING

DEPOSIT MASK

GRAZING ION-BEAM ETCH

ION-BEAM ETCH

DISSOLVE MASK

FIGURE 13. Spatial frequency doubling of a triangular grating of period Λ. After depositing a mask material at normal incidence, the peaks are removed by ion-beam etching at near grazing incidence. A grating of period $\Lambda/2$ is then obtained by ion-beam etching (L. F. Johnson(64)).

faces, with columns of cross-section smaller than the wavelength of the light used for recording. Figs. 11.11 and 11.12 of ref. 2 show a schematic cross-section of a textured Si storage medium, and a micrograph of reflected light from recorded "spots" on a textured Ge surface, respectively. The advantage of this recording medium is that different materials can, in principle, be used and therefore no limitation exist imposed by the use of special absorbing films. Details related to the production of textured optical storage media and necessary laser characteristics can be found elsewhere(66).

3.2.2 Surface Acoustical Devices. Surface gratings also provide the foundation for surface acoustic wave (SAW) devices based on resonant reflection of surface acoustic waves(71). Ion bombardment-induced etching can be used to develop grooves on the surface of substrates, in such a way that the grooves' depth can be varied in a controlled manner, to produce a desired characteristic for SAW reflective array grating filters. High resolution electrode arrays with 0.5μm period and low insertion loss have been fabricated for SAW filters operating at 1.89 GHz(72). Recently, SAW resonators have been fabricated in a ZnO/SiO$_2$ layered medium on a Si substrate, with the objective of producing monolithic integration(73). The etching techniques described in previous sections are applicable to processing of SAW devices, therefore no further details are necessary here.

3.3 Solar Energy Conversion Technology

Optically selective absorbing surfaces are desirable for manufacturing photothermal solar energy convertors. Such selective surfaces strongly absorb solar radiation but lose little of the absorbed energy by reradiation. The antireflection necessary for good solar energy convertors' performance has been obtained in the past mostly by the use of antireflection coatings. (2). These, however, are only effective over a limited wavelength range and involve some complexity and stability problems always associated with additional layers(2). A novel highly competitive alternative, which virtually eliminates the reflectivity losses without the use of coatings, involves the use of textured surfaces. These consist of a dense array of sharp cones or columnar structures, which can be obtained either by ion bombardment or chemical vapor deposition. A detailed description of surface texturing for this kind of application has been presented in recent reviews(1,2) and will not be repeated here.

3.4 Film Deposition Technology

3.4.1 Ion Bombardment and Vapor Incidence Angle Effects on Film Topography and Some Properties. The study of film formation has evolved into a very dynamic and rapidly expanding field of research, which has been stimulated by a continuously growing number of applications to different technologies. It is completely out of the scope of this paper to present a detailed account of previous and present work in this field of research. Extensive reviews have been published on some topics of interest to this summer school, namely, (a) techniques for film deposition(74), (b) surface morphology in vapor-deposited films(75,76,77), (c) ion bombardment effects on film formation(77-82), and (d) films application to different technologies(83,84).

The topic of major interest to this school is that related to the surface morphology of deposited films whether affected or not by ion bombardment. A general observation is that crystalline and amorphous thin films produced by vapor deposition exhibit a columnar microstructure(75); this consists of a network of low density material that surrounds an array of parallel columns of higher density. The columns are generally oriented along the direction of atoms arrival to the substrate. Therefore, the columns will be normal or at oblique angles with respect to the surface, depending on the angle of arrival of atoms. Many techniques have been used to study the columnar microstructure(75), such as microfractography, transmission electron microscopy, and small angle electron and X-ray scattering. The effect of this structure on the physical properties of thin films if often remarkable. Magnetic, optical, electrical and mechanical effects have been reported(75). The surface topography, specific surface area, oxygen uptake, and adhesion to the substrate are also known to be affected by the existence of the columnar structure(75).

Many investigators have noted that the formation of columns depends upon the deposition conditions—substrate temperature, deposition rate, angle of incidence of the vapor atoms, and vacuum ambient—as well as upon the material itself. In every case, however, it appears that columnar structures are formed only when the mobility of the deposited atoms is limited(75). Columns are observed, for example, in films of high melting point materials (Cr, Be, Si, and Ge), in compound materials of high binding energy (CdTe, CaF$_2$, and PbS), and in non-noble metals evaporated in the presence of oxygen (Al, Fe, and Ni-Fe). Amorphous films (Si, Ge, SiO, and rare earth-transition metal alloys) whose existence in the amorphous state depends upon a limited atomic mobility, universally exhibit a columnar structure when deposited at sufficiently low temperatures. An increase in substrate temperature leads to the eventual elimination of the columnar structure(75).

A hypothesis widely accepted is that columns can be formed only if atoms being deposited on the substrate shield or "shadow" unoccupied sites, surrounding them, from direct impingement of subsequently arriving atoms, and if, furthermore, post-impingement migration is inhibited by low substrate temperature. A detailed description of this mechanism and computer modeling, which agrees well with experimental observations, can be found elsewhere(75).

Simultaneous ion bombardment of the growing film helps to eliminate the columnar structure(79). Fig. 14 shows, schematically, a model proposed to explain the ion bombardment-induced effect. Micrographs of ion bombardment-assisted deposited films shows good agreement with the proposed model (Fig. 2 of ref. 79 and Fig. 5 of ref. 82). The morphological changes observed in the figures mentioned above represent further examples of the influence of secondary effects (redeposition) in the topograhical evolution of ion bombarded surfaces. Fig. 15 shows an example of properties alteration of films,

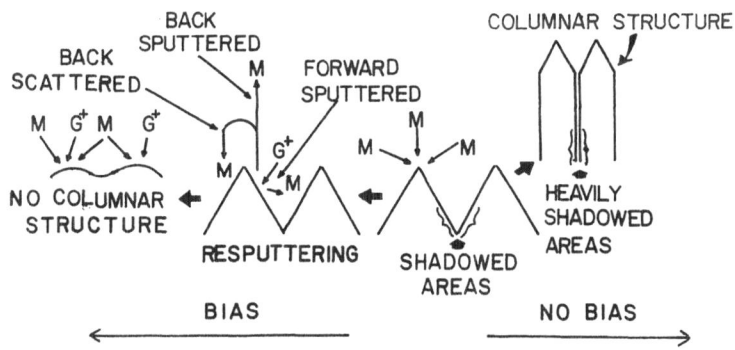

FIGURE 14. Model of growth modification by ion bombardment during deposition. Ion bombardment may produce erosion of the column tops and filling of the valleys, due to redeposition of sputtered atoms from the top. An additional process may occur when the films are grown by sputter deposition in r.f. discharges, i.e., sputtered atoms may redeposit on the growing film, due to backscattering with plasma species (Bland, et al.(79)).

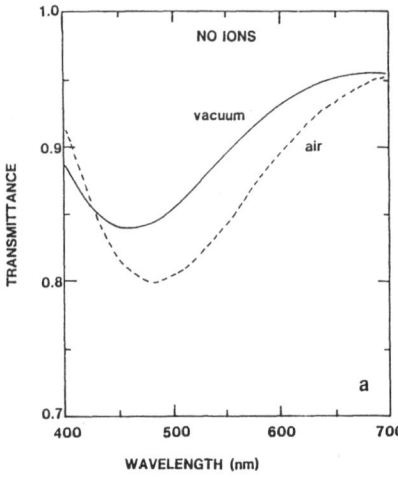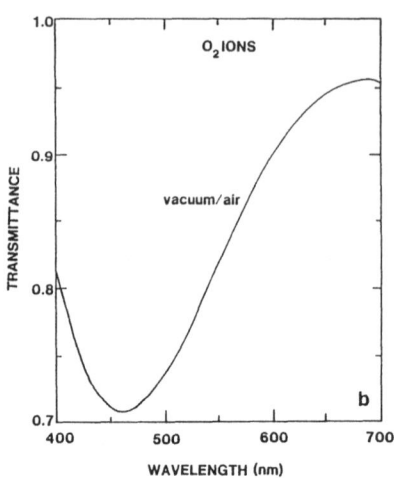

WAVELENGTH (nm)

FIGURE 15. Transmittance over the visible spectrum of a ZrO_2 film deposited: (a) without and (b) with simultaneous O^+ ion bombardment. Transmittances measured in vacuum and air are represented by solid and broken lines, respectively. The change in the film refractive index has been attributed to absorption of water vapor into the microvoids of the unbombarded film, the presence of water being inferred from hydrogen presence detected via nuclear reaction analysis (P. J. Martin, et al.(82)).

due to ion bombardment.

3.4.2 Examples of Novel Technological Applications of Films. Numerous applications of films have been described in the literature (as an example, see ref. 85), in a more extensive and comprehensive form than could be done within the limitations of the present review. However, a whole new era of possible novel applications of films, in the new frontier of space, may have been opened with the successful initial missions of the Space Shuttle. Therefore, it seems appropriate to mention here, as an example, studies on the capabilities of sputtered coating for protection of spacecraft polymers. Early Shuttle flights have demonstrated that materials such as polyimide (Kapton®), carbon coatings, and some paints undergo weight loss and changes in optical properties when exposed in low earth orbit(86). The survival of those coatings as solar cell blankets or thermal blankets may be in jeopardy when exposed to oxygen-induced chemical erosion during Shuttle flights(87). Micrographs of Kapton samples exposed during the Shuttle flight STS-8 have shown that the material became mat in appearance; this due to light absorption in a dense forest of microscopic fibrils or cone-like structures (Fig. 16), which appear to be the result of oxidation by a directed low energy (~5eV) atomic oxygen flux(88), which impacted on the exposed Shuttle's surfaces when in flight in low earth orbit.(87) Provided this is the case, researchers may have obtained first evidence for this kind of surface topography development by other means than energetic ion (>1keV) bombardment.

Ion beam sputter deposited thin films of Al_2O_3(700Å), SiO_2(650Å) and >96% SiO_2/<4% polytetrafluoroethylene (PTFE) were found to be effective in preventing oxidation, therefore erosion, of Kapton when exposed to oxygen in a Shuttle flight. Evidently the space frontier is being opened to the use of films.

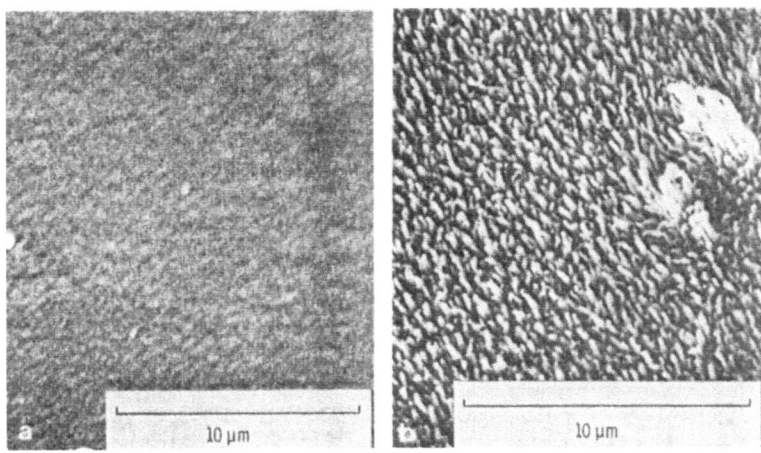

FIGURE 16. Unprotected Kapton® polyimide: (a) before and (b) after space exposure (courtesy of Banks, et al.(87)).

4. POTENTIAL TECHNOLOGICAL APPLICATIONS OF ETCHED SURFACES AND DEPOSITED FILMS

4.1 Etched Surfaces

Potential technological applications of textured surfaces have been discussed in detail in a recent review(2). Therefore, that material will not be repeated here. Suffice to say that those technologies previously identified for which textured surface may be useful are:

(a) Fusion Technology: a potential advantage of using textured surface relates to the fact that they present, under certain conditions(89), proper to plasma edge parameters, a reduced physical sputtering, H^+ and He^+ ion backscattering, and blistering(89). Further work is, however, necessary to fully assess the characteristics and behavior of those surfaces when exposed to fusion plasmas.

(b) Incandenscent Lamp Manufacturing: the efficiency of visible light emitted from the refractory metal filament has been shown to increase when the surface of the filament is textured by ion bombardment. Details about this phenomenon can be found in a recent review(2).

(c) Micromechanical Device Fabrication: tiny valves, nozzles, pressure sensors and other mechanical systems can be formed on Si wafers by chemical etching, in such a way that a complete microchromatograph can be built on a single Si wafer five centimeters in diameter(90). Ion bombardment-induced etching methods, described in section 3.1.1, may be a good alternative to chemical etching to produce these devices.

(d) Biomedicine (Implantology): the ability to alter the morphology or the chemical composition of surfaces of different materials by ion bombardment can be advantageously used to modify biological materials and orthopedic implants, in order to improve their performance when implanted in humans or animals (see ref. 91).

4.2 Deposited Films

The growth rate of III-V and II-VI compounds, due to vapor phase epitaxy (VPE), is considerably different depending on whether the growth direction is <111>A or <111>B. This effect has been used to produce interesting and potentially useful structures such as ink jet nozzles, gratings, etc. (Fig. 17) by VPE of GaAs onto a substrate of GaAs(84) coated with a 200nm SiO_2 or Al_2O_3 film, and patterned with parallel stripes. According to a tentative hypothesis, if slow-growing (111) planes are adjacent to the edges of the stripes, V-grooves would be formed over the stripes (Fig. 17a), whereas if fast-growing (111) planes are adjacent to the stripes, tunnels would be produced (Fig. 17b). Further work is in progress to better understand and control the mechanisms of VPE for this particular application. Details can be found elsewhere(84).

5. EROSIONAL EFFECTS IN ION-BEAM BASED SURFACE ANALYSIS TECHNIQUES

Rather comprehensive reviews on the influence of ion bombardment-induced surface topography on ion-beam based surface analysis techniques have been published recently(1,2). Therefore, that material will not be repeated here. Suffice to say that topography effects are important, and should be considered, due to their deleterious effects, in:

(a) Depth Profiling by SIMS: it has been observed that the sputter-etching necessary for profiling elements in solids can produce, under certain circumstances, undesirable surface roughness (cones, steps, ridges, etc.) which may alter to a large extent the depth resolution during the measurements(1,2).

(b) Auger Surface Analysis: the yield of Auger electrons from analyzed samples can change due to variations in the orientation of the local surface normal, an effect that occurs as a consequence of surface topography development(2).

(c) Rutherford Backscattering (RBS) Analysis: theoretical and experimental work have shown that oscillations that may appear in back-scattering spectra, obtained under certain conditions, may be related to a particular surface topography(1,2).

On the other hand, the development of certain surface topographies may be beneficial in other analytical techniques, for example:

(d) Field ion emission and electron microscopy: it has been shown that multipoint electron emitters, which consist of surfaces covered with dense arrays of cones/pyramids or needles, produce much higher emission than needles with smooth surfaces, used as single emitters (1). Multipoint emitters are now being used in commercial instruments.

(e) Surface enhanced Raman scattering spectroscopy: the analysis of molecules by Raman scattering spectroscopy may benefit from a recent discovery, which has shown that the intensity of the Raman light inelastically scattered from molecules adsorbed onto metal surfaces may increase by factors of 10^5-10^6 with respect to that measured for molecules in the free state(92). Although the mechanism for this effect is not yet clear, surface roughness seems to be a necessary condition for the adsorbed molecules to yield such high light intensities. Surface covered with dense array of cones/pyramids are being investigated in relation to their potential application in this analytical technique(1,2).

(f) Measurement of momentum of low energy particle beams: momentum transfer to a solid surface of superthermal to approximately 100eV particles is frequently used as a diagnostic means for determining parameters of particle flows. For example, ballistic pendula have

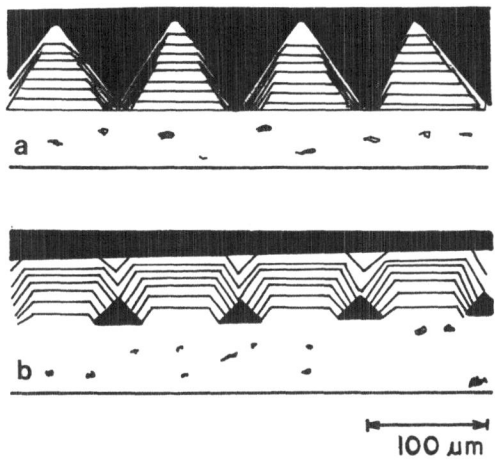

FIGURE 17. (a) Grooves and (b) tunnels formed by vapor phase epitaxy (VPE) of GaAs on a GaAs substrate (J. C. Marinace(84)).

been used to measure particle fluxes in shock-tube-produced molecular beams, laser-produced plasmas, or in accelerated cluster-ion beams(93). One way to obtain reliable results of momentum transfer is to use perfectly absorbing surfaces, such that reflection of particles is eliminated. This ideal situation can be approximated by using pendula with surfaces covered with dense arrays of cones, grooves, needles, etc.(93).

Perhaps future work will show that surface texturing also affects other material analysis techniques.

6. CONCLUSIONS

As shown in this and previous reviews(1,2), particle bombardment-induced erosional/depositional phenomena occurring during particle-surface interaction are relevant to different technologies and material analysis techniques. It has been demonstrated that these phenomena may be used advantageously in the processing of surfaces for applications to micro-electronics, surface optical and acoustical technologies, solar energy conversion technology, film deposition technology, and others less publicized. The mentioned phenomena may also be useful for applications in emerging technologies like fusion, micromechanical device fabrication, incandescent lamp manufacturing, cold welding, biomedicine, and others not yet identified.

One of the noticeable characteristics of this field of research is its interdisciplinary nature, which results from the fact that different disciplines (atomic collisions physics, solid state physics, chemical physics, etc.), within the broad fields of physics, chemistry, and materials science, have to be intermixed in order to understand and later try to control, for the purpose of technological applications, the complex phenomena occurring during particle-solid interactions. The rapid growth of technological applications of erosional/depositional phenomena suggests that challenging, exciting, and probably fruitful times can be expected in this field of research.

418

ACKNOWLEDGEMENTS

The author is grateful to all scientists who kindly contributed with figures and photographs included in this review.

REFERENCES

1. O. Auciello. Ion Interaction with Solids: Surface Texturing, Some Bulk Effects, and Their Possible Applications. J. Vac. Sci. Technol. 19 (1981)841-867.
2. O. Auciello. Ion Bombardment Modification of Surfaces: Fundamentals and Applications, Chs. 1 and 11. Eds. O. Auciello and R. Kelly (Elsevier Science Publishers, The Netherlands, 1984).
3. J. M. E. Harper, J. J. Cuomo, and H. R. Kaufman. Technology and Applications of Broad-Beam Ion Sources Used in Sputtering; Part II: Applications. J. Vac. Sci. Technol. 21(1982)737-756.
4. J. M. E. Harper, J. J. Cuomo, R. J. Gambino, and H. R. Kaufman. Ion Bombardment Modification of Surfaces: Fundamentals and Applications, Ch. 4. Eds. O. Auciello and R. Kelly (Elsevier Science Publishers, The Netherlands, 1984)pp.127-162.
5. A. R. Bayly. Secondary Processes in the Evolution of Sputter-Topographies. J. Mater. Sci. 7(1972)404-412.
6. I. H. Wilson. The Topography of Sputtered Semiconductors. Rad. Eff. 18, (1973)95-103.
7. G. Carter and M. J. Nobes. Ion Bombardment Modification of Surfaces: Fundamentals and Applications, Ch. 5. Eds. O. Auciello and R. Kelly (Elsevier Science Publishers, 1984)pp.163-224.
8. R. E. Chapman. Redeposition: a Factor in Ion-Beam Etching Topography. J. Mater. Sci. 12(1977)1125-1133.
9. O. Auciello, R. Kelly, and R. Iricibar. On the Problem of the Stability of Pyramidal Structures on Bombarded Copper Surfaces. Rad. Eff. Lett. 43 (1979)37-42.
10. O. Auciello and R. Kelly. On the Relative Stability of Different Topographical Features Developed on Bombarded Copper Surfaces. Rad. Eff. Lett. 43(1979)187-192.
11. O. Auciello and R. Kelly. Further Experimental Evidence on the Importance of Tertiary Effects in the Evolution of Pyramids on Bombarded Copper. Rad. Eff. Lett. 43(1979)117-123.
12. O. Auciello, R. Kelly, and R. Iricibar. New Insight into the Development of Pyramidal Structures on Bombarded Copper Surfaces. Rad. Eff. 46 (1980)105-118.
13. O. Auciello and R. Kelly. On the Role of the Primary Beam and of Scattered or Sputtered Particles in the Faceting of Cones on Bombarded Surfaces. Nucl. Inst. Meth. 182/183(1981)267-273.
14. G. W. Lewis, J. S. Colligon, F. Paton, M. J. Nobes, G. Carter, and J. L. Whitton. The Life Cycle of Copper Cones. Rad. Eff. Lett. 43(1979)49-54.
15. G. Carter, M. J. Nobes, G. W. Lewis, and J. L. Whitton. Combined Sputtering Yield and Surface Topography Development Studies on Si. Proc. Symp. Sputtering, Eds. P. Varga, G. Betz, and F. P. Viehböck (Inst. für Allgemeine Phys. Tech. Univ. Wien, 1980)pp.604-613.
16. H. I. Smith. In Proceedings of the Symposium on Etching for Pattern Definition, Electrochemical Society, 1976, Eds. M. G. Hughes and M. J. Read, pp. 133.
17. J. Belson and I. H. Wilson. Flux Density Equations for Topographical Evolution of Features on Ion Bombarded Surfaces. Rad. Eff. 51(1980)27-34.
18. J. Belson and I. H. Wilson. Equations for Redeposition of Sputtered Flux

onto Surface Asperities. In Symposium on Sputtering, Eds. P. Varga, G. Betz, and F. P. Viehböck (Inst. für Allgemeine Phys. Tech. Univ. Wien, 1980)pp.574-583.

19. J. Belson and I. H. Wilson. Theory of Redeposition of Sputtered Flux onto Surface Asperities. Nucl. Instr. Meth. 182/183(1980)275-281.

20. I. H. Wilson, S. S. Todorov, and D. S. Karpuzov. Profile Evolution During Ion Beam Etching of Germanium Targets. Nucl. Instr. Meth. 209/210(1983) 549-554.

21. R. Smith, S. S. Makh, and J. M. Walls. Surface Morphology During Ion Etching: The Influence of Redeposition. Phil. Mag. A47(1983)453-481.

22. R. Smith, M. A. Tagg, and J. M. Walls. Deterministic Model of Ion Erosion, Reflection and Redeposition. Vacuum 34(1984)175-180.

23. R. Kelly and O. Auciello. On the Origin of Pyramids and Cones on Ion-Bombarded Copper Surfaces. Surf. Sci. 100(1980)135-153.

24. R. S. Robinson and S. M. Rossnagel. Monte Carlo Model of Topography Development During Sputtering. J. Vac. Sci. Technol. A1(1983)426-429.

25. A. Güntherschülze and W. Tollmien. Neve Unterscuchungen über die Kathodenzerstäubung der Glimmentladung. Z. Phys. 119(1942)685-695.

26. O. Auciello (unpublished, 1980).

27. O. Auciello and R. Kelly. The Evolution of Pyramidal Structures on Surfaces Bombarded at 60°. In Symposium on Sputtering, Eds. P. Varga, G. Betz, and F. P. Viehböck (Inst. für Allgemeine Phys. Tech. Univ. Wien, 1980)594-603.

28. O. Auciello and R. Kelly. The Evolution of Pyramidal Structures on Surfaces Bombarded at Oblique Angles. Rad. Eff. 66(1982)195-210.

29. G. W. Lewis, G. Carter, M. J. Nobes, and S. A. Cruz. The Development of Tailed-Cones on Non-Normal Incidence Ion Bombarded Solids. Rad. Eff. Lett. 58(1981)119-124.

30. I. H. Wilson, J. Belson, and O. Auciello. Secondary Effects in Ion Bombardment-Induced Surface Erosion. In Ion Bombardment Modification of Surfaces: Fundamentals and Applications, Ch. 6 Eds. O. Auciello and R. Kelly (Elsevier Science Publishers, The Netherlands, 1984)pp.225-297.

31. H. H. Andersen and H. L. Bay. Sputtering Yield Measurements. Sputtering by Particle Bombardment I, Ed. R. Behrisch (Topics in Applied Physics, Springer Verlag, 1981)pp.145-218.

32. R. Kelly and O. Auciello. On the Origin of Pyramids and Cones on Ion-Bombarded Copper Surfaces. Surf. Sci. 100(1980)135-153.

33. M. Hou and M. T. Robinson. The Conditions for Total Reflection of Low-Energy Atoms from Crystal Surfaces. Appl. Phys. 17(1978)371-375.

34. D. Ghose, D. Basu, and S. B. Karmohapatro. Cone Formation on Argon-Bombarded Copper. J. Appl. Phys. 54(1983)1169-1171.

35. N. N. Efremow, M. W. Geis, D. C. Flanders, G. A. Lincoln, and N. P. Economou. Ion Beam-Assisted Etching of Diamond. J. Vac. Sci. Technol. B3(1985)416-418.

36. R. L. Kubena, R. L. Seliger, and E. H. Stevens. High Resolution Sputtering Using a Focused Ion Beam. Thin Solid Films 92(1982)165-169.

37. T. Kato, H. Morimoto, K. Saitoh, and H. Nakata. Submicron Pattern Fabrication by Focused Ion Beams. J. Vac. Sci. Technol. B3(1985)50-53.

38. P. G. Glöersen. Ion Beam Etching. J. Vac. Sci. Technol. 12(1975)28-35.

39. H. Dimigen and H. Lüthje. An Investigation of Ion Etching. Philips Tech. Rev. 35(1975)199-208.

40. R. E. Lee. Microfabrication by Ion Beam Etching. J. Vac. Sci. Technol. 16(1979)164-170.

41. V. M. Donnelly, D. E. Ibbotson, and D. L. Flamm. Fundamental Aspects of Plasma-Surface Interactions and The Etching Process. In "Ion Bombardment Modification of Surfaces: Fundamentals and Applications," Ch. 8, Eds.

O. Auciello and R. Kelly (Elsevier Science Publishers, 1984)323-359.

42. H. W. Lehmann, L. Krausbauer, and R. Widmer. Redeposition - A Serious Problem in r.f. Sputter Etching of Structures with Micrometer Dimensions. J. Vac. Sci. Technol. 14(1977)281-284.

43. O. Auciello, A. A. Haasz, and P. C. Stangeby. Synergism in Materials Erosion Due to Multispecies Impact. Rad. Eff. 89(1985)63-101.

44. O. Auciello. Recent Progress in Understanding Ion Bombardment-Induced Synergism in the Erosion of Carbon Due to Multispecies Impact. Nucl. Instr. Meth. B(in press).

45. D. L. Flamm and V. M. Donnelly. The Design of Plasma Etchants. Plasma Chem. and Plasma Processing 1(1981)317-363.

46. L. F. Johnson, K. A. Ingersoll, and D. Kahng. Planarization of Patterned Surfaces by Ion Beam Erosion. Appl. Phys. Lett. 40(1982)636-638.

47. G. D. Boyd, L. A. Coldren, and F. G. Storz. Directional Reactive Ion Etching at Oblique Angles. Appl. Phys. Lett. 36(1980)583-585.

48. D. E. Prober, M. D. Feuer, and N. Giordano. Fabrication of 300A Metal Lines with Substrate-Step Techniques. Appl. Phys. Lett. 37(1980)94-96.

49. M. K. Sullivan and G. A. Kolb. Direct Photoetching of Evaporated Germanium and Its Use in Mask Fabrication. Electrochem. Techn. 6(1968)430-434.

50. L. L. Sveshnikova, V. I. Donin, and S. M. Repinskii. Initiation of Bromine-Silicon Reaction by a High-Power Argon Laser. Sov. Tech. Phys. Lett. 3(1977)223-224.

51. I. M. Beterov, V. P. Chebotaev, N. I. Yurshima, and B. Ya. Yurshin. Effect of the Laser Radiation Intensity on the Kinetics of the Heterogeneous Photochemical Reaction Between Single-Crystal Germanium and Bromine Gas. Sov. J. Quantum Electron. 8(1978)1310-1315.

52. (a) J. Ehrlich, R. M. Osgood, Jr., and T. F. Deutsch. Laser-Induced Microscopic Etching of GaAs and InP; Appl. Phys. Lett. 36(1980)698-700. (b) Laser Chemical Technique for Rapid Direct Writing of Surface Relief in Silicon; Appl. Phys. Lett. 38(1981)1018-1020. (c) Laser Photochemical Microalloying for Etching of Aluminum Thin Films; Appl. Phys. Lett. 38 (1981)399-401.

53. T. J. Chuang. Infrared Laser-Induced Reactions at Metal and Semiconductor Surfaces. J. Vac. Sci. Technol. 18(1981)638-642.

54. T. F. Deutsch, D. J. Erlich, and R. M. Osgood, Jr. Laser Photodeposition of Metal Films with Microscopic Features. Appl. Phys. Lett. 35(1979)175-177.

55. S. D. Allen and M. Bass. Laser Chemical Vapor Deposition of Metals and Insulators. J. Vac. Sci. Technol. 16(1979)431.

56. R. J. von Gutfeld, E. E. Tynan, R. L. Melcher, and S. E. Blum. Appl. Phys. Lett. 35(1979)651-653.

57. T. J. Chuang. Infrared Laser-Induced Reaction of SF_6 with Silicon Surfaces. J. Chem. Phys. 72(1980)6303-6304.

58. T. J. Chuang. Multiple Photon Excited SF_6 Interaction with Silicon Surfaces. J. Chem. Phys. 74(1981)1453-1460.

59. P. Sigmund. Sputtering by Ion Bombardment: Theoretical Concepts. In "Sputtering by Particle Bombardment," Ed. R. Behrisch (Springer Verlag, 1981)9-71.

60. H. R. Deppe, B. Hasler, and J. Höepfner. Investigations on the Damage Caused by Ion Etching of SiO_2 Layers at Low Energy and High Dose. Solid State Electron. 20(1977)51-55.

61. A. Yariv. Guided-Wave Optics. Scientific American (Jan. 1979) 64-72.

62. L. F. Johnson. Evolution of Grating Profiles Under Ion-Beam Erosion. Applied Optics 18(1979)2559-2574.

63. L. F. Johnson and K. A. Ingersoll. Interference Gratings Blazed by Ion-Beam Erosion. Appl. Phys. Lett. 35(1979)500-503.

64.L. F. Johnson and K. A. Ingersoll. Generation of Surface Gratings With Periods <1000Å. Appl. Phys. Lett. 38(1981)532-534.

65.L. F. Johnson. Ion Beam Microstructure Fabrication in Optical, Magnetic and Surface Acoustical Technologies. In "Ion Bombardment Modification of Surfaces: Fundamentals and Applications," Ch. 9, Eds. O. Auciello and R. Kelly (Elsevier Science Publishers, 1984)361-397.

66.H. G. Craighead and R. E. Howard. Microscopically Textured Optical Storage Media. Appl. Phys. Lett. 39(1981)532-534.

67.H. G. Craighead, R. E. Howard, J. E. Sweeney, and D. M. Tennant. J. Vac. Sci. Technol. 20(1982)316-319.

68.H. G. Craighead, R. E. Howard, P. F. Liao, D. M. Tennant, and J. E. Sweeney. Textured Germanium Optical Storage Medium. Appl. Phys. Lett. 40(1982)662-664.

69.R. G. Zech, SPIE 177,56. Optical Information Storage. Proc. of the "Symposium on Optical Storage Materials," 1980. Proc. of "Outlook for Optical and Video Disc Systems and Applications," Florida, 1981.

70.R. A. Bartolini. Media for High Density Optical Recording. J. Vac. Sci. Technol. 18(1981)70-74.

71."Surface Wave Filters, Design, Construction and Use"; Ed. H. Matthews (John Wiley and Sons, NY, 1977).

72.M. Itoh, H. Gokan, S. Esko, and K. Asakawa. Fabrication Process for Surface Acoustic Wave Filters Having 0.5μm Finger Period Electrodes. J. Vac. Sci. Technol. 20(1982)21-25.

73.S. J. Marti, S. S. Schwartz, R. L. Gunshor, and R. F. Pierret. Surface Acoustic Wave Resonators on a ZnO-on-Si Layered Medium. J. Appl. Phys. 54(1983)561-569.

74.Thin Film Processes, Eds. J. L. Vossen and W. Kern (Academic, New York, 1978).

75.A. G. Dirks and H. J. Leamy. Columnar Microstructure in Vapor-Deposited Thin Films. Thin Solid Films 47(1977)219-233.

76.J. A. Thornton. Influence of Apparatus Geometry and Deposition Conditions on the Structure and Topography of Thick Sputtered Coatings. J. Vac. Sci. Technol. 11(1974)666-670.

77.J. M. E. Harper, J. J. Cuomo, R. J. Gambino, H. R. Kaufman. Modification of Thin Film Properties by Ion Bombardment During Deposition. In "Ion Bombardment Modification of Surfaces: Fundamentals and Applications." Eds. O. Auciello and R. Kelly (Elsevier Science Publishers, 1984)127-159.

78.J. Amano and R. P. W. Lawson. Thin-Film Deposition Using Low-Energy Ion Beams, Mg^{+}Ion-Beam Deposition and Analysis of Deposits. J. Vac. Sci. Technol. 14(1977)695-698.

79.R. D. Bland, G. J. Kominiak, and D. M. Mattox. Effect of Ion Bombardment During Deposition on Thick Metal and Ceramic Deposits. J. Vac. Sci. Technol. 11(1974)671-674.

80.E. H. Hirsch and I. K. Varga. Thick Film Annealing by Ion Bombardment. Thin Solid Films 69(1980)99-105.

81.T. Takagi. Role of Ions in Ion-Based Film Formation. Thin Solid Films 92(1982)1-17.

82.P. J. Martin, H. A. MacLeod, R. P. Netterfield, C. G. Pacey, and W. G. Sainty. Ion-Beam Assisted Deposition of Thin Films. Applied Optics 22 (1983)178-184.

83.D. M. Mattox. Preparation of Thin Films for Solar Energy Utilization. J. Vac. Sci. Technol. 17(1980)370-373.

84.J. C. Marinace. Tunnels in Semiconductor Epitaxy. IBM J. Res. Develop. 23(1979)459-461.

85.Proceedings of the 31st National Symposium of the American Vaccum Society. J. Vac. Sci. Technol. A3, Part I and II(1985).

86. L. J. Leger. Oxygen Atom Reaction with Shuttle Materials at Orbital Altitudes, NASA Tech. Memo TM-58246(1982).
87. B. A. Banks, M. J. Mirtich, S. K. Rutledge, and D. M. Swee. Sputtered Coatings for Protection of Spacecraft Polymers. NASA Tech. Memo 83706 (1984).
88. D. C. Ferguson. Laboratory Degradation of Kapton in a Low Energy Oxygen Ion Beam, NASA Tech. Memo TM-8353(1984).
89. O. Auciello. A Critical Review on the Origin, Stability, Relative Sputtering Yield and Related Phenomena of Textured Surfaces Under Ion Bombardment. Rad. Eff. 60(1982)1-26.
90. J. B. Angell, S. C. Terry, and P. W. Barth. Silicon Micromechanical Devices. Scientific American (April 1983)44-55.
91. B. A. Banks. Ion Bombardment Modification of Surfaces in Biomedical Applications. In "Ion Bombardment Modification of Surfaces: Fundamentals and Applications," Eds. O. Auciello and R. Kelly (Elsevier Science Publishers, 1984)399-434.
92. T. E. Furtak and J. Reyes. A Critical Analysis of Theoretical Models for the Giant Raman Effect from Adsorbed Molecules. Surf. Sci. 93(1980)351-382.
93. W. Keller, R. Klingelhöfer, B. Krevet, H. O. Moser, and R. Ries. Influence of Surface Roughness on the Momentum Transfer by 350-keV Hydrogen-Cluster Ions. Rev. Sci. Instr. 55(1984)468-471.

THE INFLUENCE OF INCIDENCE ANGLE ON DAMAGE PRODUCTION IN In[+] ION IMPLANTED Si

S KOSTIC, G KIRIAKIDIS[*] M J NOBES and G CARTER
University of Salford, Department of Electronic and Electrical Engineering, U.K.

[*]University of Crete, Department of Physics, Heraklion, Crete, Greece.

INTRODUCTION AND EXPERIMENTAL TECHNIQUE

One set of models (1) of ion irradiation enhanced reactive gas etching of materials, including Si, suggests that increased reaction rates occur as a result of surface damage production by the irradiation. If this is the case then etching rates should depend upon the incident angle of irradiation since this should modify the surface damage density as a result of varying energy deposition density near the surface. Although theoretical predictions of surface damage can be made in the limit of low ion fluence (2) these do not apply for high incidence angles (relative to the surface normal) since collision cascades are incompletely developed near a free surface, nor at high fluence where individual disordering events accumulate and overlap (3). Numerical simulations are possible, and some such data will be presented in this paper but, of course, experimental observation is the only really valid method. Whereas some results of the dependence on incident angle upon ion ranges and range distributions have been published (4) no such data exists for the damage production. This paper reports the first such results.

In selecting the system for study and the damage analysis method we were constrained by techniques available to us. For reactive gas etching purposes one should ideally use low keV incident ion energies but the damage is then confined to depths very close to the surface and no currently available techniques, for semiconductors, allow such near surface analysis. It was therefore necessary to employ higher ion energies (80 keV was chosen) so that the damage was sufficiently depth distributed to allow analysis. Secondly the damage generated by irradiation should be retained in the solid and not anneal (either thermally or athermally). This, ideally, requires low temperature conditions but, since this was not available to us, the next alternative is to employ a system in which annealing is apparently minimal.

Apart from its intrinsic interest for reactive etching purposes Si is a good candidate material for study since it is known (3) that, even at room temperature, when irradiated with heavy ions, Si becomes locally amorphised by such ion impact and this disorder is stable. Consequently In ions were implanted into Si.

Finally the analysis technique should be as quantitative and with as high depth resolution as possible. The only technique which fulfils these requirements is Rutherford backscattering in the low incidence and/or exit angle mode (5) coupled with incidence or exit channelling. This technique of damage analysis was therefore chosen using 2 MeV He[+] ions with normal incidence and 97° scattering geometry. The disadvantage of this technique is that it is relatively insensitive and only detects displaced atom densities in Si greater than about 10^{15} cm^{-2}. For a depth of Si of several hundred Å this represents several percent of complete atomic

424

randomisation. As heavy ion implantation fluence increases, the damage
created, initially in isolated amorphous zones, begins to overlap and
forms the beginning of a complete amorphous layer and as this process
occurs the measured damaging rate falls below the actual production rate.

In order to compromise between the (relatively) poor sensitivity and the
demand not to enter the damage overlap regime unless desired, the In$^+$ ion
fluence must be carefully chosen. The problem is further compounded when
it is realised that even if, for normal incidence to the surface, no
overlap conditions occur, as incidence angle is increased towards grazing,
the damage depth profile is expected to move towards the surface and
compress, increasing the damage density and, consequently, the propensity
for cascade or amorphous region overlap.

Advised by earlier studies (3) of damage density depth distributions for
normal ion incidence and simple calculations assuming a Gaussian depth
distribution (4) it was concluded that, for the majority of incidence
angles, an ion fluence of 10^{13} cm^{-2} would produce little or no damage
overlap, whereas for the higher dose of 5×10^{14} cm^{-2} major overlap would
occur. These were the ion fluences employed.

RESULTS AND DISCUSSION

As noted earlier the analysis technique is relatively insensitive and,
since, even without implantation there will always be a disordered (oxidised)
region at the Si surface, a surface peak, corresponding to this state, must
always be subtracted from the implanted sample RBS/channelling data. This
becomes increasingly difficult for high incidence angles as the disorder
shifts towards the surface but a careful channel by channel subtraction
does allow reasonable data extraction.

Figure 1 shows deconvoluted damage-depth distributions following 10^{13} cm^{-2}
implantation and Figure 2 the equivalent data for 5×10^{14} cm^{-2} implantation.

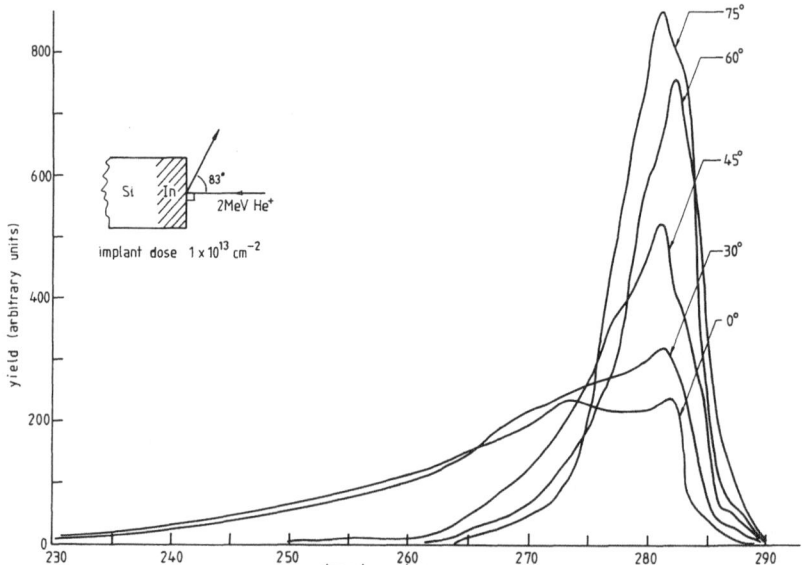

FIGURE 1. RBS spectra of samples implanted with a 80KeV In$^+$ beam to a dose of 1×10^{13} cm^{-2} at various
implant angles, where the virgin Si spectrum has been subtracted.

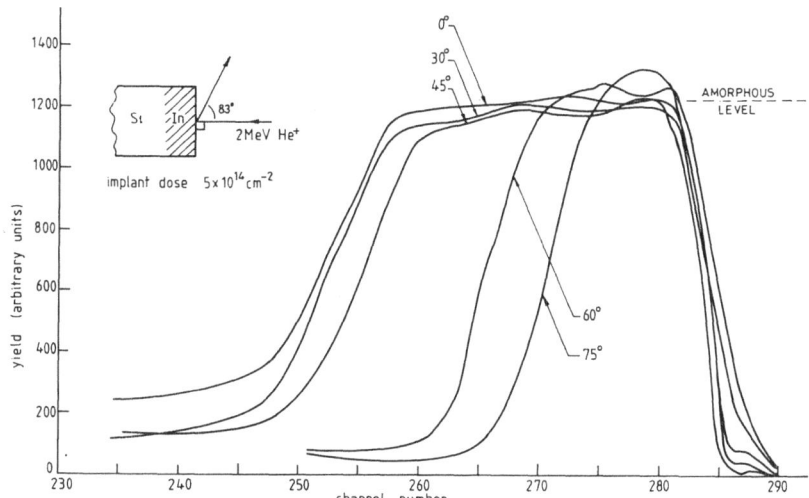

FIGURE 2. RBS spectra of samples implanted with a 80KeV In⁺ beam to a dose of 5x10¹⁴ cm⁻² at various implant angles where the virgin Si spectrum has been subtracted.

Figure 1 clearly reveals the shift towards the surface and the depth compression, together with an increase in the maximum density, of disorder with increasing incidence angle. Figure 2 shows, again, the depth compression of disorder, with increasing incidence angle and here, it is clear from the plateaux nature of the data, that the Si is always amorphised, for all incidence angles at this fluence to depths which decrease with increasing incidence angle. Comparison of the maximum backscattering signals in Figures 1 and 2 indicates that even at the lower ion fluence some disorder overlap probably would have occurred for the highest incidence angles.

From the data of Figures 1 and 2 the mean value $\bar{L}$ and the standard deviation ΔL of the damage depth distributions may be determined and these parameters normalised to their values for normal incidence are plotted in Figures 3 and 4 as a function of ion incidence angle for both ion fluences. It is clear from both figures that $\bar{L}$ and ΔL remain approximately constant for incidence angles up to 40-50° and then decrease rather rapidly.

The total depth integrated, disorder may also be obtained from Figures 1 and 2, and this parameter N_d is plotted, for both fluences, as a function of ion incidence angle in Figure 5. It is clear that, for the lower fluence, the total disorder remains substantially constant up to the largest incidence angle (80°) employed suggesting that the cascades remain still contained within the substrate for 80 keV In. The reduction in the total disorder for the higher fluence is indicative of the decreasing depth over which total amorphisation has occurred with increasing incidence angle.

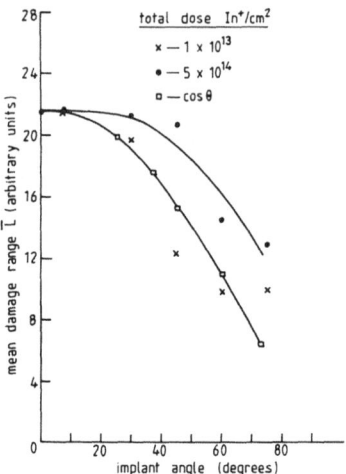

Calculated mean range of damage in Si
provided by two different doses of In.

FIGURE 3.

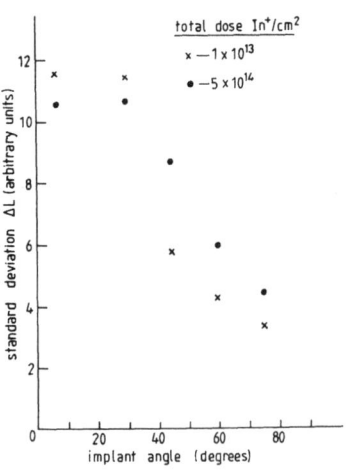

Calculated straggling of damage in Si
produced by two different doses of In.

FIGURE 4.

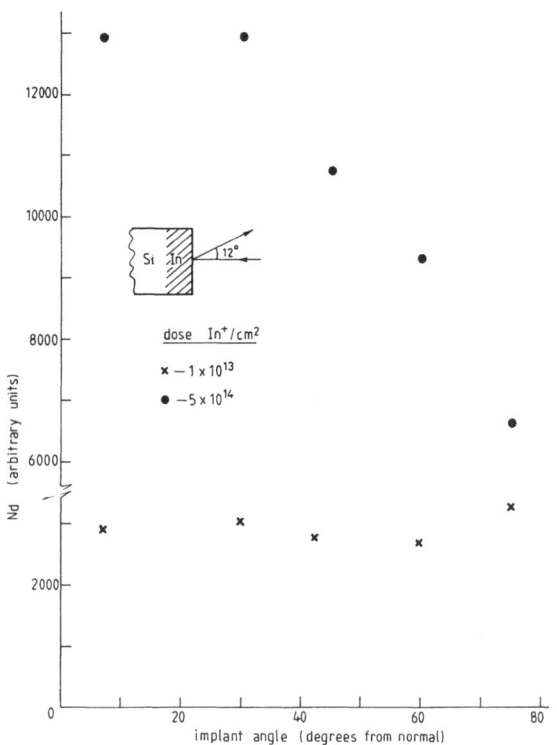

FIGURE 5.

Change in Nd with variation of the incident beam angle for two given doses of In into Si.

If the damage production function is of a Gaussian form with parameters $\bar{X}_D$ = mean damage depth, ΔX_D = standard deviation parallel to direction of ion incidence and ΔY_D = standard deviation transverse to direction of ion incidence then simple theoretical considerations indicate that:

$$\bar{L} = \bar{X}_D \cos\theta \qquad \text{and} \qquad (1)$$

$$\Delta L^2 = \Delta X_D^2 \cos^2\theta + \Delta Y_D^2 \sin^2\theta \qquad (2)$$

where θ is the angle of ion incidence relative to the surface normal.

The first two damage moments were calculated from linear cascade theory (6) and the values of $\bar{L}$ and ΔL calculated from equations (1) and (2). These values are also plotted in Figures 3 and 4, again normalised to their values for normal incidence. In addition, numerical computations were made of the number of i-v pairs created by 80 keV In implantation using a MARLOWE Code (7), as a function of ion incidence angle with results shown in Figure 6. The similarity of the distributions, at each selected incidence angle, with the experimental low fluence data of Figure 1 is clear, with major changes in depth distribution occurring only for incidence angles greater than about 40°. This study also allowed evaluation of the depth integrated disorder density and it was found that for incidence angles up to 80°, no more than a 20% reduction in defect production occurred, in line with the results of Figure 5 for the low fluence case. The comparison of predicted and low fluence L̄ data shows rather good agreement (i.e. a cosθ projection behaviour) which is always expected to occur independent of an assumed form of the depth distribution function. However, comparison of measured ΔL values with the predictions of equation (2) is rather poor, which we believe, is largely due to the

difficulty in obtaining such data from RBS spectra as seen in Figures 1 and 2.

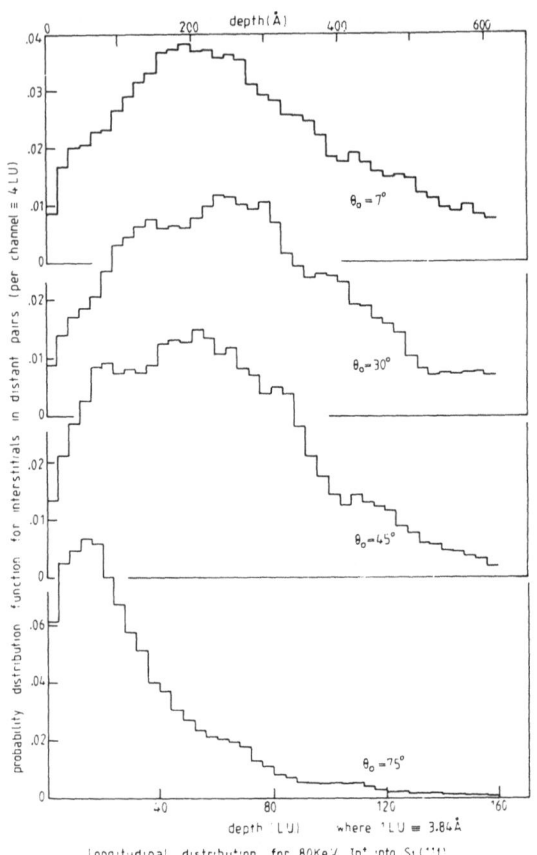

FIGURE 6.   Longitudinal distribution for 80KeV In⁺ into Si(¹¹¹)

IMPLICATIONS FOR REACTIVE GAS ETCHING

The results of this study were obtained for low fluence, where the data reasonably reflects the defect generation rate depth profiles and at high fluences where disorder is completely overlapped and layer amorphisation occurs. In neither case is substantial surface sputtering involved. When this does occur during implantation the steady state surface damage will be given by the integral under the damage depth distribution (3,8). If erosion is slow the surface damage, as a function of incident angle, will tend towards the overlapped damage form (high fluence) of the data of Figure 5 whilst if the erosion rate is high the surface damage will tend towards low fluence data of Figure 5. The erosion rate, under ion bombardment enhanced reactive gas etching conditions, will generally be a function of the reactive gas flux and if this is varied, at constant ion flux, then if ion bombardment induced damage production is an important process in the etch process, there should be major changes in etch rate with incident ion angle as the reactive gas flux is varied. It is recommended that studies of this type may be useful in determining the role of substrate surface damage in enhanced etching.

REFERENCES

1. Donnelly VM, Ibbotson DE and Flamm DL: Chapter 8 'Ion Bombardment Modification of Surfaces' (Eds O Auciello and R Kelly, Elsevier, Amsterdam) (1984).
2. Winterbon KB: 'Ion Implantation Range and Energy Deposition Distributions', Vol 2; Low incident ion energies, IFI/Plenum, New York.
3. Carter G, Webb R and Collins R: Rad Effects 37, 21 (1978).
4. Grant WA, Williams JS and Dodds D: Rad Effects 29, 189 (1976); and Furukawa S and Matsumura H: Appl Phys Lett 22, No 3 (1973).
5. Williams JS: Nucl Instrum & Meth 126, 205 (1975).
6. Jimenez-Rodriguez JJ: Private communication.
7. Karpuzov DS: Private communication.
8. Carter G and Nobes MJ: Vacuum 32, 593 (1982).

STRIATION PRODUCTION BY GRAZING ION INCIDENCE ON Si

S KOSTIC, W BEGEMANN,[*] G CARTER, D G ARMOUR AND M J NOBES

University of Salford, Department of Electronic and Electrical Engineering
U.K.
*University of Bielefeld, Physics Department, F.R.G.

1. INTRODUCTION AND EXPERIMENTAL

Our own (1,2) and other earlier studies (3,4) of the development of surface topography during ion bombardment of Si suggested that different observations by different investigators could possibly be ascribed to surface contamination and residual vacuum condition effects. We therefore decided to conduct studies of the topography developed on Si under relatively high vacuum conditions. Cleaved segments of (100) oriented Si wafers were bombarded to fluences of about 5×10^{17} cm^{-2} with approximately 5 keV Ar$^+$ ions, without beam aperturing, at a base pressure of about 10^{-9} torr. Bombardment was at normal incidence to the surface. The ion beam flux density profile was non uniform so that ion fluence decayed quite rapidly towards the periphery of the bombarded area.

Following bombardment any morphological changes to the Si surface were examined by Scanning Electron microscopy.

2. RESULTS

After bombardment to 5×10^{17} ions cm^{-2} no measurable change in the morphology of the plane area of the Si had occurred. The Si had, however, been sputtered since 1) local areas covered with inadvertent contaminant became elevated (due to contaminant protection) above their immediate sputtered surroundings and 2) the bombarded area was depressed (eroded) relative to the unbombarded surroundings.

However part of the (spatially) inhomogeneous ion beam frequently intercepted part of a cleaved edge which was fractured at a steep angle to the plane surface, and along these edges a well developed morphology developed.

The form of this morphology was always of a striated or ripple-like habit which lies parallel to the projection of the ion flux on the fracture surface as shown in the examples of Figs. 1, 2 and 3.

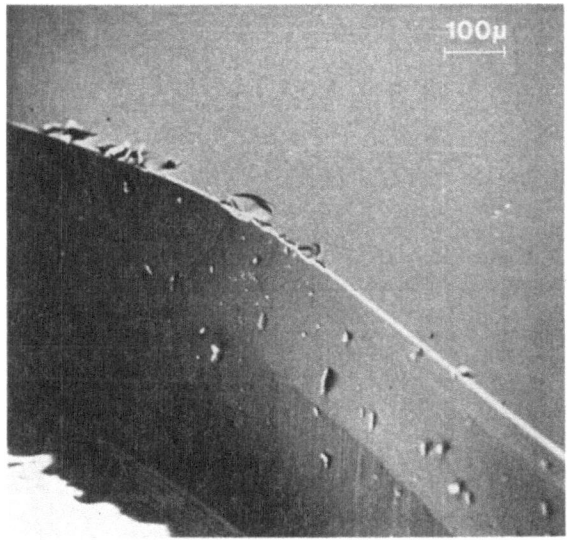

FIGURE 1. Striations produced in two different regions of the fracture boundary of the Si.

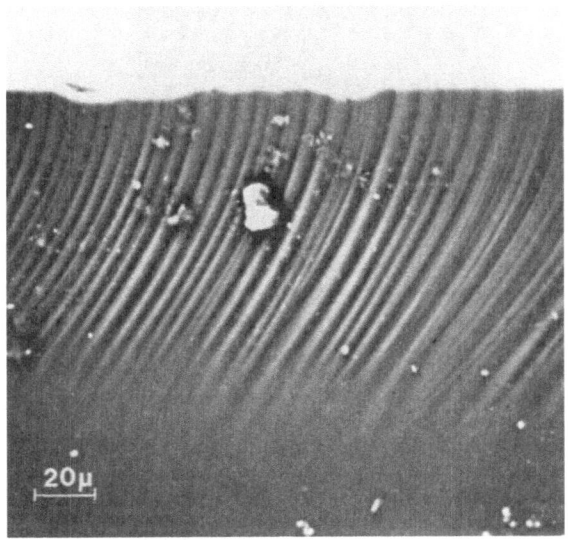

FIGURE 2. Striations produced in the upper part of the fracture boundary intersecting the plane surface of the Si.

FIGURE 3. Intersecting and crossed striation patterns at the lower part of a fracture surface.

FIGURE 4. A typical unbombarded area of the fracture surface – plane surface of the Si.

As a comparison Figure 4 shows a region of a fracture surface which was unbombarded. This region, typical of all fracture edges examined is featureless, showing no signs of incipient striations or corrugations at the fracture-wafer surface interface.

Figures 1 and 2 show that the ripples do not always cover the complete fracture surface, appearing to terminate where there are slope changes in the non planar fracture surface. Even when such ripples do intersect the wafer surface they do not extend into that surface.

Figure 3 indicates that, sometimes, the striations are, composed of two intersecting and superimposed ripple patterns.

3. DISCUSSION

The lack of topography development for normal incidence bombardment is not surprising since our earlier work (1,2) has revealed very slow feature development for such conditions.

The development of topography for near grazing incidence is not surprising either since, again, our earlier work (1,2) reveals similar behaviour. What is surprising, however, is that 1) the observed topography must have initiated and developed at fluences much less than 5×10^{17} ions cm^{-2} because of the rapid decrease of flux away from the centre of the beam spot, 2) the topography appears to be rather sensitive to the incidence angle near grazing incidence and 3) there appears to be no correlation with initial features on the surface or at the edge of the surface.

Many authors have displayed micrographs showing the topography of Si (5) when etched by inert gas or reactive species bombardment through defining masks which show similar striation behaviour to that reported here. It is usually believed that this morphology results from corrugation or roughness in the mask. The present study, where no masking occurred, nor where there appeared to be any initial face-edge roughness, reveals that flux non uniformity (i.e. mask supression of flux) is not a necessary condition for striation development. The existence of contaminant (e.g. deliberate masking or inadvertent deposits on the wafer surface) does not therefore, as we have argued previously (1,2) appear to be generally necessary to initiate feature developments. The reasons for such development are unclear but could arise, for example, from redeposition of sputtered atom flux, thermal atomic migration or, we believe most probably from facetting of the Si to produce surfaces of equal and locally maximum sputtering yield relative to the yield of the initial quasi-planar fracture surface. Variations in local orientation (slope) of the initial surface would lead to different, and even no sets of equivalent surfaces.

It might be noted that similar striation structures have also been observed (6,7) on some high angle elevation boundaries of etch pits in Ar^+ bombarded single crystal Cu and not on other neighbouring boundaries of the same pits. The phenomenon therefore appears to be of some general nature for near grazing incidence conditions on a variety of substrate materials.

434

4. REFERENCES

1. M J Nobes, G W Lewis, G Carter and J L Whitton, Nucl Instrum and Meth
    194, 509 (1982).
2. G Carter, M J Nobes, I Abril and R Garcia-Molina, Surface and Inter-
    face Analysis 7, 41 (1985).
3. W Hauffe. Doctoral Thesis. Tech Univ Dresden (1971).
4. P R Boudewijn, H W P Akerboom, C W T Bulle-Liewma and J Haisma,
    Surface and Interface Analysis 7, 49 (1985).
5. See for example; "Ion bombardment modification of surfaces", Eds
    O Auciello and R Kelly (Elsevier, Amsterdam) (1984).
6. G Carter, M J Nobes and J L Whitton, Appl Phys A38, 1 (1985).
7. G Carter, M J Nobes, G W Lewis and J L Whitton, Vacuum 33, 373 (1980).

INFLUENCE OF LATTICE DEFECTS PRODUCED BY ION IMPLANTATION ON ELECTRICAL AND
OPTICAL PROPERTIES OF BULK-BARRIER-DIODES

N. GEORGOULAS

Democritus University of Thrace, Xanthi, Greece.

1. INTRODUCTION

Bulk-Barrier-Diodes (BBD) have recently gained growing interest for the ap-
plication as photodiodes with internal gain |3|, mixers |5| and as high sen-
sitivity temperature sensors |4|. Previous work |3| considers only the inter-
band transition of optically excited carriers, since the number of lattice
defects is minimized after the diode annealing |6|.

The present work studies the influence of lattice defects produced during
ion implantation on the BBD electric and optical properties and investigates
whether such an influence can be employed into a useful application.

Section 2 gives a brief description of BBD and an explanation of the appearance and distribution of lattice defects during ion implantation. Experimental results are reported in section 3 and discussed in section 4.

2. DIODE DESCRIPTION

Fig. 1 shows schematically the structure, the space charge density and the
corresponding band diagram of a BBD. The diode has been prepared by a double
implantation of P and B ions at Siemens research centre in West Germany. In
contrast to a similarly structured transistor, the middle layer is sufficiently thin as to be depleted of carriers for all bias voltage values.The
current transport is controlled exponentially by the potential barrier height
caused by the space charge of the thin middle n-layer. The current-voltage
characteristic of BBD is given by |1,2|.

When the middle layer is depleted of carriers, the potential barrier height
(for V = 0) is proportional to the space charge density and to the square of
the layer thickness. Therefore, changing the value of a middle layer parameter such as space charge density N_B or layer thickness d would change the value of the barrier height and consequently the current. One of the causes for
such a change is the presence of lattice defects in the middle layer provided that they can be ionized under the influence of external factors such as
light and temperature. In addition to change in electrical behaviour these
lattice defects can also change the optical behaviour as they can extend the
optical sensitivity region into the infrared ($\lambda > 1.1 \mu m$). In this case, carrier
optical stimulation occurs from the intermediate level (traps) of the energy
gap.

Lattice defects appear during the ion implantation stage. Their profile
$N_T(x)$ is proportional to the doping profile N(x) |8,9|.

$$N_T(x) = K \cdot N(x) , \qquad (1)$$

where K is the number of defects formed by each implanted ion. The position
of the defects in the energy gap depends on the ion energy, dose and mass.
Small implantation doses usually form vacancies or associations of vacancies
and chemical impurities such as B and P |7,8|. Lattice defects can be elimina-
ted |6| by an annealing process (800°C for 30 min).

436

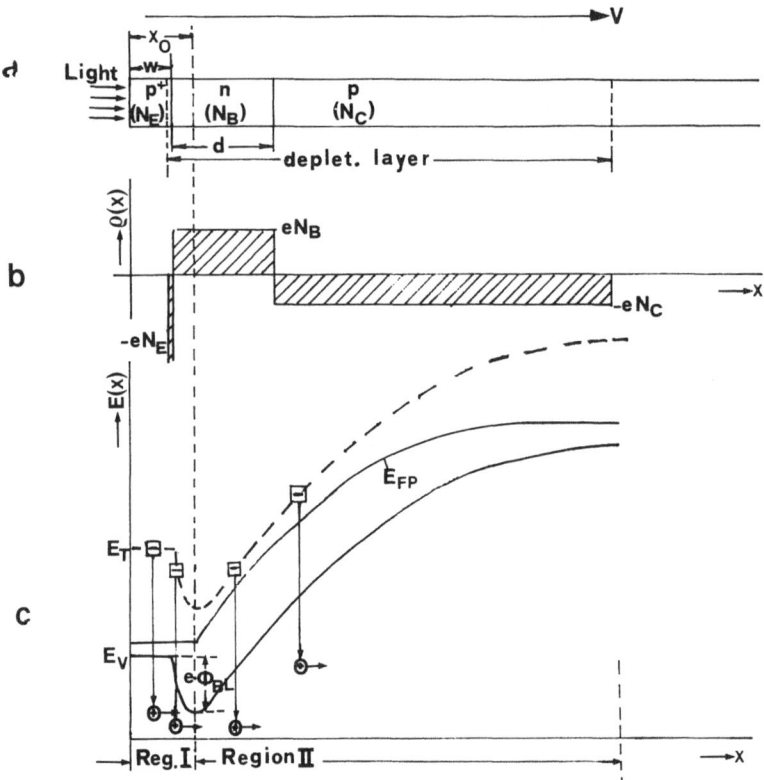

FIGURE 1. a) Schematic diagram of the BBD structure. b) Space charge density. c) The corresponding energy band diagram. Note: The band diagram has been drawn qualitatively for small potential barrier and traps level near the middle of the energy gap.

3. MEASUREMENTS

In the study of lattice defects and their influence on the bulk-barrier-diode properties a large number of different diodes were investigated. The study parameters were the annealing temperature of the sample, the type, and the dose of implanted ions. Two representative diodes 222/3 and 520/2 were chosen. They differ in the type of implanted ions causing the lattice defects. The most important data are:

222/3	520/2

p-type   :$\rho=18$ Ωcm.   p-type   :$\rho=18$ Ωcm
n-type   :Phosphorus -implantation,   n-type   :Phosphorus - implatation,
            3×10^{12}cm^{-2}, 150 keV. 3×10^{12}cm^{-2}, 150 keV, d≈200 nm,
            d ≈ 200 nm, annealing   annealing temper. 800°C, 30min
            temperature 800°C, 30min.   p$^+$-type :Boron-implantation, 1×10^{14}cm^{-2},
p$^+$-type :Boron-implantation,   30 keV, w≈100 nm, annealing
            1×10^{14} cm^{-2}, 30 keV,   temp. 800°C, 30 min.
            w≈100 nm, no annealing.   Si-implantation, 1×10^{12} cm^{-2},
                                       150 keV, no annealing.

Fig. 2a illustrates the measured internal quantum efficiency as a function of wavelength with the bias voltage as parameter for the sample 222/3. Fig.

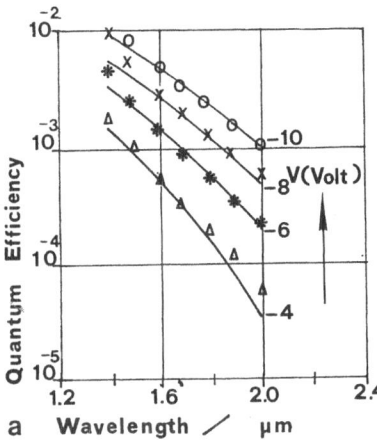
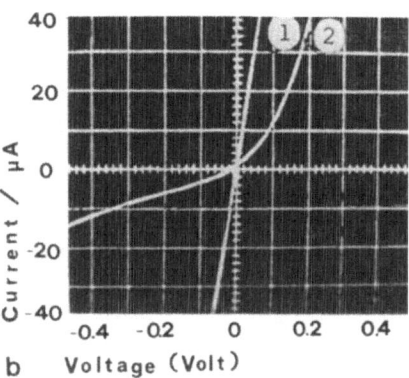

FIGURE 2. a) Quantum efficiency of a BBD (sample no.222/3) as function of wavelength. Points: measured values, solids lines: calculated, same fitting for all curves. b) (I-V)-characteristics of a BBD (sample no.520/2. ① : under illumination, ② : without illumination.

2b illustrates the I-V characteristic of a BBD specially prepared with high defect density in the middle layer (sample no.520/2) to be converted to an ohmic characteristic when the diode is illuminated with light of sufficient power (tungsten lamp, 40 watt).

4. EXPERIMENTAL RESULTS AND DISCUSSION

The measurements showed that the cause of the BBD IR-sensitivity is the defects formed during the ion implantation. A verification is the dissapearance of IR-sensitivity in all diodes which have been annealed following the double implantation.

For the explanation of the IR-sensitivity the following assumptions were made:

- the diode is reverse biased
- excitation of carriers occurs from a certain energy level.
- consider defects like acceptors; they are neutral prior to illumination and become negatively charged after the optical excitation of holes.
- the incident photons in the emitter region are absorbed only by traps, because their number in this region is much larger than the number of free carriers (s. eqn. 1).

Fig. 1c illustrates the phenomenon of optical excitation and carrier transport. Three different mechanisms can be seen:

1. Excitation of carriers from deep traps and emission over the barrier. This mechanism considers defects from region I only. The photogenerated holes must have sufficient energy to overcome the barrier.
2. Photogeneration in the space charge region. In region II, which almost coincides with the space charge region, the photogenerated holes can drift easily towards the collector under the influence of an electric field.

3. Current controlled by barrier modulation. It is possible to control the height of the pontential barrier by negatively charged traps in the middle layer. Their presence in this layer decreases the potential barrier and increases the current. The requirement is that the recombination time of the traps must be greater than the transit time of the excited holes in the middle layer.

All the above transport mechanisms can occur simultaneously. Their contribution to photocurrent depends on the position of the defect profile in the diode. The last two mechanisms require high defects density in the middle and depletion layer.

The appearance of the defects in the diode 222/3 is due to B ion implantation. Brice (|9|) gives a range of 63 nm and a standard deviation of 40 nm for this diode. Hence the defects are primarily distributed in the p^+- and at the beginning of the n-zone. Consequently the mechanism responsible for the photocurrent may be that of photoemission. This mechanism considers doping and defect distribution, trap energetic level, Fermi-Dirac distribution, optical absorption cross section, mean free path and emission probability of the excited carriers. The quantum efficiency η is given |6| by:

$$\eta = \frac{1-R}{2} \int_0^{x_o} \alpha_T(h\nu,x) \cdot \exp\left[-\alpha_T(h\nu,x) \cdot x\right] \cdot \exp(-\frac{x_o-x}{l}) \, (1 - \sqrt{\frac{e \cdot \Delta\Phi_B}{h\nu-E_T}}) \cdot dx \qquad (2)$$

where: $\Delta\Phi_B = \Phi_{BL} - \frac{1}{e} E_V(x)$,

$\alpha_T(h\nu,x) = N_T(x) \left|1 - f_T(E_T,x)\right| \sigma(h\nu)$,

Φ_{BL} : potential barrier (s.Fig 1c),

$E_V(x)$: valence band as a function of x,

$f_T(E_T,x)$: Fermi-Dirac distribution,

$\sigma(h\nu)$: capture cross section,

R : reflectivety index,

$h\nu$: photon energy,

l : mean free path,

x_o : distance between the surface (x=0) and the point corresponding to the maximum of the potential barrier.

Fig.2a shows the measured (points) and the calculated (solid lines, using eqn.2) quantum efficiency for the diode 222/3. It is obvious from this figure that there is a good agreement between the measured and the theoretical results. The fitting curves shown in fig. 2a are obtained using the following parameters values: $E_T-E_V = 0.45$ eV, $l = 56$ nm, $\sigma(0.689$ eV$) = 2.10^{-13}$cm^2, $d = 200$ nm and $N_B = 2.8 \cdot 10^{16}$ cm^{-3}.

Regarding the 520/2 diode, it should be noted that this diode was bombarded with Si atoms (150 keV) after the annealing process of the double ion implantation. For this implantation energy of 150 keV Brice gives a range of 120 nm and a standard deviation of 90 nm, which means that the defect density in the n-zone is high. The I-V characteristics of the diode 520/2, illustrated in Fig.2b, validate the mechanism of barrier modulation by negatively charged traps. In this diode, the trap density in the n-zone is higher than the donor density, so that, following the optical stimulation, the excited holes drift to the collector and the remaining negative charge of the traps eliminates the positive charge of the donors and, furthermore, the potential barrier. As a result the p^+-n-p structure is converted to an ohmic contact, p^+-p, whereas the diode 222/3 under the same illumination conditions kept its exponential characteristic.

CONCLUSIONS

It has been shown that the BBD can be used as an IR-photodetector. The three different mechanisms described in this work can lead to three different types of IR-photodetectors. The photoemission of deep traps for an IR-detector with controlled sensitivity limit by the potential barrier, the photogeneration in the space charge region for a conventional IR-photodiode and finally the barrier modulation by ionized traps for an IR-photodetector with high internal gain.

ACKNOWLEDGMENT

The author wishes to thank Dr.H.Mader,Siemens,Munich for supplying him with the implanted wafers.

REFERENCES

1. Shannon J.M.: A Majority-Carrier Camel Diode, Appl. Phys. Lett. 35 (1), Jul. 1979, pp 63-65.
2. Mader H.: Electrical Properties of Bulk-Barrier-Diodes, IEEE Trans. Electron Devices, Vo. ED-29, No.11,1982, pp 1766-1771.
3. Georgoulas N.: The Camel Diode as Photodetector with High Internal Gain, IEEE Electron Device Letters, EDL-3, No.3, pp 61-63, 1982.
4. Schaffer H.: Ein empfindlicher Silizium-Temperature Sensor, AEÜ, Band 38, Heft 1, pp 69-72, 1984.
5. Malik R.T., Dixon S.: A Subharmonic Mixer Using a Planar Doped Barrier Diode with Symmetric Conductance, Electr. Device Letters, Vol. EDL-3, No.7, pp 205-207, 1982
6. Georgoulas N.: Infrarotempfindlichkeit von Bulk-Barrier-Dioden. Dissertation, Technische Universität München, 1981.
7. Matsui, P.Baruch: Lattice Defects in Semiconductors, Ryukiti R. Hasiguti, University of Tokyo Press (1968), p 282.
8. H.Ryssel,I.Ruge: Ionenimplantation, B.G.Teubner, Stuttgart (1978).
9. D.K.Brice: Ion Implantation Range and Energy Deposition Distributions, Vol. 1 incident Ion Energies, New-York (1975).

INFLUENCE OF THE TARGET ENVIRONMENT ON ANGULAR DISTRIBUTION OF SPUTTERED
PARTICLES

W. HUANG and J. ONSGAARD, Fysisk Institut,
DK-5230 Odense M, Denmark
and
M. SZYMONSKI
Instytut Fizyki, Uniwersytet Jagiellonski
Reymonta 4, 30-059 Krakow, Poland.

ABSTRACT

Angular distributions of sputtered Ag atoms have been measured under various target environments by a combination of a collector technique and Auger
electron spectroscopy. The obtained results show that sputtering under Ultra
High Vacuum (UHV) conditions leads to an isotropic, cosine-like distribution, while the high partial pressure of impurities causes the distribution
to be strongly outward-peaked. Ion beam-induced-texturing effects are also
investigated.

INTRODUCTION

The commonly accepted methods for measurement of angular distributions of
sputtered particles rely on a careful analysis of the sputter-deposited material. For this purpose different surface sensitive techniques have been
used, in particular Rutherford backscattering spectroscopy (1). In a forth-
coming paper (2), we report on angular distribution measurements with Auger
electron spectroscopy for analysing the collected, sputtered material. The
present contribution is a continuation of that work in which we have studied
the influence of the target environment during ion bombardment on the angular distribution of sputtered particles. Thus, we have measured such distributions under good vacuum as well as under well controlled impurity conditions.

EXPERIMENTAL

A schematic view of the experimental set-up, built at Odense University,
is shown in figure 1. The target was mounted on the horizontal arm of the
rotatable manipulator, installed in the UHV system. 5 keV Ar^+ and N_2^+ ion
beams were delivered by a differentially pumped Leybold ion gun. The beam
densities, measured with a removable Faraday cup, were 0.2-0.3 $\mu A/mm^2$. The
ion beam under normal incidence was defined by the aperture on the collector (Fig. 1b). The base pressure in the system was in the low part of the
10^{-10} Torr range.
Sputtered material was collected on a stainless steel foil with an evaporated thin layer of aluminum. The foil could be cylindrically shaped during the sputter deposition and formed planar for the Auger analysis of the
deposit. Both the sample and the collector were sputter cleaned prior to
measurements. The sample, 5.5×5 cm^2 in size, was cut from a polycrystalline
silver bar (99.999% Ag). It was finally polished by SiO_2 paste (0.35-0.4
micron grain). Scanning electron micrographs taken in advance and after
sputtering showed the target surface as rather flat in a magnification of
36,000. No specific check of the presence of texture structure before sput-

tering was made. The sample cleaness as well as the composition of the collector could be monitored by the same Auger electron spectrometer.

In this contribution, we report on the angular distribution of Ag atoms sputtered from the polycrystalline silver sample. The partial pressure of the working gas in the chamber was varied from 3×10^{-8} Torr (we shall call this "UHV results") to 5×10^{-5} Torr and the corresponding distributions were compared for argon and nitrogen. Doses up to 4×10^{17} Ar$^+$ (7×10^{17} N$_2^+$) ions/cm^2 were used. This assured that the thickness of the sputter deposited film ($<3\times10^{15}$ Ag atoms/cm^2) did not exceed the mean escape depth of Auger electrons. From the point of view of texture effects, a small dose is preferable too, minimizing an influence of these effects on the angular spectra (1). Beam-induced texturing effects (4×10^{17} Ar$^+$ ions/cm^2) were observed in our earlier measurements (2). In order to reduce the effects, the target, after a fluence of 1.5×10^{17} Ar$^+$ ions/cm^2, was moved to a new position without changing the measurement geometry, whereupon a hitherto unirradiated area was exposed to the ion beam. Here, we have deliberately assigned the term "dose" to the total sputtering ion density and the term "fluence" to the ion density on a particular spot.

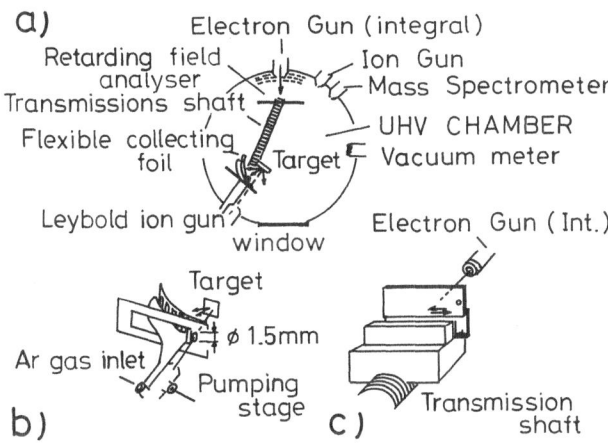

Fig. 1 Experimental set-up

The concentration of deposited Ag atoms as a function of ejecting angle θ, measured with respect to the surface normal, was determined by peak to peak heights of the characteristic Ag-MNN transitions (347.5 eV and 353.4 eV), measured as a function of the position on the foil.

The data presented in this report were obtained with a collector of radius of 22 mm.

RESULTS AND DISCUSSION

Polar plots of the Auger peak to peak heights of sputter-deposited Ag are presented in figure 2 and 3. The experimental points have been normalized in order to give the same integrated yield ($2\pi\int Y(\theta)\sin\theta d\theta$) as a cosine distribution represented in the figures.

It is seen from Fig. 2 that the Ag atom angular distribution caused by sputtering with Ar$^+$ ions is close to cosine, if the partial pressure of the

442

rest gas during sputtering is within the 10^{-8} Torr range, while in the 10^{-5} Torr range, the distribution is over -cosine.

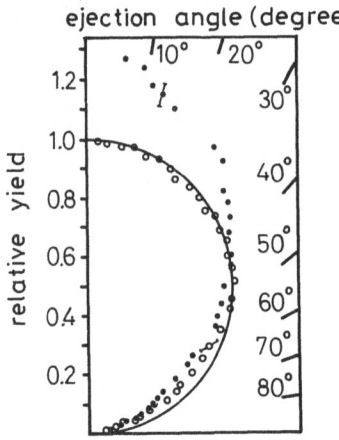

ejection angle (degree)

Fig. 2. Polar plots of angular distribution of Ag atoms sputtered from polycrystalline silver target by 5keV Ar$^+$ ion bombardment.
o:P=6.8×10^{-8} Torr
   fluence ~1.5×10^{17} Ar$^+$ ions/cm^2.
•:P=4×10^{-5} Torr,
   fluence ~2×10^{17} Ar$^+$ ions/cm^2,
The particles were deposited on a stainless steel catcher foil.

The strong back-peaked tendency with a pressure increase in the target environment was also pronounced in the case of nitrogen as shown in Fig. 3. These data are determined with less accuracy than the corresponding Ar$^+$ results because a background subtraction was performed in order to take a residual amount of Ag on the collector into consideration. Also a slight texturing effect could explain the deviation from a strict cosine-behavior for the case of P=1×10^{-7} Torr.

The UHV results presented here, as well as those shown elswhere (2), are consistent with the theoretical predictions (3-5). The relevant sputtering theory based on the linear collision cascade predicts an isotropic distribution for the sputtered flux. Recent computer simulations have shown that for low keV energies, the angular distribution were close to cosine too (5).

This indicates that over -cosine angular distributions of sputtered particles, frequently found previously (6,7), might be due to disturbances caused by adsorbed impurity layers on the surface. Target atoms originating from layers deeper than the adsorbate layer are preferentially ejected into a narrow cone around the surface normal (8,9). Our experimental data taken under 4×10^{-5} Torr partial pressure of the residual gas (shown in Figs. 2 and 3) support this explanation.

On the other hand, a high dose causing heavy damage to the surface, can restrict glancing ejection angles in a similar manner as discussed in ref. (8,9) despite the vacuum conditions. Therefore, it is not surprising that the distributions measured by Besocke et al. (10) were over-cosine.

In our earlier report (2) on Ar$^+$ ion sputtering the same Ag sample, a hump in the distribution pattern in the 30°-45° interval was interpreted as a result of beam-induced-texturing effects. Similar phenomena were observed when the N$_2^+$ ion beam fluence exceeded 3×10^{17}N$_2^+$ ions/cm^2, while for Ar$^+$ ion fluence case, it was a little lower.

Beam-induced-texturing effects corresponding to the closed packes direction were found by Tsuge et al. (11) in the case of polycrystalline fcc materials. Moreover, the distributions were symmetric about the normal incident ion beam. Our measurements prove that the hump in 30°-45° of the distribution patterns is independent of azimuthal angle, taking target surface normal as the polar axis. For fcc structured polycrystalline material, the <111> direction of the grains, ordered as a consequence of the high ion fluences, coincides with the direction of the surface normal. The <111> directions of those grains randomly lie on a cone with a polar angle θ=37.5°

ejection angle (degree)

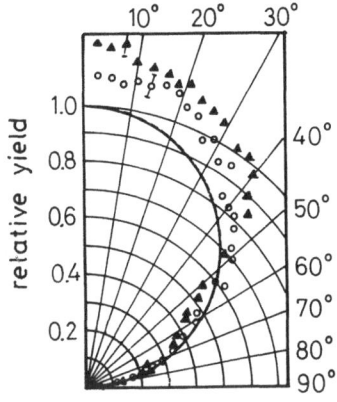

Fig. 3. Polar plots of angular
distribution of Ag atoms sputtered from polycrystalline
silver target by 5 keV N_2^+ ion
bombardment.
Fluence: 3×10^{17} N_2^+ ions/cm^2
o   : P=1.2 $\times$ 10^{-7} Torr
▲   : P=4$\times 10^{-5}$ Torr.

related to the cone axis. The present obtained results confirm this
interpretation(1).

ACKNOWLEDGEMENTS

Fruitful discussions with Prof.
Peter Sigmund are gratefully acknowledged. We are most grateful
to Dr. Hans A. Bøye for taking the
scanning electron micrographs. Two
of us, W.H. and M.S. acknowledge
the support from the Danish Natural
Science Research Council.

REFERENCES

1. Andersen H.H., Stenum B., Sørensen T., and Whitlow H.J., Nucl. Instr.
   Meth. B6 (1985) 459.
2. Szymonski M., Huang W., and Onsgaard J., submitted to Nucl. Instr. Meth.
   B.
3. Sigmund P., Phys. Rev. 184 (1969) 383; Phys. Rev. 187 (1969) 768.
4. Sigmund, P., in <<Sputtering by Particle Bombardment I>>, ed. R. Behrisch
   (Springer-Berlin-Heidelberg-New York 1981) p.9.
5. Biersack J.B., and Eckstein W., Appl. Phys. A34 (1984) 73.
6. Perovic B., and Cobic B., in (Ionization Phenomena in Gases II), ed. H.
   Maecker (North-Holland, Amsterdam 1962) p. 1165.
7. Andersen, H.H., Stenum B., Sørensen T., and Whitlow H.J., Nucl. Instr.
   Meth. 209/210 (1983) 487.
8. Sigmund, P. Oliva A., and Falcone G., Nucl. Instr. Meth. 194 (1982) 541.
9. Okutani T., Shikata M., and Shimizu R., Sur. Sci. 99 (1980) L410.
10. Besocke K., Berger S., Hofer W.O., and Littmark, Rad. Effects 66 (1982)
    35.
11. Tsuge H., and Esho S., J. Appl. Phys. V52,7 (1981) 4391.

SIMULATION OF NEAR ATOMIC SCALE SPUTTER INDUCED MORPHOLOGY

JOEL A. KUBBY and BENJAMIN M. SIEGEL

Department of Applied and Engineering Physics and the National Research and Resource Facility for Submicron Structures, Cornell University, Ithaca, New York 14853

On a length scale of ion penetration range $R_p(E)$, the prediction of a surfaces topographical development due to atom and ion bombardment as a result of erosion by sputtering and diffusional processes, that can be radiation enhanced, depends on parameters that are instantaneous contour dependent. On this length scale an analytic treatment results in an intractable mathematical description of the temporal topographic growth requiring computer simulation for detailed prediction. A recursive computer simulation, that advances points on a contour a short distance along the surface normal direction to form a new and slightly modified contour, is compared to TEM micrographs that show this contour development on the length scale where these second order effects first become important. Simulation results qualitatively confirm the experimental findings that topography development is influenced on this length scale by the spatial distribution of sputtering as well as curvature dependent migrational processes.

1. INTRODUCTION

A previous experimental investigation (1) of sputter induced morphology has shown morphological detail that is attributed to mechanisms which become important on the reduced length scale of ion penetration depth. The mechanism of sputter yield reduction due to the spatial distribution of sputter etch effects (2,3) has been discussed in regard to the life cycle (4), and stability (5) of surface morphological features that result due to ion bombardment induced erosion. The distribution of the sputter yield effect has also been discussed (6,7) in regard to its influence on the shape of the sputter yield curve $Y(\theta)$ where 'downstream' (2) effects may be important in turning the curve over at θ_p, the maximum in the sputter yield curve. This process tends to sharpen protuberant surface features as it reduces the sputter yield at the apex of the protuberance within a distance of the average energy deposition depth. A process that competes with this sharpening mechanism is the blunting action of surface diffusion, that can be radiation enhanced, as discussed by Carter et al (8,9). Consideration of this surface smoothing action, in competition with edge formation due to erosion, led to the Carter et al (9) intuitive prediction that the equilibrium for a sputtered hemispherical end cap of a cylindrical shank would be largely of conical section (with semi-vertical angle θ_p) but gently radiused at both tip and shank. This rod configuration is similar to the standard structure discussed by Witcomb (10). We have used this configuration experimentally to study the influence of both the sharpening mechanism due to yield reduction and the blunting action of the surface diffusion. Computer simulation of these effects are compared to TEM micrographs (fig. 1) of the cone apex region showing the initiation of morphology on this length scale. The qualitative results show that both mechanisms can be important to explain morphological development on a near atomic length scale.

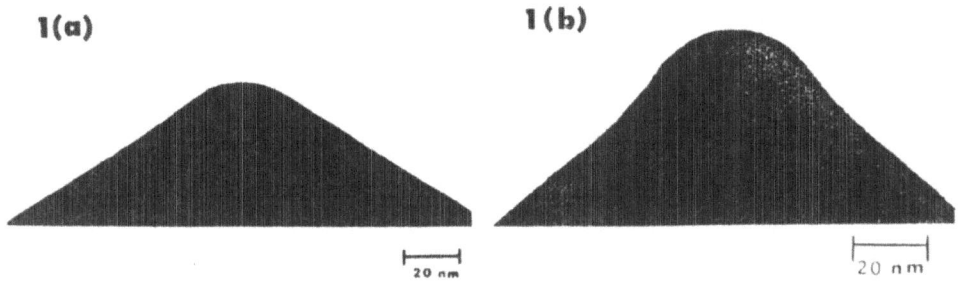

1(a)

1(b)

20 nm

20 nm

Figure 1(a) Transmission Electron Micrograph of cone apex region when second
   order effects first become appreciable. The profile resembles a
semi-circular arc fitted smoothly to conical asymptotes. The initial
development of concave curvature is just evident to the left of the cone
apex.

   (b) Transmission Electron Micrograph showing morphology that relates
to local sputter yield reduction. To either side of the apex region high
negative radius of curvature 'ears' have developed.

2. THEORY
     To predict the topographical development of the surface under ion
bombardment induced sputter erosion, the sputter yield given by Sigmund (2)
is used. Here, the average number of sputtered atoms per bombarding ion
from a surface element da is;

$$Y(\underline{r})da = \xi F_d(\underline{r})da \qquad (1)$$

where $Y(\underline{r})da$ is the average number of target atoms sputtered from a surface
element da at a vector distance $\underline{r}$ from the point of impact of the bombarding
ion, ξ a constant characterizing the target surface, and $F_d(\underline{r})$ the energy
deposited in atomic motion per unit volume at $\underline{r}$. This geometry is
illustrated in figure 2 for an ion of mass M_1 and energy E_1 incident on a
target composed of atoms of mass M_2 that at time t is described by the set
of points $(x(t),y(t))$. The intersection of the energy deposition contours
with the target surface illustrates the distribution of the sputter effect
on a length scale of the projected ion path length $R_p(E)$ as shown in
figure 3. The total number of atoms sputtered at a point $\underline{r}$ is due to the
contribution to the sputter yield of all ions impinging on the target
surface, of which those within a cascade length are most important.
Sigmund (2) sums the effect at $\underline{r}$ of ions incident at various points $\underline{r}'$ over
the target surface to find the total number of atoms sputtered from the area
da at $\underline{r}$;

$$N_s(\underline{r})da = J\xi da \iint_A dx'dy' f_d(\underline{r}-\underline{r}') \qquad (2)$$

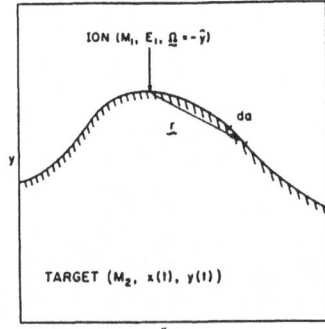

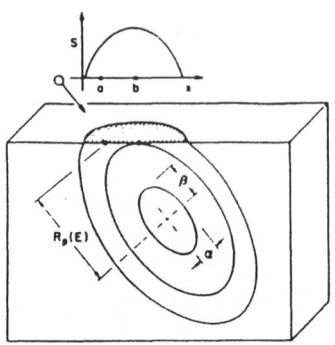

Figure 2: Sputter yield geometry. An ion of mass M_1 is incident on a target from above with energy E_1. The target is composed of atoms with mass M_1 and has a profile, at time t, described by the set of points (x(t),y(t)). The bombarding ion causes $Y(\underline{r})$da target ions to be sputtered at a vector distance $\underline{r}$ from the point of impact.

Figure 3: The intersection of the atomic collision cascade volume with the target surface. Contours of equal energy deposition are drawn in centered about the projected range $R_p(E)$ of the incident ion with widths α and β along and perpendicular to the direction of incidence. An ion incident at point a sputters target atoms mainly at the point b, shown projected on to the inset sputter yield curve Y, as this is the point of maximal energy deposition at the target surface.

where Jdx'dy' is the number of ions hitting an area dx'dy' on the target surface per unit time for a flux of J ions/cm²sec incident along the $-\hat{y}$ direction.

The rate of recession of a surface element da along the direction normal to the surface, $\rho_n(\underline{r})$, would be given by the product of the decrease in the number of atoms per unit area, per unit time, and the atomic volume Ω;

$$\rho_n(\underline{r}) = \Omega N_s(\underline{r}) \tag{3}$$

The deposited energy distribution is approximated by a Gaussian as discussed by Sigmund (2);

$$F_d(x-x',y-y') = (v/2\pi\alpha\beta)e^{-(y-y'+d)^2/(2\alpha^2)-(x-x')^2/(2\beta^2)} \tag{4}$$

where v is the total energy deposited in atomic motion, d the average energy deposition depth with α and β the longitudinal and lateral widths of the distribution. All three lengths are of the order of the ion range $R_p(E)$ as shown in figure 3. A surface that is instantaneously described by y=f(x) would have an erosion rate at the point (x,y) given by;

$$p_n(x,y) = (J\Omega vN\xi/2\pi\alpha\beta)\int_{-\infty}^{\infty} dx' e^{-(y-f(x')+R_p(E))^2/(2\alpha^2)-(x-x')^2/(2\beta^2)} \tag{5}$$

Sigmund (2) considered the ratio of the number of atoms sputtered at the apex of a cone to that further down the slope and predicted a reduced sputter yield at the top region of a narrow cone, as most of the cascades will miss the top of the cone and hit somewhere underneath. As a result of this partial cascade development near the cone apex, Sigmund (2) predicted that a cone should have an apical angle that is smaller than the value θ_p predicted by first order erosion theory (11). The exclusion of local flux has a sharpening effect around the apex region of a cone.

A competing process to the cone sharpening mechanism of Sigmund (2) is the smoothing action of surface diffusion discussed by Carter et al'(8,9) that has the effect of blunting high curvature surface features. The mathematical analysis for surface diffusion has been disucssed for the two dimensional case of a contour described by $y=f(x)$ by Mullins (12). For a segment of the surface with curvature $K=1/R$ where;

$$K(x,y) = -y''/(1+y'^2)^{3/2} \tag{6}$$

the speed of movement ρ_n of the surface element along its normal direction would be;

$$\rho_n(x,y) = (D_S \gamma \Omega^2/A_0 kT) \; (\partial^2 K/\partial s^2) \tag{7}$$

where $D_S = D_0 \exp(-Q/kT)$ is the coefficient of surface diffusion at temperature T, γ the surface free energy per unit area, Ω the atomic volume, A_0 the surface area per atom, kT the thermal energy, and s the arc length along the profile. Substituting for the curvature K and using the relation $\partial/\partial s = (\partial x/\partial s)(\partial/\partial x)$ with $\partial s/\partial x = (1+y'^2)^{1/2}$, the speed of a surface element along the normal direction is given by;

$$\rho_n = -(D_S \gamma \Omega^2/A_0 kT)(1+y'^2)^{-1/2}\partial/\partial x\left\{(1+y'^2)^{-1/2}\partial/\partial x(y''/(1+y'^2)^{3/2})\right\} \tag{8}$$

Although thermally stimulated diffusion is generally negligible below 1/3 of a substances melting temperature, Carter et al (8,9) point out that bombardment may significantly enhance surface diffusion by the creation of a vacancy density in excess of the thermal equilibrium value.

3. SIMULATION

An initial contour $y=f_0(x)$ is chosen that is qualitatively similar to the TEM shadow micrograph of figure 1(a) of the cone apex region when second order effects first become noticeable. Here a semi-circular arc is fit smoothly (continuity of first derivative) to linear conical asymptotes giving the profile shown in figure 4(a) as the heavy line marked (a). Both the apex radius of curvature as well as the cone half angle are variable; the apex radius of curvature has been set equal to the average ion path length, the cone half angle is set equal to the angle formed at the apex region of a surface under erosion as described by first order erosion theory (11) as the apex radius of curvature approaches zero.

The dynamic morphological development of this initial contour under the influence of Sigmund's (2) cone sharpening mechanism is found by dividing the curve into 100 coordinates $(x(t),y(t))$ that are equally spaced along the x axis. At each point along the curve an integration is performed over the surface (and extended beyond the region of interest by a few characteristic lengths to allow the integral to build up to a steady state value) to find

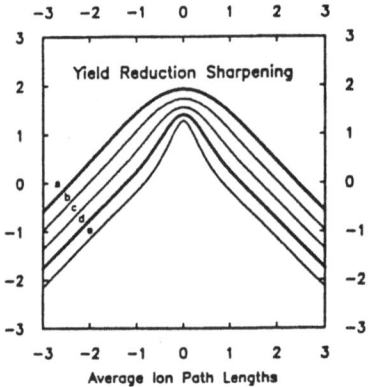
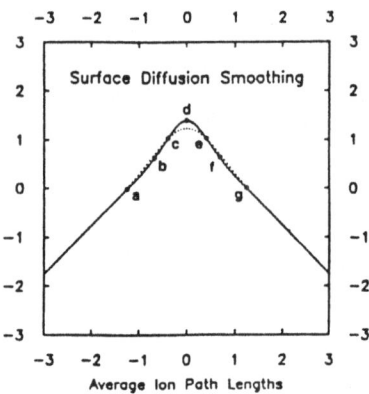

Figure 4(a) Sharpening of the cone due to the reduction of sputter yield in the apex region. The target profile is initially a semi-circular arc with conical asymptotes as shown by the heavy line a. As erosion proceeds, the cone apex sharpens approaching the limiting case, line e, as described by Sigmund. The heavy line marked d is an intermediate contour with appreciable curvature at the apex.

(b) Contour d from the previous figure has been reproduced here as the solid line. The points a,b,c,e,f, and g mark the boundary of segments with positive and negative curvature. The maximal curvature occurs at the point d in the segment ade. Down the slope from either a or g no second order effects are expected as there is no surface curvature beyond these points. The dotted line shows the development of the contour due to surface migrational smoothing. The maximum smoothing affect, which is driven by surface curvature, occurs at the point d.

the magnitude of the normal erosion rate $\rho_n(x(t),y(t))$ as given by equation 5. The motion of each surface point is then determined by moving a distance $\rho_n \partial t$ along the normal direction as shown in figure 5 for a fixed time step ∂t (or local ion density step $J\partial t$) and a new surface is constructed from the individual points $(x(t+\partial t),y(t+\partial t))$. This technique is similar to that used by Catana et al (13) in relation to the equilibrium topography of sputtered amorphous solids under first order erosion theory.

A cubic spline is then fit through all of the points of the new contour to describe the next profile $y=f_{i+1}(x)$. The previous erosion step has carried each surface point into a new normal erosion rate ρ_n so that the next iteration proceeds with the new normal speed $\rho_n(x(t+\partial t),y(t+\partial t))$ found by integration of equation 5 all over the new surface $y=f_{i+1}(x)$. This process is continued iteratively to find the dynamic development of surface morphology shown in figure 4(a) where every other contour has been plotted. The step $J\partial t$ is chosen so that only small geometric changes occur between contours. The sequence of profiles shows an apex radius of curvature that decreases with continued localized erosion and the development of 'ears' with maximal concave (negative) curvature within an average ion path length to either side of the apex. The sequence approaches the limiting behavior, that has an analytic solution, as discussed by Sigmund (2).

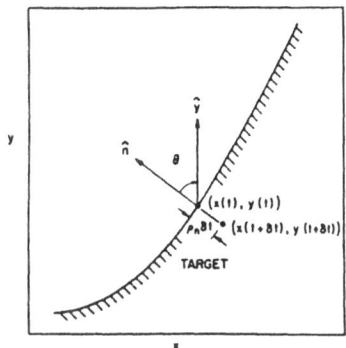

Figure 5: With an erosion rate
ρ_n along the direction normal
to the target surface, a point
$(x(t),y(t))$ will move a distance
$\rho_n \partial t$ to the new position
$(x(t+\partial t),y(t+\partial t))$ after a short
erosion time ∂t.

As the apex radius of curvature decreases, the driving force for surface diffusion (gradients in chemical potential) increases in accordance with equation 8 that has the effect of blunting the high curvature profile developing at the cone apex. If an intermediate contour from figure 4(a) is chosen, such as the heavy line (d) in this figure, the profile can be modified to take into account the action of surface diffusion. This contour has been re-drawn as the solid line that appears in figure 4(b). The contour as modified by equation 8 is shown in figure 4(b) by the dotted line. The main effect due to surface diffusion is the modification of the high curvature arc cde to the blunter segment ce of the dotted contour. A small amount of material has accumulated in the concave ears (negative radius of curvature) segments abc and efg on either side of the apex region. Surface diffusion is a competing process to the sharpening caused by sputter erosion and any amount of it will inhibit edge formation as discussed by Carter et al (9).

4. EXPERIMENTAL
The experimental conditions under which the erosion was carried out have been described previously (14). A high voltage discharge source was used to produce an Ar^+ ion beam of nominally 10keV energy. However the energy spread is estimated to be on the order of the acceleration energy for this source, and no mass analysis was performed on the ion beam that is estimated at up to 50 percent atomic composition. The target was mounted in a bulk specimen holder for imaging, with the variation in specimen position about the optimal planar object position estimated to give no more than a factor of 2 correction for the specified magnification. This difference in specimen height could not be compensated via the focusing current of the TEM as no calibration curve was available on either the JEOL 200 CX Temscan or JEOL 1200 EX microscopes used that related lens current to magnification. For these reasons the TEM images are only considered on a qualitative basis.

5. RESULTS
The TEM micrograph of figure 1(a) is digitized and used as the initial contour for simulation as shown by the first heavy line in figure 6(a). After a spline fit, the contour is modified iteratively in accordance with equation 5 giving the sequence of contours shown as light lines. An intermediate contour, that had similar radius of curvature in the 'ears' to either side of the apex as the TEM micrograph of figure 1(b), has been drawn

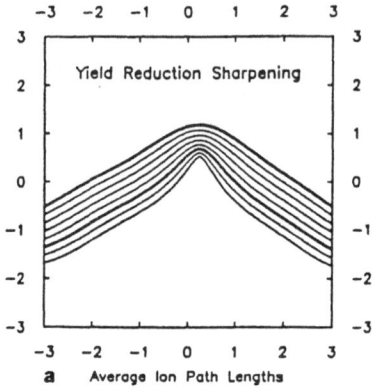 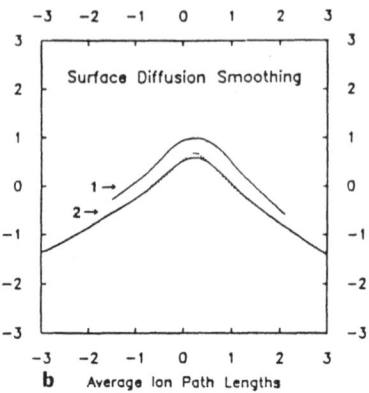

Figure 6(a) Yield reduction sharpening as applied to the contour shown in
the TEM shadow micrograph of figure 1(a). The micrograph is digitized and
appears at the initial heavy line in this figure and is used as the
initial contour $y=f_0(x)$ in the simulation. The light lines show the
intermediate contours $f_i(x)$ used to predict further development. The
second heavy line has 'ears' with curvature similar to the 'ears' on
either side of the cone apex shown in the micrograph 1(b).

   (b) The intermediate contour of the previous figure has been
re-drawn here as the dotted line. After surface migrational smoothing it
appears as the solid line marked 2. The micrograph from 1(b) is digitized
and appears as the solid line 1 for comparison.

in with a heavy line and reproduced as the dotted line in figure 6(b). The
solid line marked 2 in figure 6(b) is the dotted contour as modified by the
surface diffusion smoothing of equation 8.

6. DISCUSSION

   Comparison of line 1 in figure 6(b), the digitized profile of the TEM
micrograph of figure 1(b), with the simulation results drawn as the solid
line 2 in figure 6(b), shows reasonable qualitative agreement between the
two. The 'ears' in each curve occur near the point taken as an average
ion path length to either side of the apex, and show similar curvature at
the apex region, although the micrograph of figure 1(b) has less curvature
here than the simulation. A more realistic simulation would have a contour
that is continuously modified by surface diffusional smoothing as the
profile is sharpened due to localized erosion rather than the single
expost-facto surface diffusion step taken here.
   The macroscopic equilibrium end-form expected under first order erosion
theory (11) for this target configuration has been discussed by Witcomb (10)
and is a conical end cap (of half angle θ_p) to a rod with sides parallel
to the ion beam direction. This solution would obtain in the regions $(-\infty, a]$
and $[g, \infty)$ of figure 4(b). However the internal region bounded by the points
a and g has segments of both positive and negative curvature that are on the
order of average ion path length so that second order effects must be
included; the question of an equilibrium morphology within this region must
be handled with regard to these effects.

Carter et al (3) has discussed the question of whether an equilibrium morphology is expected to develop under the localized erosion theory. Using erosion slowness analysis similar to that used to predict morphology development under a spatially variable ion flux density J(x,y), Carter et al (3) conclude that in general no steady state end forms should be expected when local sputter variations merely moderate the standard sputter yield-projectile incidence angle function by multiplication.

However, under the influence of both erosional sharpening and surface migrational blunting, Carter (8) points out the interesting analogy to the case of a field emitter that is subject to the sharpening action of 'build-up' due to the surface stress $F^2/8\pi$ of an applied electrostatic field F, and the thermally activated surface migration driven by gradients in chemical potential due to surface curvature. Operation of a field emitter about the point where electrostatic stresses balance surface tension stresses allow the emitter to be heated to maintain electrical stability, while avoiding surface migrational dulling. Relating the morphological changes to the parameters of field and temperature that affect the rate of sharpening and dulling, Barbour et al (15) were able to determine the surface tension and migrational constants for tungsten. A similar quantitative study here would give useful information on the activation processes that are important, at a given temperature (8), for surface and volume diffusion under ion bombardment.

Future experimental work following the dynamic morphology development as a function of dose, corrected for secondeary ions and electrons, will determine if a steady state end-form on the length scale of ion penetration depth develops. Quantitative results will require a mass and energy analyzed beam as well as a TEM with magnification calibrated to lens current to see if the dimension of apex structure scales in accordance with ion range $R_p(E)$. A target stage with the ability to lower the target temperature should enhance the significance of Sigmund's (2) second order theory when migrational processes are dominated by thermally activated surface diffusion. Future computational work will include migrational smoothing continuously throughout the erosion process.

The authors wish to express their gratitude to the staff of the Materials Science Center: John Hunt, Margret Craft, and Ray Coles for the provision of the ion mill as well as assistance and training on the SEM and TEM, Jane Jorgenson and Marcia Kelley for the drawings and photography, Doug Neuhauser for technical support, and Bonni Jo Davis for typing the manuscript. We would also like to thank David McQueeney for the computer graphics. The simulation was done using his language LTPLOT, a powerful tool for the manipulation and plotting of data.

This work was supported by DARPA under ONR contract N00014-80-C-0587.

REFERENCES

1. Joel A. Kubby and Benjamin M. Siegel, Proc. of the 11th Inter. Conf. on Atomic Collisions in Solids, Georgetown University, Washington D.C., August (1985), to be published in Nucl. Instr. and Meth. B.
2. Peter Sigmund, J. Mater. Sci. 8, (1973) 1545.
3. G. Carter, M.J. Nobes, and R.P. Webb, J. Mater. Sci. 16, (1981) 2091.
4. G.W. Lewis, J.S. Colligon, F. Paton, M.J. Nobes, G. Carter and J.L. Whitton, Rad. Eff. Lett. 43, (1979) 49.
5. O. Auciello, Roger Kelly, and R. Iricibar, Rad. Eff. Lett. 43, (1979) 37.
6. I.H. Wilson, S. Chereckdjian, and R.P. Webb, Nucl. Instr. and Meth. B7, (1985) 735.

452

7. I.H. Wilson and R.P. Webb, Proc. 11th Inter. Conf. on Atomic Collisions in Solids, Georgetown University, Washington D.C., August (1985), to be published in Nucl. Instr. and Meth. B.
8. G. Carter, J. Mater. Sci. $\underline{11}$, (1976) 1091.
9. G. Carter, J.S. Colligon, and M.J. Nobes, Rad. Eff. $\underline{31}$, (1977) 65.
10. M.J. Witcomb, J. Mater. Sci. $\underline{10}$, (1975) 669.
11. G. Carter, B. Navinsek, and J.L. Whitton, in Sputtering by Particle Bombardment II, ed. by R. Behrisch (Springer-Verlag, Berlin 1983) pg. 257.
12. W.W. Mullins, J. Appl. Phys. $\underline{28}$, (1957) 333.
13. Cristina Catana, J.S. Coligon, and G. Carter, J. Mater. Sci. $\underline{7}$, (1972) 467.
14. Joel A. Kubby and Benjamin M. Siegel, Proc. 29th Inter. Symp. on Electron, Ion, and Photon Beams, to be published in J. Vac. Sci. Technol., Jan. (1986).
15. J.P. Barbour, F.M. Charbonnier, W.W. Dolan, W.P. Dyke, E.E. Martin, and J.K. Trolan, Phys. Rev. $\underline{117}$, (1960) 1452.

EPITAXY OF ERBIUM ON TUNGSTEN

G. KOZŁOWSKI, A. CISZEWSKI

Institute for Experimental Physics, University of Wrocław,
50-205 Wrocław, ul. W. Cybulskiego 36, Poland.

The process of the growth of erbium crystals by evaporation
from vapour onto a surface of tungsten under UHV conditions
was investigated. The experiment was carried out by using the
technique of field electron emission microscopy [1,2].
A feature of this technique is a hemispherical form of the
substrate with a tip curvature radius of order of thousand Å.
The substrate is usually a single crystal, the spherical surface of which exposes several low index crystallographic
planes. The atom flux impinging onto such a surface is accommodated on it, and the condensation process mostly occurs in a
different way on different regions of the substrate surface
due to a significant anisotropy of the electronic structure
of the surface. The aim of this work was to examine which crystallographic type of the surface of tungsten is preferential
for the epitaxial growth of erbium crystals and to find favourable conditions for growing big crystals of erbium the
size of which would be comparable with the size of the substrate.

The experiment was performed using a glass field emission
microscope connected to a sputter - ion pumped metal vacuum
system. The base pressure in the microscope was about 10^{-8} Pa
estimated from the rate of the specimen surface contamination.

Erbium was deposited from vapour. The evaporator was a
conical tungsten coil basket which was heated by electrical
current through W leads, which in turn were separately degass-
able. This kind of source was stable and regulation of vapour
flux was possible in a wide range. The source was thoroughly
vacuum degassed before filling with erbium. The Er used as
the vapour source was of 99.9% purity.

The temperature of the substrate during the evaporation of
erbium, the other important factor of crystal growth, was sta-
bilized and carefully measured by means of an electronic tip-
temperature controller of the resistance of the support loop.
The loop resistance was then related to the temperature by
using a conventional method [1]. The specimen was thermally
cleaned. Field emission observations and photography were made
with the tip held near 78 K. The measurement of the absolute
magnitude of the erbium flux during deposition was not possible
in this experiment. The flux magnitude was conventionally controlled and estimated based on the heating current through the
tungsten coil of the erbium source.

Changes in the structure of the deposited layer of erbium
were observed with temperature increasing from room temperature to about 1000 K. The vapour flux of erbium was assumed

to be a constant parameter in this observation. In this way, the changes in the structure of the adsorption layer were examined as a function of the layer growth temperature for a few values of the vapour flux of deposition. The sequence of the changes was similar in all cases except that the higher the flux magnitude of erbium vapour the higher temperature range needed to observe structure features typical for a given stage of structure development.

Fig. 1 shows the most typical structures of crystalline adsorption layers of erbium obtained on a tungsten substrate at various temperatures for a constant value of the vapour flux of erbium. A field emission microscopy /FEM/ pattern of a clean tungsten surface is shown in fig. 1a. The first crystalline form of the erbium layer /fig. 1b/ appears when the temperature of the substrate is in excess of 500 K. A stage preceding it shows an FEM pattern which is characteristic of an amorphous structure, and a number of grains indicate a marked roughness of layer. The layer of fig. 1b shows grainy features also. However the emerging symmetry of the grains gives evidence for the hcp structure of erbium [3]. The grains appear mostly on the edges of the low index planes (011) , (112) and sometimes the (001) plane of tungsten. They appear first on that part of the substrate surface which is closer to the impinging vapour flux from the erbium source. The distribution of the grains becomes more uniform with increasing temperature of the substrate. Then, a stage which appears in the temperature range of 650 K to 800 K shows a significant uniformity of layers. The latter have a definite crystallographic structure. Mostly these are either crystal twins, such as the ones shown in figs. 1c and 1d, or polycrystals obtained often at a substrate temperature close to the lower limit of the temperature range. In the case of the crystal twins it is not possible to determine their epitaxial affinity in relation to the substrate of tungsten. It can be assumed, however, from the symmetry observed in FEM patterns that the twins are cooriented self-epitaxial crystals of erbium. We were not successful in observation of the original stage of the crystal twin growth and so one cannot discern between two possible mechanisms of growth. This could occur either via coalescence of two nuclei which had appeared simultaneously on the tungsten surface or through supersaturation of a growing single crystal of erbium resulting in the appearance of another nucleus which would initiate the growth of the other crystal of erbium under more favourable conditions. Fig. 1c shows an epitaxial crystal of erbium obtained at a substrate temperature of about 800 K. The (0001) crystal plane of erbium, which has a 6-fold symmetry axis, is perpendicular to the emitter axis. The crystal of erbium covers the entire area of electron emission of the substrate. A crude estimation of the thickness of the crystal gives a magnitude of several hundred Å. The deposition of erbium at temperatures higher than 850 K did not result in obtaining crystalline erbium layers for the vapour flux of erbium used to obtain the layers shown in figs. 1b - e. The field electron emission pattern in this case was indicative of the symmetry of tungsten, and the emission characteristics

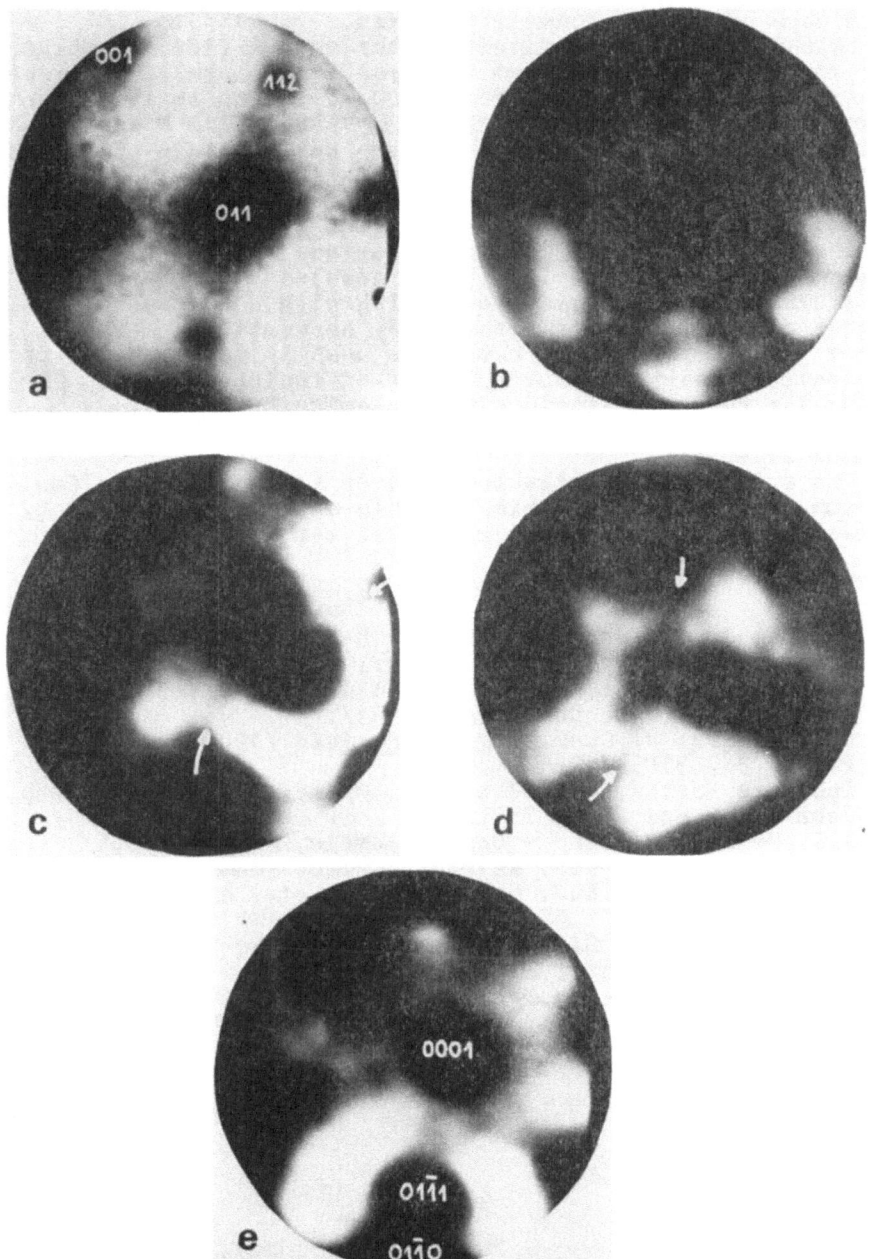

FIGURE 1. Structures of crystalline adsorption layers of erbium on tungsten FEM emitter tip; a/ pattern of clean tungsten, (011) - oriented, bcc structure; b-e/ erbium crystalline layers on W obtained at different temperatures. Note the hcp structure of the crystal form in e. Arrows indicate the grain boundary.

were ones of erbium covered tungsten.

Depending on the magnitude of the vapour flux of erbium, the epitaxial erbium layers appeared in a range of substrate temperatures of 800 K to about 1000 K. The epitaxial orientation $\{0001\}$ Er $\|$ $\{011\}$ W with $\langle 11\bar{2}0 \rangle$ Er $\|$ $\langle 11 \rangle$ W was found which is typical also for other rare earth metals of the hcp structure [4,5,6], The nuclei initiating the epitaxy of erbium appeared mostly on the edges of the (011) plane of tungsten, on a precovered surface. The erbium crystals obtained had a uniformly developed surface which could remain clean for long time periods. This enabled the use of such crystals in experiments, which require dealing with a surface of a high degree of cleanness, not possible by conventional methods. These erbium crystals have been used in measurements of the activation energy for surface self-diffusion of erbium [7] by using the method of field electron emission [8,9].

ACKNOWLEDGEMENT

The authors would like to thank Dr Allan J. Melmed for the erbium used in this experiment. This work was supported by the Polish Academy of Sciences under Project No MR.I.9.

REFERENCES

1. R.H. Good, Jr. and E.W. Müller, Handbuch der Physik, Springer - Verlag, Berlin /1956/;
   R. Gomer, Field Emission and Field Ionization, Harvard University Press, Cambridge /1961/.
2. A.J. Melmed, J. Chem. Phys., 38, 1444 /1963/, J. Appl. Phys., 36, 3585 /1965/.
3. The Rare Earths, edited by F.H. Spedding and A.H. Daane, John Wiley and Sons, Inc., New York and London /1961/.
4. A.J. Melmed, J. Less - Common Metals, 8, 320 /1965/.
5. A. Ciszewski and A.J. Melmed, Surfece Sci. 145 L471 /1984/.
6. A. Ciszewski and A.J. Melmed, J. Crystal Growth, accepted for publication.
7. G. Kozłowski, A. Ciszewski and W. Święch, Journal de Physique, accepted for publication.
8. A.J. Melmed, J. Appl. Phys., 38, 1885 /1967/.
9. A. Ciszewski, A.J. Melmed, Surface Sci. 145, L509 /1984/.

Ref. IEF 85.54mk

GROWTH OF THIN FILMS BY UHV ION BEAM SPUTTERING DEPOSITION TECHNIQUE

C. PELLET, C. SCHWEBEL, F. MEYER, G. GAUTHERIN
Institut d'Electronique Fondamentale, Unité associée au CNRS 22, Université Paris-Sud, bâtiment 220 - 91405 ORSAY CEDEX (France)

1. INTRODUCTION

To date the ion beam sputtering deposition (IBSD) technique has been used to deposit metallic, dielectric, superconductor and semi-conductor films [1]. Many promising results have been obtained, for example good mechanical characteristics of the films. In the demanding domain of microelectronics, a thorough knowledge, theoretical as well as experimental, of the sputtering and deposition processes seems to be necessary for determining the potential for using IBSD technique to deposit thin films [2]. We have studied IBSD underlying mechanism with the aim of forming epitaxial thin films. This paper reports some results of this study.

2. EXPERIMENTAL SET-UP

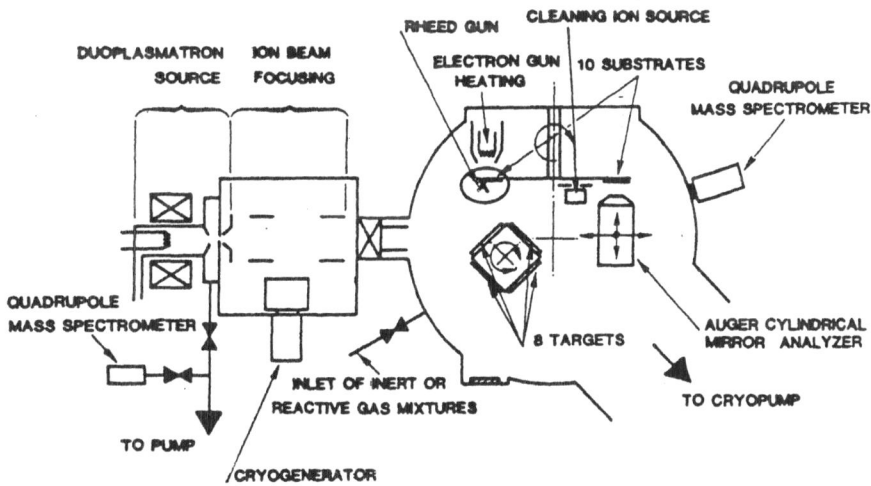

Figure 1 - Schematic drawing of the experimental set-up.

The experimental set-up previously described [2] consists of three main parts : the duoplasmatron source, the intermediate chamber (focusing chamber) and the deposition chamber. The latter is equipped with four diagnostic techniques: reflection high energy electron diffraction (RHEED), Auger electron spectrometer (AES), quadrupole mass spectrometer (QMS), and secon-

458

dary ion mass spectrometer (SIMS). The ultimate pressure, in this chamber, is 10^{-5} Pa.

3. DEPOSITION PARAMETERS

Neon, argon, krypton and xenon ion beams of 10-20 KeV energy and 50-500 μA/cm^2 maximum current density at the target are used. For these conditions, the pressure in the deposition chamber is 10^{-5} Pa of rare gas. Boron or antimony doped silicon targets and gallium arsenide targets have been used.

4. PARTICLES EMITTED FROM THE TARGET

4.1. Doped silicon target

The angular distribution of the particles emitted from the target (silicon, doping element, rare gas) has been determined by thickness measurement and composition analysis of films deposited on nine substrates. These substrates are attached to a semicircular substrate holder, as schematically illustrated in Figure 2. The ion beam impinged on the target surface at an incidence angle of 45°.

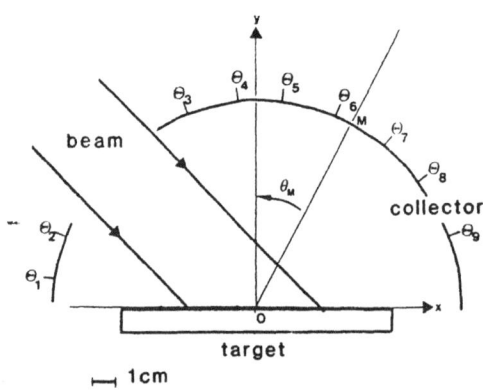

4.1.1. Silicon

For the energy and the current density of the primary ion beam, the angular distribution of sputtered silicon is found to follow a $\cos^2\theta$ law (fig. 3).

Figure 2 - Semicircular substrate holder

4.1.2. Dopant (SIMS measurement)

The dopant is transferred from the target to the film with constant efficiency (50 % in the case of boron) in any emission direction (fig. 4). Compared with MBE results |3|, the transfer efficiency value of Table 1 is very good. Since these doping elements with thermal energy possess very low sticking probability |4| one can suppose that our results can be attributed to the suprathermal energy of the sputtered particle.

Dopant	Target	Film
B	4.10^{16} cm^{-3}	2.10^{16} cm^{-3}
Sb	3.10^{17} cm^{-3}	2.10^{17} cm^{-3}

Table 1. Dopant concentration in the target and in the epitaxial film.

4.1.3. Rare gas

The major impurity in our films is the rare gas. Figure 3, shows the angular distribution of xenon incorporated in the films at room temperature. The maximum concentration $1,5.10^{20}$ cm^{-3} is found near the direction of the ion beam specular reflection. Since rare gas atoms with thermal energy possess low sticking probability even at room temperature [5], this rare gas concentration can only be attributed to the trapping of energetic atoms. In addition, since $M_t/M_g < 1$ (where M_t is the atomic mass of the target material and M_g is the atomic mass of the working gas) the observed distribution cannot result from one collision backscattering process. It seems that the main mechanism is the backscattering of primary ions due to multiple collisions at small angles [6].

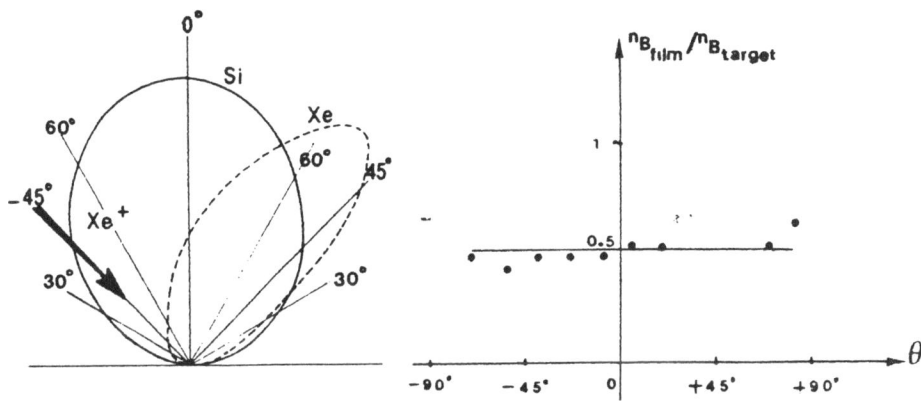

Figure 3 – Angular distribution of sputtered silicon and trapped xenon.

Figure 4 – Yield of boron concentration in the films and in the target versus the ejection angle.

4.2. Gallium arsenide target

The results obtained have been previously published [7]. We merely recall that epitaxial GaAs films with target stoichiometry (to within electron microprobe sensitivity) have been deposited at 520°C.

5. GROWTH OF SILICON EPITAXIAL THIN FILM [2]

The beginning of single crystal growth occurs at 250° C and good crystalline structure is obtained above 700° C as it can be seen on the RHEED pattern (fig. 5). Most of the deposits have been obtained from a silicon target doped with boron (4.10^{16} cm^{-3}). For a 500 nm thick layer with this target we observe at room temperature a net carrier concentration of 2.10^{16} cm^{-3}, the mobility of the deposited Si has been measured and the results are shown in Figure 6.

460

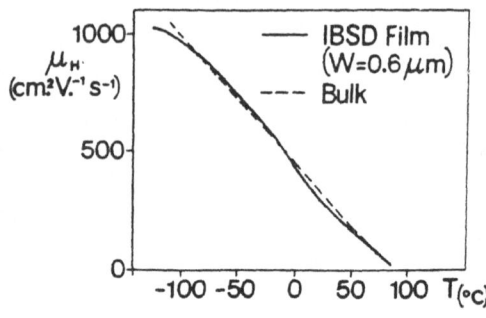

Figure 5 - RHEED pattern of silicon
epitaxial thin film deposited at
710° C.

Figure 6 - Hall mobility versus
temperature (film thickness
W = 0.6 μm).

REFERENCES

1. Klabunde J. (ed) : Thin films from free atoms and particles : Academic
   press Inc. (in press).

2. Schwebel C., Meyer F., Gautherin G., Pellet C. : submitted to J. Vac.
   Sci. Technol.

3. Bean J.C. : Appl. Phys. Lett. 33, 654, 1978.

4. Ota Y. : J. Electrochem. Soc. Vol. 126, n° 10, 1761, 1979.

5. Comas J., Wolicki E.A. : J. Electrochem. Soc. 117, 1197, 1970.

6. Pellet C., Schwebel C. : to be published.

7. Schwebel C., Vapaille A., Bouchier D., Gautherin G., Meyer F. :
   Third international congress cathodic sputtering and related applications, september 11-14, 1979.

MICROSTRUCTURAL ANALYSIS OF ELECTRONICALLY CONDUCTIVE CERAMIC COATINGS SYNTHESIZED BY REACTIVE ION PLATING AND SPUTTER DEPOSITION.

L. D. STEPHENSON**, V. F. Hock*, J. M. RIGSBEE**, D. TEER***, AND D. ARNELL***.

*U.S. ARMY Corps of Engineers, CONSTRUCTION ENGINEERING RESEARCH LABORA-TORY, CHAMPAIGN, ILLINOIS, USA, 61820.
**Department of Metallurgy and Mining Engineering, University of Illinois, 1304 West Green Street, Urbana, Illinois, USA 61801.
***Department of Aeronautrical and Mechanical Engineering, University of Salford, Salford M54WT, U. K.

1. INTRODUCTION

Electrically conducting oxide (ceramic) coatings for composite anodes have been used in recent years for various electrochemical processess, particularly in the electrolytic production of chlorine, chlorine oxide compounds, fuel cells, and cathodic protection systems (1,2). The composite anode usually consists of a self passivating conductive metal base (e.g., titanium or niobium) and a coating of transition metal oxides and noble metal oxides of the platinum group (3).

This paper presents the results of an applied research program designed to fabricate and characterize the chemistry, structure and electrophysical properties (electrical conductivity, dissolution rate, adhesion) of ion plated Ru-Ti-O coatings and Nb-Ti-O coatings on niobium substrates. Initial studies, which employed a substrate-biased sputter deposition process at the University of Salford, England, indicated the feasibility of fabricating conductive metal oxide compounds. In more recent studies ion plating has been used as the deposition process. Bombardment of the substrate by energetic ions and neutrals has been shown to produce: (a) a clean and reactive substrate surface by "sputter cleaning" prior to coating (4,5,6,7); (b) heating of the surface (8,9); (c) development of a chemically graded coating/substrate interface because of recoil implantation and diffusion induced by ion mixing (10); and (d) improved adhesion from the combined effects of a, b, and c above.

2. ION PLATING OF CONDUCTIVE OXIDE COATINGS

2.1 Experimental Procedure

Investigations were carried out on Nb-Ti-O films and Ru-Ti-O films deposited on niobium, 1018 mild steel, and glass substrates. The metals were electron beam evaporated with the evaporation rates controlled by varying the power input and raster frequency of the individual electron beams. The microstructure and elemental composition of the deposited oxide films have been studied by scanning electron microscopy with x-ray energy dispersive spectroscopy (SEM/EDX), x-ray diffraction and Auger electron spectroscopy (AES). Conductivities of the oxide films deposited on glass substrates were measured by a four-point-probe technique. The dissolution rates of the oxide films were determined by testing the coated anodes in a 3.5% NaCl solution and in ordinary tap water. In each case, a 25mA current (current density = 25 A/m^2) was passed through the solution with the anode biased positive relative to a graphite cathode. The dissolution rates were based on an averaged amount of constituent ions in solution as determined by mass spectroscopy.

462

2.2 Results and Discussion

2.2.1 "Niobium-doped" titanium oxide coatings.

Composition-depth profiles of as-deposited coatings using AES revealed a uniform composition of approximately 3.5 at.% Nb, 30.5 at.% Ti, 66.0 at.% oxygen upon sputtering over an estimated first 0.5 µm layer of the ion plated coating, as shown in Figure 1.

The results of x-ray diffraction scans of the as-deposited coatings are shown in Figure 2. Most of the peaks can be indexed as belonging to the rutile (TiO_2) structure, although some of the peaks are consistent with the anatase (TiO_2) structure and possibly one of the TiO phases. The broadening of these peaks indicates that the coating is likely microcrystalline. It was difficult to positively identify any of the x-ray diffraction peaks as belonging to either an intermetallic oxide compound containing Ti, Nb, and O (e.g. $Nb_xTi_{1-x}O_2$) or to one of the oxides of niobium, although it is possible that small amounts of these phases may actually exist in the ion plated structure. No evidence of metallic Nb or Ti was found by x-ray diffraction. On the basis of the composition and structural analysis to date, it is likely that these coatings are composed of oxides of titanium (chiefly rutile) with niobium atoms substituted in the crystal lattice. The term "niobium-doped" titanium oxide is used in this report since the true nature of the structure is not yet fully established. More advanced techniques, such as analytical electron microscopy (AEM), in which microdiffraction techniques may be employed, are needed to ascertain the accuracy of the results.

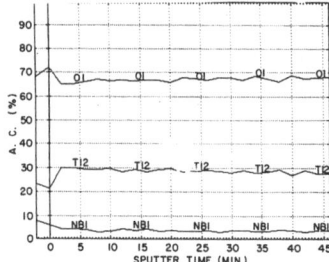

Fig. 1. AES composition-depth profile of ion plated Nb-Ti-O coating.

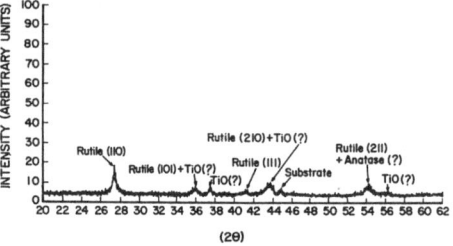

Fig. 2. X-ray diffraction scan of ion plated Nb-Ti-O coating.

Figure 3 is an SEM micrograph from a fracture specimen showing the surface and cross section of the "niobium-doped" titanium oxide film on a 1018 steel substrate. The morphology can be characterized as a dense columnar structure 10 µm thick. The apparent gain size is less than 1 micron.

The resistivity of the "niobium-doped" titanium oxide coating deposited on glass slides was 0.08 ohm-cm, which is quite satisfactory

Fig. 3. SEM micrograph of ion plated Nb-Ti-O coating.

for its intended application. The measured coating adhesion was found to be 51.3 MPa.

Composite anodes, formed by ion plating the "niobium-doped" titanium oxide films onto niobium substrates were found to have low dissolution rates, 0.329 g/A/yr when tested in 3.5% NaCl solutions and 0.054 g/A/yr when tested in ordinary tap water. Thus, anodes with highly adherent, electrically conductive "niobium-doped" titanium oxide coatings can be produced by reactive ion plating.

2.2.2 Mixed oxides of ruthenium and titanium (Ru-Ti-O). Figure 4 shows an AES composition-depth profile of the as-deposited coating containing mixed oxides of Ru and Ti. After 60 minutes of sputtering the substrate peaks appeared, indicating that the entire coating had been sputtered through and analyzed. The composition of this coating can be seen to vary with depth, forming layers that are alternately Ti-rich and Ru-rich with an average composition of approximately 20 at.% ruthenium, 20 at.% titanium, and 60 at.% oxygen. These results were somewhat curious as the production of this layered structure was unintentional. An attempt was made to hold all process parameters constant during ion plating (as in the case of the "niobium-doped" titanium oxide coatings), however, it is apparent that the evaporation rates and/or plasma characteristics actually varied during the ion plating experiments.

A typical x-ray diffraction scan over the as-deposited coating is shown in Figure 5. Analysis of this data is difficult, due to very low intensities and peak broadening. The visible peaks can be indexed as belonging to either the RuO_2 structure or the rutile (TiO_2) structure. Based on the fact that the relative concentrations of Ti and Ru are each approximately 50 at.%, the coating is most likely a mixed oxide of rutile (TiO_2) and RuO_2. The peak broadening evident in Figure 5 indicates that the structure of the coating is microcrystalline. Figure 6 is an SEM micrograph from a fracture specimen showing the surface and cross section of

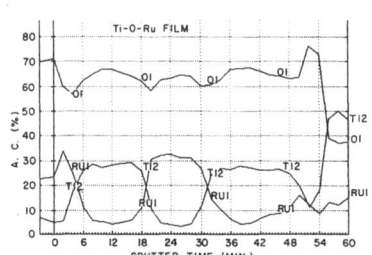

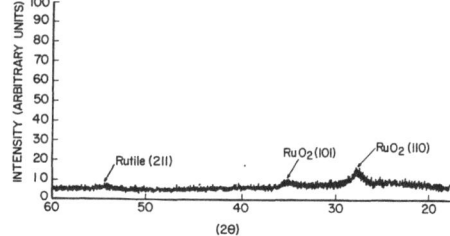

Fig. 4. AES composition-depth profile of ion plated Ru-Ti-O coating.

Fig. 5. X-ray diffraction of an ion plated Ru-Ti-O coating.

the as-deposited RuO_2 - TiO_2 film. The cross section indicates that the coating structure is dense and apparently columnar. It will be necessary to employ AEM to determine the true nature of the apparently complex crystallographic structure and morphology of the coating.

The resistivity of the RuO_2 - TiO_2 coating on glass was 0.0018 ohm-cm. The measured coating adhesion was found to be 53.3 MPa.

Composite anodes formed by ion plating the RuO_2 - TiO_2 film onto niobium were found to have very low dissolution rates, 0.016 g/A/yr in 3.5% NaCl solutions and 0.003 g/A/yr in ordinary tap water. Anodes with RuO_2 - TiO_2 ion plated coatings can thus be expected to have a longer lifetime than "niobium-doped" titanium oxide coated anodes. Furthermore, it has been ascertained that the mixed oxide RuO_2 - TiO_2 coating will exhibit high conductivity over a much greater composition range (30-100 at.% RuO_2) as compared to the composition range of 1-5 at.% Nb for the "niobium doped" titanium oxide coatings (11,12). Based on composition, it is more feasible to produce a conductive coating composed of a mixed oxide of ruthenium and titanium. However, a disadvantage of the RuO_2 - TiO_2 coating is the high cost of the ruthenium, about 15 times greater than the cost of niobium and titanium.

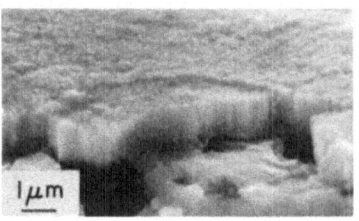

Fig. 6. SEM micrograph of ion plated Ru-Ti-O coating.

3. SPUTTER DEPOSITION OF Nb-Ti-O COATINGS
3.1 Experimental Procedure
    Composite mangetron sputtering targets were produced by placing an array of niobium inserts onto a titanium target with the relative areas of titanium and niobium controlling the alloy composition of the coating. The sputter target had Nb over 15% of its area. The samples consisted of 1.3 cm diameter titanium cylinders mounted on an electrode which could be RF biased. The distance between the magnetron sputtering target and the substrate was 3.8 cm. The samples were first sputter cleaned in an R.F. biased argon plasma. The magnetron sputtering target was then energized, and oxygen introduced into the chamber. Deposition was continued for 10 hours.

3.2 Results and Discussion
    AES surveys over the topmost layers of the sputter deposited Nb-Ti-O coatings, revealed an approximate composition of 18 at.% titanium, 12 at.% niobium, 65 at.% oxygen along with 5 at.% carbon contamination as shown in Figure 7.
    X-ray diffraction analysis showed that these sputter deposited coatings were most likely a mixture of Nb_2O_5 and TiO_2, rather than the "niobium-doped" TiO_2 as in the case of the ion plated coatings. Some broadening of the x-ray diffraction peaks was noted, indicating a microcrystalline grain structure, these findings are consistent with the elemental compositions determined by AES.
    Figure 8 is an SEM micrograph showing the surface and cross-section of the sputter deposited Nb-Ti-O coating. A dense columnar structure 35 μm thick is indicated by these micrographs.

4. CONCLUSIONS
    It has been shown that electrically conductive metal oxide (ceramic) coatings can be produced by ion plating and reactive sputter deposition.

The systems employed in these investigations were Nb-Ti-O coatings and Ru-Ti-O coatings on selected substrates. Cathodic protection anodes, fabricated by depositing the coatings on Nb substrates, can be expected to have long lifetimes as a result of their low resistivity, good adherence, and low dissolution rates.

The ion plated Nb-Ti-O coatings, the ion plated Ru-Ti-O coatings, and the sputter-deposited Nb-Ti-O coatings all had dense columnar morphologies with microcrystalline structures. The ion plated Ru-Ti-O coatings appeared to consist of mixtures of RuO_2 and TiO_2. The sputter-deposited Nb-Ti-O coatings appeared to be composed of mixtures of Nb_2O_5 and TiO_2, rather than "niobium-doped" titanium oxide as in the case of the ion plated Nb-Ti-O, possibly because more niobium was deposited than was intended.

Future work will involve a more detailed analysis of the asdeposited films and in-situ plasma diagnostics correlated with processing conditions. In addition, anodes produced by these methods will be field tested.

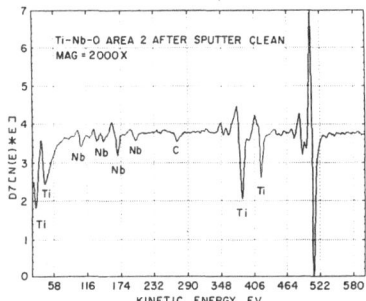

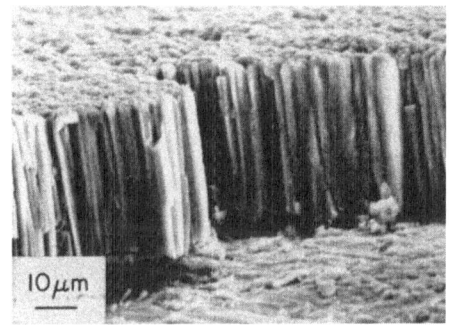

Fig. 7. AES survey of sputter deposited Nb-Ti-O coating.

Fig. 8. SEM micrograph of sputter deposited Nb-Ti-O coating.

5. ACKNOWLEDGMENTS

The authors wish to recognize the partial financial support and technical expertise of Dr. Rothwarf, U.S. Army Research Development and Standardization Group, London, UK and Dr. M. Di Lullo, NATO Scientific Affiars Division, Bruxelles, Belgium. In addition, the contributions of Dr. R. Quattrone, Dr. A. Kumar, Mr. J. Boy and Mr. J. Givens (USA-CERL), and the Center for Microanalysis of Materials, which is supported by DOE under contract DE-AC02-76ER01198, University of Illinois, Urbana-Champaign are also recognized. Two authors (LDS and JMR) further acknowledge the partial financial support of the U.S. Army Research Office under Contract No. DAAG 29-83-K-0151.